W9-BFO-045

Confidence interval for
$\mu_1 - \mu_2$

$$\bar{x}_1 - \bar{x}_2 \pm ts_\rho \sqrt{\frac{1}{n_1} + \frac{1}{n_2}}$$

Statistical test of H_0:
$\mu_1 - \mu_2 = 0$

$$t = \frac{\bar{x}_1 - \bar{x}_2}{s\rho \sqrt{\frac{1}{n_1} + \frac{1}{n_2}}}$$

Confidence interval for π

$$\hat{\pi} \pm z\sigma_{\hat{\pi}}$$

Statistical test of H_0: $\pi = \pi_0$

$$z = \frac{\hat{\pi} - \pi_0}{\sigma_{\hat{\pi}}}$$

Confidence interval for $\pi_1 - \pi_2$

$$\hat{\pi}_1 - \hat{\pi}_2 \pm z\sigma_{\hat{\pi}_1 - \hat{\pi}_2}$$

Statistical test of H_0:
$\pi_1 - \pi_2 = 0$

$$z = \frac{\hat{\pi}_1 - \hat{\pi}_2}{\sigma_{\hat{\pi}_1 - \hat{\pi}_2}}$$

Test of H_0: $\sigma^2 = \sigma_0^2$

$$\chi^2 = \frac{(n-1)s^2}{\sigma_0^2}$$

Test of H_0: $\sigma_1^2 = \sigma_2^2$

$$F = \frac{s_1^2}{s_2^2}$$

Least squares line

$$\hat{y} = \hat{\beta}_0 + \hat{\beta}_1 x$$

Slope of least squares line

$$\hat{\beta}_1 = \frac{S_{xy}}{S_{xx}}$$

Intercept of least squares line

$$\hat{\beta}_0 = \bar{y} - \hat{\beta}_1 \bar{x}$$

Test of H_0: $\beta_1 = 0$

$$t = \frac{\hat{\beta}_1}{\sqrt{s_\epsilon^2/S_{xx}}}$$

Population correlation coefficient ρ

Sample correlation coefficient

$$\hat{\rho} = \frac{S_{xy}}{\sqrt{S_{xx}S_{yy}}}$$

Paired-difference test

$$t = \frac{\bar{d}}{s_d/\sqrt{n}}$$

Chi-square test of independence

$$\chi^2 = \sum \frac{(O-E)^2}{E}$$

Spearman's rank correlation
coefficient

$$\hat{\rho}_s = 1 - \frac{6(\Sigma d^2)}{n(n^2 - 1)}$$

UNDERSTANDING STATISTICS

The Duxbury Advanced Series in Statistics and Decision Sciences

Applied Nonparametric Statistics, Second Edition, Daniel
Applied Regression Analysis and Other Multivariable Methods, Second Edition, Kleinbaum, Kupper, and Muller
Classical and Modern Regression with Applications, Second Edition, Myers
Elementary Survey Sampling, Fourth Edition, Scheaffer, Mendenhall, and Ott
Introduction to Contemporary Statistical Methods, Second Edition, Koopmans
Introduction to Probability and Its Applications, Scheaffer
Introduction to Probability and Mathematical Statistics, Bain and Engelhardt
Linear Statistical Models, Bowerman and O'Connell
Probability Modeling and Computer Simulation, Matloff
Quantitative Forecasting Methods, Farnum and Stanton
Time Series Analysis, Cryer
Time Series Forecasting: Unified Concepts and Computer Implementation, Second Edition, Bowerman and O'Connell

The Duxbury Series in Statistics and Decision Sciences

Applications, Basics, and Computing of Exploratory Data Analysis, Velleman and Hoaglin
A Course in Business Statistics, Second Edition, Mendenhall
Elementary Statistics, Fifth Edition, Johnson
Elementary Statistics for Business, Second Edition, Johnson and Siskin
Essential Business Statistics: A Minitab Framework, Bond and Scott
Fundamentals of Biostatistics, Third Edition, Rosner
Fundamentals of Statistics in the Biological, Medical, and Health Sciences, Runyon
Fundamental Statistics for the Behavioral Sciences, Second Edition, Howell
Introduction to Probability and Statistics, Seventh Edition, Mendenhall
An Introduction to Statistical Methods and Data Analysis, Third Edition, Ott
Introductory Business Statistics with Microcomputer Applications, Shiffler and Adams
Introductory Statistics for Management and Economics, Third Edition, Kenkel
Mathematical Statistics with Applications, Fourth Edition, Mendenhall, Wackerly, and Scheaffer
Minitab Handbook, Second Edition, Ryan, Joiner, and Ryan
Minitab Handbook for Business and Economics, Miller
Operations Research: Applications and Algorithms, Winston
Probability and Statistics for Engineers, Third Edition, Scheaffer and McClave
Probability and Statistics for Modern Engineering, Second Edition, Lapin
Statistical Experiments Using BASIC, Dowdy
Statistical Methods for Psychology, Second Edition, Howell
Statistical Thinking for Behavioral Scientists, Hildebrand
Statistical Thinking for Managers, Second Edition, Hildebrand and Ott
Statistics: A Tool for the Social Sciences, Fourth Edition, Ott, Larson, and Mendenhall
Statistics for Business and Economics, Bechtold and Johnson
Statistics for Management and Economics, Sixth Edition, Mendenhall, Reinmuth, and Beaver
Understanding Statistics, Fourth Edition, Ott and Mendenhall

FIFTH EDITION

UNDERSTANDING STATISTICS

Lyman Ott
Merrell Dow Research Institute

William Mendenhall
Professor Emeritus, University of Florida

PWS-KENT Publishing Company
Boston

PWS-KENT
Publishing Company

20 Park Plaza
Boston, Massachusetts 02116

Sponsoring Editor: Michael Payne
Assistant Editor: Marcia Cole
Production Editor: Susan L. Krikorian
Production Services: Sara Hunsaker, Ex Libris
Text and Cover Designer: Susan L. Krikorian
Compositor: Weimer Typesetting, Inc.
Manufacturing Coordinator: Margaret Sullivan Higgins
Cover Printer: Henry N. Sawyer Company, Inc.
Text Printer/Binder: R. R. Donnelley & Sons Company

Cover art: *Ariel's Conation* by Michael Lasuchin. Reprinted by permission of the artist.

© 1990 by PWS-KENT Publishing Company. All rights reserved. No part of this book may be reproduced, stored in a retrieval system, or transcribed in any form or by any means, electronic, mechanical, photocopying, recording, or otherwise, without the prior written permission of the publisher, PWS-KENT Publishing Company.

PWS-KENT Publishing Company is a division of Wadsworth, Inc.

Printed in the United States of America

1 2 3 4 5 6 7 8 9 — 94 93 92 91 90

Library of Congress Cataloging-in-Publication Data

Ott, Lyman.
    Understanding statistics / Lyman Ott, William Mendenhall. — 5th ed.
        p.   cm.
    Includes bibliographical references.
    1. Statistics. I. Mendenhall, William. II. Title.
QA276.12.O87   1990
001.4'22—dc20
                                                         89-16136
                                                              CIP

PREFACE

This text is designed for a one-quarter or one-semester introductory course in statistics. It can be used in those colleges and universities that teach a general course in statistics appropriate for students majoring in, or intending to major in, many different areas. The approach, examples, and exercises provide a basic knowledge of statistical concepts that will be useful in business and in the biological, social, and physical sciences.

The focus of this edition is similar to that for the fourth edition. One of the primary objectives of an introductory statistics course is to develop in the student an understanding and appreciation of the role of statistics in society. The fifth edition of *Understanding Statistics* emphasizes that statistics, as a subject, is the study of making sense of data. Thus, after completing an introductory course in statistics, students should have a *basic understanding* of how to make sense of data. In this text, we approach the study of statistics by considering four steps in making sense of data: (1) gathering methods, (2) methods for summarizing data, (3) methods for analyzing data, and (4) ways to report the results of analyses. Most texts focus on the summarization and analysis steps; we emphasize that all problems involve the collection, summarization, analysis, and reporting steps of making sense of data. We do not, however, spend a great deal of time on teaching statistical calculations. Rather, with calculators and computers available to do the calculations required in analyzing data, we try to focus on the inferences that are made after the calculations have been done.

The use of this text does not require that students have access to a computer and standard statistical software. However, we do discuss the use of calculators and computers in statistics (Chapter 2) and, in an optional section at the end of each chapter, we give reusable programs and output to show how two software systems (SAS and Minitab) can be used to perform some of the calculations of that chapter. Exercises with computer output also provide opportunities for students to understand and interpret the results of a statistical analysis presented as computer output.

To summarize, the fifth edition provides an introduction to statistics with strong emphasis on making sense of data. From this text, students gain an understanding and appreciation of the role of statistics in society and of the steps and methods used in making sense of data. Since computers are becoming so much a part of our lives, an optional section is included in each chapter to show where computer software can be of assistance in performing the calculations of statistics. Important features of this edition include the following:

- Organizational and textual changes to emphasize the four steps in making sense of data: collecting data, summarizing data, analyzing data, and reporting data.

- A new chapter on methods for gathering data (Chapter 3). This chapter discusses some of the survey methods and designs of scientific studies used in collecting data.

- A new chapter on the reporting of results of statistical analyses (Chapter 18).

- A brief discussion of observational studies (Section 5.7).

- An early discussion of methods for summarizing data from more than one variable (Chapter 5).

- New sections on discussing assumptions (Sections 8.7, 9.5, and 15.4).

- Discussion of inferences about β_0 (Section 13.2).

- A new section on predicting y for a given value of x (Section 13.4).

- Updated and expanded computer output for examples and exercises.

- Expanded and improved exercise sets and examples.

A special note of appreciation is extended to Susan Reiland, who provided an extensive prerevision review of this edition. We are also indebted to Will Sullivan for redoing the computer output and to Jim Stegeman for preparation of the solutions manual. Special thanks are also due to Barbara and Robert Beaver for preparing the study guide.

Many students and professors who used previous editions of this text have contacted us with suggestions, and these are greatly appreciated. The authors especially want to acknowledge the reviewers for this edition:

Maria Betkowski
Middlesex City College

Howard A. Bird
St. Cloud State University

James R. Case
Hiram College

James Finch
University of San Francisco

Ester Guerin
Seton Hall University

Donald Hotchkiss
Iowa State University

Harold Nemer
Riverside City College

Franklin Sheehan
San Francisco State University

Franklin D. Shobe
St. Andrews Presbyterian College

Karen H. Smith
West Georgia College

Joseph Walker
Georgia State University

Mary Woods
University of Louisville

Karl Zilm
Lewis and Clark Community College

The success of this book through the years is in many ways a team effort. For this edition, we gratefully thank our secretarial assistant, Phyllis Switzer, who transformed many drafts into polished word-processing documents. Thanks are also due A. Hald, E. S. Pearson, the Biometrika Trustees, and the Chemical Rubber Company for permission to reproduce tables, and to the SAS Institute, and Minitab, Inc., for allowing us to include output from their software systems.

We also want to acknowledge the cooperative efforts of our editor, Michael Payne, his assistant editor, Marcia Cole, his editorial assistant, Susan Hankinson, and the production staff of PWS-KENT. Finally, we mention the ongoing support and encouragement of our families.

Lyman Ott
William Mendenhall

CONTENTS

CHAPTER 7   **SAMPLING DISTRIBUTIONS**   **209**

PART 5   **STEP THREE: METHODS FOR ANALYZING DATA**   **245**

CHAPTER 8   **INFERENCES ABOUT μ**   **247**

UNDERSTANDING
STATISTICS

PART 1

STATISTICS: MAKING SENSE OF DATA

CHAPTER 1

WHAT IS STATISTICS?

1.1 Introduction ▪ **1.2** Why study statistics? ▪ **1.3** Some current applications of statistics ▪ **1.4** What do statisticians do? ▪ **1.5** A note to the student ▪ Summary ▪ Key terms ▪ Exercises

1.1 INTRODUCTION

What is statistics? Is it the addition of numbers? Is it graphs, batting averages, percentages of passes completed, percentages of unemployment, and, in general, numerical descriptions of society and nature?

Statistics, as a subject, is the study of making sense of data. Almost everyone—including corporate presidents, marketing representatives, social scientists, chemists, and consumers—deal with data. These data could be in the form of quarterly sales figures, expenditures for goods and services, pulse rates for patients undergoing therapy, contamination levels in samples of surface water, or census data. In this text, we approach the study of statistics by considering the four steps in making sense of data: (1) gathering data, (2) summarizing data, (3) analyzing data, and (4) reporting the results of analyses. The text is divided into these four major sections to reflect this focus. Within each section, there is one or more chapters dealing with that aspect of making sense of data.

Before we jump into the study of statistics, let's consider three instances in which the application of statistics could help to solve a practical problem.

1. Suppose that a manufacturer of light bulbs produces roughly a half-million bulbs per day. Because of some customer reactions to its product, the firm wishes to determine the fraction of bulbs produced on a given day that are defective. It can solve the problem in two ways. The half-million bulbs could be inserted into sockets and tested, but the cost of this solution would be substantial and could greatly increase the price per bulb. A second method for determining the fraction of defective bulbs is to select 1000 bulbs from the half-million produced and test each one. The fraction of bulbs defective in the 1000 tested could be used to estimate the fraction defec-

tive in the entire day's production. We will show in later chapters that the fraction defective in the bulbs tested will probably be quite close to the fraction defective for the entire half-million bulbs. Also, we will be able to tell you by how much you might expect this esti-mate to differ from the fraction of defective bulbs produced on any given day.

2. A similar application of statistics is brought to mind by the frequent use of the Gallup poll, the Harris poll, and other public opinion polls. How can these pollsters presume to know the opinions of more than 100 million Americans? They certainly cannot reach their con-clusions by contacting every voter in the United States. Rather, as we have suggested in the light bulb example, they sample the opinions of a small number of voters, perhaps as few as 1500, to estimate the reaction of every voter in the country. The amazing result of this process is that the fraction of those people contacted who hold a particular opinion will match very closely the fraction of voters hold-ing that opinion in the total population at that point in time. Most students find this assertion difficult to believe, but we will supply con-vincing supportive evidence in subsequent chapters.

3. Another example of a statistical problem is taken from the field of medicine. Suppose a research physician wishes to investigate the ef-fect of a new drug on the stimulation of a patient's heart. The phy-sician is really interested in the effect of the drug on all future heart patients who might be treated with the drug. He selects fifty heart patients and treats each with the drug. The increase in the pulse rate is recorded for each over a period of time. After observing the effect of the drug on the fifty patients, the physician may infer that the drug will have a similar effect on all heart patients in the future.

These problems illustrate the four steps in making sense of data. First, each problem involved a data-gathering stage using sampling. A group (sample) of light bulbs was selected from the day's production, a sample of people was obtained from the entire voting population in the United States, and a sample of fifty heart patients was obtained. Then a measurement was obtained for each element (bulb, voter, or patient) in the sample. These data are then used to solve the problem.

Next, in order to make sense of the data collected, someone would have to summarize and analyze it. In the light bulb example the fraction of defective bulbs could be computed for those bulbs tested. Based on this value and the number of bulbs tested (1000), one could accurately predict the fraction of defective bulbs in the entire production of a half-million bulbs. Similarly, for the voter opinion poll, one could compute the fraction of the sample voters who favored each of the candidates. Based on the results for the 1500 sample voters, one could accurately predict the voting pattern for the entire voting public in the United States, at that point in time. In the study of fifty heart patients, measurements on variables such as exercise capacity, oxygen consumption, and quality of life could be used to predict the effect (efficacy) of the compound on other patients, who would be candidates for similar treatment.

Finally, having collected, summarized, and analyzed the data, it would be important to report the results in unambiguous terms to interested persons. For the light bulb example, management and technical staff would need to know the quality of their production batches. Based on this information they could determine whether adjustments in the process are necessary. The results of the statistical analyses cannot be presented in ambiguous terms; decisions must be made from a well-defined knowledge base. The results of the voter opinion poll example would be of vital interest to political candidates, campaign managers, and potential campaign contributors and might lead to major shifts in campaign and funding strategies. Finally, the results of the heart patient study would interest the physician treating the patients, the company developing the compound, the Food and Drug Administration, and the medical community in general. The results must be presented clearly so that informed decisions can be made for the future development of the compound in the treatment of heart patients.

Definition 1.1

population

A **population** is the set of all measurements of interest to the sample collector.

Definition 1.2

sample

A **sample** is any subset of measurements selected from the population.

FIGURE 1.1 Population and Sample

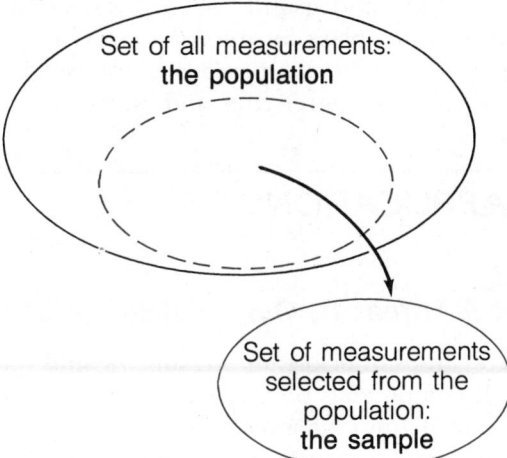

1.2   WHY STUDY STATISTICS?

We can think of two good reasons for taking an introductory course in statistics. One reason is that you need to know how to evaluate published numerical facts. Every person is exposed to manufacturers' claims for products; to the results of sociological, consumer, and political polls; and to the published achievements of scientific research. Many of these results are inferences based on sampling. Some of the inferences are valid; others are invalid. Some are based on samples of adequate size; others are not. Yet all these published results bear the ring of truth. Some people say that statistics can be made to support almost anything (particularly statisticians). Others say it is easy to lie with statistics. Both statements are true. It is easy, purposely or unwittingly, to distort the truth by using statistics when presenting the results of sampling to the uninformed.

A second reason for studying statistics is that your profession or employment may require you to interpret the results of sampling (surveys or experimentation) or to employ statistical methods of analysis to make inferences in your work. For example, practicing physicians receive large amounts of advertising describing the benefits of new drugs. These advertisements frequently display the numerical results of experiments that compare a new drug with an older one. Do such data really imply that the new drug is more effective, or is the observed difference in results due simply to random variation in the experimental measurements?

Recent trends in the conduct of court trials indicate an increasing use of probability and statistical inference in evaluating the quality of evidence. The use of statistics in the social, biological, and physical sciences is essential because all these sciences make use of observations of natural phenomena, through sample surveys or experimentation, to develop and test new theories. Statistical methods are employed in business when sample data are used to forecast sales and profit. In addition, they are used in engineering and manufacturing to monitor product quality. The sampling of accounts is a new and useful tool to assist accountants in conducting audits. Thus statistics plays an important role in almost all areas of science, business, and industry; persons employed in these areas need to know the basic concepts, strengths, and limitations of statistics.

1.3   SOME CURRENT APPLICATIONS OF STATISTICS

Acid Rain: A Threat to Our Environment

The accepted causes of acid rain are sulfuric and nitric acids; the sources of these acidic components of rain are hydrocarbon fuels, which spew sulfur and nitric oxide into the atmosphere when burned. The effects of acid rain are many. Some of these effects are listed here:

- Acid rain, when present in spring snow melts, invades breeding areas for many fish and this prevents successful reproduction. Forms of life that depend on ponds and lakes contaminated by acid rain begin to disappear.

- In areas surrounded by affected bodies of water, nutrients are leached from the soil affecting the reproduction of the soil, especially in forests.

- Man-made structures are also affected by acid rain. Experts from the United States estimate that acid rain has caused nearly $15 billion of damage to buildings and other structures thus far.

Solutions to the problems associated with acid rain will not be easy. The National Science Foundation (NSF) has recommended that we strive for a 50% reduction in sulfur-oxide emissions. Perhaps that is easier said than done. High-sulfur coal is a major source of these emissions, but in states dependent on coal for energy, a shift to lower sulfur coal is not always possible. Rather, better scrubbers must be developed to remove these contaminating oxides from the burning process before they are released into the atmosphere. Fuels for internal combustion engines are also major sources of the nitric and sulfur oxides of acid rain. Clearly, better emission control is needed for automobiles and trucks.

Reducing the oxide emissions from coal-burning furnaces and motor vehicles will require greater use of existing scrubbers and emission control devices as well as the development of new technology to allow us to use available energy sources. Developing alternative, cleaner energy sources is also important if we are to meet NSF's goal. Statistics and statisticians will play a key role in monitoring atmosphere conditions, testing the effectiveness of proposed emission control devices, and developing new control technology and alternative energy sources.

Determining the Effectiveness of a New Drug Product

The development and testing of the Salk vaccine for protection against poliomyelitis (polio) provide an excellent example of how statistics can be used in solving practical problems. Most parents and children growing up before 1954 can recall the panic brought on by the outbreak of polio cases during the summer months. Although relatively few children fell victim to the disease each year, the pattern of outbreak of polio was unpredictable and caused great concern because of the possibility of paralysis or death. The fact that very few of today's youth have even heard of polio demonstrates the great success of the vaccine and the testing program that preceded its release on the market.

It is standard practice in establishing the effectiveness of a particular drug product to conduct an experiment (often called a *clinical trial*) with human subjects. For some clinical trials assignments of subjects are made at random, with half receiving the drug product and the other half receiving a solution or tablet (called a *placebo)* that does not contain the medication. One statistical problem concerns the determination of the total number of subjects to be included in the clinical trial. This problem was particularly important in the testing of the Salk vaccine because data from previous years suggested that the incidence rate might be less than 50 cases for every 100,000 children. Hence a large number of subjects had to be included in the clinical trial in order to detect a difference in the incidence rates for those treated with the vaccine and those receiving the placebo.

With the assistance of statisticians it was decided that a total of 400,000 children should be included in the Salk clinical trial begun in 1954, with half of them randomly assigned the vaccine and the remaining children assigned the placebo. No other clinical trial had ever been attempted on such a large group of subjects. Through a public school inoculation program, the 400,000 subjects were treated and then observed over the summer to determine the number of children contracting polio. Although less than 200 cases of polio were reported for the 400,000 subjects in the clinical trial, more than three times as many cases appeared in the group receiving the placebo. These results together with some statistical calculations were sufficient to indicate the effectiveness of the Salk polio vaccine. However, these conclusions would not have been possible if the statisticians and scientists had not planned for and conducted such a large clinical trial.

The development of the Salk vaccine is not an isolated example of the use of statistics in the testing and developing of drug products. In recent years the Food and Drug Administration (FDA) has placed stringent requirements on pharmaceutical firms to establish the effectiveness of proposed new drug products. Thus statistics has played an important role in the development and testing of birth control pills, rubella vaccines, chemotherapeutic agents in the treatment of cancer, and many other preparations.

Applications of Statistics in Our Courts

Libel suits related to consumer products have touched each one of us; you may have been involved as a plaintiff or defendant in a suit or you may know of someone who was involved in such litigation. Certainly we all help to fund the costs of this litigation indirectly through increased insurance premiums and increased costs of goods. The testimony in libel suits concerning a particular product (automobile, drug product, and so on) frequently leans heavily on the interpretation of data from one or more scientific studies involving the product. This is how and why statistics and statisticians have been pulled into the courtroom.

For example, epidemiologists have used statistical concepts applied to data to determine whether there is a statistical "association" between a specific characteristic, such as the use of a brand-name tampon, and a disease condition, such as toxic shock syndrome. An epidemiologist who finds an association should try to determine whether the observed statistical association from the study is due to random variation or whether it reflects an actual association between the characteristic and the disease. Arguments in courtrooms about the interpretations of these types of associations involve data analyses using statistical concepts as well as a clinical interpretation of the data.

The Energy Crisis: A Search for New Sources and a Search for Oil

The OPEC oil crisis of 1973–1974 brought to American's attention a problem that is with us today and will continue to plague us for decades: a shortage of energy. The United States is confronted by staggering annual demands for energy with supplies that may not meet current and future demands, especially if a major supplier "interrupts" service. Such an interruption by OPEC in 1974 led to an energy rush and subsequent supply problems.

Possible sources of energy needed to supply the present and future requirements of the United States include vast coal and oil shale reserves, nuclear reactors, new oil and natural gas reserves, solar energy, and alternative new fuels. For example, methanol (wood alcohol) and ethanol (grain alcohol) may be major contributors to octane boost as leaded fuels are phased out. These alcohols are also likely candidates to reduce our dependence on foreign crude oil.

In which of these resources should we, the American public, invest the capital necessary for development? Which source will yield a given amount of energy at minimum cost? What unfavorable impact will each have on the environment or the quality of life? Which might yield dangerous side effects? These questions and others must be answered by experimentation. Statisticians will assist in designing experiments and in interpreting experimental data.

Opinion and Preference Polls

Public opinion, consumer preference, and election polls are commonly used to assess the opinions or preferences of a segment of the public for issues, products, or candidates of interest. And we, the American public, are exposed to the results of these polls on a daily basis in newspapers, in magazines, on the radio, and on television. For example, the results of polls related to the following subjects were printed in local newspapers over a two-day period.

- consumer confidence related to future expectations about the economy

- preferences for candidates in upcoming elections and caucuses
- attitudes toward cheating on federal income tax returns
- preference polls related to specific products (for example, foreign vs. American cars, Coke vs. Pepsi, McDonald's vs. Wendy's)
- reactions of North Carolina residents toward arguments about the morality of tobacco
- opinions of voters toward proposed tax increases and proposed changes in the Defense Department budget

A number of questions can be raised about these polls. How many people were polled? What questions were asked? Was each person asked the same question? How were people chosen or selected for the poll? Can we believe the results of these polls? Do these results "represent" how the general public feels?

Opinion and preference polls are an important, visible application of statistics for the consumer. We will discuss this topic in more detail in Chapter 10. Hopefully after studying this material, you will have a better understanding of how to interpret the results of these polls.

1.4   WHAT DO STATISTICIANS DO?

What do statisticians do? In the context of making sense of data, statisticians are involved with all aspects of gathering, summarizing, and analyzing data, and reporting the results of their analyses. There are both good and bad ways to gather data. Statisticians apply their knowledge of existing survey techniques and scientific study designs or they develop new techniques to provide a guide to good methods of data collection. We will explore these ideas further in Chapter 3.

Once the data are gathered, they must be summarized before any meaningful interpretation can be made. Statisticians can recommend and apply useful methods for summarizing data in graphical, tabular, and numerical forms. Intelligent graphs and tables are useful first steps in making sense of the data. Also, measures of the average (or typical) value and some measure of the range or spread of the data help in interpretation. These topics will be discussed in detail in Chapters 4 and 5.

The objective of statistics is to make an inference about a population of interest based on information obtained from a sample of measurements from that population. The analysis stage of making sense of data deals with making inferences. For example, a market research study reaches only a few of the potential buyers of a new product, but probable reaction of the set of potential buyers (population) must be inferred from the reactions of the buyers included in the study (sample). If the market research study has been carefully planned and executed, the reactions of those included in the sample should agree reasonably well (but not necessarily exactly) with the population. The reason we can say this is because the basic concepts of probability allow us to make an inference about the population of interest that includes our best guess plus a statement of the probable error in our best guess.

We will illustrate how inferences are made by way of an example. Suppose an auditor samples 2000 financial accounts from a set of more than 25,000 accounts and finds that 84 (4.2%) are in error. What can be said about the set of 25,000 accounts? That is, what inference can we make about the percentage of accounts in error for the population of 25,000 accounts based on information obtained from the sample of 2000 accounts? We will show (in Chapter 10) that our best guess (inference) about the percentage of accounts in error for the population is 4.2%, and this best guess should be within ±.9% of the actual unknown percentage of accounts in error for the population. The plus-or-minus factor is called the probable error of our inference. Anyone can make a guess about the percentage of accounts in error; concepts of probability allow us to define the (probable) error of our guess.

In dealing with the analyses of data, statisticians can apply existing methods for making inferences; some theoretical statisticians engage in the development of new methods with more advanced mathematics and probability theory. Our study of the methods for analyzing sample data will begin in Chapter 8, after we discuss the basic concepts of probability and sampling distributions in Chapters 6 and 7.

Finally, statisticians are involved with communicating the results of their analyses as the final stage in making sense of data. The form of the communication varies from an informal conversation to a formal report. The advantage of a more formal verbal presentation with visual aids or study report is that the communication can make use of the graphical, tabular, and numerical displays as well as the analyses done on the data to help convey the "sense" found in the data. Too often this is lost in an informal conversation. The report or communication should convey to the intended audience what can be gleaned from the sample data, and it should be conveyed in as nontechnical terms as possible so that there can be no confusion as to what is being inferred. More information about the communication of results is presented in Chapter 18.

1.5 A NOTE TO THE STUDENT

We think with words and concepts. A study of the discipline of statistics requires the memorization of new terms and concepts (as does the study of a foreign language). Commit these definitions, theorems, and concepts to memory.

Also, focus on the broader concept of making sense of data. Do not let details obscure these broader characteristics of the subject. The teaching objective of this text is to identify and amplify these broader concepts of statistics.

SUMMARY

The discipline of statistics and those who apply the tools of that discipline deal with making sense of data. As such, statisticians are involved with methods of data collection, data summarization, and data analyses, as well as communicating the results of its analyses.

KEY TERMS

population (*def*)
sample (*def*)

EXERCISES

1.1 Selecting the proper diet for shrimp or other sea animals is an important aspect of sea farming. A researcher wishes to estimate the mean weight of shrimp maintained on a specific diet for a period of six months. One hundred shrimp are randomly selected from an artificial pond and each is weighed.

(a) Identify the population of measurements that is of interest to the researcher.

(b) Identify the sample.

(c) What characteristics of the population would be of interest to the researcher?

(d) If the sample measurements are used to make inferences about certain characteristics of the population, why would a measure of the reliability of the inferences be important?

1.2 Radioactive waste disposal as well as the production of radioactive material in some mining operations are creating a serious pollution problem in some areas of the United States. State health officials decided to investigate the radioactivity levels in one suspect area. Two hundred points were randomly selected in the area and the level of radioactivity was measured at each point. Answer questions (a), (b), (c), and (d) in Exercise 1.1 for this sampling situation.

1.3 A social researcher in a particular city wishes to obtain information on the number of children in households that receive welfare support. A random sample of 400 households was selected from the welfare rolls of the city. A check on welfare recipient data provided the number of children in each household. Answer questions (a), (b), (c), and (d) of Exercise 1.1 for this sample survey.

1.4 **Class Exercise** Search issues of your local newspaper to locate the results of a recent Harris or Gallup survey.

(a) Identify the items that will be observed in order to obtain the sample measurements.

(b) Identify the measurement made on each item.

(c) Clearly identify the population associated with the survey.

(d) What characteristic(s) of the population is (are) of interest to the pollster?

(e) Does the article explain how the sample was selected?

(f) Does the article include the number of measurements in the sample?

(g) What type of inference is made concerning the population characteristics?

(h) Does the article tell you how much faith you can place in the inference about the population characteristic?

PART 2

STEP ONE: METHODS FOR GATHERING DATA

CHAPTER 2

USING CALCULATORS AND COMPUTERS IN STATISTICS

2.1 Introduction ▪ **2.2** Calculators ▪ **2.3** Computers and statistical software ▪ **2.4** A brief introduction to Minitab and SAS ▪ Summary ▪ Key terms

2.1 INTRODUCTION

Before we begin any formal development of the subject matter of statistics we should mention two very important tools of the trade: electronic, hand-held calculators and, more importantly, computers. As the lay connotation of statistics implies, the field of statistics involves data: we will be summarizing data and using data to make inferences in the form of estimates, predictions, and decisions. The better we are able to perform the calculations required in summarizing data or in drawing inferences from data, the more time we can spend on actually understanding them — that is, understanding the summarizations and inferences made from the data.

2.2 CALCULATORS

A calculator can be very helpful to a student taking a first course in statistics. It is useful in doing standard arithmetic operations such as addition ($+$), subtraction ($-$), multiplication ($\times$), division ($\div$), as well as for squaring numbers and taking square roots. Other features offered by many hand-held calculators include the ability to take logarithms, to compute two statistical quantities presented in Chapter 5 (a mean $\bar{x}$ and a standard deviation s), and to do minor amounts of programming. If you don't have a calculator, perhaps your professor could refer you to a store that carries inexpensive, quality calculators suitable for a beginning course in statistics. The calculator should be able to do at least the following: add, subtract, multiply, divide, calculate squares and square roots, and save the sum of a sequence of numbers.

2.3 COMPUTERS AND STATISTICAL SOFTWARE

software

The availability and accessibility of computers and the development of computer **software** (programs or program systems developed for computers) that is relatively easy to use have had a great impact on the field of statistics. The calculations for large data sets can become complex and time-consuming when using a hand-held calculator. This is where a computer can be of tremendous help, even for a student in an introductory course. Specific programs or more general software systems can be used to perform statistical calculations and analyses almost instantaneously with entering the data into the computer. Also, it is not necessary to have knowledge of computer programming in order to make use of most statistical software for planned analyses. The major **statistical software systems** (such as Minitab, SAS®, BMDP, and SPSS^X) have user's manuals that give detailed instructions as to how data are to be entered and how you can perform the desired calculations and analyses. Other systems that have been developed specifically for interactive use at a computer terminal provide program "prompts" to lead the user through the desired analysis.

statistical software systems

We will spend some time introducing Minitab and SAS, since these are the two software systems used to illustrate calculations and analyses throughout the text. However, if you have access to a computer and are interested in using it, ask your professor how to obtain an account, what statistical software systems are available, and where to learn how to use these software systems. We will not cover statistical packages for microcomputers in this text. However, most of what we illustrate here and throughout the text can be duplicated using packages such as *Systat, Statgraphics, PC SAS,* and the *Student Edition of Minitab* developed for microcomputers.

Before we introduce Minitab and SAS, though, we want to say a few words about the use of computers for calculation and analysis. First, if used properly, computers can be a tremendous aid to us, especially for processing large data sets that require repetitive calculations—the so-called number-crunching operations. But, computers are mindless beasts; they can do only what they have been told (programmed) to do, no matter how absurd the result may be.

Second, the oft-used acronym from computer technology GIGO (Garbage In, Garbage Out) aptly states another problem area. Inferences made from data (whether based on analyses done by hand or by a computer) are only as good as the data upon which they are based. If we have erroneous data or apply the wrong procedures (analyses), the inferences that we make will probably be wrong.

Third, there is a general misconception that computers always provide numerically accurate results. While it is true that the major software systems have been thoroughly tested, we should still review the computer calculations just to make certain that they make sense. Don't believe everything that is on a neatly printed computer output. Computers are not a substitute for thinking; the user must still correctly formulate the problem and interpret the computer output. It follows that sound judgment and an understanding of statistical concepts are required in order to properly analyze data.

Finally, this isn't a text on computer usage, and we won't spend much time on the mechanics of using particular packages; these are best learned by doing. However, if you do have access to a particular software package, your professor could help you apply it to the business of statistics—making sense of data.

2.4   A BRIEF INTRODUCTION TO MINITAB AND SAS

We will provide a brief introduction to how data can be entered into Minitab or SAS and how to obtain a printout of the data. For more instruction and details about these two software systems (as well as for SPSS[X] and BMDP), see Lefkowitz (1984) and the user's manual and documentation for each of the software systems (see the references at the end of this text).

Now let's consider entry of data into Minitab. Suppose that we want to enter the sales volume ($000) and profit before tax (PBT in $000) for each of fourteen pharmacies. These data are shown in Table 2.1.

TABLE 2.1   *Sales and Profit Before Tax (PBT) for Pharmacies*

Pharmacy	Sales ($000)	PBT ($000)
1	38	1.3
2	20	2.1
3	48	2.2
4	44	2.6
5	56	3.3
6	39	4.0
7	65	4.1
8	84	4.2
9	82	5.5
10	105	5.7
11	126	7.0
12	52	7.5
13	80	7.7
14	101	7.9

After you access a mainframe version or PC version of Minitab at a terminal, these data can be entered into Minitab and printed out using the following Minitab program:

```
READ INTO C1, C2, C3
 1          38       1.3
 2          20       2.1
 3          48       2.2
 4          44       2.6
 5          56       3.3
 6          39       4.0
 7          65       4.1
 8          84       4.2
 9          82       5.5
10         105       5.7
11         126       7.0
12          52       7.5
13          80       7.7
14         101       7.9
PRINT C1, C2, C3
STOP
```

The first command of this program (READ . . .) tells Minitab that the data follow, and it specifies that the data should be placed into Columns 1, 2, and 3. Next are the 14 lines of data. Following that, the PRINT statement of the program tells Minitab to print out the data from Columns 1–3. The STOP statement designates the end of the Minitab program. The printout from this program is shown here.

```
MTB>    READ INTO C1, C2, C3
DATA>    1          38       1.3
DATA>    2          20       2.1
DATA>    3          48       2.2
DATA>    4          44       2.6
DATA>    5          56       3.3
DATA>    6          39       4.0
DATA>    7          65       4.1
DATA>    8          84       4.2
DATA>    9          82       5.5
DATA>   10         105       5.7
DATA>   11         126       7.0
DATA>   12          52       7.5
DATA>   13          80       7.7
DATA>   14         101       7.9
DATA> PRINT C1, C2, C3
14 ROWS READ
 1        38       1.3
 2        20       2.1
 3        48       2.2
 4        44       2.6
 5        56       3.3
 6        39       4.0
 7        65       4.1
 8        84       4.2
 9        82       5.5
10       105       5.7
11       126       7.0
```

```
12        52        7.5
13        80        7.7
14       101        7.9
MTB> STOP
```

As shown in our example, each Minitab command is listed, followed immediately by the results of that command. After the READ statement, the data are displayed; the entire data set is also listed after the PRINT statement. Each value should be checked carefully to ensure that the data were entered correctly.

Entry into SAS is about as simple as that for Minitab. A SAS program or job has two kinds of commands: DATA commands and PROC commands. The DATA commands are used to enter or modify data sets, while the PROC commands are used to perform calculations and analyses on data sets. A SAS program for entering and printing the data of Table 2.1 is shown here:

```
DATA SALESPBT;
INPUT PHARMACY SALES PBT;
CARDS;
   1         38        1.3
   2         20        2.1
   3         48        2.2
   4         44        2.6
   5         56        3.3
   6         39        4.0
   7         65        4.1
   8         84        4.2
   9         82        5.5
  10        105        5.7
  11        126        7.0
  12         52        7.5
  13         80        7.7
  14        101        7.9
;
PROC PRINT DATA=SALESPBT;
```

For this program, we have one DATA command (which includes everything from the line "DATA SALESPBT" through the last data line) followed by a semicolon (which denotes the end of the data) and one PROC command. Note that each SAS statement ends with a semicolon.

For this example we chose the name SALESPBT as the data set name. A data set name can have up to eight characters (letters, digits, or an underscore), must begin with a letter, and cannot end with the underscore character. The INPUT statement indicates the names we have chosen for the three numeric variables for which data are being entered; a variable name can be from one to eight characters (letters or digits) and must start with a letter. The CARDS statement indicates that the data for the three variables follow.

The PROC PRINT command of this SAS program follows the last data line and tells SAS to print out the data for the data set named SALESPBT. Output for this SAS program is shown here:

OBS	PHARMACY	SALES	PBT
1	1	38	1.3
2	2	20	2.1
3	3	48	2.2
4	4	44	2.6
5	5	56	3.3
6	6	39	4.0
7	7	65	4.1
8	8	84	4.2
9	9	82	5.5
10	10	105	5.7
11	11	126	7.0
12	12	52	7.5
13	13	80	7.7
14	14	101	7.9

Remember that the use of computers in an introductory statistics course is not a necessity; rather, we think that using computers allows us to focus more on the understanding of statistics by recognizing the important role that computers can play in the processing and analysis of large data sets.

SUMMARY

In this chapter we introduced you to two very important tools of the trade: electronic calculators and computers. Specifically, we indicated that calculators and computers are especially helpful in performing the time-consuming calculations that are part of a statistical evaluation of data. Two widely used software systems, Minitab and SAS, were used to illustrate how data can be entered into a computer and then listed. These same two systems will be discussed in more detail in the "Using Computers" section of subsequent chapters, where we show how Minitab and SAS can solve some of the problems discussed in each chapter. The programs used in these illustrations can be readily modified to apply the same analysis to other data sets; in this sense they are "reusable" programs.

KEY TERMS

software   statistical software systems

CHAPTER 3

USING SURVEYS AND SCIENTIFIC STUDIES TO GATHER DATA

3.1 Introduction ▪ **3.2** Surveys ▪ **3.3** Scientific studies ▪ **3.4** Observational studies ▪ Summary ▪ Key terms

3.1 INTRODUCTION

As we've mentioned previously, the first step in making sense of data is to gather data on one or more variables of interest. But intelligent data gathering doesn't just happen, it takes a conscious, concerted effort focused on the following steps:

- Specifying the objective of the data-gathering exercise
- Identifying the variable(s) of interest
- Choosing an appropriate design for the survey or scientific study
- Collecting the data

To specify the objective of the data-gathering exercise you must understand the problem under study. For example, if the management of a large manufacturing company is considering whether to institute a new incentive pay plan for its production workers, it might want to determine the attitudes of the production supervisors toward the proposed plan. This, then, could be the objective of a data-gathering exercise.

To identify the variable(s) of interest you must examine the objective of the data-gathering exercise. For the production incentive plan, the variable of interest would be the attitude of the production supervisors. Measurements would consist of preferences (favor, oppose) accompanied by comments as to why a production supervisor may favor or oppose the new incentive plan.

Once the objective is determined and the variable(s) of interest specified, you must choose how to collect the data. In statistics, data can be gathered by way of a survey, a study, or a combination of the two. Survey theory and the theory of experimental designs for scientific studies provide good methods for collecting data. Usually surveys are passive where the aim is to gather (survey) data on existing conditions, attitudes, or behaviors. Thus the management of the manufacturing company would use a survey to sample the opinions of the production supervisors on the merits of a new incentive plan. Scientific studies, on the other hand, tend to be more active: the person conducting the study would tend to deliberately vary certain conditions in order to reach a conclusion. For example, if a plant manager is interested in the effect of noise level in his manufacturing plant, he could vary the noise level and certain controlled conditions in order to see, directly, the gains or losses in productivity.

In this chapter we will consider some of the survey methods and designs for scientific studies. We will also make a distinction between a scientific study and an observational study.

3.2   Surveys

Information from surveys affects almost every facet of our daily lives. These surveys determine such government policies as the control of the economy and the promotion of social programs. Opinion polls are the basis of much of the news reported by the various news media. Ratings of television shows determine which shows are to be available for viewing in the future.

One usually thinks of the U.S. Census Bureau as contacting every household in the country. Actually, in the 1980 census only 14 questions were asked of all households. Information on an additional 42 questions was obtained from only a sample of households. The resulting information is used by many agencies and individuals for manifold purposes. For example, the federal government uses it to determine allocations of funds to states and cities; it is used by businesses to forecast sales, to manage personnel, and to establish future site locations; it is used by urban and regional planners to plan land use, transportation networks, and energy consumption. It is used by social scientists to study economic conditions, racial balance, and other aspects of the quality of life.

The U.S. Bureau of Labor Statistics (BLS) routinely conducts over twenty surveys. Some of the best known and most widely used are the surveys that establish the consumer price index (CPI). The CPI is a measure of price change for a fixed market basket of goods and services over time. It is used as a measure of inflation and serves as an economic indicator for government policies. Businesses have wage rates and pension plans tied to the CPI. Federal health and welfare programs, as well as many state and local programs, tie their bases of eligibility to the CPI. Escalator clauses in rents and mortgages are based on the CPI. So we can see that this one index, determined on the basis of sample surveys, plays a fundamental role in our society.

Many other surveys from the BLS are crucial to society. The monthly Current Population Survey establishes basic information on the labor force, employment, and unemployment. The consumer expenditure surveys collect data on family expenditures for goods and services used in day-to-day living. The Establishment Survey collects information on employment hours and earnings for nonagricultural business establishments. The survey on occupational outlook provides information on future employment opportunities for a variety of occupations, projecting to approximately ten years ahead. Other activities of the BLS are addressed in the *BLS Handbook of Methods* (1982).

Opinion polls are constantly in the news, and the names of Gallup and Harris have become well known to everyone. These polls, or sample surveys, reflect the attitudes and opinions of citizens on everything from politics and religion to sports and entertainment. The Nielsen ratings determine the success or failure of TV shows.

Businesses conduct sample surveys for their internal operations, in addition to using government surveys for crucial management decisions. Auditors estimate account balances and check on compliance with operating rules by sampling accounts. Quality control of manufacturing processes relies heavily on sampling techniques.

One particular area of business activity that depends on detailed sampling activities is marketing. Decisions on which products to market, where to market them, and how to advertise them are often made on the basis of sample survey data. The data may come from surveys conducted by the firm that manufactures the product or may be purchased from survey firms that specialize in marketing data.

Sampling Techniques

simple random sampling

The basic design (**simple random sampling**) consists of selecting a group of n sampling units in such a way that each sample of size n has the same chance of being selected. Thus we can obtain a random sample of n eligible voters in the bond-issue poll by drawing names from the list of registered voters in such a way that each sample of size n has the same probability of selection. (The details of simple random sampling can be found in Chapter 4 of Scheaffer et al., 1986.) At this point we merely state that a simple random sample will contain as much information on the community preference as any other sample survey design, provided all voters in the community have similar socioeconomic backgrounds.

Suppose, however, that the community consists of people in two distinct income brackets, high and low. Voters in the high bracket may have opinions on the bond issue that are quite different from the opinions of voters in the low bracket. Therefore, to obtain accurate information about the population, we want to sample voters from each bracket. We can di-

stratified random sample

vide the population elements into two groups, or strata, according to income and select a simple random sample from each group. The resulting sample is called a **stratified random sample.** (See Chapter 5 of Scheaffer et al., 1986.)

ratio estimation

Note that stratification is accomplished by using knowledge of an auxiliary variable, namely, personal income. By stratifying on high and low values of income, we increase the accuracy of our estimator. **Ratio estimation** is a second method for using the information contained in an auxiliary variable. Ratio estimators not only use measurements on the response of interest, but they incorporate measurements on an auxiliary variable. Ratio estimation can also be used with stratified random sampling.

cluster sampling

Although individual preferences are desired in the survey, a more economical procedure, especially in urban areas, may be to sample specific families, apartment buildings, or city blocks rather than individual voters. Individual preferences can then be obtained from each eligible voter within the unit sampled. This technique is called **cluster sampling.** Although we divide the population into groups for both cluster sampling and stratified random sampling, the techniques differ. In stratified random sampling we take a simple random sample within each group, whereas in cluster sampling we take a simple random sample of groups and then sample all items within the selected groups (clusters). (See Chapters 8 and 9 of Scheaffer et al., 1986, for details.)

systematic sample

Sometimes, the names of persons in the population of interest are available in a list, such as a registration list, or on file cards stored in a drawer. For this situation an economical technique is to draw the sample by selecting one name near the beginning of the list and then selecting every tenth or fifteenth name thereafter. If the sampling is conducted in this manner, we obtain a **systematic sample.** As you might expect, systematic sampling offers a convenient means of obtaining sample information; unfortunately, we do not necessarily obtain the most information for a specified amount of money. (Details are given in Chapter 7 of Scheaffer et al., 1986.)

The important point to understand is that there are different kinds of surveys that can be used to collect sample data. For the surveys discussed in this text, we will be dealing with simple random sampling and methods for summarizing and analyzing data collected in such a manner. More complicated surveys lead to even more complicated problems at the summarization and analyses stages of statistics.

Data Collection Techniques

Having chosen a particular sample survey, how does one actually collect the data? The most commonly used methods of data collection in sample surveys are personal interviews and telephone interviews. These methods, with appropriately trained interviewers and carefully planned callbacks, commonly achieve response rates of 60% to 75%, and sometimes even higher. A mailed questionnaire sent to a specific group of interested persons can achieve good results, but generally the response rates for this type of data collection are so low that all reported results are suspect. Frequently, objective information can be found from direct observation rather than from an interview or mailed questionnaire.

personal interviews

Data are frequently obtained by **personal interviews.** For example, we can use personal interviews with eligible voters to obtain a sample of the public sentiments toward a community bond issue. The procedure usually requires the interviewer to ask prepared questions and to record the respondent's answers. The primary advantage of these interviews is that people will usually respond when confronted in person. In addition, the interviewer can note specific reactions and eliminate misunderstandings about the questions asked. The major limitations of the personal interview (aside from the cost involved) concern the interviewers. If they are not thoroughly trained, they may deviate from the required protocol, thus introducing a bias into the sample data. Any movement, facial expression, or statement by the interviewer can affect the response obtained. For example, a leading question such as "Are you also in favor of the bond issue?" may tend to elicit a positive response. Finally, errors in recording the responses can also lead to erroneous results.

telephone interviews

Information can also be obtained from persons in the sample through **telephone interviews.** With the advent of wide-area telephone service lines (WATS lines), an interviewer can place any number of calls to specified areas of the country for a fixed monthly rate. Surveys conducted through telephone interviews are frequently less expensive than personal interviews, owing to the elimination of travel expenses. The investigator can also monitor the interviews to be certain that the specified interview procedure is being followed.

A major problem with telephone surveys is the establishment of a frame that closely corresponds to the population. Telephone directories have many numbers that do not belong to households, and many households have unlisted numbers. A few households have no phone service, although lack of phone service is now only a minor problem for most surveys in the United States. A technique that avoids the problem of unlisted numbers is random-digit dialing. In this method a telephone exchange number (the first three digits of the seven-digit number) is selected, and then the last four digits are dialed randomly until a fixed number of households of a specified type are reached. This technique seems to produce unbiased samples of households in selected target populations and avoids many of the problems inherent in sampling a telephone directory.

Telephone interviews generally must be kept shorter than personal interviews because respondents tend to get impatient more easily when talking over the telephone. With appropriately designed questionnaires and trained interviewers, telephone interviews can be as successful as personal interviews.

self-administered questionnaire

Another useful method of data collection is the **self-administered questionnaire,** to be completed by the respondent. These questionnaires usually are mailed to the individuals included in the sample, although other distribution methods can be used. The questionnaire must be carefully constructed if it is to encourage participation by the respondents.

The self-administered questionnaire does not require interviewers, and thus its use results in a savings in the survey cost. This savings in cost is usually bought at the expense of a lower response rate. Nonresponse can be a problem in any form of data collection, but since we have the least contact with respondents in a mailed questionnaire, we frequently have the lowest rate of response. The low response rate can introduce a bias into the sample because the people who answer questionnaires may not be representative of the population of interest. To eliminate some of the bias, investigators frequently contact the nonrespondents through follow-up letters, telephone interviews, or personal interviews.

direct observation

The fourth method for collecting data is **direct observation.** For example, if we were interested in estimating the number of trucks that use a particular road during the 4–6 PM rush hours, we could assign a person to count the number of trucks passing a specified point during this period. Possibly, electronic counting equipment could also be used. The disadvantage in using an observer is the possibility of error in observation.

Direct observation is used in many surveys that do not involve measurements on people. The U.S. Department of Agriculture, for instance, measures certain variables on crops in sections of fields in order to produce estimates of crop yields. Wildlife biologists may count animals, animal tracks, eggs, or nests in order to estimate the size of animal populations.

A closely related notion to direct observation is that of getting data from objective sources that are not affected by the respondents themselves. For example, health information can sometimes be obtained from hospital records, and income information from employer's records (especially for state and federal government workers). This approach may take more time but can yield large rewards in important surveys.

EXERCISES

3.1 An experimenter wants to estimate the average water consumption per family in a city. Discuss the relative merits of choosing individual families, dwelling units (single-family houses, apartment buildings, etc.), and city blocks as sampling units.

3.2 A forester wants to estimate the total number of trees on a tree farm that possess diameters exceeding 12 inches. A map of the farm is available. Discuss the problem of choosing what to sample and how to select the sample.

3.3 A safety expert is interested in estimating the proportion of automobile tires with unsafe treads. Should he use individual cars or collections of cars, such as those in parking lots, in his sample?

3.4 An industry is composed of many small plants located throughout the United States. An executive wants to survey the opinions of the employees on the vacation policy of the industry. What would you suggest she sample?

3.5 A state department of agriculture desires to estimate the number of acres planted in corn within the state. How might one conduct such a survey?

3.6 A political scientist wants to estimate the proportion of adult residents of a state who favor a unicameral legislature. What could be sampled? Also, discuss the relative merits of personal interviews, telephone interviews, and mailed questionnaires as methods of data collection.

3.7 Discuss the relative merits of using personal interviews, telephone interviews, and mailed questionnaires as methods of data collection for each of the following situations:

(a) A television executive wants to estimate the proportion of viewers in the country who are watching her network at a certain hour.

(b) A newspaper editor wants to survey the attitudes of the public toward the type of news coverage offered by his paper.

(c) A city commissioner is interested in determining how homeowners feel about a proposed zoning change.

(d) A county health department wants to estimate the proportion of dogs that have had rabies shots within the last year.

3.8 A Yankelovich, Skelly, and White poll taken in the fall of 1984 showed that one-fifth of the 2207 people surveyed admitted to having cheated on their federal income taxes. Do you think that this fraction is close to the actual proportion who cheated? Why? (Discuss the difficulties of obtaining accurate information on a question of this type.)

3.3 SCIENTIFIC STUDIES

The subject of experimental designs for scientific studies cannot be given much justice in the beginning of an introductory course in statistics, since entire courses at the undergraduate and graduate levels are needed to get a comprehensive understanding of the methods and concepts of experimental design. Even so, we will attempt to give you a brief overview of the subject because much data requiring summarization and analyses arise from scientific studies involving one of a number of experimental designs. We will work by way of examples.

A multinational oil company has been developing an unleaded gasoline that can be competitively priced while delivering higher gasoline mileage. A number of blends of gasolines have been proposed and undergone initial testing. One blend in particular appears to yield good gasoline mileage and can be produced economically. An additional test is planned to obtain an accurate estimate of the gasoline mileage (miles per gallon) for the blend under normal road conditions. To do this a standard car model is chosen and each of ten cars of this model type is driven over a predetermined course and the miles per gallon recorded. (Later, in Chapter 8, we will show how these data could then be summarized to provide an estimate of the miles per gallon for the blend under these fixed road conditions.)

To change the problem slightly, suppose the oil company had three different gasoline blends for further testing in the gasoline mileage test conducted under normal road conditions. For this study the company could take nine standard model cars and randomly assign three cars to each gasoline blend and test the cars under the road conditions dictated by the experiment. There would be a recorded gasoline mileage for each car, and three cars per gasoline blend. The methods presented in Chapters 4 and 15 could be used to summarize and analyze the sample mileage data in order to make comparisons (inferences) among the three gasoline blends. One possible inference of interest could be the selection of the best gasoline blend from a gasoline mileage standpoint. Which blend performed better? Can the best performing blend in the sample data be expected to provide better gasoline mileage if the same study were repeated?

Experimental Designs

completely randomized design

The experimental design for this scientific study is called a **completely randomized design.** Table 3.1 displays a completely randomized design for the gasoline blend study.

In general a completely randomized design is used when one is interested in comparing t "treatments" (in our $t = 3$ case the treatments were gasoline blends). For each of the treatments we obtain a sample of

TABLE 3.1 *Completely Randomized Design of Gasoline Blends*

Blend 1	Blend 2	Blend 3
3 cars	3 cars	3 cars

observations, and the samples are not necessarily of the same size for the different treatments. The sample of observations from a treatment is assumed to be the result of a simple random sample of observations from the hypothetical population of possible values that could have been obtained for that treatment. In our example, the sample of three gasoline mileages obtained from blend 1 was considered to be the outcome for a simple random sample of three observations selected from the hypothetical population of possible mileages for standard model cars using gasoline blend 1. The same reasoning applies for the samples from blends 2 and 3.

This experimental design could be changed to accommodate the study of the same three blends using each of, say, three different drivers. Because of individual driving habits, not all drivers get the same gasoline mileage for the same blend of gasoline; so it would be desirable to have each of the drivers test each of the blends over the specified road course. Here we avoid having the comparison of blends distorted by differences among drivers. The experimental design is called a **randomized block design** because we have "blocked" out any differences among drivers in order to get a precise comparison of the three blends. See Table 3.2.

randomized block design

TABLE 3.2 Randomized Block Design of Gasoline Blends

Driver 1	Driver 2	Driver 3
Blend 1	Blend 2	Blend 1
Blend 3	Blend 3	Blend 2
Blend 2	Blend 1	Blend 3

Note that each driver tests each blend, and the order of testing the blends is randomized for each driver (i.e., driver 2 will test blend 2 first, then blend 3, and finally blend 1).

What happens if the order of testing influences a driver's performance and the first blend tested generally receives a higher mileage rating than the ones tested second or third. Then blend 1 could (possibly) look better than blends 2 and 3 simply because it was tested first in two of the three drivers (see Table 3.2).

A variation on the previous randomized block design, called a **Latin square design** eliminates the order of testing as a factor affecting the comparison of treatments (blends). A Latin square design for our example is shown in Table 3.3.

Latin square design

Note that with this design, each blend is tested first once, second once, and third once, *and* each driver tests all blends.

The randomized block and Latin square designs are both extensions of the completely randomized design where the objective is to compare *t* treatments. The analysis of data collected according to a completely ran-

TABLE 3.3 Latin Square Design of Gasoline Blends

Order of Testing	Driver 1	Driver 2	Driver 3
1st	Blend 1	Blend 3	Blend 2
2nd	Blend 2	Blend 1	Blend 3
3rd	Blend 3	Blend 2	Blend 1

domized design and the inferences made from these analyses are discussed further in Chapter 15. A special case of the randomized block design is presented in Chapter 14, where the number of treatments is $t = 2$, and the analysis of data and the inferences from these analyses are discussed. More details about the analysis and interpretation of data from completely randomized designs, randomized block designs, and Latin square designs (not discussed in this text) are presented in Ott (1988).

Factorial Experiments

Suppose that we want to examine the effects of two (or more) variables (factors) on a response. For example, suppose that an experimenter is interested in examining the effects of two independent variables, nitrogen and phosphorus, on the yield of a crop. For simplicity we will assume that two levels have been selected for the study of each factor: 40 and 60 pounds per plot for nitrogen, 10 and 20 pounds per plot for phosphorus. For this study the experimental units are small, relatively homogeneous plots that have been partitioned from the acreage of a farm.

one-at-a-time
approach

One approach for examining the effects of two or more factors on a response is called the **one-at-a-time approach.** To examine the effect of a single variable, an experimenter varies the levels of this variable while holding the levels of the other independent variables fixed. This process is continued until the effect of each variable on the response has been examined. See Table 3.4 for an example.

TABLE 3.4 Factor-Level Combination for a One-at-a-Time Approach

Combination	Nitrogen	Phosphorus
1	60	10
2	40	10
3	40	20

TABLE 3.5 Yields for the Three Factor-Level Combinations

Observation (yield)	Nitrogen	Phosphorus
145	60	10
125	40	10
160	40	20
?	60	20

Hypothetical yields corresponding to the three factor-level combinations of our experiment are given in Table 3.5. Suppose the experimenter is interested in using the sample information to determine the factor-level combination that will give the maximum yield. From the table we see that crop yield increases when the nitrogen application is increased from 40 to 60 (holding phosphorus at 10). Yield also increases when the phosphorus setting is changed from 10 to 20 (at a fixed nitrogen setting of 40). Thus it might seem logical to predict that increasing both the nitrogen and phosphorus applications to the soil will result in a larger crop yield. The fallacy in this argument is that our prediction is based on the assumption that the effect of one factor is the same for both levels of the other factor.

We know from our investigation what happens to yield when the nitrogen application is increased from 40 to 60 for a phosphorus setting of 10. But will the yield also increase by approximately 20 units when the nitrogen application is changed from 40 to 60 at a setting of 20 for phosphorus?

To answer this question we could apply the factor-level combination of 60 nitrogen–20 phosphorus to another experimental plot and observe the crop yield. If the yield is 180, then the information obtained from the three factor-level combinations would be correct and useful in predicting the factor-level combination that produces the greatest yield. However, suppose the yield obtained from the high settings of nitrogen and phosphorus turns out to be 110. If this happens, the two factors nitrogen and phosphorus are said to **interact.** That is, the effect of one factor on the response does not remain the same for different levels of the second factor, and the information obtained from the one-at-a-time approach would lead to a faulty prediction.

interaction

The two outcomes just discussed for the crop yield at 60–20 setting is displayed in Figure 3.1 along with the yields at the three initial design points. Figure 3.1 (top) illustrates a situation with no interaction between the two factors. The effect of nitrogen on yield is the same for both levels of phosphorus. In contrast, Figure 3.1 (bottom) illustrates a case in which the two factors nitrogen and phosphorus do interact.

FIGURE 3.1   Yields of the Three Design Points and Possible
Yield at a Fourth Design Point

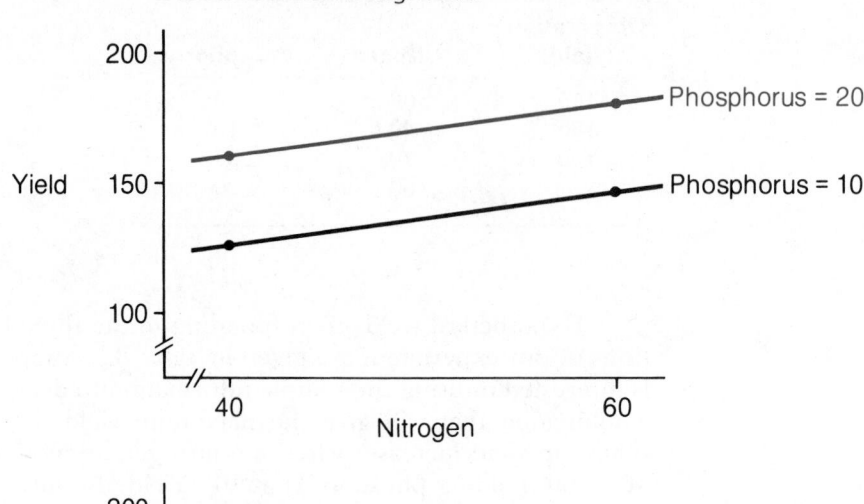

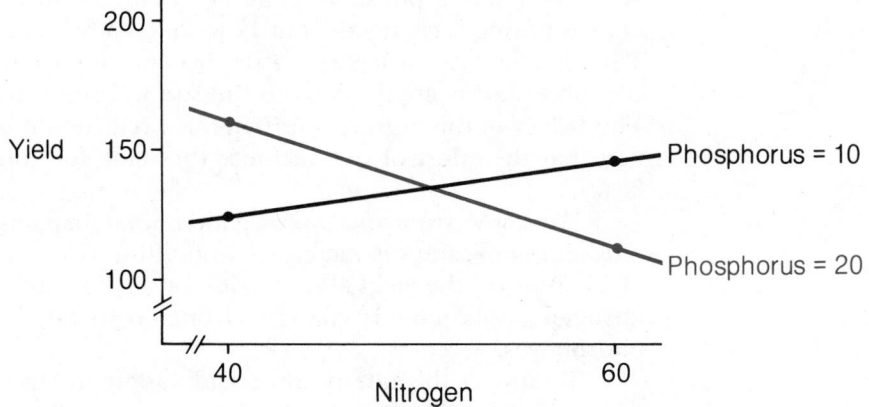

We have seen that the one-at-a-time approach to investigating the effect of two factors on a response is suitable only for situations in which the two factors do not interact. Although this was illustrated for the simple case in which two factors were to be investigated at each of two levels, the inadequacies of a one-at-a-time approach are even more salient when trying to investigate the effects of more than two factors on a response.

Factorial experiments are useful for examining the effects of two or more factors on a response, whether or not interaction exists. As before, the choice of the number of levels of each variable and the actual settings of these variables is important. But assuming we have made these selections with help from an investigator knowledgeable in the area being examined, we must decide which factor-level combinations we will observe.

Definition 3.1

factorial
experiment

A **factorial experiment** is an experiment in which the response is observed at all factor-level combinations of the independent variables.

Using our previous example, if we are interested in examining the effect of two levels of nitrogen at 40 and 60 pounds per plot and two levels of phosphorus at 10 and 20 pounds per plot on the yield of a crop, we must decide how to prepare plots to observe yield. A 2 × 2 factorial experiment for this example is shown in Table 3.6. The four factor-level combinations are assigned at random to the experimental units.

TABLE 3.6   2 × 2 Factorial Experiment for Crop Yield

Factor-Level Combinations	
Nitrogen	*Phosphorus*
40	10
40	20
60	10
60	20

Similarly, if we wished to examine nitrogen at the two levels 40 and 60 and phosphorus at the three levels 10, 15, and 20, the 2 × 3 factorial experiment would have the factor-level combinations shown in Table 3.7.

TABLE 3.7   2 × 3 Factorial Experiment for Crop Yield

Factor-Level Combinations	
Nitrogen	*Phosphorus*
40	10
40	15
40	20
60	10
60	15
60	20

The examples of factorial experiments presented in this section have concerned two independent variables. However, the procedure applies to any number of factors and levels per factor. Thus if we have four different factors at two, three, three, and four levels, respectively, we could formulate a $2 \times 3 \times 3 \times 4$ factorial experiment by considering all $2 \cdot 3 \cdot 3 \cdot 4$ = 72 factor-level combinations. For more information on the analyses and inferences of data obtained from a factorial experiment, see Ott (1988).

Combination Designs

Not all designs can be classified as either a block design or a factorial experiment. Sometimes the objectives of a study are such that we wish to investigate the effects of certain factors on a response while blocking out certain other extraneous sources of variability. Such experiments may require a design that is a combination of a block design and a factorial experiment. This can be illustrated with the following example.

An investigator wants to examine the effects of two factors (factors A and B each measured at three levels) on a response y. It is determined that $r = 2$ observations are desired at each factor-level combination, but only nine observations can be done each day. Since nine observations can be obtained each day, it is possible to run a complete replication of the 3×3 factorial experiment on two different days to get the desired number of observations. The design is shown in Table 3.8.

TABLE 3.8 A Block Design Combined with a Factorial Experiment

Day 1				Day 2			
	Factor B				*Factor B*		
Factor A	1	2	3	*Factor A*	1	2	3
1				1			
2				2			
3				3			

Note that this design is really a randomized block design where the blocks are days and the treatments are the nine factor-level combinations of the 3×3 factorial experiment. Other more complicated combinations of block designs and factorial experiments are possible. As with sample surveys, though, we will deal only with the simplest experimental designs in this text. The point we want to make is that there are many different experimental designs that can be used in scientific studies for designating the collection of sample data. Each has certain advantages and disadvantages, but we will use only the completely randomized design and the randomized block design in this text. Other more advanced courses in statistical methods use more complicated designs as the plans for gathering sample data in experimental studies.

3.4   OBSERVATIONAL STUDIES

*observational
study*

Before leaving the subject of sample data collection, we will draw a distinction between an **observational study** and a scientific study. In experimental designs for scientific studies, the observation conditions are fixed or controlled. For example, with a randomized block design, an observation is obtained on each treatment in every block. Similarly, with a factorial experiment, an observation is obtained at each factor-level combination. These "controlled" experiments are very different from observational studies, which are sometimes used in place of controlled studies because it is not feasible to do a proper scientific study. This can be illustrated by way of an example.

Much research and public interest centers on the effect of cigarette smoking on lung cancer and cardiovascular disease. One possible experimental design would be to randomize a fixed number of individuals (say 1000) to each of two groups — one group would be required to smoke cigarettes for the duration of the study (say 10 years) while those in the second group would not be allowed to smoke throughout the study. At the end of the study, the two groups would be compared for lung cancer and cardiovascular disease. Even if we ignore ethical questions, this type of study would be impossible to do. Because of the long duration, it would be difficult to follow all participants and make certain that they follow the study plan. And it would be difficult to find nonsmoking individuals willing to take the chance of being assigned to the smoking group.

Another possible study would be to sample a fixed number of smokers and a fixed number of nonsmokers to compare the groups for lung cancer and for cardiovascular disease. Assuming one could obtain willing groups of participants, this study could be done in a *much shorter* period of time.

What has been sacrificed? Well, the fundamental difference between an observational study and a scientific study lies in the inference(s) which can be drawn. For a scientific study comparing smokers to nonsmokers, assuming the two groups of individuals followed the study plan, the observed differences between the smoking and nonsmoking groups could be attributed to the effects of cigarette smoking because individuals were randomized to the two groups; hence the groups were assumed to be comparable at the outset.

This type of reasoning does not apply to the observational study of cigarette smoking. Differences between the two groups in the observation could not necessarily be attributed to the effects of cigarette smoking because, for example, there may be hereditary factors that predispose people to smoking and cancer of the lungs and/or cardiovascular disease. Thus differences between the groups might be due to hereditary factors, smoking, or a combination of the two. Typically the results of an observational study are reported by way of a statement of association. For our example, if the observational study showed a higher frequency of lung cancer and cardiovascular disease for smokers relative to nonsmokers, it would be stated that this study showed that cigarette smoking was associated with an increased frequency of lung cancer and cardiovascular disease. It is a careful rewording so as not to infer that cigarette smoking causes lung cancer and cardiovascular disease.

Many times, however, an observational study is the only type of study that can be run. Our job is to make certain that we understand the type of study run and hence understand how the data were collected. Then we can critique inferences drawn from an analysis of the study data.

SUMMARY

The first step in making sense of data involves intelligent data gathering. This involves specifying the objectives of the data-gathering exercise, identifying the variables of interest, and choosing an appropriate design for the survey or scientific study. In this chapter we discussed various survey designs and experimental designs for scientific studies. Armed with a basic understanding of some design considerations for conducting surveys or scientific studies, one can address how data are to be collected on the variables of interest in order to address the stated objectives of the data-gathering exercise.

We also drew a distinction between observational and scientific studies in terms of the inferences (conclusions) that can be drawn from the sample data. Differences found between treatment groups from an observational studies are said to be *associated with* the use of the treatments; on the other hand, differences found between treatments in a scientific study are said to be *due to* the treatments. In the next chapter, we will examine the methods for summarizing the data we collect.

KEY TERMS

simple random sampling
stratified random sampling
ratio estimation
cluster sampling
systematic sample
personal interviews
telephone interviews
self-administered questionnaire

direct observation
completely randomized design
randomized block design
Latin square design
one-at-a-time approach
interaction
factorial experiment (*def*)
observational study

PART 3

STEP TWO: METHODS FOR SUMMARIZING DATA

4

GRAPHICAL METHODS OF DATA DESCRIPTION

4.1 Introduction ■ **4.2** Organizing the data ■ **4.3** Pie charts ■ **4.4** Bar charts ■ **4.5** Frequency histograms ■ **4.6** Frequency polygons ■ **4.7** Comments concerning histograms ■ **4.8** Stem-and-leaf plots ■ **4.9** Using computers ■ Summary ■ Key terms ■ Supplementary exercises ■ Exercises from the data base

4.1   INTRODUCTION

In Chapters 2 and 3 we dealt with gathering data — the first step in making sense of data. The second step in this process is to summarize the data that have been gathered so that meaningful interpretations can be made. Imagine having to describe the annual incomes for all families registered in the last census or having to describe the sales data by quarter for the steel industry broken up by domestic consumption, exports, inventory change, and imports. Because it is almost impossible to understand and describe such detailed data, it is necessary to summarize the "raw" data.

The first step in summarizing data is to graph the data. Take a "look" at the data to see if you can get a better understanding of what the data say. Are there any trends? Do most of the values fall near some central value? Do the data have considerable variability? Sometimes answers to these questions and others will follow directly from the data plots (graphs) that you construct. In this chapter we will examine how to use graphical displays to summarize data. In Chapter 5 we will discuss a second way to summarize data by focusing on the average or typical value in a data set and spread or variability of the measurements.

4.2   ORGANIZING THE DATA

When a data set involves many measurements, the data must be organized prior to being presented in graphical form. The data should be arranged in such a way that each measurement can fall into one and only one category. This procedure eliminates any ambiguity that might arise in placing measurements into categories and aids in the interpretation of the data.

For example, administrative officials of many universities require the parents of students applying for financial aid to file a financial report. Suppose a particular university required parents to pick one of the following gross income categories:

less than $15,000
$15,000–$34,999
$35,000–$54,999
$55,000–$74,999
$75,000 or more

Clearly, the adjusted gross income for a family filing a joint return or the combined adjusted gross income for separate returns will fall into one and only one income category. However, if the income categories had been defined as

less than $15,000
$15,000–$35,000
$35,000–$55,000
$55,000–$75,000
$75,000 or more

there could be some confusion as to which category should be checked for families with adjusted gross incomes falling on the boundary points.

Having organized the data according to the guideline suggested, there are a number of ways to describe the data in graphical form. These are presented in the next few sections.

4.3   PIE CHARTS

pie chart

A **pie chart** is often used to show how a number of objects are apportioned to a group of categories. For example, Table 4.1 provides a summary of the employment status of 1000 adults and shows the number falling into each of the three categories of employment. From the table we see that 122, or 12.2%, had no jobs; 536, or 53.6%, had one job; and 342, or 34.2%, held more than one job. The percentages of adults falling into each of three employment categories are shown graphically in Figure 4.1.

TABLE 4.1   Employment Data

No job	One job	More than one job
122	536	342
(12.2%)	(53.6%)	(34.2%)

FIGURE 4.1   Employment Pie Chart

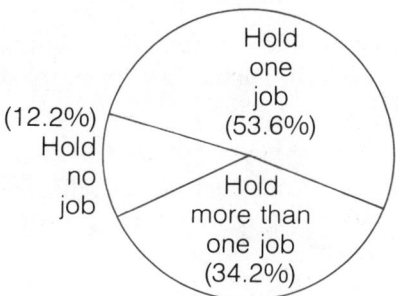

Figure 4.1 was constructed by partitioning a circle to match the percentage of each employment category, much as one might slice a pie. Thus 12.2% of the circle was assigned to those adults holding no job. Similarly, 53.6% of the circle was assigned to those holding one job. The remainder of the circle (34.2%) was allocated to those holding more than one job.

Figure 4.2 depicts the employment growth of major occupational groups in the state of Ohio between the years 1980 and 1990. As can be seen, the major increase will be in the clerical, service, professional and technical, and operatives occupational groups. The pie chart gives a rapid overview of the *relative* growth rates of the categories shown in the pie. More details would be needed to ascertain the number of jobs available in 1990 for these same occupation groups.

In summary, the pie chart can be used to display the percentages associated with each category of the variable. The following guidelines should help you to obtain a clear presentation using a pie chart.

FIGURE 4.2   Employment Growth for 1980–1990 by Major Occupational Groups

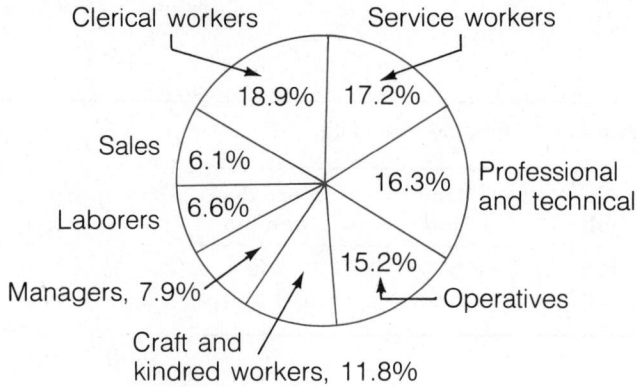

Guidelines for Constructing a Pie Chart

1. Choose a small number of categories (say 5–10) for the data being summarized. Too many categories make the pie chart difficult to interpret.
2. Partition the circle to match the percentages for the categories.
3. Whenever possible, construct the pie chart so that percentages are in either ascending or descending order. This helps in interpreting the data.

4.4 BAR CHARTS

bar chart

The **bar chart** is another graphical method for showing how data fall into a group of categories. Figure 4.3 displays the number of workers in the greater Cincinnati area for the largest foreign investors. The estimated total work force is 680,000.

Bar charts are relatively easy to construct using the guidelines given here.

Guidelines for Constructing Bar Charts

1. Label the vertical axis with the number of objects falling into the categories; label the categories along the horizontal axis.

2. Construct a rectangle over each category, with the height of the rectangle equal to the number of objects in that category. The base of each rectangle should be of the same width.

3. Leave space between each category on the horizontal axis to distinguish between the categories and to clarify the presentation.

FIGURE 4.3   Number of Workers by Major Foreign Investors

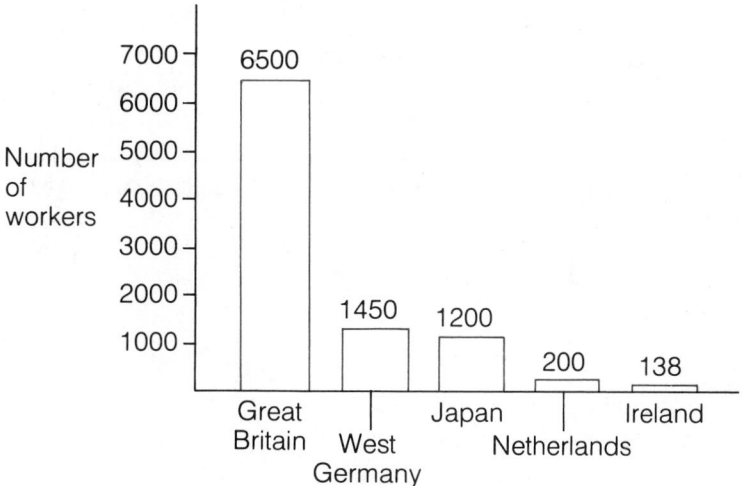

Sometimes bar charts are placed on their sides so that the bars run horizontally rather than vertically; bar charts are also used to display percentage data. The bar chart in Figure 4.4 offers a comparison of the percentages of black-occupied and white-occupied housing units classified by location in metropolitan and nonmetropolitan areas.

There are variations of the bar chart, many of which you may have seen in printed form. Figure 4.5 shows the number of active drill rigs in the United States from 1971 to 1987, while Figure 4.6 compares the sources of world crude oil from 1971 to 1986. From Figure 4.6 it is clear that there have been major shifts in the sources of crude oil from 1971 to 1986. Also it appears that the United States is headed toward increased dependence on oil imports. This latter interpretation seems to gain additional credence when we see the drop in active drilling rigs over recent years (see Figure 4.5), the projected increase in United States petroleum demand from 1986 to 1990 illustrated in Figure 4.7, and the projected drop in United States production shown in Figure 4.8.

FIGURE 4.4 Percentage Distribution of the Population by
Metropolitan and Nonmetropolitan Residence

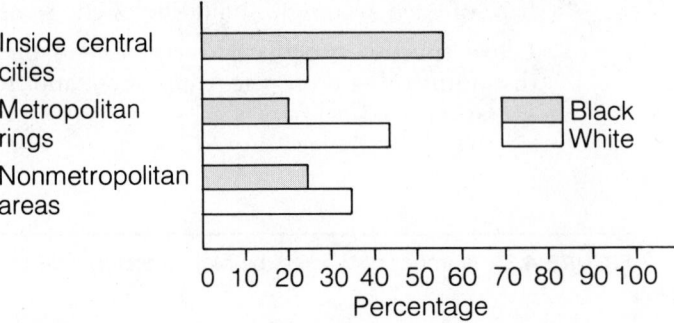

Source: U.S. Bureau of the Census, 1980, p. 18.

FIGURE 4.5 Active Drilling Rigs in the United States, 1971
to 1987

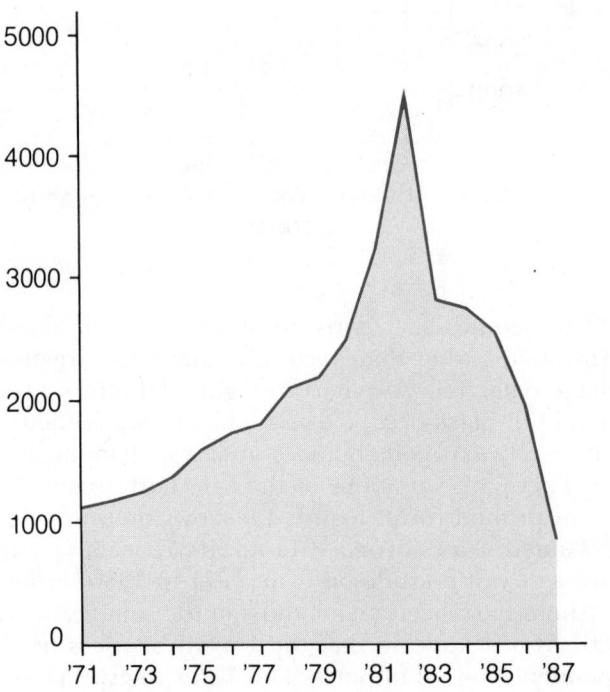

FIGURE 4.6   *Sources of World Crude Oil, 1971 and 1986.
Leading suppliers' proved reserves in billions
of barrels.*

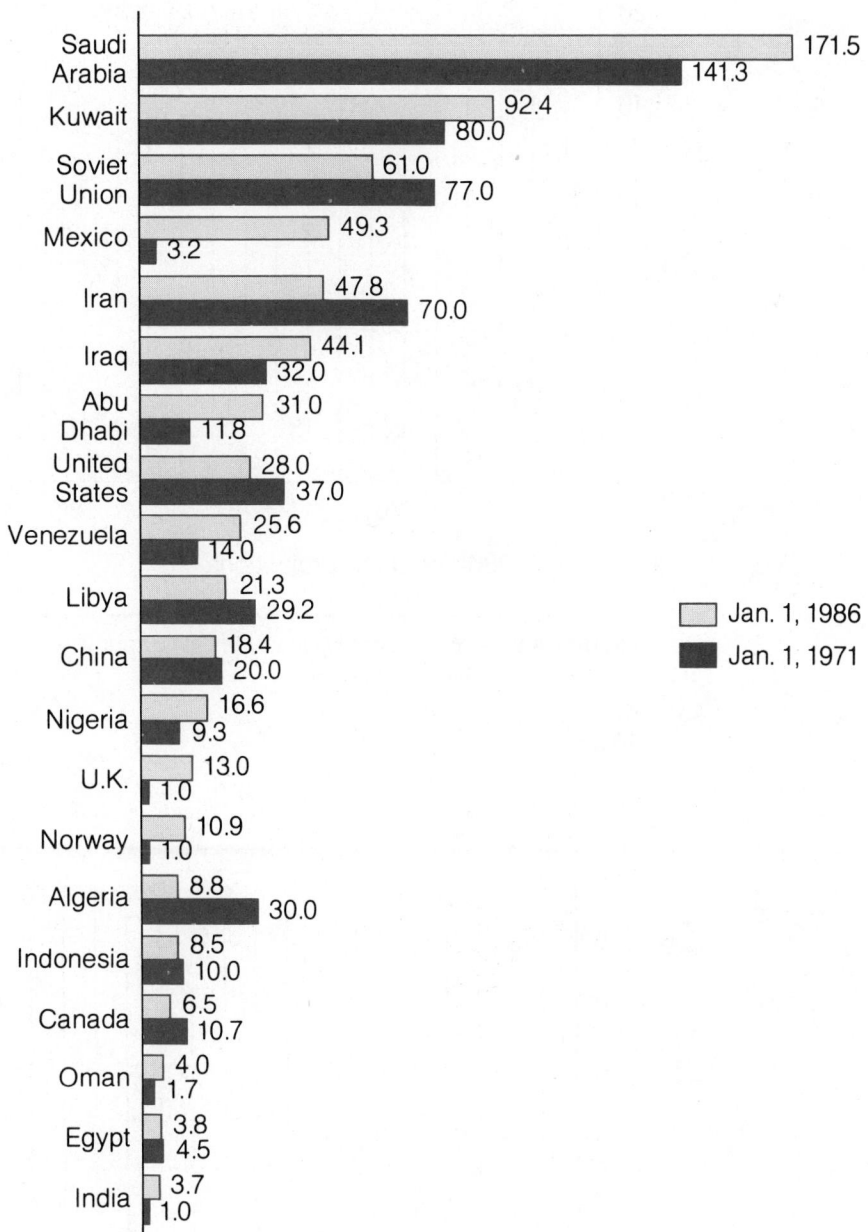

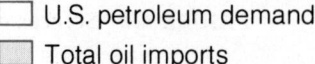

☐ U.S. petroleum demand*
▨ Total oil imports

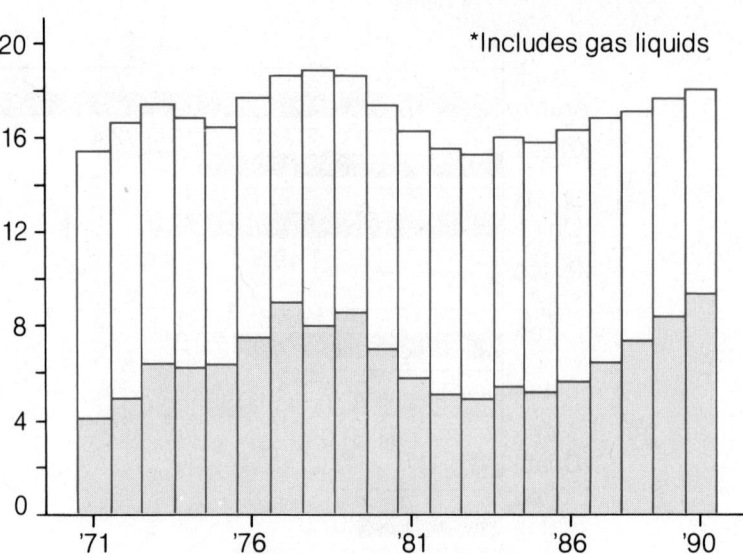

Note: 1986–1990 are projections

FIGURE 4.8 United States Daily Crude Oil Output in
Millions of Barrels

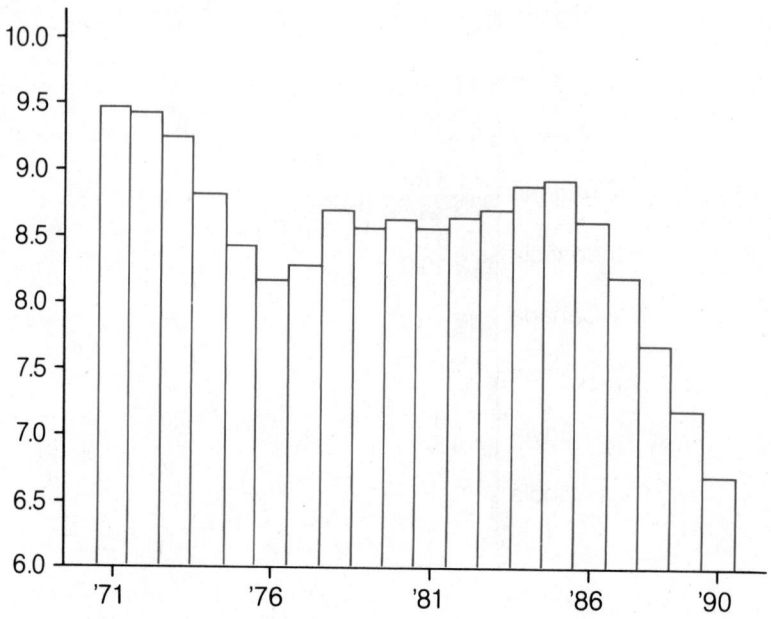

Note: 1986–1990 are projections

EXERCISES

 4.1 The accompanying data give a breakdown of the total United States oil consumption, in percentages, by the various purposes for which the oil is used. Describe these data by using a pie chart.

Use	Percentage of total
Gasoline	43.1
Industrial fuel oil	12.0
Heating oil	17.0
Jet fuel	6.9
Diesel fuel	5.4
Petrochemical	3.6
Other	12.0
Total	100.0

 4.2 Use the consumption figures in Exercise 4.1 to construct a bar graph. Which presentation, the bar graph or pie chart, seems to provide a clearer display of the data?

4.3 The U.S. Bureau of the Census publishes the *Statistical Abstract of the United States*. This reference was used to obtain the listing in the accompanying table. Use these data to construct a pie chart for the percentage of males by years of school completed.

| | Persons 65 or older, 1980 (million) | |
Years of schooling	Male	Female
8 years or less	45.3	41.6
1–3 years of high school	15.5	16.7
4 years of high school	21.4	25.8
1–3 years of college	7.5	8.6
4 years or more of college	10.3	7.4
Total	100.0	100.1*

*This column sums to 100.1 rather than 100.0 due to rounding errors.

 4.4 Refer to Exercise 4.3. Construct a bar chart for the female population by years of schooling.

 4.5 Can we combine the male and female data for each group in Exercise 4.3, and construct a bar chart for the combined population?

 4.6 Can we combine the male and female data for each group in Exercise 4.3 and construct a pie chart for the combined population?

§ 4.7   A large study of employment trends, based on a survey of 45,000 businesses, was conducted by The Ohio State University. Assuming an unemployment rate of 5% or less, it is predicted that 2.1 million job openings will be created between 1980 and 1990. This employment growth is shown next by major industry groups.

Industry group	Percentage of job openings 1980–1990
Service	33.2%
Manufacturing	25.0%
Retail trade	17.9%
Finance, insurance, real estate	6.6%
Wholesale trade	4.8%
Construction	4.6%
Transportation	3.9%
Government	2.7%
Other	1.3%

Construct a pie chart to display these data.

§ 4.8   From the same study described in Exercise 4.7, data were obtained on the job openings between 1980 and 1990. Use the data to construct a bar chart.

Occupational groups	Percentage of job openings 1980–1990
Clerical workers	20.9%
Sales	7.3%
Managers	9.5%
Professional and technical	16.3%
Laborers	3.7%
Service workers	18.1%
Operatives	13.1%
Craft and kindred workers	11.1%

4.5 FREQUENCY HISTOGRAMS

The frequency histogram offers a third graphical method for describing a set of measurements. We will illustrate the frequency histogram for the following situation.

Most of us are so accustomed to drinking water treated in a municipal water treatment plant that we may not concern ourselves with the quality of the water we drink. If you were to visit a water treatment plant, you would observe that many different measurements are obtained throughout the day to monitor the quality of water sent into the system. For example, 1000 1-milliliter samples of water are taken from the water flow and sent through a filtering process. If, after filtering, one or more colonies of coliform bacteria appear, the water is unacceptable. However, since the testing process is so delicate, the appearance of a colony could mean there was a laboratory error in the analysis (something as trivial as coughing over the water). Thus more samples would have to be obtained to verify the findings.

In addition to checking for the presence of coliform bacteria, workers obtain measurements on the chlorine residual in the water leaving the treatment plant. A reading of 2.2 parts per million (ppm) could be an acceptable value for a particular community. Similarly, readings on the hardness, turbidity, acidity, and color also are made in monitoring the quality of water sent throughout the system. And for those communities that add fluoride to their water as a tooth decay preventative (especially for children), the level of fluoride must also be monitored in the system.

The regulations of the board of health in a particular state specify that the fluoride level must not exceed 1.5 ppm. The 25 measurements in Table 4.2 represent the fluoride levels for a sample of 25 days. Although fluoride levels are measured more than once per day, these data represent the early morning readings for the 25 days sampled.

TABLE 4.2 *Fluoride Levels (ppm) for a Sample of 25 Days*

.75	.86	.84	.85	.97
.94	.89	.84	.83	.89
.88	.78	.77	.76	.82
.72	.92	1.05	.94	.83
.81	.85	.97	.93	.79

Note from Table 4.2 that the recorded fluoride levels range from .72 ppm to 1.05 ppm. Although we might examine the table closely, it would be difficult to describe precisely how the measurements are distributed along this range. For example, are most of the fluoride measurements less than .90 ppm or are most greater than .90 ppm? To answer these questions, we summarize the data in a **frequency table.**

frequency table

class interval

We begin by partitioning the range from .72–1.05 into an arbitrary number of subintervals called **class intervals.** The number of subintervals chosen depends on the number of measurements being summarized, but we generally recommend using from 5 to 20 class intervals. The more data we have, the larger the number of intervals we tend to use. The guidelines given here can be used for constructing appropriate class intervals.

range

Guidelines for Constructing Class Intervals

1. Divide the **range** of the measurements (the difference between the largest and the smallest measurements) by the approximate number of class intervals desired. Generally we will have from 5 to 20 class intervals.

2. After dividing the range by the desired number of subintervals, round the resulting number to a convenient (easy to work with) unit. This unit represents a common width for the class intervals.

3. Choose the first class interval so that it contains the smallest measurement. It is also advisable to choose a starting point for the first interval so that no measurement falls on the boundary between two subintervals. This eliminates any ambiguity in placing measurements into the class intervals. (One possibility that may work is to choose class boundaries to one more decimal place than the data, e.g., see Table 4.3.)

For the data in Table 4.2, the range is

$$\text{range} = 1.05 - .72 = .33$$

Suppose we want to have approximately 7 class intervals. Dividing the range by 7 and rounding to a convenient unit we have $.33/7 \approx .05$. Thus .05 is the **class interval width.** Since the fluoride measurements have two decimal places, it is convenient to choose the boundary points of the classes with three decimal places. Let .705 be the lower boundary of the first class, and obtain the other boundaries by adding multiples of the class width .05. Beginning with .705, the 7 classes are

class interval width

```
 .705 to  .755
 .755 to  .805
 .805 to  .855
 .855 to  .905
 .905 to  .955
 .955 to 1.005
1.005 to 1.055
```

Note that each class has a width of .05 and that the seven classes span the range of the fluoride measurements. Further, note that by choosing the lower class boundary at .705 (rather than .70 or .71), we made it impossible for a measurement to fall on a class boundary.

Since we are interested in knowing how the fluoride readings are distributed among the 7 classes, we will examine each of the 25 measurements in Table 4.2 and tally the number of measurements falling in each class. The number of measurements falling into a given class is called the **class frequency.** These frequencies are shown in Table 4.3. Notice that the sum of the frequencies equals the total number of measurements n, and that this will always be true.

class frequency

TABLE 4.3   *Frequency Table for the Fluoride Data*

Class	Class boundaries	Class frequency	Relative frequency
1	.705–.755	2	$^2/_{25}$
2	.755–.805	4	$^4/_{25}$
3	.805–.855	8	$^8/_{25}$
4	.855–.905	4	$^4/_{25}$
5	.905–.955	4	$^4/_{25}$
6	.955–1.005	2	$^2/_{25}$
7	1.005–1.055	1	$^1/_{25}$
Total		$n = 25$	1

relative frequency

The **relatively frequency** of a class is defined as the frequency of the class divided by the total number n of measurements (total frequency). Thus if we let f_i equal the frequency for class i, then

$$\text{relative frequency} = \frac{f_i}{n} \quad \text{for class } i$$

For example, the relative frequency for class 6 in Table 4.3 can be found as follows: The total number of measurements is $n = 25$, and the frequency for class 6 is $f_6 = 2$. Hence

$$\text{relative frequency} = \frac{2}{25} \quad \text{for class } 6$$

A frequency table can be presented as a graph called a frequency histogram. We mark the class boundaries (.705, .755, and so on) along the horizontal axis. Frequencies (2, 4, 8, and so on) are labeled along the vertical axis. Rectangles are then constructed over each subinterval, with the height of the rectangle equal to the class frequency. The frequency histogram for the data in Table 4.3 is given in Figure 4.9.

relative frequency histogram

Sometimes the results of a frequency table are presented graphically by using a **relative frequency histogram.** The only difference between the frequency histogram and the relative frequency histogram is that the ver-

FIGURE 4.9   Frequency Histogram for the Fluoride Data

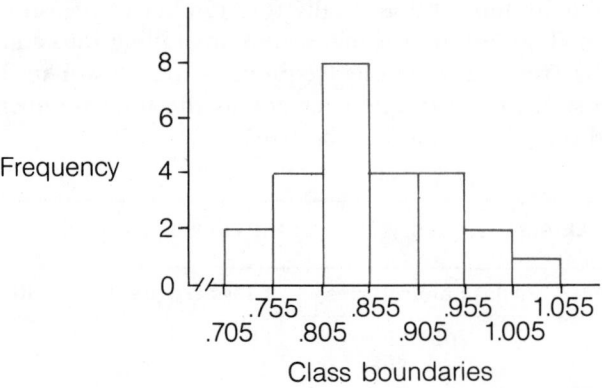

tical axis in the relative frequency histogram is scaled for relative frequency rather than for frequency. The relative frequency histogram for
Table 4.3 is presented in Figure 4.10. Very little distinction is made between these two histograms since they become the same figures if drawn
to the same scale. We frequently refer to either one as simply a histogram.

FIGURE 4.10   Relative Frequency Histogram for the Fluoride
Data

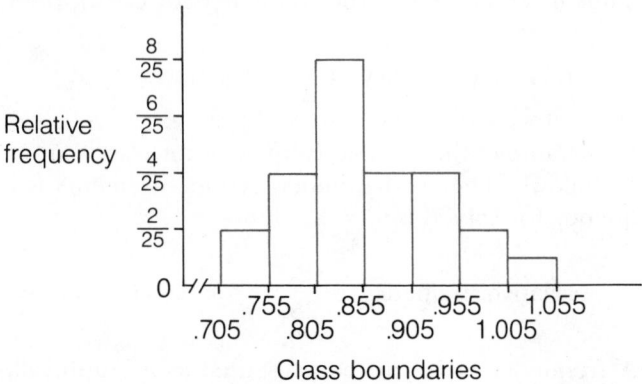

A relative frequency histogram describing a set of 1000 recorded
fluoride readings is shown in Figure 4.11. Note that we have used more
intervals to obtain a better description of the data. Also note that the relative frequency histogram begins to approach a smooth curve when both
the number of measurements and the number of class intervals are
increased.

How do you decide on the number of classes to be used to construct
a histogram for a set of data? There is no exact answer to this question,

FIGURE 4.11 *Histogram for 1000 Fluoride Readings*

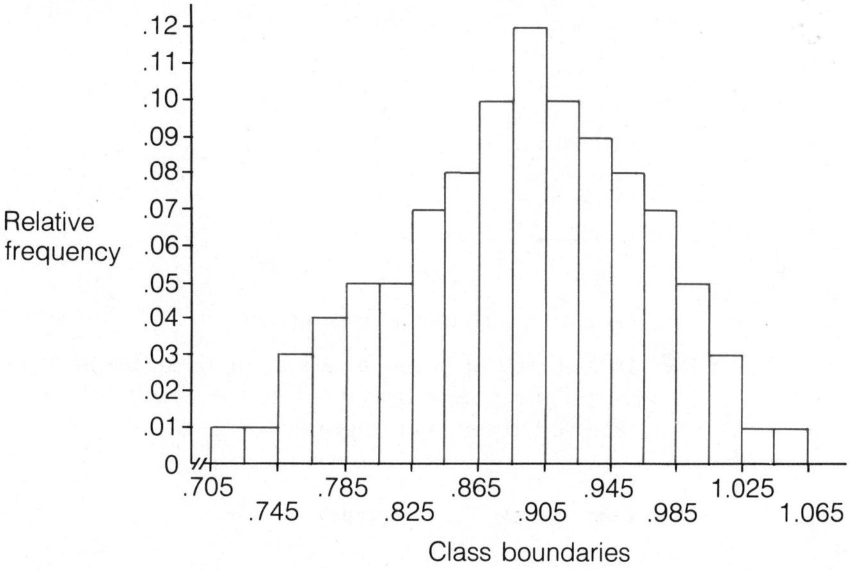

but we can obtain a partial answer if we consider what the histogram is supposed to accomplish. Ideally we want to choose the number of classes that will provide the best graphical description of the way the data are distributed over the scale of measurement. Particularly we would like to see the rise and fall of the relative frequencies over the interval of measurement and the approximate location of the maximum relative frequency. These characteristics, which are visible in the histograms of Figures 4.10 and 4.11, would be lost if the number of classes were either very small or very large. For this reason most histograms are constructed by using between 5 and 20 classes, the smaller number being reserved for small sets of data. More than 7 class intervals would probably be too many to properly characterize the 25 measurements of Table 4.2, but 18 classes were successfully used to characterize the 1000 measurements in Figure 4.11. Thus for large sets of data you can use a larger number of class intervals, say from 15 to 20.

EXERCISES

4.9 During a crackdown, the number of speeding citations delivered by each of the 30 members of a police department was recorded. These data are summarized in the following frequency table.

Number of citations	Frequency
−.5–4.5	2
4.5–9.5	3
9.5–14.5	5
14.5–19.5	8
19.5–24.5	9
24.5–29.5	2
29.5–34.5	1

Construct a frequency histogram.

$ 4.10   In a survey of prices of a loaf of white bread at regional stores, we obtained the following data. Use the results of this survey to construct a relative frequency histogram.

Cost (cents)	Frequency
89.5–94.5	5
94.5–99.5	10
99.5–104.5	14
104.5–109.5	13
109.5–114.5	15
114.5–119.5	21
119.5–124.5	16
124.5–129.5	7
129.5–134.5	2

(a) Compute the relative frequency for each class interval.

(b) Construct a relative frequency histogram.

$ 4.11   The degree of job satisfaction among employees in any job classification is difficult to quantify. Attempts have been made to develop questionnaires (sometimes called *instruments*) composed of specific questions related to a variable of interest. Each respondent is asked to quantify his or her answer to each question on a scale (perhaps from 0 to 5). In this way a respondent obtains a total score (often called *index score*) for the entire questionnaire. It is hoped that different numerical scores will differentiate the variable of interest, such as the degree of job satisfaction among employees. On the following table one such instrument was used to measure the degree of job satisfaction among a sample of 219 nurses. (A high index score indicates a high degree of job satisfaction.) The data are summarized in the accompanying frequency table. Construct a frequency histogram for these data.

Degree of job satisfaction among 219 nurses:

Index score	f	Index score	f
19.5–23.5	2	47.5–51.5	43
23.5–27.5	1	51.5–55.5	38
27.5–31.5	1	55.5–59.5	43
31.5–35.5	3	59.5–63.5	24
35.5–39.5	7	63.5–67.5	13
39.5–43.5	7	67.5–71.5	4
43.5–47.5	33		
		Total	219

Source: Marshall, unpublished paper.

4.12 Refer to the data of Exercise 4.11. Compute the relative frequency for each class and construct a relative frequency histogram. Note that the shape of the relative frequency histogram should be the same as that of the frequency histogram of Exercise 4.11.

4.13 Many metropolitan areas throughout the country have experienced staggering increases in size over the past decade. It has been estimated that nearly 70% of the entire United States population lives in 264 metropolitan areas, and by the year 2000 this percentage could go as high as 83%. To further emphasize the crowded conditions in these areas, the average number of people per square mile for the 264 metropolitan areas is 400. The accompanying table lists the 50 most crowded and the 50 least crowded metropolitan areas from the original list of 264.

Most crowded		Least crowded	
Metropolitan area	*People per square mile*	*Metropolitan area*	*People per square mile*
1. Jersey City	12,963	1. Reno	19
2. New York	7,206	2. Laredo	22
3. Paterson-Clifton-Passaic	2,400	3. Richland-Kennewick, Wash.	31
4. Boston	2,351	4. Great Falls, Mont.	31
5. Meriden, Conn.	2,332	5. Billings	33
6. Nassau-Suffolk, N.Y.	2,096	6. Yakima, Wash.	34
7. Newark, N.J.	2,039	7. Las Vegas	35
8. Bridgeport	2,029	8. Duluth-Superior	36
9. Chicago	1,877	9. Tucson	38
10. New Brunswick-Perth Amboy-Sayreville, N.J.	1,871	10. Bakersfield, Calif.	40
		11. Riverside-San Bernardino-Ontario	42
11. Anaheim-Santa Ana-Garden Grove	1,816	12. Fargo-Moorhead	43
		13. Abilene, Tex.	45

(continued)

(continued)

Most crowded		Least crowded	
Metropolitan area	*People per square mile*	*Metropolitan area*	*People per square mile*
12. Los Angeles-Long Beach	1,728	14. Eugene-Springfield, Ore.	47
13. Stamford	1,706	15. Fort Smith	47
14. New Britain	1,670	16. San Angelo, Tex.	47
15. Norwalk	1,449	17. Pueblo, Colo.	49
16. Cleveland	1,359	18. Texarkana	56
17. Philadelphia	1,356	19. St. Cloud, Minn.	62
18. Trenton	1,333	20. Alexandria, La.	66
19. San Francisco-Oakland	1,254	21. Albuquerque	68
20. Lowell, Mass.	1,219	22. Provo-Orem	68
21. Providence-Warwick-Pawtucket	1,212	23. Fresno	69
22. Detroit	1,132	24. Midland, Tex.	70
23. New Haven-West Haven	1,109	25. Fayetteville-Springdale, Ark.	71
24. Brockton, Mass.	1,098	26. Salinas-Seaside-Monterey, Calif.	75
25. Honolulu	1,056	27. Killeen-Temple, Tex.	76
26. Washington	1,034	28. Wichita Falls, Tex.	76
27. Long Branch-Asbury Park, N.J.	965	29. Amarillo	80
28. Milwaukee	964	30. Salt Lake City-Ogden	82
29. Baltimore	917	31. Tallahassee	86
30. Bristol, Conn.	885	32. Tuscaloosa	87
31. Springfield-Chicopee-Holyoke	856	33. Colorado Springs	88
32. Buffalo	849	34. Bloomington-Normal, Ill.	89
33. Lawrence-Haverhill, Mass.	848	35. Sherman-Denison, Tex.	89
34. Waterbury, Conn.	844	36. Williamsport, Pa.	93
35. San Jose	819	37. Florence, Ala.	94
36. Fall River, Mass.	807	38. Tulsa, Okla.	97
37. Pittsburgh	788	39. Santa Barbara-Santa Maria-Lompoc	97
38. New Bedford, Mass.	783	40. Lynchburg, Va.	97
39. Akron	752	41. Salem, Ore.	98
40. Hartford	698	42. Pine Bluff, Ark.	98
41. Gary-Hammond-East Chicago	675	43. Bryan-College Station, Tex.	99
42. Worcester, Mass.	667	44. Lawton, Okla.	100
43. Cincinnati	644	45. Odessa, Tex.	101
44. Louisville	623	46. Topeka	102
45. Miami	621	47. Sioux City	103
46. Lewiston-Auburn, Maine	604	48. Wilmington, N.C.	103
		49. Tyler, Tex.	104

Most crowded		Least crowded	
Metropolitan area	*People per square mile*	*Metropolitan area*	*People per square mile*
47. Fitchburg- Leominster, Mass.	581	50. Phoenix	106
48. Nashua, N.H.	560		
49. Norfolk-Virginia Beach-Portsmouth	548		
50. New Orleans	532		

Note immediately that Jersey City and New York have population densities (number of people per square mile) that far exceed the densities of the remaining 48 cities in the list of the 50 most crowded metropolitan areas. Since it would be difficult to include Jersey City and New York on the same graph with the remaining 48 because of the extremely high densities in these two cities, graph only the remaining 48. Use a frequency histogram to describe the population densities for the 48 cities from Paterson-Clifton-Passaic to New Orleans. Begin the first subinterval at 531.5 and construct each subinterval with a width of 190.

4.14 Refer to Exercise 4.13. Construct a frequency histogram for the 50 least crowded cities among the original 264 cities. Use approximately 10 subintervals with an interval width of 9. Begin the first interval at 18.5.

4.15 Refer to Exercise 4.13. Construct a relative frequency histogram for the most crowded metropolitan area data (excluding Jersey City and New York). Use the same subintervals as in Exercise 4.13.

4.16 Refer to Exercise 4.13. Construct a relative frequency histogram for the 50 least crowded metropolitan areas, using the same subintervals as in Exercise 4.14. Note that the shape of the relative frequency histogram is identical to that for the frequency histogram of Exercise 4.14.

4.17 The length of time an outpatient must wait for treatment is a variable that plays an important role in the design of outpatient clinics. The waiting times (in minutes) for 50 patients at a pediatric clinic are as follows:

```
 35   22   63    6   49   19   15   83   46   19
 16   31   24   29   36   68   42   57   64    8
 23   47   21   51    7   40   19   46   16   32
108   33   55   32   22   36   25   27   37   58
 39   10   42   28   72   13   51   45   77   16
```

Construct a relative frequency histogram for these data.

4.6   FREQUENCY POLYGONS

frequency polygon

Class relative frequencies can also be portrayed in the form of a **frequency polygon.** The only difference between a frequency polygon and a frequency histogram is that the frequency associated with each class is indicated by a dot placed over the midpoint of the class interval. The dots are then joined by straight lines. The frequency polygon for the fluoride data of Table 4.3 is presented in Figure 4.12. Note that we actually used two additional classes, one at each end. This device provides a neater graph, with endpoints that fall off to zero on the horizontal axis.

Frequency polygons are often used in news articles when it is necessary to summarize frequency data. For example, the data summarized in Figure 4.13 were published in a news article concerning the number and intensity of tropical storms. Some of these results are presented in Figure 4.13, which provides information on the frequency of hurricanes and tropical storms by month from data collected over a 73-year period.

FIGURE 4.12   Frequency Polygon for the Fluoride Data

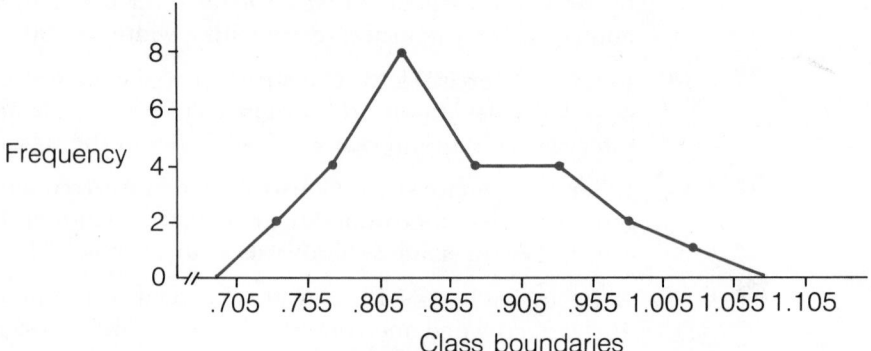

4.7   COMMENTS CONCERNING HISTOGRAMS

There are several comments that should be made concerning histograms. First, we need a brief discussion of variables.

Definition 4.1

variable

When observations on the same phenomenon vary from trial to trial, the phenomenon is called a **variable.**

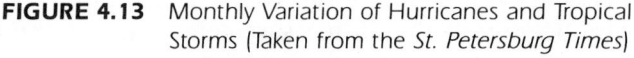

FIGURE 4.13 Monthly Variation of Hurricanes and Tropical Storms (Taken from the *St. Petersburg Times*)

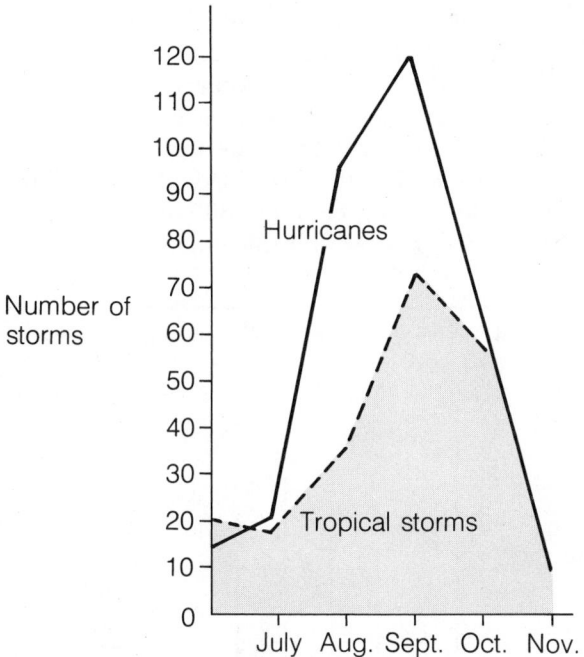

For example, the marriage license fee in a state is a variable because the fee may vary from county to county. Similarly the monthly weight gain (or loss) for an individual is a variable because it changes from month to month.

qualitative and quantitative variables

Variables can be classified as **qualitative** or **quantitative.** Values of qualitative variables vary in kind but not degree, and hence are not measurements. For example, the variable political party affiliation can be categorized as Republican, Democrat, or other, and, although we could label the categories as one, two, or three, these values are only codes and have no quantitative interpretation. Political party affiliation is a qualitative variable. In contrast, quantitative variables have actual units of measure. For example, the yield (in bushels) per acre of corn can assume specific, numerical values. Bar charts are used to display frequency data for qualitative variables; histograms are appropriate for displaying frequency data for quantitative variables.

Second, the histogram is the most important graphical technique we will present because of the role it plays in statistical inference, a subject we will discuss in later chapters. Third, the fraction of the total number of measurements in an interval is equal to the fraction of the total area under the histogram over the interval. For example, consider the relative frequency histogram for the sample of 25 fluoride readings (Figure 4.10). The fraction of the 25 measurements less than or equal to .905 is $^{18}/_{25}$. You will note that $^{18}/_{25}$ of the total area under the histogram (shaded) lies to the left of .905. See Figure 4.14.

probability

Fourth, if a single measurement is selected arbitrarily from the set of measurements, the chance (probability) that it lies in a particular interval is equal to the fraction of the total area under the histogram over that interval. For example, if 25 cards were labeled with respective fluoride readings in Table 4.2, then shuffled, the probability of choosing a card with a fluoride reading between .805 and .855 ppm is $^{8}/_{25}$, because 8 out of the 25 readings fall within this interval. From Figure 4.15 we note that

---

FIGURE 4.14   Relative Frequency Histogram for the Fluoride Data

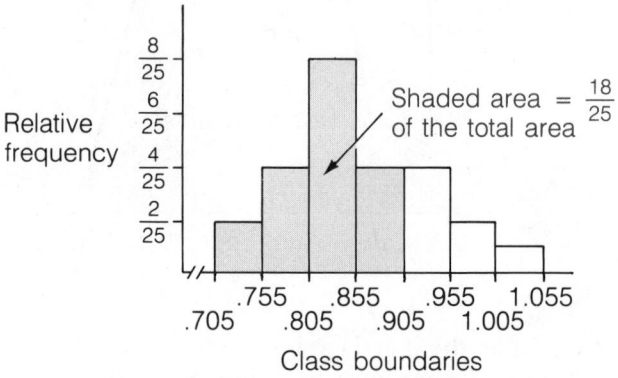

FIGURE 4.15   Relative Frequency Histogram Showing the Fraction of Fluoride Readings in the Interval .805 to .855

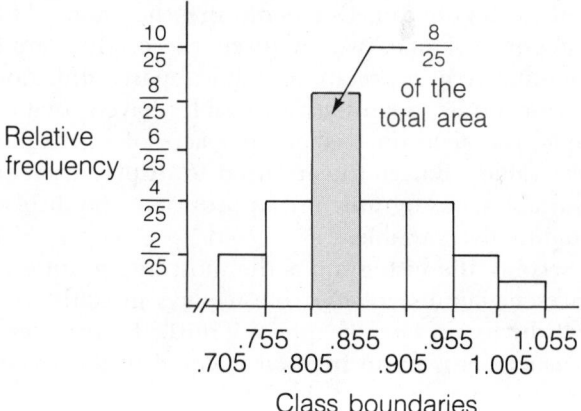

the shaded area under the relative frequency histogram over the interval .805 to .855 is equal to $^8/_{25}$ of the total area.

Fifth, we can construct a frequency histogram for any set of measurements (sample or population), but in most situations we will not know the population values. Imagine what the histogram for a population might look like. Since populations usually contain a large number of measurements, the number of classes can usually be made rather large so that the population frequency histogram becomes almost a smooth curve. The 25 fluoride readings of Table 4.2 were obtained from a sample of 25 days. As the number of measurements in a sample increases, we can select smaller class intervals. The resulting histogram will then become more regular and tend to become a smooth curve. Figure 4.16 shows three relative frequency histograms for the fluoride measurements. The first is a histogram

FIGURE 4.16 Relative Frequency Histograms for the Fluoride Data

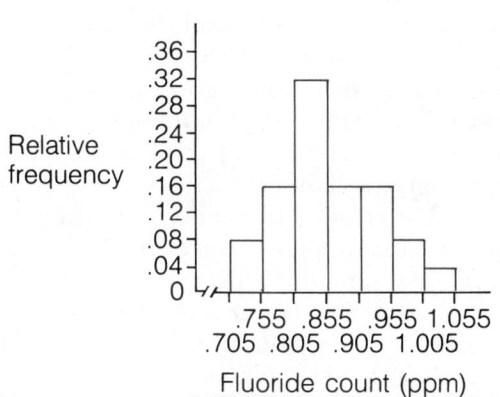

(a) Sample size: $n = 25$

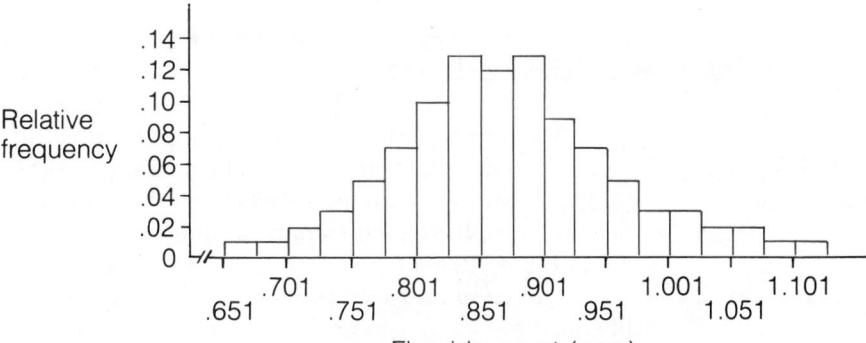

(b) Sample size: $n = 100$

(continued)

FIGURE 4.16 Relative Frequency Histograms for the

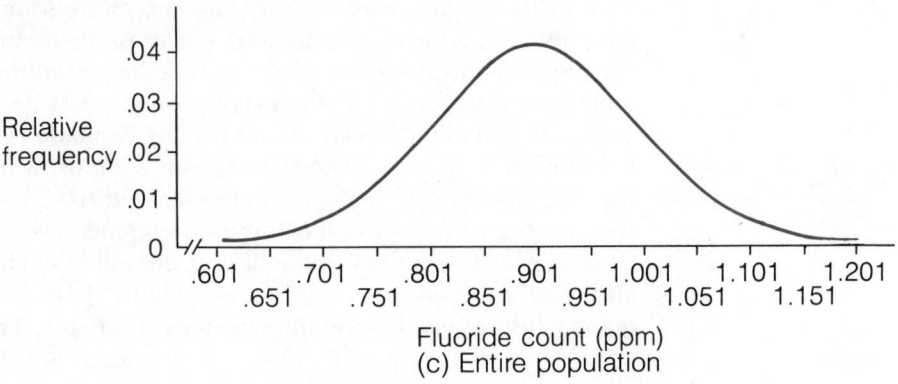

Fluoride count (ppm)
(c) Entire population

for a sample of 25 measurements; the second is a histogram for a sample of 100 measurements; and the third is a possible histogram for the entire population. Note that the scale of relative frequency (the vertical scale) will change from one figure to another. Also observe that it may be necessary to change the endpoints of the class intervals [as was done in Figure 4.16(a) and (b)] so that no measurement falls on the boundary between two classes.

The relative frequency histogram can be very useful in describing a set of measurements. For example, if the set of measurements represents a random sample of measurements from a population of interest, we can use the fraction of measurements falling in a given interval as an "estimate" of the fraction of measurements falling in the same interval for the corresponding population.

4.8 STEM-AND-LEAF PLOTS

exploratory data analysis (EDA)

The final graphical technique we will present is a display technique taken from an emerging area of statistics called **exploratory data analysis (EDA).** Professor John Tukey has been the leading proponent of this practical philosophy of data analysis aimed at exploring and understanding data (Tukey, 1977).

stem-and-leaf plot

The **stem-and-leaf plot** is a simple device to construct a histogram-like picture of a frequency distribution. It allows us to use the information contained in a frequency distribution to show the range of scores, their concentration, the shape of the distribution, whether there are any specific values or scores not represented, and whether there are any stray or extreme scores.

TABLE 4.4 Violent Crime Rates for 90 SMSAs Selected from South, North, and West

South	Rate*	North	Rate*	West	Rate*
Atlanta	843	Albany, N.Y.	270	Bakersfield	818
Augusta, Ga.	560	Allentown	171	Boise	441
Baton Rouge	792	Atlantic City	858	Colorado Springs	451
Beaumont, Tex.	668	Canton, Ill.	336	Denver	640
Birmingham	637	Chicago	585	Eugene, Ore.	351
Charlotte	707	Cincinnati	521	Fresno	925
Chattanooga	522	Cleveland	794	Honolulu	322
Columbia, S.C.	953	Detroit	860	Kansas City	869
Corpus Christi	592	Evansville, Ind.	444	Las Vegas	1148
Dallas	752	Grand Rapids	411	Lawton, Okla.	572
El Paso	621	Johnstown, Pa.	201	Modesto, Calif.	570
Fort Lauderdale	991	Kankakee, Ill.	453	Oklahoma City	618
Greensboro, N.C.	503	Kenosha, Wis.	250	Oxnard, Calif.	480
Houston	719	Lancaster, Pa.	112	Pueblo, Colo.	792
Jackson	495	Lansing	322	Sacramento	709
Knoxville	404	Lima, Ohio	346	St. Louis	766
Lexington, Ky.	370	Madison, Wis.	212	Salinas, Calif.	574
Lynchburg	184	Mansfield, Ohio	758	Salt Lake City	390
Macon, Ga.	470	Milwaukee	306	San Diego	629
Mobile	832	Newark	1000	San Francisco	1005
Monroe, La.	918	Paterson, N.J.	737	San Jose	514
Nashville	550	Philadelphia	638	Seattle	559
Newport News	437	Pittsfield, Ma.	305	Sioux City	289
Richmond	499	Racine, Wis.	527	Spokane	439
Roanoke	272	Rockford, Ill.	592	Stockton	739
Shreveport, La.	656	South Bend	436	Tacoma	578
Washington, D.C.	802	Springfield, Ill.	607	Topeka	494
West Palm Beach	1068	Syracuse	265	Tucson	616
Wichita Falls	830	Vineland, N.J.	482	Vallejo	487
Wilmington, Del.	448	Youngstown	409	Waco, Tex.	548

*Rates represent the number of violent crimes (that is, murder, forcible rape, robbery, and aggravated assault) per 100,000 inhabitants, rounded to the nearest whole number.

Source: Department of Justice, 1980, pp. 60–86.

The stem-and-leaf plot does *not* follow the organization principles stated earlier for histograms. We will use the data shown in Table 4.4 to illustrate how to construct a stem-and-leaf plot.

leading digit
trailing digits

The original scores of Table 4.4 are either three- or four-digit numbers. We will use the first or **leading digit** of scores as the stem (see Figure 4.17) and the **trailing digits** as the leaf. For example, the violent crime rate in St. Louis is 766. The leading digit is 7 and the trailing digits are 66. In the case of Las Vegas, the leading digits are 11 and the trailing digits are 48.

FIGURE 4.17   Stem-and-Leaf Plot for Violent Crime Rates of Table 4.4.

1	84 71 12
2	72 70 01 50 12 65 89
3	70 36 22 46 06 05 51 22 90
4	95 04 70 37 99 48 44 11 53 36 82 09 41 51 80 39 94 87
5	60 22 92 03 50 85 21 27 92 72 70 74 14 59 78 48
6	68 37 21 56 38 07 40 18 29 16
7	92 07 52 19 94 58 37 92 09 66 39
8	43 32 02 30 58 60 18 69
9	53 91 18 25
10	68 00 05
11	48

If our data consisted of 6-digit numbers such as 104,328, we might use the first two digits as stem numbers, the second two digits as leaf numbers, and ignore the last two digits.

For the data on violent crime, the smallest rate is 112, the largest is 1148, and the leading digits are 1, 2, 3, . . . , 11. In the same way as a class interval determines where a measurement is placed in a frequency table, the leading digit (stem of a score) determines the row in which a score is placed in a stem-and-leaf plot. This has been done in Figure 4.17 for the violent crime data. The end result is that a stem-and-leaf plot is a graph that looks much like a histogram turned sideways.

We can see that each stem defines a class interval and the limits of each interval are the largest and smallest possible scores for the class. The values represented by each leaf must be between the lower and upper limits of the interval.

The plot can be made a bit neater by ordering the data within a row from lowest to highest. (See Figure 4.18.) The advantages of such a graph over the histogram are that not only does it reflect frequencies, concentra-

FIGURE 4.18   Reordering of the Stem-and-Leaf Plot of Figure 4.17.

1	12 71 84
2	01 12 50 65 70 72 89
3	05 06 22 22 36 46 51 70 90
4	04 09 11 36 37 39 41 44 48 51 53 70 80 82 87 94 95 99
5	03 14 21 22 27 48 50 59 60 70 72 74 78 85 92 92
6	07 16 18 21 29 37 38 40 56 68
7	07 09 19 37 39 52 58 66 92 92 94
8	02 18 30 32 43 58 60 69
9	18 25 53 91
10	00 05 68
11	48

tion(s) of scores, and shape of the distribution; but it also gives actual scores from which we can determine whether there are any values not represented and whether or not there are any stray or extreme values.

In summary, to display a data set in the format of a stem-and-leaf plot:

1. Split each score or value into two sets of digits. The first (or leading) set of digits is the stem, and the second (or trailing) set of digits is the leaf.

2. List all possible stem digits from lowest to highest.

3. For each score in the mass of data write down the leaf numbers on the line labeled by the appropriate stem number.

4. If the display looks too cramped and narrow, we can stretch the display by using two lines (or more) per stem so that, for example, we place leaf digits 1, 2, 3, and 4 on the first line of the stem and leaf digits 5, 6, 7, 8, and 9 on the second line. (Exercise 4.50 shows how this is done.)

5. If too many trailing digits are present, such as in a 6- or 7-digit score, we drop the rightmost trailing digit(s) to maximize the clarity of the display.

6. The rules for developing a stem-and-leaf plot are somewhat different from the rules governing the establishment of class intervals for the traditional frequency distribution and for a variety of other procedures that we will consider in later sections of the text. Class intervals for stem-and-leaf plots are, then, in a sense slightly atypical.

EXERCISES

4.18 The final examination scores are listed for 50 students from a mathematics course.

43	61	68	72	79
51	62	69	73	79
53	63	69	73	81
55	64	69	74	82
57	65	69	74	82
58	65	70	75	85
58	66	70	76	87
59	66	70	76	89
61	67	71	77	91
61	68	71	78	96

Construct a stem-and-leaf plot for these data.

4.19   Refer to the data of Exercise 4.18. Construct a histogram for these data. Compare this histogram to the stem-and-leaf plot of Exercise 4.18.

4.9   USING COMPUTERS

In this section are several examples of computer-generated graphs from Minitab using data from this chapter. First we give a Minitab program that can be used to generate a histogram and stem-and-leaf plot for the data of Table 4.4. Compare the output shown here to the plots we did by hand earlier in the chapter. The histogram displayed in the output is fairly straightforward. The middle of each interval and the number of observations falling in each interval are given. The histogram is displayed horizontally rather than vertically, as we're used to seeing it. Each asterisk (*) in the histogram represents an observation.

The stem-and-leaf plot in the Minitab output is not as clearly labeled as the histogram. First, you can ignore the left-hand column for now; it is used to locate the central measurement (median), discussed in Chapter 5. The middle column of the display gives the stems of the stem-and-leaf plot. The last column of the output gives the leaf portion of the stem-and-leaf plot. And, as indicated by the note at the top of the output, the leaf digit unit is 10.0. The first row of our stem-and-leaf plot in Figure 4.18 is 1 | 12 71 84. If we round the leaf digits 12, 71, and 84 to the nearest 10, we have the numbers 10, 70, and 80. These digits give rise to the first row of the Minitab stem-and-leaf plot, 1 | 1 7 8. With this basic understanding, don't worry about being able to reconstruct all the leaves in a Minitab stem-and-leaf plot; you should, however, be able to describe a set of measurements from a Minitab stem-and-leaf plot. That is, from a Minitab plot you should be able to determine (approximately) concentrations of scores, the shape of the distributions, the range of scores, low and high values, and whether or not there are extreme values.

```
MTB > SET INTO C1
DATA> 843   560   792   668   637   707   522   953   592   752
DATA> 621   991   503   719   495   404   370   184   470   832
DATA> 918   550   437   499   272   656   802  1068   830   448
DATA> 270   171   858   336   585   521   794   860   444   411
DATA> 201   453   250   112   322   346   212   758   306  1000
DATA> 737   638   305   527   592   436   607   265   482   409
DATA> 818   441   451   640   351   925   322   869  1148   572
DATA> 570   618   480   792   709   766   574   390   629  1005
DATA> 514   559   289   439   739   578   494   616   487   548
DATA> END

MTB > PRINT C1

C1
    843   560   792   668   637   707   522   953   592   752   621
    991   503   719   495   404   370   184   470   832   918   550
```

```
437   499   272   656   802  1068   830   448   270   171   858
336   585   521   794   860   444   411   201   453   250   112
322   346   212   758   306  1000   737   638   305   527   592
436   607   265   482   409   818   441   451   640   351   925
322   869  1148   572   570   618   480   792   709   766   574
390   629  1005   514   559   289   439   739   578   494   616
487   548
```

```
MTB > HISTOGRAM C1

Histogram of C1   N = 90

Midpoint   Count
     100       1   *
     200       4   ****
     300      11   ***********
     400      12   ************
     500      15   ***************
     600      18   ******************
     700       7   *******
     800      11   ***********
     900       5   *****
    1000       4   ****
    1100       2   **

MTB > STEM-AND-LEAF C1

Stem-and-leaf of C1        N  = 90
Leaf Unit = 10

      3     1  178
     10     2  0156778
     19     3  002234579
     37     4  001333444557888999
    (16)    5  0122245567777899
     37     6  0112233456
     27     7  00133556999
     16     8  01334566
      8     9  1259
      4    10  006
      1    11  4

MTB > STOP
```

Next we have a SAS program, which shows how to use PROC CHART to generate a histogram for the data of Table 4.4 and how PROC UNIVARIATE with the PLOT option can be used to construct a stem-and-leaf plot for the same data. Following the listing of data in the output, we see a histogram (or as SAS calls it, a *frequency bar chart)* of the crime rate data. The midpoints of the intervals are given along the horizontal axis and class frequency along the vertical axis. Each line of five asterisks in a given interval represents one observation.

The stem-and-leaf plot for SAS (as with Minitab) is a bit more difficult to read than a SAS histogram, and the one displayed in the output looks different from the plot that we constructed in Figure 4.18. The first column of the plot gives the stems. Note that there are more stems in this plot than there are in Figure 4.18. The leaf portions of the plot are listed in the second column. According to the note at the bottom of the plot, we must multiply each number represented in the form of a stem and leaf by $10^2 = 100$. The first line of the output 11 | 5 is taken to be 11.5. If we multiply 11.5 by 100 we get 1150. Actually, the highest crime rate (1148) has been rounded to 1150 and presented in the plot. Similarly, in the third line 10 | 7 (or 10.7) becomes 1070, which corresponds to the crime rate 1068. The top and bottom lines of the stem-and-leaf plot, 1 | 1 and 1 | 78, represent 110, 170, and 180. As with Minitab, these numbers are rounded versions of the original measurements 112, 171, and 184.

Again, don't worry too much about how the leaves of a SAS stem-and-leaf plot are constructed. Just be able to describe a set of measurements from the plot. You can ignore much of the other sections of the output from this program until we cover these topics later in the text.

The SAS and Minitab programs presented in this section can be used as models for similar analyses to be performed on other data sets, particularly for histograms and stem-and-leaf plots. Additional details about Minitab statements and SAS procedures are available in the user's manuals for these two software systems (see references at the end of this text). Also Exercises 4.50, 4.52, and 4.53 offer other examples of the use of Minitab to graph and plot data.

Minitab Output

```
OPTIONS NODATE NONUMBER PS=60 LS=78;
TITLE1 "VIOLENT CRIME RATES FOR 90 SMSA'S SELECTED";
TITLE2 'FROM THE NORTH, SOUTH AND WEST';

DATA CRIME;
INPUT RATE @@;
CARDS;
      843   560   792   668   637   707   522   953   592   752
      621   991   503   719   495   404   370   184   470   832
      918   550   437   499   272   656   802  1068   830   448
      270   171   858   336   585   521   794   860   444   411
      201   453   250   112   322   346   212   758   306  1000
      737   638   305   527   592   436   607   265   482   409
      818   441   451   640   351   925   322   869  1148   572
      570   618   480   792   709   766   574   390   629  1005
      514   559   289   439   739   578   494   616   487   548
;
PROC PRINT N;
TITLE3 'LISTING OF THE DATA';

PROC CHART DATA=CRIME;
   VBAR RATE;
TITLE3 'EXAMPLE OF PROC CHART';
TITLE4 'FREQUENCY DISTRIBUTION OF THE VARIABLE CRIME RATE';

PROC UNIVARIATE PLOT DATA=CRIME;
   VAR RATE;
TITLE3 'EXAMPLE OF PROC UNIVARIATE WITH THE PLOT OPTION';
RUN;
```

SAS Output

```
                VIOLENT CRIME RATES FOR 90 SMSA'S SELECTED
                    FROM THE NORTH, SOUTH AND WEST
                         LISTING OF THE DATA

                         OBS     RATE

                          1       843
                          2       560
                          3       792
                          4       668
                          5       637
                          6       707
                          7       522
                          8       953
                          9       592
                         10       752
                         11       621
                         12       991
                         13       503
                         14       719
                         15       495
                         16       404
                         17       370
                         18       184
                         19       470
                         20       832
                         21       918
                         22       550
                         23       437
                         24       499
                         25       272
                         26       656
                         27       802
                         28      1068
                         29       830
                         30       448
                         31       270
                         32       171
                         33       858
                         34       336
                         35       585
                         36       521
                         37       794
                         38       860
                         39       444
                         40       411
                         41       201
                         42       453
                         43       250
                         44       112
                         45       322
                         46       346
                         47       212
                         48       758
                         49       306
                         50      1000
                         51       737
```

(continued)

```
        VIOLENT CRIME RATES FOR 90 SMSA'S SELECTED
           FROM THE NORTH, SOUTH AND WEST
                 LISTING OF THE DATA

              OBS      RATE

               52       638
               53       305
               54       527
               55       592
               56       436
               57       607
               58       265
               59       482
               60       409
               61       818
               62       441
               63       451
               64       640
               65       351
               66       925
               67       322
               68       869
               69      1148
               70       572
               71       570
               72       618
               73       480
               74       792
               75       709
               76       766
               77       574
               78       390
               79       629
               80      1005
               81       514
               82       559
               83       289
               84       439
               85       739
               86       578
               87       494
               88       616
               89       487
               90       548

              N = 90
```

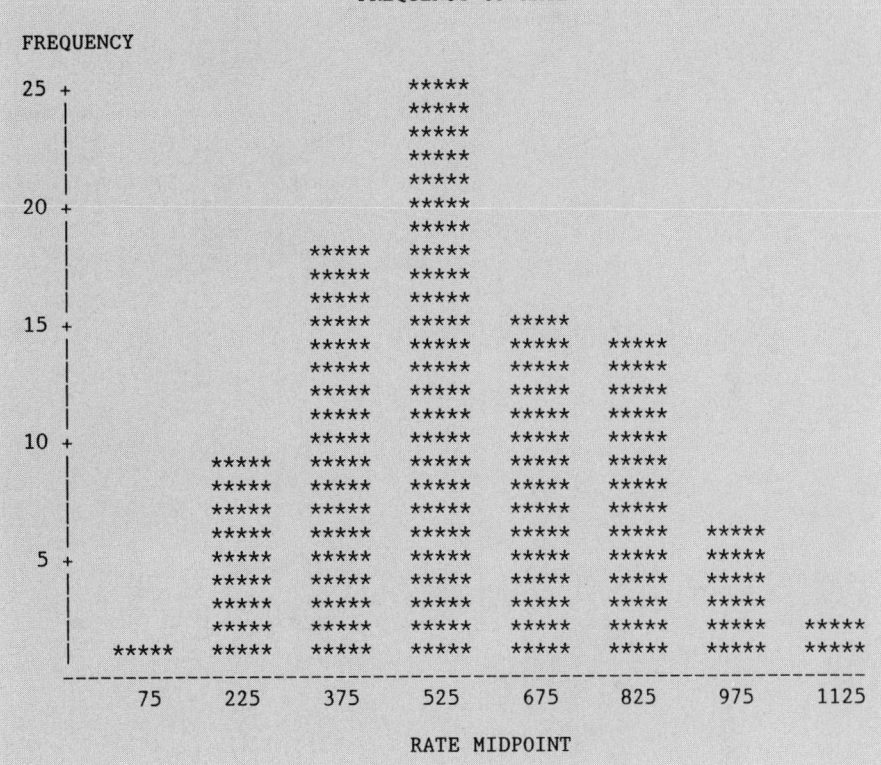

```
              VIOLENT CRIME RATES FOR 90 SMSA'S SELECTED
                  FROM THE NORTH, SOUTH AND WEST
                      EXAMPLE OF PROC CHART
             FREQUENCY DISTRIBUTION OF THE VARIABLE CRIME RATE

                          FREQUENCY OF RATE

FREQUENCY

   25 +                            *****
      |                            *****
      |                            *****
      |                            *****
      |                            *****
   20 +                            *****
      |                            *****
      |                     *****  *****
      |                     *****  *****
      |                     *****  *****
   15 +                     *****  *****  *****
      |                     *****  *****  *****  *****
      |                     *****  *****  *****  *****
      |                     *****  *****  *****  *****
      |                     *****  *****  *****  *****
   10 +              *****  *****  *****  *****  *****
      |              *****  *****  *****  *****  *****
      |              *****  *****  *****  *****  *****
      |              *****  *****  *****  *****  *****  *****
    5 +              *****  *****  *****  *****  *****  *****
      |              *****  *****  *****  *****  *****  *****
      |              *****  *****  *****  *****  *****  *****
      |              *****  *****  *****  *****  *****  *****  *****
      |       *****  *****  *****  *****  *****  *****  *****  *****
      ------------------------------------------------------------
             75    225    375    525    675    825    975   1125

                          RATE MIDPOINT
```

```
              VIOLENT CRIME RATES FOR 90 SMSA'S SELECTED
                   FROM THE NORTH, SOUTH AND WEST
              EXAMPLE OF PROC UNIVARIATE WITH THE PLOT OPTION

                        UNIVARIATE PROCEDURE

Variable=RATE


                               Moments

            N                  90    Sum Wgts             90
            Mean          573.7333   Sum               51636
            Std Dev       225.3603   Variance       50787.25
            Skewness      0.279827   Kurtosis       -0.40103
            USS           34145360   CSS             4520066
            CV            39.27962   Std Mean       23.75506
            T:Mean=0      24.15205   Prob>|T|         0.0001
            Sgn Rank        2047.5   Prob>|S|         0.0001
            Num ^= 0            90


                            Quantiles(Def=5)

            100% Max          1148    99%             1148
             75% Q3            739    95%              991
             50% Med         559.5    90%            864.5
             25% Q1            436    10%            280.5
              0% Min          112     5%              212
                                      1%              112

            Range            1036
            Q3-Q1             303
            Mode              322


                               Extremes

            Lowest      Obs        Highest     Obs
               112(      44)          991(      12)
               171(      32)         1000(      50)
               184(      18)         1005(      80)
               201(      41)         1068(      28)
               212(      47)         1148(      69)
```

```
          VIOLENT CRIME RATES FOR 90 SMSA'S SELECTED
                FROM THE NORTH, SOUTH AND WEST
         EXAMPLE OF PROC UNIVARIATE WITH THE PLOT OPTION

                      UNIVARIATE PROCEDURE

Variable=RATE


        Stem Leaf                          #          Boxplot
          11 5                             1             |
          11                                             |
          10 7                             1             |
          10 00                            2             |
           9 59                            2             |
           9 22                            2             |
           8 667                           3             |
           8 02334                         5             |
           7 567999                        6             |
           7 11244                         5          +------+
           6 67                            2          |      |
           6 12223444                      8          |      |
           5 55667778899                  11          *--+---*
           5 0001223                       7          |      |
           4 55578899                      8          |      |
           4 01144444                      8          +------+
           3 5579                          4             |
           3 01224                         5             |
           2 56779                         5             |
           2 01                            2             |
           1 78                            2             |
           1 1                             1             |
             ----+----+----+----+
        Multiply Stem.Leaf by 10**+2
```

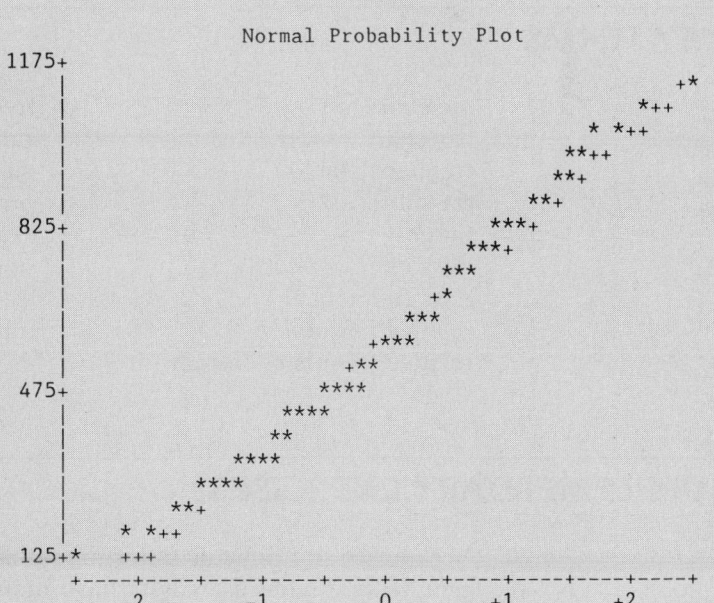

                      Normal Probability Plot

SUMMARY

We study how to describe a set of measurements for two reasons. First, the objective of statistics is to make inferences about a population based on information contained in a sample. Since populations are sets of measurements, existing or conceptual, we need some method for talking about the population or, equivalently, for describing a set of measurements. Second, graphical description of data is very useful because it is easily comprehended by both the novice and the scientist. For example, there is a need for the condensation and description of large quantities of economic or sociological data that are collected annually by various government bureaus or even every ten years by the U.S. Bureau of the Census. Graphical descriptions of these data are easily understood by the layperson and the scientist engaged in economic or sociological research.

Data can be described either graphically or numerically. Graphical methods, such as *pie charts, bar charts, histograms, frequency polygons,* and *stem-and-leaf plots,* are presented in this chapter. Numerical descriptive measures are discussed in Chapter 5.

Note the role of data description in statistical inference: it is impossible to make an inference about anything—measurements or what have you—unless you are able to describe the object of your interest. But graphical description has limitations when used for making inferences, and these will be made apparent in the next chapter.

KEY TERMS

pie chart	frequency polygon
bar chart	variable (*def*)
frequency table	qualitative and quantitative variables
class interval	probability
range	exploratory data analysis (EDA)
class interval width	stem-and-leaf plot
class frequency	leading digit
relative frequency	trailing digits
relative frequency histogram	

SUPPLEMENTARY EXERCISES

4.20   Because of the difficult times many basic industries have endured in recent years, financial analysts have monitored the influx of foreign materials. The data here show steel industry imports (in 1000s of tons) for the years 1979 to 1986.

Year	1979	1980	1981	1982	1983	1984	1985	1986
Import	17,518	15,491	19,898	16,663	17,061	26,171	23,650	19,650

(a) Would a pie chart be an appropriate graphical method for describing these data? Explain.

(b) Construct a bar graph.

4.21 Graph the data shown here in the allocation of our food dollars to the categories of the table below. Try a pie chart and a bar graph. Which seems better?

Where our food dollars go	Percent
Dairy products	13.4
Cereal and baked goods	12.6
Nonalcoholic beverages	8.9
Poultry and seafood	7.5
Fruit and vegetables	15.6
Meat	24.5
Other foods	17.5

4.22 The following table shows quarterly gross national product (GNP) and disposable personal income (DPI) for the years 1985 and 1986 (billions of dollars). Choose a form of a single bar graph to display these data on a single graph.

Year/	1985				1986			
Quarter	I	II	III	IV	I	II	III	IV
GNP	3910	3961	4017	4067	4137	4203	4266	4308
DPI	2505	2532	2503	2533	2536	2555	2579	2589

4.23 Below are data from SAT exams for selected years. Plot these data and give some interpretations to the data.

Gender, Type	Year				
	1967	1970	1975	1980	1983
Male, Math	514	509	495	491	493
Female, Math	467	465	449	443	445
Male, Verbal	463	459	437	428	430
Female, Verbal	468	461	431	420	420

Source: College Entrance Examination Board.

4.24   An hour examination in a statistics course produced the following scores:

40	61	66	74	80
52	61	68	75	82
54	63	68	75	85
57	65	70	78	88
60	66	72	79	97

(a) Set up a frequency table using 7 intervals with the lowest class boundary at 39.5.

(b) Construct a frequency histogram for the set of grouped data.

4.25   Consider the following data:

7   5   4   7   3   6   6   8   5   3   8   6   5   8   6

(a) Construct the frequency histogram.

(b) Group the data using a class interval width of 2. Construct a frequency distribution.

(c) What is the median score for this distribution?

4.26   A student bought 25 new textbooks at her college bookstore in 1979–1980 for the following prices (rounded to the nearest dollar) :

22	15	22	1	9
33	17	12	9	7
20	2	7	4	5
7	5	3	12	12
9	20	6	6	8

(a) Using 5 intervals with the lowest class boundary at 0.5, set up a frequency table.

(b) Construct a frequency histogram for the above set of grouped data.

(c) Construct a frequency polygon for the above set of grouped data.

4.27   A family's monthly expenses are listed as follows:

Housing	$528
Utilities	194
Medical expenses	21
Food	181
Transportation	187
Clothing	35
Savings	35
Miscellaneous	69

Draw a pie chart showing this information graphically. (Hint: Use percentages.)

4.28 A study was conducted to compare prices for auto repair. A test car was repaired at each of 25 different shops, with the following resulting costs (in dollars). (Values are arranged in order for convenience.)

48, 50, 52, 52, 54, 56, 57, 62, 64, 66, 67, 69, 70, 71, 71, 73, 76, 77, 77, 78, 80, 85, 89, 94, 105

Construct a relative frequency histogram for the data; use four equal width intervals. (Label axes.)

4.29 The following are IQ scores of 40 randomly selected high school students:

154	118	127	120
128	91	102	103
125	115	91	118
109	114	98	106
90	131	104	81
85	112	110	99
111	85	83	123
122	147	112	108
104	105	122	127
87	108	133	97

Construct a stem-and-leaf plot for the data set.

4.30 A psychologist has developed a new technique for improving memory. To test this method, 20 students were taught the new technique, and then asked to memorize 100 words. The following list gives the number of words memorized correctly by each student.

91	64	98	66	83	87	83	86	80	93
83	75	72	79	90	80	90	71	84	68

Construct a stem-and-leaf diagram for the data.

4.31 Standardized scores were obtained for each of 88 students (maximum score = 30).

11	29	14	28	29	18	30	4	28	23	1	27	16	7	5	26
5	16	18	1	13	8	12	26	12	28	15	15	21	17	30	24
27	10	16	16	29	6	13	13	10	11	7	5	28	11	26	27
30	29	29	16	11	3	29	15	20	13	13	28	5	30	1	17
17	6	11	9	28	27	12	24	17	22	25	24	17	7	17	14
23	13	27	10	28	28	22	28								

(a) Construct a frequency histogram.

(b) Draw a frequency polygon.

4.32 The data from Exercise 4.31 were grouped as follows:

Grouped	Frequency
1– 5	9
6–10	10
11–15	19
16–20	14
21–25	9
26–30	27

(a) Draw a frequency histogram.

(b) Superimpose a frequency polygon on the histogram of (a).

4.33 Construct a relative frequency histogram for the accompanying public welfare expenditure data.

Per capita expenditure by state ($)	Frequency
49.5–74.5	3
74.5–99.5	6
99.5–124.5	14
124.5–149.5	11
149.5–174.5	2
174.5–199.5	5
199.5–224.5	2
224.5–249.5	5
249.5–274.5	1
274.5–299.5	1

4.34 Consumers have been increasing their personal debt (through credit card and time purchases) during the past 30 years. Total consumer debt, expressed as a percentage of total personal income after taxes, is shown in the accompanying table for 1950 through 1980. Use a bar chart to display these data graphically.

Year	1950	1955	1960	1965	1970	1975	1980
Total consumer debt as a percentage of total personal income	37	53	65	75	73	73	85

4.35 The oil reserves in the Western Hemisphere are estimated to be as shown in the accompanying table. Display these data in a pie chart.

Location	Barrels (in billions)
United States	38.7
South America	22.6
Canada	8.8
Mexico	60.0

4.36 Display the data of Exercise 4.35 in a bar chart.

4.37 Residents of a condemned area in a large city were moved into a new apartment complex. At the end of the first year, a building inspector evaluated the amount of damage that had been done to each apartment. These data are shown here.

Amount of damage ($)	Number of apartments
0–999	7
1000–1999	16
2000–2999	20
3000–3999	35
4000–4999	16
5000–5999	11
6000–6999	5
7000 or greater	2

(a) Construct a frequency histogram for these data.

(b) What fraction of the apartments had damage of $5000 or more? Less than $2000?

4.38 In general, what additional information is provided by a stem-and-leaf plot compared to a frequency histogram?

4.39 A study was conducted to evaluate the absorption characteristics of an antibiotic preparation, chloramphenicol. A subject was given a .5-gram oral dose of the preparation. Urine collections were then made over the next 12 hours, and the number of milligrams of chloramphenicol excreted was recorded at each collection. The data are given in the accompanying table. Use a frequency histogram to display graphically the chloramphenicol excretion pattern for the subject.

Urine collection period (hours after medication)	Chloramphenicol excreted (mL)
0–2	94
2–4	107
4–6	70
6–8	62
8–10	31
10–12	28

4.40 Construct a frequency polygon for the data in Exercise 4.39.

4.41 Urine collections for the subject in Exercise 4.39 continued to be taken over a 48-hour period after the subject received the .5-gram oral dose of chloramphenicol. These data appear in the accompanying table. Discuss how to present these data graphically. Note that the intervals are of unequal width.

Urine collection period (hours after medication)	Chloramphenicol excreted (mL)
0–2	94
2–4	107
4–6	70
6–8	62
8–12	59
12–24	47
24–48	11

4.42 Explain the difference between a frequency histogram and a bar chart.

4.43 An investigator was interested in studying the sedative effect on rats of different doses of a drug. A small cage was constructed with several electric eyes focused at different angles. Attached to each electric eye was a counter that monitored the number of times a rat broke any of the light beams in a 15-minute period. Twenty-five rats were in injected with a specified dose of the drug, and each one was observed in the cage for a 15-minute period. These data are recorded in the accompanying table. Construct a frequency histogram for these data using five or more class intervals.

Rat	Number of times a light beam was broken in the 15-minute period	Rat	Number of times a light beam was broken in the 15-minute period
1	107	14	128
2	99	15	106
3	171	16	177
4	116	17	144
5	101	18	102
6	109	19	196
7	199	20	191
8	142	21	169
9	118	22	182
10	173	23	148
11	155	24	130
12	184	25	159
13	132		

4.44 A questionnaire circulated last year asked 94 economic forecasters to estimate the probability of an increase in the gross national product from the third quarter to the fourth quarter of the year. The results of the survey are summarized in the accompanying table. Construct a relative frequency histogram for these results.

Estimated probability of an increase in GNP	Frequency
.01–.10	38
.11–.20	18
.21–.30	15
.31–.40	15
.41–.50	8

4.45 A study was conducted among smokers to examine symptoms of breathlessness and wheeze. A total of 1827 subjects exhibiting breathlessness and wheeze were classified by age. Use the data in the accompanying table to construct a relative frequency histogram.

Age	Frequency	Age	Frequency
20–24	9	45–49	269
25–29	23	50–54	404
30–34	54	55–59	406
35–39	121	60–64	372
40–44	169		

4.46 Summarize the data from Exercise 4.45 into the following age categories: 20–29, 30–39, 40–49, and 50 and over. Construct a pie chart depicting the percentage of subjects in each of these categories.

4.47 Select a sample of 50 measurements from a population of interest to you and construct a relative frequency distribution for the data. The data could be observations on a variable measured in a chemistry or physics experiment, they could be measurements of highway traffic on a given route during a fixed period of time, or they could be measurements on the number of people waiting in a queue (such as at a ticket counter, supermarket checkout counter) at particular times. Carefully define the population before you select your sample, and make certain that each sample measurement is selected from the target population.

4.48 Construct a frequency histogram plot for the telephone data in the accompanying table (telephones per 1000 population).

State	Telephones	State	Telephones	State	Telephones
Alabama	500	Louisiana	520	Ohio	550
Alaska	350	Maine	540	Oklahoma	580
Arizona	550	Maryland	610	Oregon	560
Arkansas	480	Massachusetts	570	Pennsylvania	610
California	610	Michigan	580	Rhode Island	560
Colorado	570	Minnesota	560	S. Carolina	510
Connecticut	620	Mississippi	470	S. Dakota	540
Delaware	630	Missouri	570	Tennessee	540
Florida	620	Montana	540	Texas	570
Georgia	570	Nebraska	590	Utah	560
Hawaii	480	Nevada	720	Vermont	520
Idaho	550	New Hampshire	590	Virginia	530
Illinois	650	New Jersey	650	Washington	570
Indiana	580	New Mexico	470	W. Virginia	450
Iowa	570	New York	530	Wisconsin	540
Kansas	600	N. Carolina	530	Wyoming	580
Kentucky	480	N. Dakota	560		

4.49 Construct a stem-and-leaf plot for the data of Exercise 4.48. Interpret the data display.

4.50 Compare the stem-and-leaf plot from the Minitab computer output shown here to the one you constructed in Exercise 4.49. Notice how Minitab broke the stems into more categories to "stretch" out the plot. The stem 4 has been stretched into 2 stems, and the stem 5 has been stretched into 5 stems.

```
MTB> PRINT C1

C1
    500     350     550     480     610     570     620     630     620     570     480
    550     650     580     570     600     480     520     540     610     570     580
    560     470     570     540     590     720     590     650     470     530     530
    560     550     580     560     610     560     510     540     540     570     560
    520     530     570     450     540     580

MTB > STEM-AND-LEAF C1;
SUBC> TRIM;
SUBC> END

Stem-and-leaf of C1          N  = 50
Leaf Unit = 10

        LO  35, 45,

     4     4 77
     7     4 888
     9     5 01
    14     5 22333
    22     5 44444555
   (12)    5 666667777777
    16     5 888899
    10     6 0111
     6     6 223
     3     6 55

        HI  72,

MTB > STOP
```

🏛 **4.51**   The table gives the age at inauguration and at death for 35 Presidents.

President	Age at inauguration	Age at Death	President	Age at inauguration	Age at Death
Washington	57	67	Hayes	54	70
J. Adams	61	90	Garfield	49	49
Jefferson	57	83	Arthur	50	56
Madison	57	85	Cleveland	47	71
Monroe	58	73	B. Harrison	55	67
J. Q. Adams	57	80	McKinley	54	58
Jackson	61	78	T. Roosevelt	42	60
Van Buren	54	79	Taft	51	72
W. H. Harrison	68	68	Wilson	56	67
Tyler	51	71	Harding	55	57
Polk	49	53	Coolidge	51	60
Taylor	64	65	Hoover	54	90
Filmore	50	74	F. Roosevelt	51	63
Pierce	48	64	Truman	60	88
Buchanan	65	77	Eisenhower	62	78
Lincoln	52	56	Kennedy	43	46
A. Johnson	56	66	L. Johnson	55	64
Grant	46	63			

(a) Construct a frequency table for the age-at-inauguration data.

(b) Use a frequency histogram to graph the age-at-inauguration data.

🏛 **4.52** Refer to Exercise 4.51. Use the computer-generated stem-and-leaf plot to describe the age-at-death data.

```
MTB > SET INTO C1
DATA>  67    90    83    85    73    80    78
DATA>  79    68    71    53    65    74    64
DATA>  77    56    66    63    70    49    56
DATA>  71    67    58    60    72    67    57
DATA>  60    90    63    88    78    46    64

MTB > PRINT C1

C1
    67    90    83    85    73    80    78    79    68    71    53    65    74
    64    77    56    66    63    70    49    56    71    67    58    60    72
    67    57    60    90    63    88    78    46    64

MTB > STEM-AND-LEAF C2

Stem-and-leaf of C1        N  = 35
Leaf Unit = 1.0

     2      4 69
     3      5 3
     7      5 6678
    13      6 003344
    (6)     6 567778
    16      7 011234
    10      7 7889
     6      8 03
     4      8 58
     2      9 00

MTB > STOP
```

🏛 **4.53** Refer to Exercise 4.51. Minitab was used to compute the years lived after inauguration for each President and to graph these data using a stem-and-leaf plot. Describe the distribution of years lived following inauguration.

```
MTB > READ INTO C1 C2
DATA>    57    67
DATA>    61    90
DATA>    57    83
DATA>    57    85
DATA>    58    73
DATA>    57    80
DATA>    61    78
DATA>    54    79
DATA>    68    68
DATA>    51    71
DATA>    49    53
DATA>    64    65
```

(continued)

```
DATA>    50    74
DATA>    48    64
DATA>    65    77
DATA>    52    56
DATA>    56    66
DATA>    46    63
DATA>    54    70
DATA>    49    49
DATA>    50    56
DATA>    47    71
DATA>    55    67
DATA>    54    58
DATA>    42    60
DATA>    51    72
DATA>    56    67
DATA>    55    57
DATA>    51    60
DATA>    54    90
DATA>    51    63
DATA>    60    88
DATA>    62    78
DATA>    43    46
DATA>    55    64
DATA>END

MTB > LET C3=C2-C1

MTB > PRINT C1 C2 C3

ROW     C1    C2    C3

  1     57    67    10
  2     61    90    29
  3     57    83    26
  4     57    85    28
  5     58    73    15
  6     57    80    23
  7     61    78    17
  8     54    79    25
  9     68    68     0
 10     51    71    20
 11     49    53     4
 12     64    65     1
 13     50    74    24
 14     48    64    16
 15     65    77    12
 16     52    56     4
 17     56    66    10
 18     46    63    17
 19     54    70    16
 20     49    49     0
 21     50    56     6
 22     47    71    24
 23     55    67    12
 24     54    58     4
 25     42    60    18
 26     51    72    21
 27     56    67    11
 28     55    57     2
```

(continued)

```
      29    51    60     9
      30    54    90    36
      31    51    63    12
      32    60    88    28
      33    62    78    16
      34    43    46     3
      35    55    64     9

MTB > STEM-AND-LEAF C3

Stem-and-leaf of C3        N  = 35
Leaf Unit = 1.0

      8    0 00123444
     11    0 699
     17    1 001222
    (7)    1 5666778
     11    2 01344
      6    2 56889
      1    3
      1    3 6

MTB > STOP
```

4.54   Average salaries for classroom teachers were compiled for schools in Hamilton County, Ohio. Determine the graphical technique that you think will best describe the data.

School district	Average salary ($1000)
Cincinnati	21.6
Deer Park	18.6
Forest Hills	22.5
Greenhills	22.9
Indian Hill	24.9
Loveland	20.0
Madeira	22.5
Mariemont	19.9
Princeton	25.2
Sycamore	24.7
Wyoming	23.5

EXERCISES FROM THE DATA BASE

4.55   Refer to the clinical trial data in Appendix 1.

(a) Use the data from this study of depression to form separate stem-and-leaf plots of the HAM-D anxiety scores for the four treatment groups.

(b) Do the same for the HAM-D retardation, sleep disturbance, and total scores.

(c) Based on your plots in (a) and (b), do the four treatment groups look comparable at the end of the study? Why might they be different?

4.56   Refer to the clinical trial data.

(a) Suggest some graphical technique for comparing the data on daily consumption of tobacco for the four treatment groups prior to their receiving the assigned medication.

(b) Refer to (a). Do the same for the history of alcohol use prior to medication.

(c) Why is it important to have comparable groups before starting the study?

CHAPTER 5

NUMERICAL METHODS
OF DATA DESCRIPTION

5.1 INTRODUCTION

In Chapter 4, we indicated that the first step in summarizing data is to graph the data. In this chapter we deal with a second way to summarize data, which is to focus on the "typical" value in the data set and some measure of the spread of the data. These numerical descriptions, like graphical ones, are commonly used to convey a mental image of physical objects or phenomena. To create a mental picture of a set of data, we seek one or more numbers, called **numerical descriptive measures.**

**numerical
descriptive
measures**

There are two good reasons for this search. First, we frequently wish to discuss sets of measurements with others, and it is inconvenient to carry frequency histograms about in our pockets. Discussion is much easier if we can project a picture of the frequency distribution to the minds of our listeners by means of one or two descriptive numbers. Second, the frequency distribution is an excellent method for characterizing a population, but it possesses severe limitations when used to make inferences. We know that an irregular frequency histogram of a sample will be similar to the corresponding distribution for the population. But how similar? How can we measure the "goodness" of our inference? How can we measure the degree of dissimilarity between the histogram for the sample and the histogram for the population?

So the sample frequency histogram can be used to make an inference concerning the shape of a population frequency distribution, but it is difficult to determine how good that inference is. In contrast, numerical descriptive measures of the population can be estimated by using the sample measurements. We can say, with a measured degree of uncertainty, how close the estimate will be to the population descriptive measure.

Since the same numerical descriptive measure could be computed for either a sample or a population, it is desirable to distinguish between these two applications.

Definition 5.1

statistics

Numerical descriptive measures computed from a sample are called **statistics.**

Definition 5.2

parameters

Numerical descriptive measures of a population are called **parameters.**

The two most important types of parameters are those that locate the center of the distribution and those that describe its spread. They are called, respectively, *measures of central tendency* and *measures of variability*. We will show that the two numbers, one locating the center or "typical" value of a distribution and one quantifying the amount of variability or spread, provide good descriptions of the frequency distributions for most sets of measurements. As you might suspect, we will frequently use a descriptive measure of the sample to estimate the value of the corresponding parameter of the population.

Measures of central tendency (averages, medians, and modes) and their definitions, interpretations, and applications are presented in Section 5.2. Measures of variability and their calculation and interpretation are discussed in Sections 5.3–5.5. Section 5.6 presents an additional graphic technique that incorporates measures of central tendency and variability while Section 5.7 shows how to deal with more than one variable. Section 5.8 gives you some SAS and Minitab programs for performing the calculations of this chapter. Thus in this chapter we provide the final touches to the first step in our study of statistical inference: namely, finding a way to describe a set of measurements. We will use these descriptive measures in later chapters to make inferences about populations based on sample measurements.

5.2   MEASURES OF CENTRAL TENDENCY

Definition 5.3

measures of central tendency

Numerical descriptive measures that locate the center of a distribution of measurements are called **measures of central tendency.** The most common of these are the arithmetic mean, the median, and the mode.

The Arithmetic Mean

Perhaps the most widely used measure of central tendency is the arithmetic mean (or "average") of a set of measurements.

Definition 5.4

arithmetic mean

The **arithmetic mean** of a set of measurements is the sum of the measurements divided by the number of measurements in the set.

The arithmetic mean — or simply the mean — is used extensively in many fields of science and business. You have undoubtedly observed phrases such as the mean income for persons living in ghetto areas, the mean tensile strength of a cable, the mean velocity of the first stage of a missile, the mean increase in the cost of living index over the past six months, and the mean closing price of a group of stocks.

Example 5.1

The length of survival (in years) after discovery of a rare type of cancer was recorded for each of six cancer patients. The data are shown in Table 5.1. Calculate the mean survival time for this sample of six patients.

TABLE 5–1   Length of Survival After Discovery of Cancer

Patient	Length of survival (years)
1	1.7
2	3.2
3	2.1
4	4.6
5	1.4
6	2.8

Solution

$$\text{Mean} = \frac{\text{sum of the measurements}}{\text{number of measurements}}$$

$$= \frac{1.7 + 3.2 + 2.1 + 4.6 + 1.4 + 2.8}{6} = \frac{15.8}{6} = 2.63$$

Note that the mean, 2.63, falls somewhere near the "middle" of the set of six survival-time measurements.

The means for both a sample and a population are defined and computed in the same way, since both sets are sets of measurements. However, we use different symbols for each. The symbol $\bar{x}$ (x bar) will be used to denote the mean of a sample and the symbol μ (Greek letter mu) will denote the mean of a population.

Note that we will rarely possess all the measurements in the population, and consequently we will rarely know the value of μ. But that leads us back to the objective of statistics. We will sample the population and use the sample mean $\bar{x}$ to estimate the value of μ. We will show you how to measure the goodness (accuracy of this estimate) in Chapter 8.

$\bar{x}$ is the **sample mean.**
μ is the **population mean.**

It is convenient here to introduce some notation, which we will use in the computational formulas encountered in this and later chapters. First, let the letter x represent a measurement. If we refer specifically to a sample, x represents a measurement in that set. In the case of multiple measurements, a subscript is used to denote a particular measurement in the set. If we consider the six measurements from Example 5.1,

1.7   3.2   2.1   4.6   1.4   2.8

we let x_1 denote the first observation ($x_1 = 1.7$). In the same manner we let $x_2 = 3.2, \ldots, x_6 = 2.8$.

To indicate the sum of our measurements, we use the Greek symbol Σ (sigma). Thus Σx indicates the sum of the measurements that we denoted by the symbol x. For the data of Example 5.1, we have

$$\Sigma x = x_1 + x_2 + \cdots + x_6$$
$$\Sigma x = 1.7 + 3.2 + \cdots + 2.8 = 15.8$$
$$\bar{x} = \frac{15.8}{6} = 2.63$$

We can expand this example to include any number of measurements. If we have a sample of n measurements, which we denote by x_1, $x_2, \ldots, x_n$, the **sample mean** is given by the following formula:

sample mean

$$\bar{x} = \frac{\Sigma x}{n} = \frac{x_1 + x_2 + \cdots + x_n}{n}$$

The sample mean formula for data presented in a frequency table (called grouped data) is only slightly more complicated than the formula for ungrouped data. Since we do not know the individual sample measurements, but only the class interval to which a measurement is assigned, this formula will be an approximation to the actual sample mean. So, when the sample measurements are known, compute $\bar{x}$ using the ungrouped formula.

We will use the same symbol $\bar{x}$ to designate the sample mean for grouped data. Since we cannot reconstruct the actual sample measurements from the grouped data, we represent all values in a given class interval by the midpoint of the interval. If we set x equal to the midpoint of the interval and let f denote the frequency for that interval, fx denotes the sum of the measurements of the interval. For example, if $x = 5$ is the midpoint   of a class interval and there are $f = 10$ measurements in the interval, $fx = 10(5) = 50$ denotes the sum of the 10 measurements in the interval. Similarly Σfx denotes the sum of the measurements across all class intervals. The **sample mean for grouped data** is then

sample mean for grouped data

$$\bar{x} = \frac{\Sigma fx}{n}$$

Example 5.2

In Exercise 4.51 we listed data for the age at inauguration and the age at death for 35 of our Presidents. The frequency table for the number of years the Presidents lived following inauguration is shown here.

Class interval	Frequency
0–4	8
5–9	3
10–14	6
15–19	7

(continued)

Class interval	Frequency
20–24	5
25–29	5
30–34	0
35–39	1

Compute the sample mean using the data presented in the frequency table.

Solution   Two additional columns x and fx are helpful when computing $\bar{x}$.

Class	Frequency, f	Midpoint, x	fx
0–4	8	2	16
5–9	3	7	21
10–14	6	12	72
15–19	7	17	119
20–24	5	22	110
25–29	5	27	135
30–34	0	32	0
35–39	1	37	38
			$\Sigma fx = 510$

Adding the entries in the fx column and substituting into the formula, we find

$$\bar{x} = \frac{\Sigma fx}{n} = \frac{510}{35} = 14.57$$

Thus, on the average our Presidents have lived about 14½ years after their inauguration.

The characteristics of the sample mean are summarized here:

1. The sample mean is the arithmetic average of the measurements.
2. The sample mean lies between the largest and smallest measurements of the set.
3. The sample mean is influenced by extreme measurements. This property will be illustrated in Exercises 5.2 and 5.3.
4. There is only one mean for a set of measurements.
5. The sample mean will be used extensively for statistical inferences discussed in later chapters.

The Median

median

The **median** is a second measure of central tendency. It is computed in the same way for a sample or for a population.

Definition 5.5

median for an
odd number of
measurements

The **median for an odd number of measurements** is the middle measurement when the measurements are arranged in increasing order.

Example 5.3

A sample of 7 students was given a reading achievement test. Find the median for these test scores:

$$95 \quad 86 \quad 78 \quad 90 \quad 62 \quad 73 \quad 89$$

Solution   We must first arrange the scores in increasing order.

$$62 \quad 73 \quad 78 \quad 86 \quad 89 \quad 90 \quad 95$$

Since we have an odd number of measurements (7), the median is then the middle score—that is, 86.

Definition 5.6

median for an
even number of
measurements

The **median for an even number of measurements** is the mean of the two middle observations when the measurements are arranged in increasing order.

Example 5.4

Suppose that 3 more students out of a class of 30 took the achievement test of Example 5.3 and scored 73, 75, and 91, respectively. Determine the sample mean and median for the combined 10 test scores.

Solution   The sample mean is given by

$$\bar{x} = \frac{\Sigma x}{n}$$

$$= \frac{62 + 73 + 73 + 75 + 78 + 86 + 89 + 90 + 91 + 95}{10}$$

$$= 81.2$$

Since we have an even number of observations, the sample median is the mean of the two middle scores when the scores are arranged in increasing order. The scores, arranged in increasing order, are 62, 73, 73, 75, 78, 86, 89, 90, 91, and 95. The two middle scores are 78 and 86; hence the median is given by

$$\text{median} = \frac{78 + 86}{2} = 82$$

The median for grouped data is slightly more difficult to compute. Since the actual values of the measurements are unknown, we can determine which particular interval contains the median, but not where it is located in that interval. Let

L = lower class limit of the interval that contains the median

n = total number of measurements

cf_b = the sum of the frequencies (cumulative frequency) for all classes before the class containing the median

f_m = frequency for the class containing the median

w = interval width

Then for grouped data

$$\text{median} = L + \frac{w}{f_m}(.5n - cf_b)$$

median for grouped data

The next example illustrates how to find the **median for grouped data.**

Example 5.5

Refer to the frequency table of Example 5.2. Compute the median time from inauguration to death for 35 of our Presidents.

Solution Let the cumulative relative frequency for class j be the sum of the relative frequencies for class 1 through class j. To determine the interval that contains the median, we must find the first interval for which the cumulative relative frequency exceeds .50. This interval will contain the median. For these data, the interval from 15–19 (or more exactly, from 14.5–19.5) is the first interval for which the cumulative frequency exceeds .50, as shown in Table 5.2. So this interval contains the median. Then

$$L = 14.5 \qquad f_m = 7$$
$$n = 35 \qquad w = 5 \text{ (i.e., } 19.5–14.5)$$
$$cf_b = 17$$

and

$$\text{median} = L + \frac{w}{f_m}(.5n - cf_b)$$

$$= 14.5 + \frac{5}{7}((.5 \times 35) - 17) = 14.86$$

TABLE 5.2   Frequency Table for Example 5.2

Class	Frequency, f	Cumulative f	f/n	Cumulative f/n
0–4	8	8	.23	.23
5–9	3	11	.09	.32
10–14	6	17	.17	.49
15–19	7	24	.20	.69
20–24	5	29	.14	.83
25–29	5	34	.14	.97
30–34	0	34	0	.97
35–39	1	35	.03	1.00
	$n = 35$		1.00	

Thus approximately 50% of our Presidents have lived for almost 15 years after their inauguration.

The median seems to be the preferred measure of central tendency for describing economic, sociological, and educational data. Newspaper reports and magazines frequently refer to the median wage increase won by unions, the median income of families in the United States, the median age of persons receiving Social Security benefits, and the gap between the median income for men and the median income for women.

Why is the median so popular in the social sciences? The answer is that many of the frequency distributions of measurements in the social sciences are **skewed** (tail off rapidly to the right or left). Because the mean is greatly affected by extremely large (or small) observations and the median is not, the median is preferred in locating the center of skewed distributions. For example, suppose you have five measurements, 1, 2, 3, 4, and 20. The mean is 6 and the median is 3. Notice how the mean is affected by the largest measurement. Also, notice that it does not appear to fall near the "center" of the five measurements. Figure 5.1 shows the location of the mean and the median for a distribution skewed to the right.

skewed

FIGURE 5.1   *Relationship Between the Mean and the Median for a Distribution Skewed to the Right*

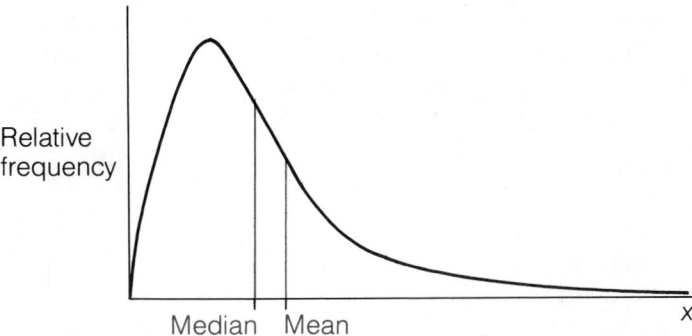

As we continue, keep in mind these four important characteristics of the median:

1. The median is the central value; 50% of the measurements lie above it and 50% fall below it.
2. The median lies between the largest and smallest measurements of the set.
3. The median is not influenced by extreme measurements. (See Exercises 5.2 and 5.3 for verification of this property.)
4. There is only one median for a set of measurements.

The Mode

A third commonly used measure of central tendency is the mode of a set of measurements.

Definition 5.7

mode

The **mode** of a set of measurements is the measurement that occurs most often in the set.

The mode is the least common of the three measures of central tendency considered in this text, but it is very useful in business planning as a measure of popularity that reflects central tendency or opinion. For example, we might talk about the most preferred stock, a most preferred refrigerator model, or the most popular manufacturer of personal computers. Identification of the mode for ungrouped data is fairly easy because we can count the number of times each measurement occurs. When dealing with grouped data we define the mode to be the midpoint of the class with the highest frequency (see Figure 5.2); this is an approximation to the actual mode of the sample measurements.

FIGURE 5.2   Location of the Mode for a Relative Frequency Histogram

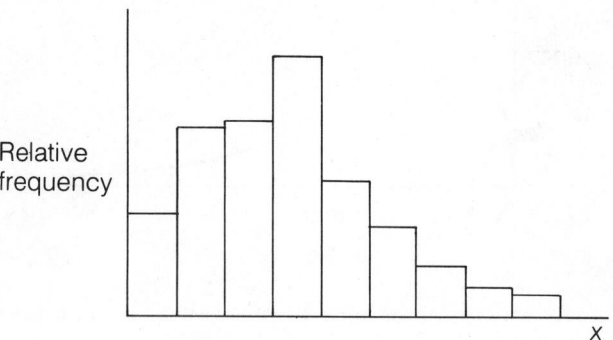

Relative frequency

X

Sometimes the frequency distribution for a set of measurements possesses more than one peak. For example, in a particular athletic shoe worn by both adults and youths, two widely differing sizes might be in great demand. In such a situation, the set of measurements is said to be bimodal. Similarly a trimodal distribution would have three modes. A variation of this situation occurs when all observations appear the same number of times. In this circumstance the mode gives no information for locating the center of the distribution, and we say that the frequency distribution possesses no mode.

Example 5.6

Refer to the frequency table for Example 5.2 and determine the mode.

Solution   The first interval has the highest frequency (8). The midpoint of this interval, 2, is the mode based on the frequency table for these data.

We can summarize the characteristics of the mode as follows:

1. The mode is the most frequent measurement in the set.
2. The mode is not influenced by extreme values. (Exercises 5.2 and 5.3 illustrate this property.)
3. There can be more than one mode for a set of measurements; if there are two modes the set of measurements is bimodal, for three modes, the set is trimodal.

We have discussed three measures of central tendency — the mean, the median, and the mode — and noted that one measure might be better than the other two in a given situation. We will concern ourselves almost exclusively with the sample mean as a measure of central tendency for the remainder of this text. The reason for this choice is that the sample mean is most widely employed in statistical inference, and the study of inference is our ultimate objective. Recall that we wish to make inferences about the population from which the sample is drawn. A sample mean is not only descriptive of the sample observations but, more importantly, it can easily be used to estimate the population mean with some degree of accuracy.

EXERCISES

5.1 Given the sample measurements 4, 9, 12, 16, 7, 9, and 6, compute the mean, median, and mode.

5.2 The following IQ scores were observed for a sample of 10 school children: 112, 109, 102, 93, 89, 111, 102, 95, 104, 103. Compute the mean, median, and mode.

5.3 Refer to Exercise 5.2. How would the mean, median, and mode be affected if the last IQ score was 173 rather than 103?

5.4 Compute the mean, median, and mode for the following sample measurements: 15, 18, 19, 21, 18, 15, and 23.

5.5 Repeat Exercise 5.4 for the sample measurements 65, 52, 71, 75, 59, and 63.

5.6 The EPA estimated the miles per gallon for 30 different car models and presented their results in the following frequency table:

Miles per gallon	Frequency
25–27	2
28–30	4
31–33	3
34–36	5
37–39	7
40–42	8
43–45	1

Use these data to compute the sample mean, median, and mode.

5.7 Refer to the data of Exercise 4.51, which was used in Example 5.2. Compute the actual mean, median, and mode for the time between inauguration and death. How do these values compare to the values computed in Examples 5.2, 5.5, and 5.6 using the frequency table?

5.8 Compute the mean and median for the exercise capacities (in seconds) shown for 11 patients suffering from congestive heart failure.

906	1320
711	1170
684	1200
837	1056
897	882
1008	

5.9 Salaries for 40 recent MBA graduates from a major university are summarized here ($000).

Interval	Frequency
24.9–29.9	6
29.9–34.9	10
34.9–39.9	15
39.9–44.9	7
44.9–49.9	2

Determine the mode, median and mean for the data shown in the frequency table. What does the relationship among the three measures indicate about the shape of the histogram for these data?

5.10 Effective tax rate (per $1000) on residential property for 10 cities from the South, North, and West are shown here.

South	Rate	North	Rate	West	Rate
Atlanta	23.8	Boston	74.6	Denver	9.1
Baltimore	26.5	Chicago	18.2	Honolulu	8.2
Dallas	24.7	Cleveland	21.4	Kansas City	12.5
Houston	21.8	Columbus	13.1	Los Angeles	12.5
Jacksonville	15.7	Detroit	36.6	Phoenix	16.8
Memphis	19.4	Indianapolis	42.8	St. Louis	21.4
Nashville	11.3	Milwaukee	34.7	San Diego	11.1
New Orleans	8.4	New York	18.4	San Francisco	12.1
San Antonio	19.5	Philadelphia	30.9	San Jose	11.2
Washington, D.C.	16.6	Pittsburgh	37.2	Seattle	15.3

Source: U.S. Bureau of the Census, 1980, p. 317.

(a) Compute the mean, median, and mode separately for the South, North, and West.

(b) Compute the mean, median, and mode for the complete set of 30 measurements.

(c) What measure or measures best summarize the center of these distributions? Explain.

5.11 Refer to Exercise 5.10. Average the three group means, the three group medians, and the three group modes, and compare your results to those of 5.10(b). Comment on your findings.

5.12 **Class Exercise** All wristwatches are subject to some degree of error. Some read fast and others slow. Consequently, if you wish an accurate estimate of the time, it would be desirable to combine the time readings from several watches in order to "average out" the errors. Try this experiment. At a particular point in time, have each person in your class (or a small subset of them) record the time on his or her wristwatch. Calculate the average of these three readings. Note that the average gives a time that counterbalances the errors caused by watches that read fast with those that read slow. Save the data; it will be used again in Exercises 5.26 and 5.34.

5.13 A recent newspaper article stated that the median wage for workers in a particular trade union is $8.48 per hour. Interpret this statistic.

5.3   MEASURES OF VARIABILITY

The importance of data variation is exemplified in a joke that is often directed at statistics and statisticians. "Have you ever heard the story about the statistician who couldn't swim and drowned in a river with an average depth of three feet?" While we admit to some discomfort every time we hear the joke, it does stress the importance of data variation. The mean (or any other measure of central tendency) only tells part of the story — or, equivalently, only partially describes a distribution of measurements.

For example, a machine manufacturing size-9 shoes "on the average" would not be considered satisfactory if the actual sizes varied from $8\frac{3}{8}$ to $9\frac{1}{2}$. A machine producing 1-inch nails "on the average" would not be very Indeed, variation of product quality is probably of far greater importance Indeed, variation of product quality is probably of far greater importance to a manufacturer than corresponding measures of central tendency.

Keep in mind that the objective of numerical description is to obtain a set of one or more measures that can be used to create a mental reconstruction of the frequency distribution of data, and that a measure of central tendency only performs one function: locating the center of the distribution. The above examples amply illustrate the need for numerical measures of data variation or spread.

The Range

The simplest measure of data variation is the range.

Definition 5.8

The **range** of a set of measurements is the difference between the largest and smallest measurements.

The range is used extensively as a measure of variability in summaries of data that are made available to the general public. We read, for instance, that the range of salaries for psychologists with the rank of assistant professor is $4,000, the range in temperature in Miami throughout the year is 50°, and the range in personal property taxes for a given state is $2,900. The range is also widely used to describe the variability in the quality of an industrial product when small samples are selected periodically from an operating production line.

Example 5.7

Compute the range of the $n = 5$ measurements shown here:

$$3.2 \quad 7.6 \quad 5.7 \quad 6.6 \quad 4.7$$

Solution   The range is the difference between the largest and the smallest measurements in a set, so

$$\text{range} = 7.6 - 3.2 = 4.4$$

Although simple to define and calculate, the range is not a sensitive measure of variability except for very small samples. In Figure 5.3, the two relative frequency distributions of fluoride readings have the same range, 6, but the data for the two distributions differ greatly in variability. The data for Figure 5.3(a) are much less variable than those for Figure

FIGURE 5.3   *Two Distributions of Fluoride Readings with the Same Range*

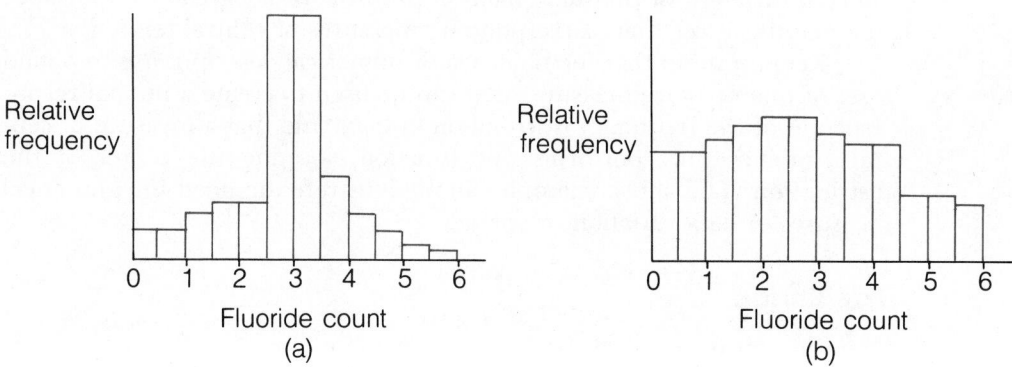

5.3(b). In Figure 5.3(a) most of the measurements are very close to the mean, while in Figure 5.3(b) the measurements are spread more evenly throughout the range.

Percentiles

Percentiles can be used to characterize the spread or variation of a (usually large) set of data, and they also can be used to give the relative standing of one measurement compared to the others in the set.

Definition 5.9

*P*th percentile

For a large set of measurements arranged in increasing order, the **Pth percentile** is the value such that $P\%$ of the measurements are less than that value and $(100 - P)\%$ are greater.

For example, the 80th percentile of a large set of measurements on a variable x is the value x such that 80% of the measurements fall below it and 20% lie above it. See Figure 5.4.

Percentiles are frequently used to describe achievement test scores and the ranking of a person in relation to the rest of the people taking an examination. Specific percentiles of interest are the 25th, 50th, and 75th percentiles, also called the lower quartile, middle quartile (median), and upper quartile, respectively (see Figure 5.5).

FIGURE 5.4   80th Percentile

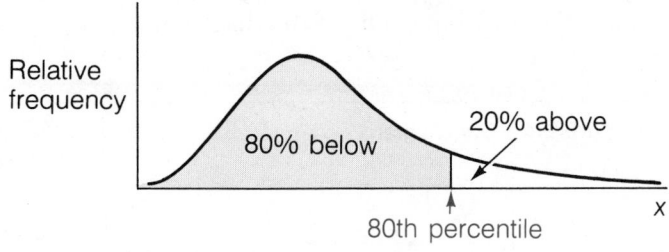

FIGURE 5.5   Lower Quartile, Median, and Upper Quartile

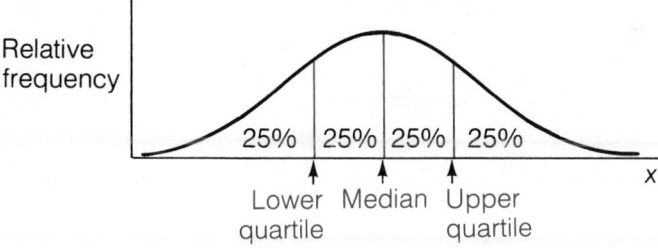

lower quartile

Definition 5.10

The **lower quartile** of a large set of data is the 25th percentile. Thus 25% of the measurements fall below the lower quartile and 75% fall above it.

upper quartile

Definition 5.11

The **upper quartile** of a large set of data is the 75th percentile. Thus 25% of the measurements fall above the upper quartile and 75% fall below it.

Specification of the median, the lower quartile, and the upper quartile provides a fairly good description of a set of measurements except that it gives no notion of the maximum or minimum measurements that we might expect in the set. Thus using percentiles requires a total of five (including the maximum and minimum) numerical descriptive measures to create a mental image of a frequency distribution. Since a small number of numerical descriptive measures would be easier to interpret, we seek a more sensitive measure to describe the variation of a set of measurements.

The Variance

The variance of a set of measurements utilizes the deviations of the measurements from their mean. To illustrate, suppose we have a set of five measurements, $x_1 = 6$, $x_2 = 7$, $x_3 = 5$, $x_4 = 3$, and $x_5 = 4$. These are shown on the **dot diagram** in Figure 5.6. (Dot diagrams are used to depict very small sets of measurements.) Each measurement is located by a dot above the horizontal axis of the diagram.

dot diagram

FIGURE 5.6   Dot Diagram

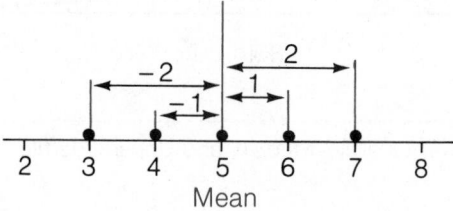

We use the mean

$$\bar{x} = \frac{\Sigma x}{n} = \frac{25}{5} = 5.0$$

deviation

to locate the center of the set. We then construct horizontal lines on Figure 5.6 to represent the distances **(deviations)** of the measurements from their mean. The larger the deviations, the greater will be the variation of the set of measurements. The deviations of the measurements are computed by using the formula $(x - \bar{x})$. In our example the deviation of x_1 from the mean is

$$(x_1 - \bar{x}) = 6 - 5 = 1.0$$

The five measurements in Figure 5.6 and their deviations from the mean are shown in columns (a) and (b), respectively, of Table 5.3.

TABLE 5.3 Deviations of the Five Measurements from Their Mean

(a) Measurement x	(b) Deviation $(x - \bar{x})$	(c) Squared deviation $(x - \bar{x})^2$
6	1	1
7	2	4
5	0	0
3	−2	4
4	−1	1
$\Sigma x = 35$	$\Sigma(x - \bar{x}) = 0$	$\Sigma(x - \bar{x})^2 = 10$

Many different measures of variability could be constructed using the deviations of the measurements from their mean. A first thought would be to use their average, but this will always equal zero. This is because the negative and positive deviations balance one another, so that their sum (and hence mean) equals zero [see column (b), Table 5.3]. A second possibility would be to ignore the minus signs and compute the average of the absolute (positive) values of the deviations. A third possibility, and one that we will employ, makes use of the squared deviations of the measurements from their mean. Recall that the deviation of a measurement x from the sample $\bar{x}$ is expressed as $(x - \bar{x})$. The square of a deviation, then, is represented as $(x - \bar{x})^2$.

Definition 5.12

variance

The **variance** of a set of n sample measurements is the sum of the squared deviations of the measurements from their mean divided by $(n - 1)$.

Because we will use information from a sample to make inferences about the population from which it was selected, it becomes convenient to draw a distinction between the variance of a set of sample measurements and the variance of a population. We use the symbol s^2 to represent the sample variance. Thus

$$s^2 = \frac{\Sigma(x - \bar{x})^2}{n - 1}$$

The corresponding population variance is denoted by σ^2 (σ is the Greek lowercase letter sigma).

s^2 is the **sample variance.**
σ^2 is the **population variance.**

Example 5.8

Calculate the variance for the $n = 5$ measurements 6, 7, 5, 3, and 4, given in Table 5.3.

Solution   Using the information in column (c) of Table 5.3, we find the sample variance to be

$$s^2 = \frac{\Sigma(x - \bar{x})^2}{n - 1} = \frac{10}{4} = 2.5$$

What can we say about the spread of a set of measurements with a variance of 2.5? The greater the variability or spread in a set of measurements, the larger is the variance; but how large is large? Although we can compare variances of sets of measurements to compare variability, it is difficult to interpret the variance for a single set of measurements.

Now we consider a measure of variability useful not only for comparison purposes but also for describing a single set of measurements.

The Standard Deviation

Definition 5.13

standard
deviation

The **standard deviation** of a set of measurements is the positive square root of the variance.

The sample standard deviation is denoted by *s* and the corresponding population standard deviation by the symbol σ.

s is the **sample standard deviation.**
σ is the **population standard deviation.**

We now state a rule that gives practical significance to the standard deviation as a measure of variability. The rule applies only to data that have mound-shaped frequency distributions—the most frequently encountered distributions in the analysis of data. Although the mathematics necessary to "prove" that the rule will always hold is very difficult and well beyond the scope of this text, it is a fact that this rule provides a reasonable characterization of many sets of measurements.

Empirical Rule

For a frequency distribution that is moundshaped, the interval

$(\bar{x} - s)$ to $(\bar{x} + s)$

contains approximately **68%** of the measurements;

$(\bar{x} - 2s)$ to $(\bar{x} + 2s)$

contains approximately **95%** of the measurements;

$(\bar{x} - 3s)$ to $(\bar{x} + 3s)$

contains **all or nearly all** of the measurements.

mound-shaped frequency distribution

An example of a **mound-shaped frequency distribution** appears in Figure 5.7. The relative frequencies are largest near the center of the dis-

Figure 5.7   Graphical Representation of the Empirical Rule

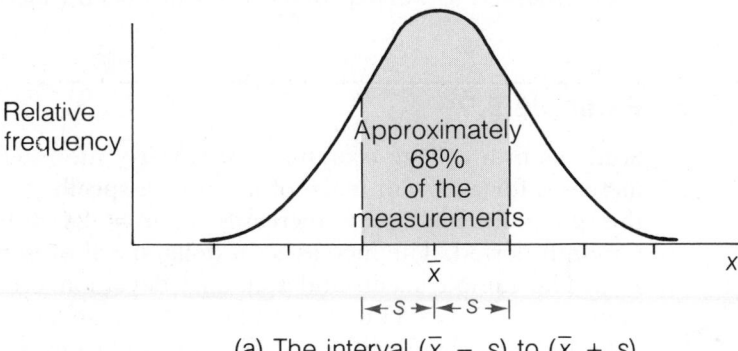

(a) The interval $(\bar{x} - s)$ to $(\bar{x} + s)$

(continued)

Figure 5.7   Graphical Representation of the Empirical Rule
(continued)

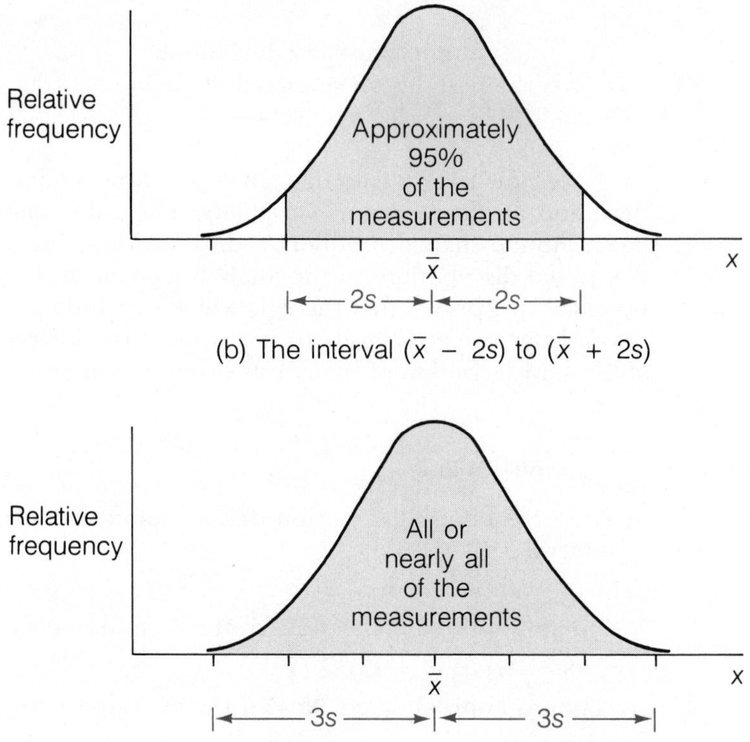

(b) The interval ($\bar{x}$ − 2s) to ($\bar{x}$ + 2s)

(c) The interval ($\bar{x}$ − 3s) to ($\bar{x}$ + 3s)

tribution and tend to decrease as you move toward the distribution tails. The exact shape of the distribution is unimportant, because the rule will adequately describe the variability for mound-shaped distributions of data encountered in real life. Because it is a rule of thumb—a rule that has been observed to work in practice—it has been called the Empirical Rule.

Example 5.9

Students in a college economics class were interested in examining price increases for a certain make of car over a specified period of time. To do this, they obtained price increases on $n = 24$ different models over a 6-month period. The increases (in dollars) are presented in Table 5.4.

The sample mean and standard deviation for this set of measurements are $\bar{x} = 118.33$ dollars and $s = 15.01$ dollars, respectively. Previous experience working with price increases indicates we can assume that the

---

TABLE 5.4 Price Increases (dollars)

100	121	130	129
150	116	120	117
154	125	110	119
130	115	125	123
90	109	100	120
92	112	115	118

---

set of 24 price increases is moundshaped. (To verify this, we could construct a relative frequency distribution for the data.) Describe the variability of the sample by using the Empirical Rule.

Solution   According to the Empirical Rule, approximately 68% of the measurements lie in the interval $(\bar{x} - s)$ to $(\bar{x} + s)$. Substituting the values given for the sample mean and standard deviation, we obtain (118.33 − 15.01) to (118.33 + 15.01), that is, 103.32 to 133.34, for the interval. From Table 5.4 we find that 18 of the 24 price increases (75%) lie in this interval. Note that the 68% specified by the Empirical Rule provides a rough approximation to the actual percentage, 75%, found in the interval $(\bar{x} \pm s)$.

The Empirical Rule also states that approximately 95% of the measurements lie in the interval $(\bar{x} - 2s)$ to $(\bar{x} + 2s)$. When we substitute the values for $\bar{x}$ and s, we obtain the interval 118.33 − 2(15.01) to 118.33 + 2(15.01), or 88.31 to 148.35. We find that 22 of the 24 price increases of Table 5.4 (92%) lie in this interval.

Similarly, the interval $(\bar{x} - 3s)$ to $(\bar{x} + 3s)$ should contain all or nearly all of the measurements. Adding and subtracting $3s = 3(15.01) = 45.03$ from $\bar{x} = 118.33$, we have the interval 73.30 to 163.36. As can be seen from Table 5.4, all of the measurements do lie in this interval.

Keep in mind that the Empirical Rule is not intended to give exact percentages in the specified intervals. Rather, it gives approximate percentages, which will be surprisingly accurate for the intervals $(\bar{x} \pm 2s)$ and $(\bar{x} \pm 3s)$ for most sets of data.

Example 5.10

To remain competitive, manufacturers must be concerned with the efficiency of their operation. One corporation conducted a study to determine the average length of time it takes for an item to be completed on an assembly line. A sample of 50 items ($n = 50$) was timed. The mean and standard deviation (in hours) for the 50 measurements were $\bar{x} = 4.8$ and $s = .42$. A relative frequency histogram for the sample measurements indicated that the distribution is mound shaped. Describe the 50 assembly-line completion times by using the Empirical Rule.

Solution The Empirical Rule tells us that approximately 68% of the measurements lie in the interval $(\bar{x} - s)$ to $(\bar{x} + s)$, that is, the interval 4.38 to 5.22. The interval $(\bar{x} - 2s)$ to $(\bar{x} + 2s)$, or 3.96 to 5.64, should contain approximately 95% of the measurements. All or nearly all of the measurements should be in the interval $(\bar{x} - 3s)$ to $(\bar{x} + 3s)$, or the interval 3.54 to 6.06. These results are summarized in Table 5.5.

TABLE 5.5 *Empirical Rule Results for Example 5.10*

k	$\bar{x} \pm ks$	Approximate percentage in interval
1	4.38 to 5.22	68
2	3.96 to 5.64	95
3	3.54 to 6.06	almost all

To increase our confidence in the Empirical Rule, let us see how well it describes the five frequency distributions of Figure 5.8. We calculated the mean and standard deviation for each of the five data sets (not given) and these are shown next to each frequency distribution. Figure 5.8(a) shows the frequency distribution for measurements made on a variable that can take values $x = 0, 1, 2, \ldots, 10$. The mean and standard deviation $\bar{x} = 5.50$ and $s = 1.49$ for this symmetric mound-shaped distribution were used to calculate the interval $(\bar{x} \pm 2s)$, which is marked below the horizontal axis of the graph. We found 94% of the measurements falling in this interval, that is, lying within two standard deviations of the mean. Note that this percentage is very close to the 95% specified in the Empirical Rule. We also calculated the percentage of measurements lying within one standard deviation of the mean. We found this percentage to be 60%, a figure that is not too far from the 68% specified by the Empirical Rule. Consequently, we think the Empirical Rule provides an adequate description for Figure 5.8(a).

Figure 5.8(b) shows another mound-shaped frequency distribution, but one that is less peaked than the distribution of Figure 5.8(a). The mean and standard deviation for this distribution, shown to the right of the figure, are 5.50 and 2.07, respectively. The percentages of measure-

FIGURE 5.8 A Demonstration of the Utility of the Empirical Rule

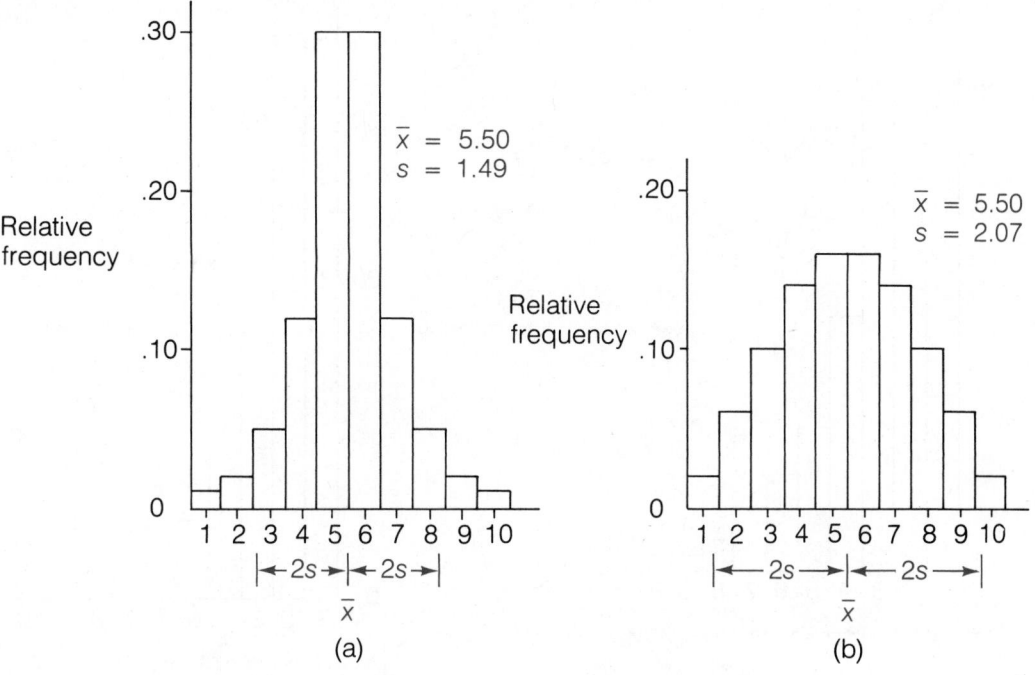

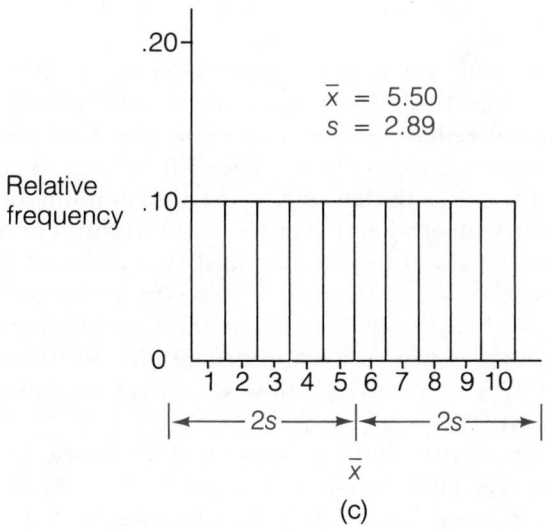

(continued)

FIGURE 5.8   A Demonstration of the Utility of the Empirical Rule (continued)

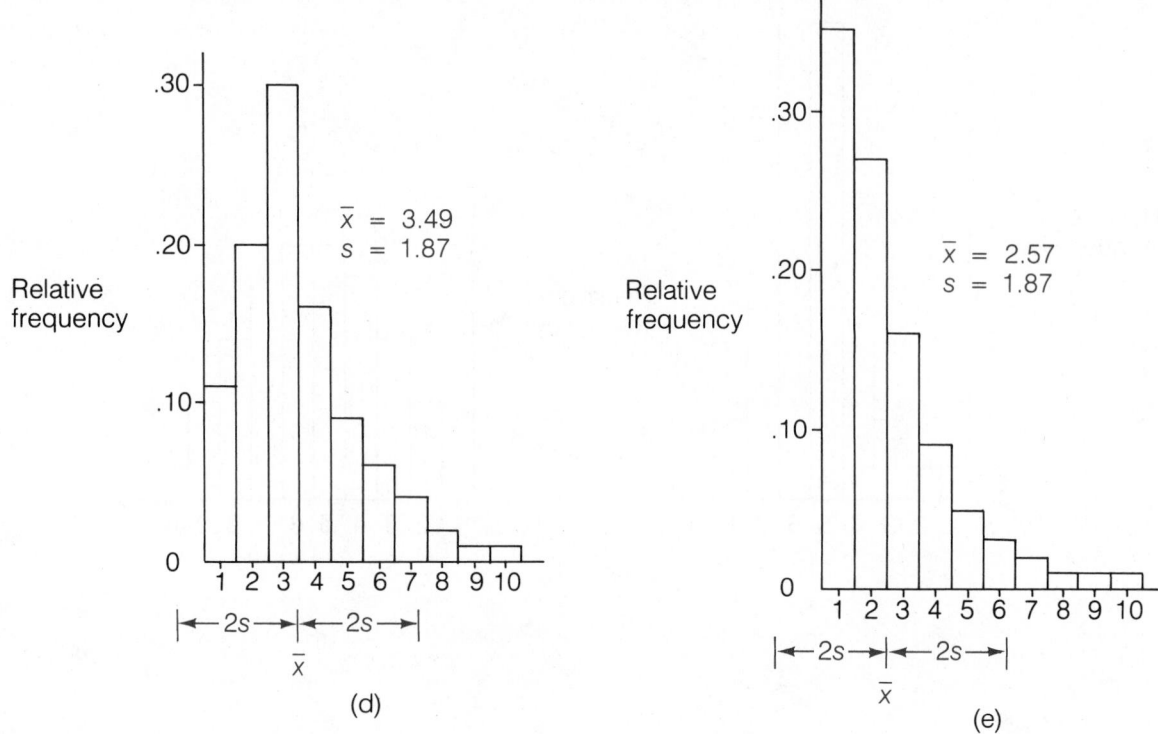

ments lying within one and two standard deviations of the mean are 64% and 96%, respectively. Once again, these percentages agree very well with the Empirical Rule.

Now let us look at three other distributions. The distribution in Figure 5.8(c) is perfectly flat, while the distributions of Figures 5.8(d) and (e) are nonsymmetric and skewed to the right. The percentages of measurements lying within two standard deviations of the mean are 100%, 96%, and 95%, respectively, for these three distributions. All these percentages are reasonably close to the 95% specified by the Empirical Rule. The percentages lying within one standard deviation of the mean (60%, 75%, and 87%, respectively) show some disagreement with the 68% of the Empirical Rule.

To summarize, you can see that the Empirical Rule accurately forecasts the percentage of measurements falling within two standard deviations of the mean for all five distributions of Figure 5.8, even for the distributions that are flat, as in Figure 5.8(c), or highly skewed to the right, as in Figure 5.8(e). The Empirical Rule is less accurate in forecasting the percentages within *one* standard deviation of the mean, but the forecast, 68%, compares reasonably well for the three distributions that might be called mound shaped, Figures 5.8(a), (b), and (d).

We have discussed four measures of variability in this section: the range, percentiles, the variance, and the standard deviation. Although each of these measures is useful in data description, the variance and the standard deviation provide us with information to

1. *compare variability* between sets of measurements.

2. *interpret the variability* of a single set of measurements by using the Empirical Rule (recall that the Empirical Rule only applies to data that have a mound-shaped frequency distribution).

In the remainder of this text we will use the standard deviation almost exclusively to measure the variability of a single set of measurements. It is easy to work with and has the same units (e.g., dollars, feet, etc.) as the x measurements.

EXERCISES

5.14 Calculate $\bar{x}$ and s for the following sample of five measurements: 3, 9, 4, 1, 3.

5.15 Given the measurements 4, 12, 9, 16, 7, 6, and 2, compute the range, variance, and standard deviation.

5.16 Refer to Exercise 5.15. Suppose the last measurement was 30 rather than 2. How would this change the range, variance, and standard deviation?

5.17 The test scores for a nationally administered college achievement test have a mound-shaped distribution with a mean of 520 and a standard deviation of 110.

(a) Use the Empirical Rule to assist you in describing this distribution of scores.

(b) The median for the scores is reported to be 510 and the 98th percentile is 795. Interpret these two statistics.

(c) The range of scores is 702. Interpret this statistic.

5.18 Students in a chemistry class were assigned the task of determining the purity of a chemical substance. Three hundred class members independently analyzed the substance. The mean and the standard deviation for the $n = 300$ measurements were 78.1% and 1.3%, respectively. Joe Smith missed his laboratory class but analyzed the substance the next day. His analysis gave a purity reading of 83.4%. Is there reason to suspect his analytical result? Why?

5.19 In manufacturing oxygen tents it is very important for the actual percentage of oxygen generated at a particular time to be close to the amount specified by a physician and indicated on the oxygen control value. To investigate a particular manufacturer's oxygen tents, 50 tents were selected and all the control values were adjusted to the same oxygen input setting. Then the atmosphere within each tent was sampled, and the difference between the actual percentage of oxygen and the valve setting was recorded for each. If the mean and the standard deviation for the sample were 1.3% and .6%, respectively, describe the distribution of the 50 readings.

5.20 Median family-income data for the United States are displayed for several years between 1970 and 1981 in the following table. Discuss and interpret these data.

Median family income	1970	1975	1979	1980	1981
Black	$ 6,229	8,779	11,574	12,674	13,266
White	$10,236	14,268	20,439	21,904	23,517
Total	$ 9,867	13,719	19,587	21,023	22,388
Total in 1981 dollars	$23,111	23,183	24,540	23,204	22,388

5.4 SHORTCUT METHOD FOR CALCULATING THE VARIANCE AND STANDARD DEVIATION

Calculation of the variance and standard deviation can be tedious if we use the definition formula and calculate each individual deviation, $(x - \bar{x})$. This method also leads to rounding error caused by rounding the values of $(x - \bar{x})$. However, there is a shortcut method that saves time and leads to more accurate calculations. We will use a sample of five measurements, 5, 7, 1, 2, 4, to introduce this shortcut method for calculating the sum of squares of deviations.

We need both the sum and the sum of squares of the x values for the shortcut formula. Recall that Σx was used to indicate the sum of the x values. Thus

$$\Sigma x = 5 + 7 + 1 + 2 + 4 = 19$$

Similarly, we use the symbol Σx^2 to indicate the sum of the squares of the x values. Thus

$$\Sigma x^2 = (5)^2 + (7)^2 + (1)^2 + (2)^2 + (4)^2$$
$$= 25 + 49 + 1 + 4 + 16 = 95$$

These quantities, Σx and Σx^2, which are given as totals in Table 5.6, are substituted into the formula (which follows) to find $\Sigma(x - \bar{x})^2$. As you will see, this formula gives us an easy way to calculate the sum of squares of deviations, the quantity that appears in the formula for s^2.

TABLE 5.6   Shortcut Calculations for $\Sigma(x - \bar{x})^2$

	x	x^2
	5	25
	7	49
	1	1
	2	4
	4	16
Total	19	95

Shortcut formula for Calculating $\Sigma(x - \bar{x})^2$

$$\Sigma(x - \bar{x})^2 = \Sigma x^2 - \frac{(\Sigma x)^2}{n}$$

(Proof of this formula is omitted.)

Substituting the 5 measurements ($n = 5$) of Table 5.6 into the shortcut formula, we arrive at the following:

$$\Sigma(x - \bar{x})^2 = 95 - \frac{(19)^2}{5}$$

$$= 95 - \frac{361}{5}$$

$$= 95 - 72.2 = 22.8$$

To check the validity of this result, we can compute, as before, the sum of the squares of the individual deviations, as shown in Table 5.7. The sample mean is $\bar{x} = 3.8$. We see that $\Sigma(x - \bar{x})^2$ is the same when using either computational procedure.

The variance and the standard deviation are found as before. Thus

$$s^2 = \frac{\Sigma(x - \bar{x})^2}{n - 1} = \frac{22.8}{4} = 5.7$$

and

$$s = \sqrt{s^2} = \sqrt{5.7} = 2.39$$

TABLE 5.7 Calculating $\Sigma(x - \bar{x})^2$

	x	$(x -)\,\bar{x}$	$(x - \bar{x})^2$
	5	1.2	1.44
	7	3.2	10.24
	1	−2.8	7.84
	2	−1.8	3.24
	4	0.2	0.04
Total	$\Sigma x = 19$	$\Sigma(x - \bar{x}) = 0$	$\Sigma(x - \bar{x})^2 = 22.8$

We can make a simple modification to our shortcut formula to approximate the sample variance for data that are summarized in a frequency table. Recall that in approximating the sample mean for grouped data, we let x and f denote the midpoint and frequency, respectively, for a class interval. With this notation the sample variance for grouped data is

$$s^2 = \frac{1}{n-1}\left[\Sigma fx^2 - \frac{(\Sigma fx)^2}{n}\right]$$

The sample standard deviation is $s = \sqrt{s^2}$.

Example 5.11

The data from Example 5.2 are shown here. Compute the sample variance and standard deviation.

Class	f	x	x^2	fx	fx^2
0–4	8	2	4	16	32
5–9	3	7	49	21	147
10–14	6	12	144	72	864
15–19	7	17	289	119	2023
20–24	5	22	484	110	2420
25–29	5	27	729	135	3645
30–34	0	32	1024	0	0
35–39	1	37	1369	37	1369
Total	35			510	10500

Solution From this table we see that $\Sigma fx = 510$ and $\Sigma fx^2 = 10500$. Substituting these values into the formula for s^2 we find

$$s^2 = \frac{1}{n-1}\left[\Sigma f x^2 - \frac{(\Sigma f x)^2}{n}\right]$$

$$= \frac{1}{34}\left[10500 - \frac{(510)^2}{35}\right]$$

$$= \frac{1}{34}[10500 - 7431.428] = 90.25$$

and $s = \sqrt{90.25} = 9.5$.

EXERCISES

5.21 To give you some practice, consider a small set of 5 measurements, say 5, 4, 1, 2, 3.

(a) So that you can see the variation in the measurements, construct a dot diagram similar to Figure 5.6.

(b) Use Σx and Σx^2 to calculate $\Sigma(x - \bar{x})^2$.

(c) Calculate x^2 and s.

(d) Because the number of measurements in the sample is so small, the frequency distribution for the sample measurements is not mound shaped. Nevertheless, note that the interval ($\bar{x} \pm 2s$) contains all the measurements. (Construct this interval on the dot diagram for the data so that you can see the location of the points within the interval.)

5.22 Repeat the instructions of Exercise 5.21 for the 6 measurements 1, 0, 3, 1, 2, 2.

5.23 Repeat the instructions of Exercise 5.21 for the 10 measurements 4, 1, 3, 5, 2, 3, 1, 4, 0, 2.

5.24 The treatment times for patients at a health clinic are as follows:

21	20	31	24	15	21	24	18	33	8
26	17	27	29	24	14	29	41	15	11
13	28	22	16	12	15	11	16	18	17
29	16	24	21	19	7	16	12	45	24
21	12	10	13	20	35	32	22	12	10

Use the shortcut formula to calculate s^2 and s. You can verify that $\Sigma x = 1016$ and $\Sigma x^2 = 24,080$ for the 50 treatment times.

5.25 Refer to Exercise 5.24. To increase your confidence in the applicability of the Empirical Rule, construct the intervals $(\bar{x} \pm s)$, $(\bar{x} \pm 2s)$, and $(\bar{x} \pm 3s)$, and count the number of treatment times falling in each of the three intervals. From these frequencies calculate the corresponding percentage of measurements falling in the three intervals. Does the Empirical Rule give a reasonable approximation to the relative frequencies you have observed?

5.26 Refer to the class experiment in Exercise 5.12. Calculate $\bar{x}$, s^2, and s, and follow the instructions of Exercise 5.25. Once again you will see that $\bar{x}$ and s describe this distribution of data and that the Empirical Rule gives an effective interpretation to s.

5.27 To assist in estimating the amount of lumber in a tract of timber, an owner decided to count the number of trees with diameters exceeding 12 inches in randomly selected 50 × 50-foot squares. Seventy 50 × 50 squares were randomly selected from the tract and the number of trees (with diameters in excess of 12 inches) were counted for each. The data were as follows:

7	8	6	4	9	11	9	9	9	10
9	8	11	5	8	5	8	8	7	8
3	5	8	7	10	7	8	9	8	11
10	8	9	8	9	9	7	8	13	8
9	6	7	9	9	7	9	5	6	5
6	9	8	8	4	4	7	7	8	9
10	2	7	10	8	10	6	7	7	8

(a) Construct a relative frequency histogram to describe these data.

(b) Calculate the sample mean $\bar{x}$ as an estimate of μ, the mean number of timber trees with diameters exceeding 12 inches for all 50 × 50-foot squares in the tract.

(c) Calculate s for the data. Construct the intervals $(\bar{x} \pm s)$, $(\bar{x} \pm 2s)$, and $(\bar{x} \pm 3s)$. Count the percentages of squares falling in each of the three intervals, and compare these percentages with the corresponding percentages given by the Empirical Rule.

5.28 A study was conducted to determine urine flow of sheep (in milliliters/minute) when infused intravenously with the antidiuretic hormone ADH. The urine flows of the 10 sheep are recorded here.

0.7   0.5   .05   0.6   0.5   0.4   0.3   0.9   1.2   0.9

(a) Determine the mean, the median, and the mode for these sample data.

(b) Suppose that the largest measurement is 6.8 rather than 1.2. How does this affect the mean, the median, and the mode?

5.29 Compute the range and standard deviation for the data of Exercise 5.28. How are the range and standard deviation affected if the largest measurement is 6.8 rather than 1.2? What about 68?

5.5 HOW TO GUESS THE STANDARD DEVIATION OF SAMPLE DATA

Many times sample data are presented without accompanying numerical descriptive measures, such as the mean and standard deviation. Although the sample mean can be computed very easily, the calculations required for obtaining s (even with the shortcut formula of Section 5.4) can be difficult and time consuming. In situations where we are interested in obtaining a rough approximation to the actual sample standard deviation, without going through the tedious calculations, we can use the formula

$$s \approx \frac{\text{range}}{4}$$

(read, "s is approximately equal to the range divided by 4"). One additional reason for using this approximation is to check our calculations of s when using the shortcut formula. Arithmetic mistakes can easily occur, so we suggest you use the range approximation as a check even when the actual calculation of s is required.

In this chapter we presented the Empirical Rule, which was useful in interpreting the variability of a mound-shaped distribution. The Empirical Rule states that approximately 95% of the measurements in a set will be within two standard deviations of their mean, or, using notation for a sample, approximately 95% of the measurements will be in the interval $(\bar{x} - 2s)$ to $(\bar{x} + 2s)$. See Figure 5.9. Since the range of the mea-

FIGURE 5.9 *Approximation to s Using the Range*

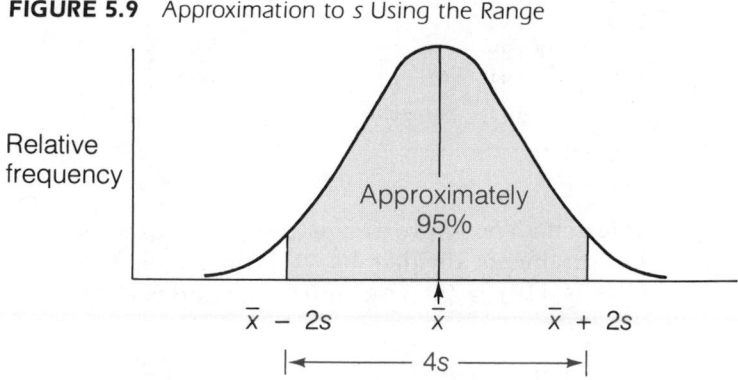

surements is approximately $4s$, one-fourth of the range will provide an approximate value for s.

> ### Approximation to s Using the Range
>
> $$s \approx \frac{\text{range}}{4}$$
>
> Note: This approximation will work best for mound-shaped distributions.

Some might wonder why we did not equate the range to $6s$ since the interval $\bar{x} \pm 3s$ should contain almost all of the measurements. This procedure would yield a value for s that is smaller than the one described previously using the range divided by 4. But, since we're bound to make an error in approximating the value of s, it is better to overestimate the value of the sample standard deviation using the range/4.

Example 5.12

Fifteen students with similar socioeconomic backgrounds were compared on an IQ test at the end of their freshman year. Their scores are presented in Table 5.8. Use the range of the observations to approximate s. Calculate s by using the shortcut formula and compare this result with the approximate value.

TABLE 5.8 IQ Scores

116	118	126
129	114	130
129	122	128
132	131	125
123	134	126

Solution We approximate the value of s by using the range of the measurements divided by 4. From Table 5.8 we see that the range is $(134 - 114) = 20$. Our approximation is then

$$\text{approximate value of } s = \frac{\text{range}}{4} = \frac{20}{4} = 5.0$$

Using the shortcut formula for $\Sigma(x - \bar{x})^2$ to calculate the exact value of s, we need to compute Σx^2 and Σx.

$$\Sigma x^2 = (116)^2 + (129)^2 + \ldots + (126)^2 = 236{,}873$$
$$\Sigma x = 116 + 129 + \cdots + 126 = 1883$$

Hence

$$\Sigma(x - \bar{x})^2 = \Sigma x^2 - \frac{(\Sigma x)^2}{n}$$
$$= 236{,}873 - \frac{(1{,}883)^2}{15}$$
$$= 236{,}873 - 236{,}379.267 = 493.733$$

The sample variance s^2 is then

$$s^2 = \frac{\Sigma(x - \bar{x})^2}{n - 1} = \frac{493.733}{14} = 35.267$$

and the standard deviation is

$$s = \sqrt{35.267} = 5.94$$

Although the approximate value of s, 5.0, differs somewhat from the actual value, 5.94, it still provides a check on our calculations. For instance, we know that our calculation of 5.94 is at least reasonable. If we had computed s to be 59.4, it would certainly not agree with our check. We would advise running this simple check every time you compute a standard deviation.

EXERCISES

5.30 Use the 10 measurements 3, 1, 0, 1, 3, 4, 2, 3, 2, 3.

(a) Use the range approximation to guess the value of s.

(b) Calculate s (use the shortcut formula) and compare this computed value of s with your approximation in part (a).

5.31 Use the 20 measurements 2, 5, 3, 4, 0, 1, 4, 2, 5, 7, 4, 4, 3, 5, 4, 6, 5, 8, 1, 3.

(a) Use the range approximation to guess the value of s.

(b) Calculate s (use the shortcut formula) and compare this computed value of s with your approximation in part (a).

5.32   Refer to the data in Exercise 5.23. Calculate the range of the measurements and then calculate an approximate value of s by using the range. Compare this result with the value of s calculated in Exercise 5.23 (or with the exercise answer given in the back of the text). The range estimate is close enough to detect gross errors in calculating s [like forgetting to divide by $(n - 1)$ or forgetting to take the square root].

5.33   Find the range for the treatment time data of Exercise 5.24 and then obtain the range approximation to s. Compare this result with the value of s calculated in Exercise 5.24.

5.34   Refer to the class experiment in Exercise 5.12. Find the range of the observations taken. Calculate the range approximation to s and compare this result with the value of s calculated in Exercise 5.26.

5.35   An experiment was conducted to investigate the effect of root temperature on the growth of soybeans. Ten soybean plants were subjected to a standard soil condition and a root temperature of 25°. Fourteen days after seed germination the weights of the exposed tops were obtained. The weights, in grams, for the 10 plants were as follows: .23, .27, .41, .25, .37, .51, .29, .34, .33, .48. The first step in the analysis of these data would require the computation of the sample mean and standard deviation.

(a) Calculate the sample mean.

(b) Approximate the value of s for these data by using the range.

(c) Calculate s and compare this value with your approximation in part (b). (You will find that $\Sigma x = 3.48$ and $\Sigma x^2 = 1.2924$ for these data.)

5.6 THE BOX PLOT

box plot

As mentioned earlier, in Chapter 4, a stem-and-leaf plot provides a graphical representation of a set of scores that can be used to examine the shape of the distribution, the range of scores, and where the scores are concentrated. The **box plot,** which builds on the information displayed in a stem-and-leaf plot, is more concerned with the symmetry of the distribution and incorporates numerical measures of central tendency and location in order to study the variability of the scores and the concentration of scores in the tails of the distribution.

Before we show how to construct and interpret a box plot, it is necessary to introduce several new terms that are peculiar to the language of exploratory data analysis (EDA). We are familiar with the definitions for the lower, middle (median), and upper quartiles of a distribution presented earlier in this chapter. The box plot uses the median and **hinges** of a distribution. Hinges are very similar to quartiles of a distribution, but owing to the method by which they are computed for sample data, the lower and upper hinges of a distribution may differ very slightly from the

hinges

lower and upper quartiles of a set of scores. Having said this and recognizing the slight distinction, we will compute hinges in this text but refer to them as the lower and upper quartiles of the sample data.

We can now illustrate a box plot by way of an example.

Example 5.13

Use the stem-and-leaf plot in Figure 5.10 for a sample of $n = 90$ violent crime rates to construct a box plot.

FIGURE 5.10 *Stem-and-Leaf Plot*

1	12 71 84
2	01 12 50 65 70 72 89
3	05 06 22 22 36 46 51 70 90
4	04 09 11 36 37 39 41 44 48 51 53 70 80 82 87 94 95 99
5	03 14 21 22 27 48 50 59 60 70 72 74 78 85 92 92
6	07 16 18 21 29 37 38 40 56 68
7	07 09 19 37 39 52 58 66 92 92 94
8	02 18 30 32 43 58 60 69
9	18 25 53 91
10	00 05 68
11	48

Solution   When the scores are ordered from lowest to highest, the median score and quartile scores are located as follows:

$$\text{median location is the } \frac{n + 1}{2}\text{th score}$$

$$\text{quartile location is the } \frac{\text{truncated median location} + 1}{2}\text{th score}$$

where the truncated median location is simply the median location with the decimal .5 omitted where present. For the distribution of $n = 90$ violent crime rates we have

$$\text{median location} = \frac{90 + 1}{2} \text{ or the 45.5th score}$$

$$\text{truncated median location} = \text{45th score}$$

and

$$\text{quartile location} = \frac{45 + 1}{2} \text{ or 23rd score}$$

Since the median location is the 45.5th score in the distribution, we average the 45th and 46th scores to compute the median. For these data the 45th score (counting from the lowest to the highest in Figure 5.10) is 559 and the 46th is 560; hence the median is

$$M = \frac{559 + 560}{2} = 559.5$$

In order to find the lower and upper quartiles for this distribution of scores, we determine the 23rd score counting in from the low side of the distribution and from the high side of the distribution, respectively. The 23rd lowest score and 23rd highest score are 436 and 739.

lower quartile, $Q_1 = 436$

upper quartile, $Q_3 = 739$

These three descriptive measures and the smallest and largest values in a data set are used to construct a *box plot* (see Figure 5.11). The box plot is constructed by drawing a box between the lower and upper quartiles with a solid line drawn across the box to locate the median. Then a straight line is drawn connecting the box to the largest value; a second line is drawn from the box to the smallest value. These straight lines are sometimes called *whiskers* and the entire graph is called a **box-and-whiskers plot.**

box-and-whiskers plot

FIGURE 5.11   Box Plot for the Data of Figure 5.10

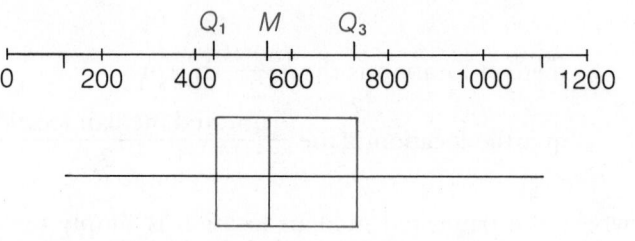

With a quick glance at a box plot, it is easy to obtain an impression about

1. the lower and upper quartiles, Q_1 and Q_3

2. the interquartile range (*IQR*), the distance between the lower and upper quartiles;

3. the most extreme (lowest and highest) values; and

4. the symmetry or asymmetry of the distribution of scores.

If we had been presented with Figure 5.11 without having seen the original data, we would have observed that

$$Q_1 \approx 450$$
$$Q_3 \approx 750$$
$$IQR \approx 750 - 450 = 300$$
$$M \approx 550$$

most extreme values: 100 and 1150

Also, because the median is closer to the lower quartile than the upper quartile and because the upper whisker is a little longer than the lower whisker, the distribution departs slightly from symmetry. This is borne out in the stem-and-leaf plot of Figure 5.10.

5.7 SUMMARIZING DATA FROM MORE THAN ONE VARIABLE

So far we have discussed graphical methods and numerical descriptive methods for summarizing data from a single variable. Frequently, more than one variable is being studied at the same time, and although we might be interested in summarizing the data on each variable separately, we might also be interested in studying relationships among the variables. For example, we might be interested in the prime interest rate and in the consumer price index, as well as in the relationship between the two. In this section we'll discuss a few techniques for summarizing data from two or more variables. These techniques will provide a brief preview and introduction to chi-square methods (Chapter 16), analysis of variance (Chapter 15), and regression (Chapters 12 and 13).

contingency
table

Consider first the problem of summarizing data from two qualitative variables. Cross-tabulations can be constructed from a **contingency table.** The rows of the table identify the categories of one variable, and the columns identify the categories of the other variable. The entries in the table are the number of times each value of one variable occurs with each possible value of the other. For example, a television viewing survey was conducted on 1500 individuals. Each individual surveyed was asked to state his or her place of residence and network preference for national news. The results of the survey are shown in Table 5.9. As you can see, 144 urban residents preferred ABC, 135 urban residents preferred CBS, and so on.

The simplest method for looking at relationships between variables in a contingency table is to do a percentage comparison based on the row totals, the column totals, or the overall total. If we calculate percentages within each row of Table 5.9, we can compare the distribution of resi-

TABLE 5.9   Data from a Survey of Television Viewing

Network Preference	Residence			
	Urban	*Suburban*	*Rural*	*Total*
ABC	144	180	90	414
CBS	135	240	96	471
NBC	108	225	54	387
Other	63	105	60	228
Total	450	750	300	1500

dences within each network preference. A percentage comparison such as this, based on the row totals, is shown in Table 5.10

TABLE 5.10   Comparing the Distribution of Residences for Each Network

Network Preference	Residence			
	Urban	*Suburban*	*Rural*	*Total*
ABC	34.8%	43.5%	21.7%	100 ($n = 414$)
CBS	28.7	50.9	20.4	100 ($n = 471$)
NBC	27.9	58.1	14.0	100 ($n = 387$)
Other	27.6	46.1	26.3	100 ($n = 228$)

Except for ABC, which has the highest urban percentage among the networks, the differences among the residence distributions are in the suburban and rural categories. The percentage of suburban preferences rises from 43.5% for ABC to 58.1% for NBC. Corresponding shifts downward occur in the rural category. Later, in Chapter 16, we will use chi-square methods to further explore relationships between two (or more) qualitative variables.

An extension of the bar graph provides a convenient method for summarizing joint data from a single *qualitative* and a single *quantitative* variable. For example, suppose that a company wants to investigate the relative effects of three different employee incentive plans (a qualitative variable) on productivity (a quantitative variable). A total of fifteen work teams are selected randomly. Of these teams, seven participate in a released-time plan, by which teams achieving certain goals are allowed to take extra time off, with pay; five participate in a bonus-pay plan; and three participate in a profit-sharing plan. The company has a standard productivity measure, and calculates the increased productivity of each work team over a three-month period. The results are as follows:

released time, R:	16.2	15.6	19.4	18.8	16.9	15.9	17.6
bonus pay, B:	12.4	15.8	14.0	9.8	10.0		
profit sharing, P:	4.6	8.0	6.0				

Do the data indicate a strong relationship between plan and productivity gain?

The data summarized in Figure 5.12 clearly indicate that productivity gains are generally largest for the released-time plan R, gains for the bonus-pay plan B are in the middle, and gains for the profit-sharing plan P are lowest. For now we will use the plot shown in Figure 5.12. Later we will use analysis of variance methods (Chapter 15) to examine the relationships between a quantitative variable and one or more qualitative variables.

FIGURE 5.12 Relationship Between Productivity and the Incentive Plans

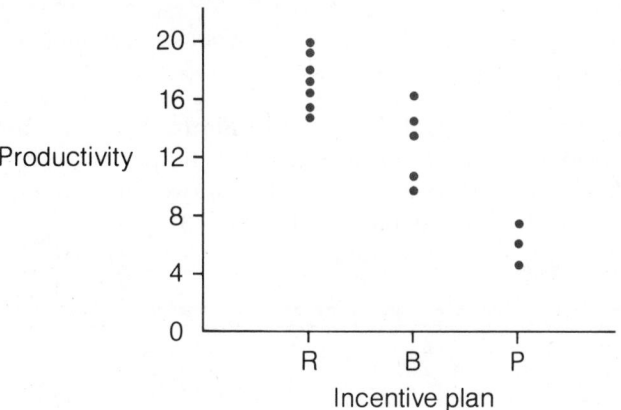

Finally we can construct data plots for summarizing the relationship between two quantitative variables. Consider the following example. A manager of a small machine shop examined the starting hourly wage y offered to machinists with x years of previous experience. The data are shown here:

y (dollars):	8.90	8.70	9.10	9.00	9.79	9.45	10.00	10.65	11.10	11.05
x (years):	1.25	1.50	2.00	2.00	2.75	4.00	5.00	6.00	8.00	12.00

Is there a relationship between x and y?

scatter plot
One way to summarize these data is to use a **scatter plot,** which is shown in Figure 5.13. Each point on the plot represents a machinist with

a particular starting wage and so many years of experience. The point circled corresponds to $y = 9.45$, $x = 4.00$.

FIGURE 5.13 Scatterplot of Starting Hourly Wage and Years Experience

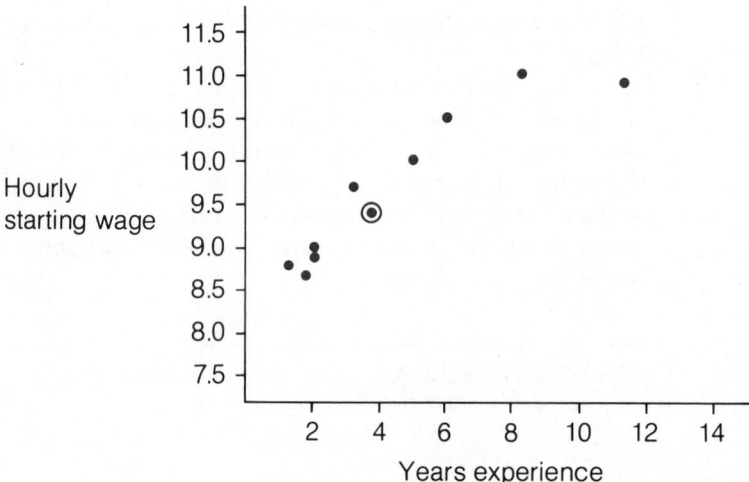

In general the data displayed in Figure 5.13 indicate that, as the years of previous experience x increase, the hourly starting wage y for machinists increases. This basic idea of relating two quantitative variables is expanded later in the book (Chapters 12 and 13).

EXERCISES

5.36 Refer to the television survey data of Table 5.9. Do a percentage comparison based on the column totals. Interpret the data.

5.37 Data on the age at the time of a job turnover and on the reason for the job turnover are displayed here for 250 job changes in a large corporation.

Reason for turnover	Age (years)				
	≤29	30–39	40–49	≥50	Total
Resigned	30	6	4	20	60
Transferred	12	45	4	5	66
Retired/fired	8	9	52	55	124
Total	50	60	60	80	250

Do a percentage comparison based on the row totals and use this to describe the data.

5.38 Refer to Exercise 5.37. What different summary would one get with a percentage comparison based on the column totals? Do this summary and describe your results.

5.39 The lengths of hospital stays were recorded for patients undergoing a particular surgical procedure at each of four hospitals. These data are shown here.

Hospital	Length of stay (days)							
A	18	20	22	22	24	26		
B	14	15	17	17	18	19	21	21
C	21	25	27	31				
D	27	33						

(a) Compute the mean stay for each hospital.

(b) Plot the sample data.

(c) Use parts (a) and (b) to describe the data. Which hospital appears to have shorter stays?

5.40 The federal government keeps a close watch on money growth versus targets that have been set for that growth. Below we list two measures of the money supply in the United States, M2 (private checking deposits, cash, and some savings) and M3 (M2 plus some investments), which are given here for 20 consecutive months.

	Money supply (in trillions of dollars)	
Month	*M2*	*M3*
1	2.25	2.81
2	2.27	2.84
3	2.28	2.86
4	2.29	2.88
5	2.31	2.90
6	2.32	2.92
7	2.35	2.96
8	2.37	2.99
9	2.40	3.02
10	2.42	3.04
11	2.43	3.05
12	2.42	3.05
13	2.44	3.08
14	2.47	3.10
15	2.49	3.10

(continued)

	Money supply (in trillions of dollars)	
Month	M2	M3
16	2.51	3.13
17	2.53	3.17
18	2.53	3.18
19	2.54	3.19
20	2.55	3.20

(a) Would a scatterplot describe the relation between M2 and M3?

(b) Construct a scatterplot. Is there an obvious relation?

5.41 Refer to Exercise 5.40. What other data plot might be used to describe and summarize these data? Make the plot and interpret your results.

5.8 USING COMPUTERS

Minitab and SAS programs for generating numerical descriptive statistics and a box plot are displayed here for the data of Table 5.4. The DE-SCRIBE statement in Minitab produces the sample size (n), the mean, median, trimmed mean (TRMEAN, which you can ignore), standard deviation, standard error of the mean, the maximum and minimum values, and the upper and lower quartiles (i.e., the 75th and 25th percentiles of the data set, shown as Q_3 and Q_1, respectively, on the printout). The BOX-PLOT statement produces a box plot as described in the chapter and also identifies the location of the more extreme values of the data set (by way of an asterisk or an "oh").

Refer to the following Minitab output to see the box plot for the data in Table 5.4. Because various numerical descriptive statistics have been computed and then shown in the output, we know that the median is 118.5 and the lower and upper quartiles are 110.5 and 125.0, respectively. Note that the box plot correctly locates these quartiles. Finally, from the box plot we see that there are two extremely large values, 150 and approximately 153 (actually the largest value is 154).

In summary, using the legend below the box plot in Minitab, we can determine the scale in order to extract desired descriptive measures of the data set.

The PROC MEANS procedure in SAS computes the sample size, mean, standard deviation, minimum, maximum, standard error of the mean, sum, and variance. The PROC UNIVARIATE procedure computes many of the same numerical descriptive measures (plus others); with the PLOT option it also gives a stem-and-leaf plot and a box plot.

The box plot for SAS is displayed in the output immediately to the right of the stem-and-leaf plot. Note that there is no scale (other than the scale for the stems of the stem-and-leaf plot) that can be used to identify the median, lower and upper quartiles, and other characteristics of the data set. However, using the summary statistics labeled QUANTILES, we can determine the median (118.5), the lower and upper quartiles ($Q1 = 111$ and $Q3 = 125$), and the largest and smallest measurements (154 and 90). Another example from SAS that shows these procedures was included in the output for Section 4.9.

The programs that are illustrated in this section can be used to analyze other data sets in a similar manner (see, e.g., Exercises 5.63 and 5.75). Additional details about the statements and procedures are available in the SAS and Minitab user's manuals.

Minitab Output

```
MTB > SET INTO C1
DATA> 100 121 130 129
DATA> 150 116 120 117
DATA> 154 125 110 119
DATA> 130 115 125 123
DATA>  90 109 100 120
DATA>  92 112 115 118
DATA> END

MTB > PRINT C1

C1
    100    121    130    129    150    116    120    117    154    125    110
    119    130    115    125    123     90    109    100    120     92    112
    115    118

MTB > DESCRIBE C1

               N      MEAN    MEDIAN    TRMEAN     STDEV    SEMEAN
C1            24    118.33    118.50    118.00     15.01      3.06

             MIN       MAX        Q1        Q3
C1         90.00    154.00    110.50    125.00

MTB > BOXPLOT C1

                              ------------
          --------------------I    +    I----                  *  *
                              ------------
          ------+---------+---------+---------+---------+---------+C1
               96        108       120       132       144       156

MTB > stop
```

(continued)

```
OPTIONS NODATE NONUMBER LS=80 PS=60;
DATA RAW;
  INPUT INCREASE @@;
  LABEL INCREASE = 'PRICE INCREASE OVER 6 MONTHS';
  LIST;
  CARDS;
  100 150 154 130  90  92 121 116 125 115
  109 112 130 120 110 125 100 115 129 117
  119 123 120 118
;
PROC PRINT N;
TITLE1 'PRICE INCREASES FOR A CERTAIN CAR MODEL OVER A 6-MONTH PERIOD';
TITLE2 'LISTING OF DATA';
PROC MEANS DATA = RAW N MEAN STD MIN MAX STDERR SUM VAR;
  VAR INCREASE;
TITLE2 'EXAMPLE OF PROC MEANS';
PROC UNIVARIATE PLOT;
  VAR INCREASE;
TITLE2 'EXAMPLE OF PROC UNIVARIATE WITH THE PLOT OPTION';
RUN;
```

SAS Output

```
PRICE INCREASES FOR A CERTAIN CAR MODEL OVER A 6-MONTH PERIOD
                    LISTING OF DATA

                   OBS    INCREASE

                     1       100
                     2       150
                     3       154
                     4       130
                     5        90
                     6        92
                     7       121
                     8       116
                     9       125
                    10       115
                    11       109
                    12       112
                    13       130
                    14       120
                    15       110
                    16       125
                    17       100
                    18       115
                    19       129
                    20       117
                    21       119
                    22       123
                    23       120
                    24       118

                   N = 24
```

```
                PRICE INCREASES FOR A CERTAIN CAR MODEL OVER A 6-MONTH PERIOD
                              EXAMPLE OF PROC MEANS

Analysis Variable : INCREASE PRICE INCREASE OVER 6 MONTHS


N Obs   N     Minimum       Maximum           Sum         Mean      Variance
--------------------------------------------------------------------------------
  24    24   90.0000000   154.0000000      2840.00    118.3333333  225.1884058
--------------------------------------------------------------------------------


                   N Obs       Std Dev      Std Error
                  ---------------------------------------
                     24      15.0062789     3.0631438
                  ---------------------------------------



                PRICE INCREASES FOR A CERTAIN CAR MODEL OVER A 6-MONTH PERIOD
                  EXAMPLE OF PROC UNIVARIATE WITH THE PLOT OPTION

                            UNIVARIATE PROCEDURE

Variable=INCREASE      PRICE INCREASE OVER 6 MONTHS

                                    Moments

                N              24   Sum Wgts            24
                Mean      118.3333   Sum               2840
                Std Dev   15.00628   Variance      225.1884
                Skewness   0.38141   Kurtosis      1.067784
                USS         341246   CSS           5179.333
                CV        12.68136   Std Mean      3.063144
                T:Mean=0  38.63133   Prob>|T|        0.0001
                Sgn Rank       150   Prob>|S|        0.0001
                Num ^= 0        24


                              Quantiles(Def=5)

                100% Max     154        99%        154
                 75% Q3      125        95%        150
                 50% Med    118.5       90%        130
                 25% Q1      111        10%        100
                  0% Min      90         5%         92
                                        1%         90

                Range         64
                Q3-Q1         14
                Mode         100


                                  Extremes

                Lowest     Obs     Highest     Obs
                    90(      5)        129(     19)
                    92(      6)        130(      4)
                   100(     17)        130(     13)
                   100(      1)        150(      2)
                   109(     11)        154(      3)
```

(continued)

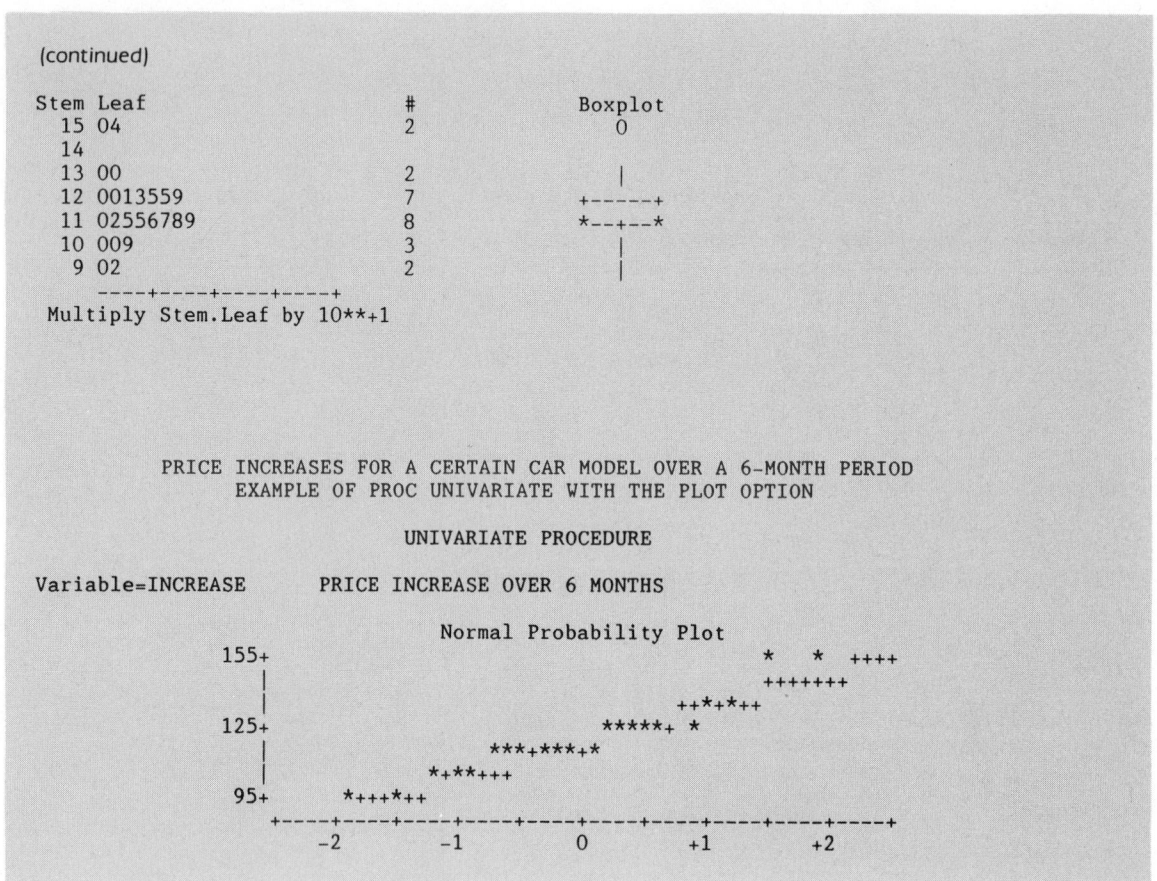

```
(continued)

Stem Leaf                        #           Boxplot
  15 04                          2             0
  14
  13 00                          2             |
  12 0013559                     7           +-----+
  11 02556789                    8           *--+--*
  10 009                         3             |
   9 02                          2             |
     ----+----+----+----+
Multiply Stem.Leaf by 10**+1
```

```
          PRICE INCREASES FOR A CERTAIN CAR MODEL OVER A 6-MONTH PERIOD
            EXAMPLE OF PROC UNIVARIATE WITH THE PLOT OPTION

                          UNIVARIATE PROCEDURE

Variable=INCREASE        PRICE INCREASE OVER 6 MONTHS

                          Normal Probability Plot
      155+                                           *    *  ++++
         |                                             ++++++
         |                                      ++*+*++
      125+                                 *****+  *
         |                        ***+***+*
         |                  *+**+++
       95+         *+++*++
         +----+----+----+----+----+----+----+----+----+----+
            -2        -1        0        +1        +2
```

SUMMARY

The objective of statistics is to make sense of data. After the data have been collected for a survey or experimental study, we can graph the measurements as a first step toward summarizing and understanding the data. (Several different graphical descriptive measures for summarizing data were discussed in Chapter 4). The second way to summarize data involves the computation of one or more *numerical descriptive measures*—numbers that convey a mental picture of the frequency distribution for a set of measurements.

The most important of these are the *mean* and *standard deviation,* which measure the center and the spread, respectively, of a frequency distribution. The standard deviation of a set of mound-shaped measurements is a meaningful measure of variability when interpreted by the Empirical Rule. Numerical descriptive measures of the sample and population are called, respectively, *statistics* and *parameters*. Numerical descriptive measures are suitable for developing both descriptions and inferences. Thus we can use a numerical descriptive measure of the sample, say the *sample mean*, to estimate a parameter such as the *population mean*. The great advantage of the numerical descriptive measure in making inferences is that we can give a quantitative measure of the "goodness" of the inference. In particular we can show that the sample mean will lie within a specified distance of the population mean with some predetermined probability.

Now that we have learned how to summarize data using graphical and numerical methods, we will turn to the analysis of data. The results of a statistical analysis of sample data are expressed in terms of inferences about the population of measurements from which the sample data were obtained—all of which are subject to a degree of uncertainty. If we decide that a population possesses a certain characteristic, we may be correct, but there is always some element of doubt. Thus we find that probability, which is a measure of uncertainty, plays a major role both in making an inference and in measuring how good it is. Probability and probability distributions (Chapter 6) and sampling distributions (Chapter 7) provide the foundation for the methods of analysis (and accompanying inferences) of Chapters 8–17.

KEY TERMS

numerical descriptive measures	range (*def*)
statistics (*def*)	*P*th percentile (*def*)
parameters (*def*)	lower quartile (*def*)
measures of central tendency (*def*)	upper quartile (*def*)
arithmetic mean (*def*)	dot diagram
sample mean	deviation
sample mean for grouped data	variance (*def*)
median	standard deviation (*def*)
median for an odd number	mound-shaped frequency
of measurements (*def*)	distribution
median for an even number	box plot
of measurements (*def*)	hinges
median for grouped data	box-and-whiskers plot
skewed	contingency table
mode (*def*)	scatter plot

KEY FORMULAS

1. Sample mean

$$\bar{x} = \frac{\Sigma x}{n}$$

2. Sample mean, grouped data

$$\bar{x} = \frac{\Sigma fx}{n}$$

3. Median, grouped data

$$median = L + \frac{w}{f_m}(.5n - cf_b)$$

4. Sample variance

$$s^2 = \frac{1}{n-1}\left[\Sigma x^2 - \frac{(\Sigma x)^2}{n}\right]$$

5. Sample variance, grouped data

$$s^2 = \frac{1}{n-1}\left[\Sigma fx^2 - \frac{(\Sigma fx)^2}{n}\right]$$

6. Sample standard deviation

$$s = \sqrt{s^2}$$

SUPPLEMENTARY EXERCISES

5.42   Given the measurements 4, 9, 12, 16, 7, 9, and 6, determine which of the following calculations are correct.

(a) $\bar{x} = 9$ (b) median $= 16$

(c) mode $= 9$ (d) range $= 8$

(e) $s^2 = 15$ (f) $s = 3$

5.43   The data for the bread price survey of Exercise 4.10 on p. 56 are shown here.

Class	Frequency
89.5–94.5	5
94.5–99.5	10
99.5–104.5	14
104.5–109.5	13
109.5–114.5	15
114.5–119.5	21
119.5–124.5	16
124.5–129.5	7
129.5–134.5	2

Determine the mean, median, and mode.

5.44   Refer to Exercises 5.43. Compute the sample standard deviation; describe the data using the Empirical Rule.

5.45   Indicate which of the following are true and which are false.

(a) The variance is the square root of the standard deviation.

(b) There is only one way for a set of numbers to have a standard deviation of zero, and that is to have all measurements equal to zero.

(c) $\Sigma(x - \bar{x}) = 0$ always.

(d) Unlike the mean, the standard deviation is not affected by extreme values.

(e) A percentile rank of 90 in the SAT would mean that a person scored as high as or higher than 90% of the people taking the SAT.

(f) A distribution with the mean = 46, median = 48, and mode = 51 is skewed to the right.

5.46   High temperatures (F°) for a period of 20 days in August are summarized here.

Class	Frequency	Class	Frequency
76–79	2	88–91	4
80–83	3	92–95	1
84–87	8	96–99	2

Compute the mean, median, mode, and standard deviation.

5.47   For (a) and (b) below, decide whether the mean or median would be a more appropriate measure of central tendency, and give a reason for your answer.

(a) The U.S. National Center for Health Statistics publishes information on the duration of marriages.

(b) The National Education Association publishes statistics on daily attendance in public elementary and secondary schools.

5.48 As part of an agricultural experiment, a new type of soybean was planted in six equal-sized fields. The yields when harvested (in hundreds of bushels) were as follows: 5, 7, 9, 8, 7, 9.

(a) Find the values of the following sample statistics (work must be shown for full credit):

Mean _____; Median _____; Range _____;

Variance _____; Standard Deviation _____.

(b) If the smallest value in the data were changed to 3, which of the above 5 measures would decrease?

5.49 Find the mean, median and mode of the following distribution:

8   5   2   3   4   5   20   3   2   5   3

(a) Is this distribution symmetric, positively skewed, or negatively skewed? Explain.

(b) Of the measures computed in (a), which one best describes the "average" of this distribution? Explain.

(c) Find the standard deviation of this distribution. Show your work and the formula you used.

5.50 Compute the arithmetic mean for each of the distributions below. Show your work. (Show the formula used in each part, and the columns and column sums used.)

(a)

x	f
7	6
8	3
9	0
10	2

(b)

Grouped x	f
5–7	3
8–10	7
11–13	5

5.51 Modified True/False Questions: If the statement is TRUE, circle the T. If the statement if FALSE, correct it by changing one (or more) of the underlined words. Do *not* change any words that are not underlined.

(a) T F If a distribution of scores is symmetric, then the median and the mode will be the same.

(b) T F The <u>mean</u> varies less from sample to sample than the <u>median</u>.

(c) T F An advantage of the <u>mean</u> is that it can be computed for distributions that contain indeterminate data.

(d) T F A disadvantage of the <u>mean</u> is that it cannot be computed for distributions of ordinally scaled data.

(e) T F A distribution always has exactly <u>one</u> median score, but it can have <u>more than one</u> mode score.

(f) T F The <u>mean</u> is the best measure of central tendency to use for distributions that are heavily skewed.

(g) T F Another name for the median is the score at the <u>second</u> quartile.

(h) T F If the <u>mean</u> of a distribution is zero, then every score in the distribution must be equal to the mean.

(i) T F The <u>interquartile range</u> of a distribution is found by subtracting the smallest score in the distribution from the largest score.

(j) T F The percentile rank of the score 15 in the following distribution is <u>70%</u>.

$$6 \quad 6 \quad 8 \quad 9 \quad 11 \quad 11 \quad 15 \quad 15 \quad 17 \quad 18$$

(k) T F The variance is the <u>square</u> of the standard deviation.

(l) T F If a distribution is <u>negatively</u> skewed, then the mean is smaller than the median.

(m) T F One measure of the dispersion of a distribution is the <u>mode</u>.

(n) T F One good way to indicate how "extreme" a score is in a distribution is to give the scores <u>deviation from the mean</u>.

5.52 Find the standard deviation of each of the following distributions of sample scores. (Show your work, including the formula you used and any new columns needed to compute the formula.)

(a)

x
4
1
5
9
3

(b)

x	f
3	7
4	1
5	2
6	4

Compute the mean, median, mode, and standard deviation.

5.53 Complete this sentence using the set of measurements 2, 5, 12, 15, 17, 23, 32, 41, and 41. The value of 17 is _____.

(a) the mean

(b) the median

(c) the mode

(d) none of the measures of central tendency

5.54 Suppose you were told that the lower quartile for the annual income of unskilled workers in the United States is $5200. What does this mean?

5.55 The mean contribution per person for a college alumni fund drive was $100.50, with a standard deviation of $38.23. What percentage of the contributions could be expected to lie in the interval $62.27 to $138.73, assuming that contributions follow a mound-shaped distribution? Explain.

5.56 The following data represent a sample of $n = 10$ measurements on the tensile strength (in pounds per square inch) of sutured wounds in 10 days after repair: 63, 56, 32, 56, 48, 45, 45, 96, 57, 71.

(a) Determine the sample mean, median, and mode.

(b) Calculate s.

(c) Do the values of $\bar{x}$ and s help you to describe this set of data?

5.57 The percentage of city telephone subscribers who use unlisted numbers is increasing. The distribution of the percentage of numbers that are unlisted in cities possesses a mean and a standard deviation that are near 14% and 6%, respectively. If you were to pick a city at random, is it likely that the percentage of unlisted numbers would exceed 20%? Explain. Within what limits would you expect the percentage to fall?

5.58 Experiments are conducted on small animals as a prelude to experimentation with larger animals or humans. The object of an experiment was to investigate the effect of propranolol in reducing hypertension in rats. The hypertension was induced by cold exposure. Two groups of rats were used, an untreated (or control) group and another that received the drug dosage. The extent of hypertension in a given rat was monitored by recording its blood pressure.

To illustrate, the blood pressures of 5 rats in the control group, which were exposed for 6 weeks to a 5°C temperature, were 152, 179, 182, 176, and 149. Corresponding blood pressures for 7 rats treated with propranolol were 113, 111, 143, 151, 109, 111, and 168. The first step in the analysis of the data requires the computation of the sample means and variances. (In a later exercise we will show how we decide whether the drug was effective in reducing hypertension.)

(a) Calculate $\bar{x}$ and s^2 for the $n = 5$ rats in the control group. (For this sample, $\Sigma x = 838$ and $\Sigma x^2 = 141,446$.)

(b) Calculate $\bar{x}$ and s^2 for the $n = 7$ rats receiving the propranolol. (For this sample, $\Sigma x = 906$ and $\Sigma x^2 = 120{,}776$.)

5.59 Calculate $\bar{x}$ and s for the following measurements: 0, 4, 1, 3, 3, 2, 4, 0, 0, 1, 3, 1, 4, 5, 0, 1, 2, 4, 3, 1. Find the percentage of measurements that fall in the interval $(\bar{x} \pm 2s)$ and compare this result with the Empirical Rule.

5.60 For the 20 measurements 0, 4, 1, 0, 0, 2, 1, 1, 0, 3, 0, 1, 0, 2, 0, 1, 2, 2, 1, 0:

(a) Use the range approximation to guess the value of s.

(b) Calculate s and compare this computed value with the approximation in part (a).

(c) Find the percentage of measurements falling in the interval $(\bar{x} \pm 2s)$, and compare this result with the Empirical Rule.

5.61 The following data summarize by state the number of people below the poverty level in the United States.

State	Number below poverty level (1000)	State*	Number below poverty level (1000)
Alabama	684	Missouri	592
Alaska	39	Montana	95
Arizona	331	Nebraska	159
Arkansas	417	Nevada	67
California	2611	New Hampshire	78
Colorado	289	New Jersey	699
Connecticut	262	New Mexico	222
Delaware	69	New York	2344
Dist. of Columbia	115	North Carolina	827
Florida	1245	North Dakota	1108
Georgia	869	Oklahoma	391
Hawaii	92	Oregon	291
Idaho	118	Pennsylvania	1213
Illinois	1284	Rhode Island	94
Indiana	523	South Carolina	479
Iowa	266	South Dakota	107
Kansas	232	Tennessee	760
Kentucky	657	Texas	2055
Louisiana	777	Utah	154
Maine	140	Vermont	56
Maryland	409	Virginia	594
Massachusetts	547	Washington	410
Michigan	1004	West Virginia	276
Minnesota	370	Wisconsin	388
Mississippi	600	Wyoming	37

*The data for Ohio were not available at this time.

Construct a frequency histogram for these data.

5.62   Refer to Exercise 5.61. Compute the mean, median, and mode.

5.63   The following output displays a stem-and-leaf plot and a box plot. Discuss the characteristics of the data (Hint: Each horizontal mark of the box plot is 50.0, with the first tick mark at 0.00.)

```
MTB > SET INTO C1
DATA> 684     39    331    417   2611   289   262    69   115   1245
DATA> 869     92    118   1284    523   266   232   657   777    140
DATA> 409    547   1004    370    600   592    95   159    67     78
DATA> 699    222   2344    827   1108   391   291  1213    94    479
DATA> 107    760   2055    154     56   594   410   276   388     37
DATA> END

MTB > STEM-AND-LEAF OF C1;
SUBC> TRIM;
SUBC> END;

Stem-and-leaf of C1          N  = 50
Leaf Unit = 10

     9     0 335667999
    15     1 011455
    22     2 2366789
    (4)    3 3789
    24     4 0117
    20     5 2499
    16     6 0589
    12     7 67
    10     8 26
     8     9
     8    10 0
     7    11 0
     6    12 148

        HI  205, 234, 261,

MTB > BOXPLOT OF C1



            -------------
      --I    +    I-------------              *    *    0
            -------------
      +---------+---------+---------+---------+---------+------C1
      0        500      1000      1500      2000      2500

MTB > STOP
```

5.64   The maximum daily temperature in the vicinity of a major resort area averages 83°F during the summer months. The standard deviation for maximum daily temperatures in this area is 7°F (assume the measurements are mound shaped). Describe the variability of the set of maximum daily temperatures by using the Empirical Rule.

5.65 Concerned with the amount of heating oil consumed over the past several years, a homeowner finds that the following amounts (in gallons) per week have been used during the winter months over the past five years: 110, 115, 105, 96, 85. Find the mean and standard deviation for the homeowner's average weekly fuel consumption over the past five winters. Would a bill for 140 gallons be unexpected? Why?

5.66 An industrial concern uses an employee screening test with an average score $\mu = 65$ and a standard deviation $\sigma = 10$. Only those applicants with the highest 2.5% of the scores receive further consideration. If the test score distribution is approximately mound shaped, what score is required to qualify an applicant for further consideration?

5.67 The intelligence quotient (IQ) expresses intelligence as a ratio of the mental age to the chronological age multiplied by 100. Thus the average (when mental age equals chronological age) is 100. Construct a relative frequency histogram for the following IQ scores.

100	103	99	101	100	120	109	82
101	112	95	118	118	89	114	113
92	137	130	94	87	93	111	96
93	98	101	96	84	86	89	90

5.68 Refer to Exercise 5.67.

(a) Calculate $\bar{x}$ and s.

(b) Use the range approximation for s to check your calculations.

(c) Find the number of scores in the intervals $(\bar{x} \pm s)$, $(\bar{x} \pm 2s)$, and $(\bar{x} \pm 3s)$. Compare the percentages of measurements in these intervals with the percentages specified by the Empirical Rule.

5.69 The number of persons who volunteered to give a pint of blood at a central donor center was recorded for each of twenty successive Fridays. The data are shown here:

320	370	386	334	325	315	334	301	270	310
274	308	315	368	332	260	295	356	333	250

(a) Construct a stem-and-leaf plot.

(b) Construct a skeletal box plot and interpret the results.

5.70 Monthly readings for the FDC Index, a popular barometer of the health of the pharmaceutical industry, are shown here. As can be seen, the Index has several components—one for pharmaceutical companies, one for diversified companies, one for chain drugstores, and another for drug and medical supply wholesalers.

Month	Pharmaceuticals	Diversified	Chain	Wholesaler
January	123.1	154.6	393.3	475.5
February	122.4	146.0	407.6	504.1
March	125.2	169.2	405.0	476.6
April	136.1	156.7	415.1	513.3
May	149.3	177.0	418.9	543.5
June	145.7	158.1	443.2	552.6
July	162.4	156.6	419.1	526.2
August	168.0	178.6	404.0	516.3
September	155.6	170.4	391.8	482.1
October	177.0	162.9	410.9	484.0
November	196.6	182.4	459.8	522.6
December	195.2	195.4	431.9	536.8

(a) Plot these data on a single graph.

(b) Discuss trends within each component and any apparent relationships among the separate components of the FDC Index.

5.71 Refer to Exercise 5.70. Compute the percentage change for each month of each component of the Index. (Assume that the percent changes in January were 12.3, −.7, 12.1, and 16.1, respectively, for the four components.) Plot these data. Are they more revealing than the original measurements were?

$ 5.72 Closing New York Stock Exchange (NYSE) prices for the components (as of March 1987) of the Dow Jones Industrial Average (DJIA) are shown here:

	Components of Dow Jones Industrial Average	
	Percent of DJIA	*Closing NYSE stock price 7/1/88*
Allied-Signal	2.81%	$35.000
Alcoa	2.39	$52.625
American Can (Primerica Corp)	3.93	$27.250
American Express	3.25	$27.000
AT&T	1.35	$26.500
Bethlehem Steel	1.06	$24.125
Chevron	2.17	$45.250
duPont	3.70	$92.250
Eastman Kodak	2.82	$45.500
Exxon	3.06	$44.625
General Electric	4.12	$43.875

(continued)

**Components of Dow Jones
Industrial Average**

	Percent of DJIA	*Closing NYSE stock price 7/1/88*
General Motors	4.22	$79.750
Goodyear	1.90	$64.250
Inco	0.85	$33.750
IBM	8.99	$126.625
Intl. Harvester (Navistar)	0.53	$ 6.625
Intl. Paper	2.96	$48.500
McDonald's	4.48	$64.250
Merck	8.10	$55.250
Minnesota Mining	5.18	$65.750
Owens-Illinois	3.35	$20.750
Phillip Morris	5.49	$83.375
Procter & Gamble	3.94	$77.000
Sears, Roebuck	2.22	$36.625
Texaco	1.72	$46.625
Union Carbide	4.95	$22.750
United Technologies	2.73	$38.625
U.S. Steel	1.40	$31.625
Westinghouse	2.00	$14.625
Woolworth	3.61	$51.875

(a) Compute the actual range of the stock prices.

(b) The DJIA is actually a weighted average, so only a certain percentage of the actual NYSE price is part of the DJIA for each stock. The weighted average can be written as

$$\bar{y}_w = \sum \frac{w_i y_i}{n}$$

where y_i is the closing price for stock i, and w_i is the weight attached to stock i. Using the weights (percentage of DJIA) listed in the above table, compute the DJIA for this particular day.

(c) Refer to part (b). Why might the DJIA be a weighted average, rather than a simple average?

5.73 As one part of a review of middle-manager selection procedures, a study was made of the relation between hiring source (promoted from within, hired from related business, hired from unrelated business) and the three-year job history (additional promotion, same position, resigned, dismissed). The data for 120 middle managers follow.

Job History	Within firm	Related business	Unrelated business	Total
Promoted	13	4	10	27
Same position	32	8	18	58
Resigned	9	6	10	25
Dismissed	3	3	4	10
Total	57	21	42	120

(a) Calculate job-history percentages within each source.

(b) Would you say that there is a strong dependence between source and job history?

5.74 Class Exercise Select an area of your undergraduate major that utilizes experimental data. Typical data sources would be chemistry, biology, psychology, geology, or physics laboratories. Or you might seek data contained in the social science or business journals. Either by experimentation or by use of a professional journal, select a sample of at least $n = 25$ observations on some experimental variable.

(a) Define the population from which your sample was drawn.

(b) Construct a relative frequency histogram for the data.

(c) Calculate $\bar{x}$ and s for the data.

(d) Do the data appear to be mound shaped and thereby satisfy the requirements of the Empirical Rule?

(e) What fraction of the observations lie within two standard deviations of $\bar{x}$? Do these results agree with the Empirical Rule?

5.75 Describe the data of Exercise 5.67 using the computer-generated box plot shown here.

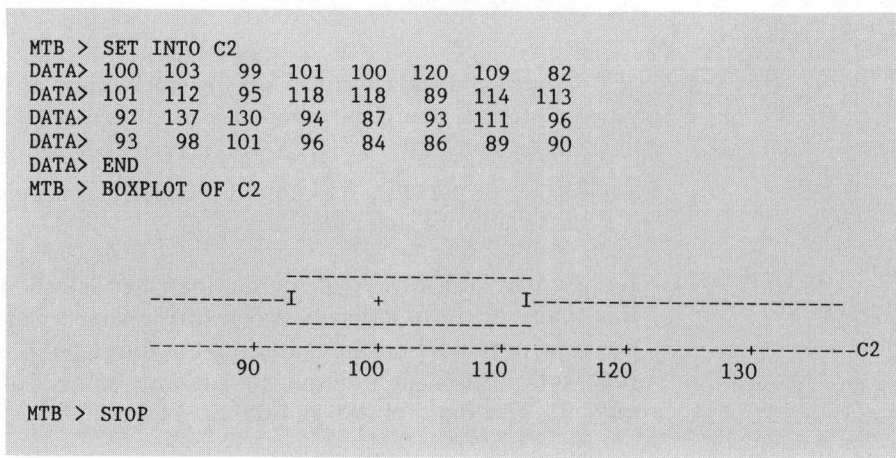

```
MTB > SET INTO C2
DATA> 100   103    99   101   100   120   109    82
DATA> 101   112    95   118   118    89   114   113
DATA>  92   137   130    94    87    93   111    96
DATA>  93    98   101    96    84    86    89    90
DATA> END
MTB > BOXPLOT OF C2


                          -------------------
             -----------I       +          I----------------------------
                          -------------------
             --------+---------+---------+---------+---------+--------C2
                    90        100       110       120       130

MTB > STOP
```

(a) Define the population from which your sample was drawn.

(b) Construct a relative frequency histogram for the data.

(c) Calculate $\bar{x}$ and s for the data.

(d) Do the data appear to be mound shaped and thereby satisfy the requirements of the Empirical Rule?

(e) What percentage of the observations lie within two standard deviations of $\bar{x}$? Three standard deviations? Do these results agree with the Empirical Rule?

5.76 Examine the population of all telephone subscribers in your community (see your local telephone directory). Randomly select a page from the directory and repeat the process until you have selected $n = 30$ pages. Let x be the number of subscribers per page and find x for each of the $n = 30$ pages.

(a) Construct a relative frequency histogram for your data.

(b) Calculate $\bar{x}$ and s.

(c) Find the percentage of observations in the intervals $(\bar{x} \pm s)$ and $(\bar{x} \pm 2s)$. Does the Empirical Rule provide a satisfactory description of the variability of the data?

5.77 **Class Exercise** Suppose you were an environmental scientist and you were required to measure the dissolved oxygen content (which is a measure of pollution) at a particular point in a lake. Why would a single chemical determination be unsatisfactory?

The answer to this question can be found by recalling your own experiences. If you have ever constructed something, measuring and cutting pieces to be assembled, you may have noticed that they frequently did not fit. This problem is often caused by errors in measuring the component parts. Similar random-measurement errors occur in almost all experimentation (if the measuring instruments are accurate enough to detect the variation).

Most errors caused by inaccurate measuring instruments are reduced (not eliminated) by using not one measurement but the mean of several measurements to characterize the true value of the quantity being measured.

To illustrate, have 10 people in your class measure some object in the room (for instance, the length of the room). Notice the variation in the recorded measurements. Calculate the mean for the 10 sample measurements ($n = 10$). Note that the mean falls near the center of the set of measurements and that it tends to offset overly large measurements by small ones.

If we were to measure an object repeatedly, millions and millions of times, a population of measurements would be generated. It is likely that the mean of this population, μ, would coincide with the true length of the object. Viewed in this manner, the 10 classroom measurements represent a sample, and the sample mean $\bar{x}$ estimates μ. In later chapters we will learn how to evaluate the error of this estimate — that is, the difference between the estimate $\bar{x}$ and the true mean μ.

5.78 In order for a health clinic to be capable of handling the desired patient load, designers need to know something about the demand for services in the area in which the clinic is to be located. This would include information on the patient arrival rate as well as the length of time to treat a patient. Both the arrival rate and the treatment time will vary in a random manner. The arrival rate varies because of the random occurrence of outbreaks of flu and other common illnesses; the time of treatment varies because treatment time will depend on a particular patient's illness. Thus the demand for physician and nurse time will vary in a random manner. By studying the frequency distributions of patient arrival rates and treatment times, the designer can specify the number of doctors, nurses, technicians, orderlies, and physical equipment needed to meet the demand. These numbers will affect the length of time a patient must wait in the clinic before receiving attention.

To answer some of the questions about clinic treatment times, a designer acquired data from an established clinic in a locale that possessed similar characteristics to the proposed new clinic location. The treatment times for 50 patients, randomly selected from the clinic's records, are as follows:

21	20	31	24	15	21	24	18	33	8
26	17	27	29	24	14	29	41	15	11
13	28	22	16	12	15	11	16	18	17
29	16	24	21	19	7	16	12	45	24
21	12	10	13	20	35	32	22	12	10

(a) What is the average treatment time for the sample of 50 patients? [You can verify that the sum of the 50 measurements ($n = 50$) is 1016.] Interpret this statistic.

(b) Find the median treatment time. Interpret this statistic.

(c) Construct a relative frequency histogram for the data. Use as class intervals 5.5–10.5, 10.5–15.5, 15.5–20.5, and so on. Mark the mean and median on the horizontal axis. Note that both measures do a reasonably good job of locating the center of distribution of treatment times.

(d) Must the sample mean and sample median be equal? Explain.

5.79 Canned salmon has a mean sodium content of 1190 milligrams per serving with standard deviation of 44 milligrams per serving. If the distribution for the sodium content of canned salmon servings is mound shaped, within what range do 95% of all such servings have sodium content? All or almost all?

5.80 During the years of 1970–1983, there was a total of 21 minority participants in the national finals of the Miss American Pageant held in Atlantic City, New Jersey, each September. The distribution of amount won by the minority participants is as follows:

Amount won:	1,000	2,000	3,000	5,500	6,500	7,000	17,500	29,000
Number	11	3	2	1	1	1	1	1

Determine the mean amount won by a typical minority participant. Would the Empirical Rule be appropriate for making inferential statements? Is the mean or the median more representative of the group's average amount won? What is the third quartile for the group?

5.81 During the years 1800 to the present, the reigns of Catholic Pontiffs (Popes) were as follows:

Pius VII	1800–1823	Benedict XV	1914–1922
Leo XII	1823–1829	Pius XI	1922–1939
Pius VIII	1829–1830	Pius XII	1939–1958
Gregory XVI	1831–1846	John XXIII	1958–1963
Pius IX	1846–1878	Paul VI	1963–1978
Leo XIII	1878–1903	John Paul I	1978–1978
St. Pius X	1903–1914	John Paul II	1978–present

Determine the range and quartiles for the length of reign for the Pontiffs listed. Does the distribution of length of reign appear to be mound shaped?

5.82 Professional football is concerned with the number of no-shows, that is, the number of ticket holders who do not attend its games. The number of no-shows for 14 randomly selected games during the 1987 professional football season was as shown below:

1,801	6,700	3,761	9,226	5,391	2,766	2,894
11,857	379	21,010	5,617	4,160	2,029	18,269

Determine the mean and standard deviation for the number of no-shows for professional football games based on this sample of 14 games. Would using the range approximation for standard deviation have provided a good approximation?

EXERCISES FROM THE DATA BASE

5.83 Refer to the clinical trial data base in Appendix 1, and compute the mean, range, and standard deviation of the HAM-D total score for the 4 treatment groups.

5.84 Again refer to the clinical trial data base. Complete a profile of the 4 treatment groups by computing the mean, range, and standard deviation for the anxiety, retardation, sleep disturbance, and total score (see Exercise 5.83) from the HAM-D scale. Are there any obvious differences among the groups following the treatment period?

5.85 Combine the OBRIST scores for the 4 treatment groups of the clinical trial data base in Appendix 1.

(a) Construct a histogram and determine the mean and median from the histogram.

(b) Compute the actual sample mean and median for the combined data, and compare these values to the approximations obtained from part (a).

PART 4

STEP THREE: BACKGROUND FOR ANALYZING DATA

CHAPTER 6

PROBABILITY AND PROBABILITY DISTRIBUTIONS

6.1   INTRODUCTION

In previous chapters we learned that a population is a set of measurements and that it can be described by a set of numerical measures called parameters. Typical population parameters are its mean μ and standard deviation σ. In most applications of statistics we will not know μ or σ (or other population parameters) but will attempt to make inferences about them based on information contained in a sample. Because a sample will represent only a subset of the measurements needed to calculate the exact value of a parameter, an inference based on a sample will almost always be subject to error.

To illustrate, suppose that a nutritionist wished to estimate the mean weight gain of one-year-old white mice placed on a specified diet for one month. Ten mice, selected from the conceptual set of all one-year-old white mice, are placed on the diet; the mean weight gain for the sample of 10 is $\bar{x} = 4.3$ grams. Note that it is highly unlikely that the sample mean weight gain, $\bar{x} = 4.3$, will equal μ (the mean weight gain that would be obtained if all one-year-old white mice were placed on the diet), but it is probably close. What is the probability that the sample mean weight gain $\bar{x}$ will lie within .1 gram of the population mean μ? This example highlights the uncertainty that surrounds every inference.

Probability is the language of uncertainty. The concepts and theorems of probability allow us to specify probable errors when making inferences in the analysis of data. The material in this chapter and Chapter 7 is fundamental background for the remainder of this text.

As you will see, not only is probability used to measure the reliability of an inference but it also plays a fundamental role in all statistical inference-making procedures. In summary, the elements of probability discussed in Chapters 6 and 7 provide the basis for measuring the uncertainty of all statistical inferences (estimates, predictions, decisions, etc.) made from the methods for analyzing data (Chapters 8–17).

6.2   INTERPRETATIONS OF PROBABILITY

Data are obtained either by observation of uncontrolled events in nature or by controlled experimentation. To simplify our terminology, we need a word that will apply to any method of data collection.

Definition 6.1

experiment

An **experiment** is the process by which an observation (or measurement) is obtained.

Typical examples of experiments are the following:

1. Measuring the weight gain for a mouse placed on a specific diet for one month.
2. Determining a test grade.
3. Making a measurement of daily rainfall.
4. Interviewing a voter to obtain the voter's preference prior to an election.
5. Inspecting a light bulb to determine whether or not it is defective.
6. Tossing a coin, and observing the face that appears.
7. Counting the number of divorces in a geographical area during a given year.

Note from Experiments 4, 5, and 6 that an observation need not be numerical.

outcome
event

Each possible, distinct result from an experiment is called an **outcome;** an **event** is a collection of one or more outcomes from an experiment. Events are usually denoted by capital letters. To see how the terminology is used, consider the following experiments.

Experiment 1   Toss two coins and observe the upper face of each coin.

Examples of events associated with this experiment are as follows:
   A: Two heads show.
   B: Two tails show.
   C: At least one tail shows.
   D: One head and one tail show.

Experiment 2   Sample 100 prospective voters and record each voter's preference, for or against a local referendum.

Examples of events associated with this experiment are as follows:
   A: Exactly 50 voters favor the referendum.
   B: Fewer than 20 voters favor the referendum.
   C: All the voters favor the referendum.

Observations of phenomena can result in many different outcomes, some of which are more likely than others. Numerous attempts have been made to give a precise definition for the probability of an outcome. We will cite a few of these.

classical interpretation of probability

The first interpretation of probability, called the **classical interpretation of probability,** arose from games of chance. Typical probability statements of this type are "the probability that a flip of a balanced coin will show 'heads' is 1/2," and "the probability of drawing an ace when a single card is drawn from a well-shuffled standard deck of 52 cards is 4/52." The numerical values for these probabilities arise from the nature of the games. A coin flip has two possible outcomes (a head or tail); the probability of a head should then be 1/2 (1 out of 2). Similarly, there are 4 aces in a standard deck of 52 cards, so the probability of drawing an ace in a single draw is 4/52 or 4 out of 52.

The probability of an event E under the classical interpretation of probability is computed by taking the ratio of the number N_E of outcomes favorable to event E in the total number N of possible outcomes.

$$P(\text{Event } E) = \frac{N_E}{N}$$

The applicability of this interpretation depends on the assumption that all outcomes are equally likely. If this assumption does not hold, the probabilities indicated by the classical interpretation of probability will be in error.

relative frequency concept

A second interpretation of probability is called the **relative frequency concept** of probability. If an experiment is repeated a large number of times and event E occurs 30% of the time, then .30 should be a very good approximation to the probability of event E. Symbolically, if an experiment is conducted n different times and if event E occurs on n_E of these trials, then the probability of event E is approximately

$$P(\text{Event } E) \approx \frac{n_E}{n}$$

We say "approximately" because we think of the actual probability P(event E) as the relative frequency of the occurrence of event E over a very large number of observations or repetitions of the phenomenon. The fact that we can check probabilities that have a relative frequency interpretation (by simulating many repetitions of the experiment) makes this interpretation very appealing and practical.

The third interpretation of probability can be used for problems in which it is difficult to imagine a repetition of an experiment. These are "one-shot" situations. For example, the director of a state welfare agency who estimates the probability that a proposed revision in eligibility rules will be passed by the state legislature would not be thinking in terms of a **subjective probability** long series of trials. Rather, the director would use a personal or **subjective probability** to make a one-shot statement of belief regarding the likelihood of passage of the proposed legislative revision. The problem with subjective probabilities is that they can vary from person to person and that they cannot be checked.

Of the three interpretations presented, the relative frequency concept seems to be the most reasonable one since it provides a practical interpretation of the probability for most events of interest. Even though we will never run the necessary repetitions of the experiment to determine the exact probability of an event, the fact that we could check the probability of an event gives meaning to the relative frequency concept. Throughout the remainder of this text we will lean heavily on this interpretation of probability.

EXERCISES

6.1 Indicate which interpretation of the probability statement seems most appropriate.

(a) The National Angus Association has stated that there is a 60/40 chance that wholesale beef prices will rise by the summer; that is, a .60 probability of an increase and a .40 probability of a decrease or no change in price.

(b) The quality control section of a large chemical manufacturing company has undertaken an intensive process-validation study. From this study the quality control section claims the probability is .998 that the shelf life of a newly released batch of chemical will exceed the minimal time specified.

(c) A new blend of coffee is being contemplated for release by the marketing division of a large corporation. Preliminary marketing survey results indicate that 550 of a random sample of 1000 potential users rated this new blend better than a brand-name competitor. The probability of this happening is approximately .001 assuming that there is no difference in consumer preference for the two brands.

(d) The probability of receiving a busy signal when attempting to access the company WATS line during the 3:00-5:00 P.M. time frame is .58.

(e) The probability that it will rain tomorrow is .30.

(f) Within a city, the probability of selecting a random household within which the head of the household is unemployed is .12.

6.2 **Class Experiment** Give your own personal probability for each of the following situations. It would be instructive to tabulate these probabilities for the entire class. In which cases did you get large disagreements?

(a) The federal budget will be balanced in the next fiscal year.

(b) You will receive a B or higher in this course.

(c) Two or more individuals in the classroom will have the same birthday.

(d) The Washington Redskins will win the Super Bowl next year.

(e) The total production of Florida oranges next year will exceed this year's production.

6.3 FINDING THE PROBABILITY OF AN EVENT

In the preceding section we discussed three different interpretations of probability. We will use the classical interpretation and the relative frequency concept to illustrate the computation of the probability of an outcome or an event. Consider an experiment that consists of tossing two coins, a penny and a dime, and observing the upturned faces. There are four possible outcomes:

TT: Tails for both coins
TH: A tail for the penny, a head for the dime
HT: A head for the penny, a tail for the dime
HH: Heads for both coins

What is the probability of observing the following event: exactly one head from the two coins?

This probability can be obtained easily if we can assume that all four outcomes are equally likely. In this case, that seems quite reasonable. Then, by the previous sections, there are $N = 4$ possible outcomes and $N_E = 2$ of these (outcomes TH and HT) are favorable for the event of interest, observing exactly one head. Hence, by the classical interpretation of probability,

$$P(\text{exactly 1 head}) = \frac{2}{4} = \frac{1}{2} = .5$$

Since the event of interest has a relative frequency interpretation, we could also obtain this same result empirically, by using the relative frequency concept. Suppose that a penny and a dime are tossed 2000 times, with the results shown in Table 6.1. Note that this approach yields approximate probabilities that are in agreement with our intuition. That is, intuitively we might expect these outcomes to be equally less likely and that each would occur with a probability equal to 1/4 or .25. This assumption was made for the classical interpretation.

TABLE 6.1   Results of 2000 Tossings of a Penny and a Dime

Outcome	Frequency	Relative frequency
TT	474	474/2000 = .237
TH	502	502/2000 = .251
HT	496	496/2000 = .248
HH	528	528/2000 = .264

If we wish to find the probability of observing exactly one head when tossing two coins, we have, from Table 6.1,

$$P(\text{exactly 1 head}) \approx \frac{502 + 496}{2000} = .499$$

This is very close to the theoretical probability, which we have shown to be .5.

No matter what the event of interest, the probability of an event, say event A, will always satisfy the property

$$0 \leq P(A) \leq 1$$

Relations between two events and probabilities associated with these event relations will be discussed in the next section.

6.4   EVENT RELATIONS

Suppose that A and B represent two experimental events and that you are interested in a new event, the event that either A or B occurs. For example, suppose that we toss a pair of dice and define the following events:

A: A total of 7 shows
B: A total of 11 shows

Then the event "either A or B occurs" is the event that you toss a total of either 7 or 11 with the pair of dice.

Notice that for this example the events A and B are mutually exclusive. That is, if you observe event A (a total of 7), you could not at the same time observe event B (a total of 11). Thus if A occurs, B cannot occur (and vice versa).

Definition 6.2

mutually exclu-
sive events

Two events A and B are said to be **mutually exclusive** if (when the experiment is performed a single time) the occurrence of one of the events excludes the possibility of the occurrence of the other event.

The concept of mutually exclusive events is used to specify a second property that the probabilities of events must satisfy. When two or more events are mutually exclusive, then the probability that any one of the events will occur is the sum of the event probabilities. That is, if two events A and B are mutually exclusive, the probability that either event A or B occurs is

$$P(\text{either } A \text{ or } B) = P(A) + P(B)$$

For example, when we toss a pair of dice, the sum S of the numbers appearing on the dice can assume any one of the values $S = 2, 3, 4, \ldots, 11, 12$. On a single toss of the dice we can observe only one of these values. Therefore, the values $2, 3, \ldots, 12$ represent mutually exclusive events. If we want to find the probability of tossing a sum less than or equal to 4, this probability is

$$P(S \leq 4) = P(2) + P(3) + P(4)$$

For this particular experiment the dice can fall in 36 equally likely different ways. For example, we can observe a 1 on die No. 1 and a 1 on die No. 2, denoted by the symbol $(1, 1)$; we can observe a 1 on die No. 1 and a 2 on die No. 2, denoted by $(1, 2)$. In other words, for this experiment the possible outcomes are

(1, 1)	(2, 1)	(3, 1)	(4, 1)	(5, 1)	(6, 1)
(1, 2)	(2, 2)	(3, 2)	(4, 2)	(5, 2)	(6, 2)
(1, 3)	(2, 3)	(3, 3)	(4, 3)	(5, 3)	(6, 3)
(1, 4)	(2, 4)	(3, 4)	(4, 4)	(5, 4)	(6, 4)
(1, 5)	(2, 5)	(3, 5)	(4, 5)	(5, 5)	(6, 5)
(1, 6)	(2, 6)	(3, 6)	(4, 6)	(5, 6)	(6, 6)

As you can see, only one of these events, (1, 1), will result in a sum equal to 2. Therefore, we would expect a 2 to occur with a relative frequency of 1/36 in a long series of repetitions of the experiment, and we let $P(2) = 1/36$. The sum $S = 3$ will occur if we observe either of the outcomes (1, 2) or (2, 1). Therefore, $P(3) = 2/36 = 1/18$. Similarly, we find $P(4) = 3/36 = 1/12$. It follows that

$$P(S \leq 4) = P(2) + P(3) + P(4) = \frac{1}{36} + \frac{1}{18} + \frac{1}{12} = \frac{1}{6}$$

A third property of event probabilities is stated in terms of an event and its complement.

Definition 6.3

complement

The **complement** of an event A is the event that A does not occur. The complement of A is denoted by the symbol $\overline{A}$.

Thus if we define the complement of an event A as a new event, namely, "A does not occur," it follows that

$$P(A) + P(\overline{A}) = 1$$

For example, refer again to the two-coin-toss experiment. If, in many repetitions of the experiment, the proportion of times you observe event A, "two heads show," is 1/4, then it follows that the proportion of times you observe the event $\overline{A}$, "two heads do not show," is 3/4. Thus $P(A)$ and $P(\overline{A})$ will always sum to 1.

The three properties that the probabilities of events must satisfy can be summarized as follows:

Properties of probabilities

If A and B are any two mutually exclusive events associated with an experiment, then $P(A)$ and $P(B)$ must satisfy the following properties:

1. $0 \leq P(A) \leq 1$ and $0 \leq P(B) \leq 1$.

2. $P(\text{either } A \text{ or } B) = P(A) + P(B)$.

3. $P(A) + P(\bar{A}) = 1$ and $P(B) + P(\bar{B}) = 1$.

6.5 CONDITIONAL PROBABILITY AND INDEPENDENCE

Consider the following situation: The examination of a large number of insurance claims, categorized according to type of insurance and according to whether the claim was fraudulent, produced the results shown in Table 6.2. Suppose you are responsible for checking insurance claims—in particular, for detecting fraudulent claims—and you examine the next claim that is processed. What is the probability of the event F, "the claim is fraudulent"? To answer the question, you examine Table 6.2 and note that 10% of all claims are fraudulent. Thus, assuming that the percentages given in the table are reasonable approximations to the true probabilities of receiving specific types of claims, it follows that $P(F) = .10$. Would you measure the risk that you face a fraudulent claim with the probability .10? We think not, because you have additional information that may affect the assessment of $P(F)$. For example, you would know the type of policy you were examining (i.e., fire, auto, or other).

TABLE 6.2 Categorization of Insurance Claims

	Type of policy			
Category	*Fire*	*Auto*	*Other*	*Total*
Fraudulent	6%	1%	3%	10%
Nonfraudulent	14%	29%	47%	90%
Total	20%	30%	50%	100%

Suppose that you have the additional information that the claim was associated with a fire policy. Checking Table 6.2, we see that 20% (or .20) of all claims are associated with a fire policy and that 6% (or .06) of all claims are fraudulent fire policy claims. Therefore, it follows that the probability that the claim is fraudulent, given that you know the policy is a fire policy, is

$$P(F|\text{fire policy}) = \frac{\genfrac{}{}{0pt}{}{\text{proportion of claims that are fraudulent}}{\text{fire policy claims}}}{\genfrac{}{}{0pt}{}{\text{proportion of claims that are}}{\text{against fire policies}}}$$

$$= \frac{.06}{.20} = .30$$

conditional probability

This probability, $P(F|\text{fire policy})$, is called a **conditional probability** of the event F, that is, the probability of event F given the fact that the event "fire policy" has already occurred. This tells you that 30% of all fire policy claims are fraudulent. The vertical bar in the expression $P(F|\text{fire policy})$ represents the phrase "given that," or simply "given." Thus the expression is read, "the probability of the event F given the event fire policy."

unconditional probability

The probability $P(F) = .10$, called the **unconditional probability** of the event F, gives the proportion of times a claim is fraudulent, that is, the proportion of times event F occurs in a very large (infinitely large) number of repetitions of the experiment (receiving an insurance claim and determining whether the claim is fraudulent). In contrast, the conditional probability of F, given that the claim is for a fire policy, $P(F|\text{fire policy})$, gives the proportion of fire policy claims that are fraudulent. Clearly the conditional probabilities of F, given the types of policies, will be of much greater assistance in measuring the risk of fraud than the unconditional probability of F.

Suppose that the probability of an event A is the same regardless of whether event B has or has not occurred. That is, suppose

$$P(A|B) = P(A)$$

Then we say that the occurrence of event A is not dependent on the occurrence of event B or, simply, that A and B are independent events.

Definition 6.4

independent events

Two events A and B are **independent events** if

$$P(A|B) = P(A) \qquad \text{or if} \qquad P(B|A) = P(B)$$

Note: You can show that if $P(A|B) = P(A)$, then $P(B|A) = P(B)$, and vice versa.

EXERCISES

6.3 A coin is to be flipped three times. List the possible outcomes in the form (result on Toss 1, result on Toss 2, result on Toss 3).

6.4   In Exercise 6.3, assume that each one of the outcomes has probability 1/8 of occurring. Find the probability of

(a) *A:*   Observing exactly 1 head

(b) *B:*   Observing 1 or more heads

(c) *C:*   Observing no heads

6.5   For Exercise 6.4,

(a) Compute the probability of the complement of event *A*, event *B*, and event *C*.

(b) Determine whether events *A* and *B* are mutually exclusive.

6.6   Determine the following conditional probabilities for the events of exercise 6.4.

(a) $P(A|B)$ (b) $P(A|C)$ (c) $P(B|C)$

6.7   Refer to Exercise 6.6. Are events *A* and *B* independent? Why or why not? What about *A* and *C*? What about *B* and *C*?

6.8   The emergency room of a hospital has two back-up generators, either of which can supply enough electricity for basic hospital operations. If we define events *A* and *B* as follows:

Event *A:*   Generator 1 works properly
Event *B:*   Generator 2 works properly

describe the following events in words

(a)   Complement of *A* ($\overline{A}$)   (b)   *B|A*   (c)   Either *A* or *B*

6.9   A die is to be rolled, and we are to observe the number face-up. Find the probabilities for these events:

(a) *A:*   Observe a 6

(b) *B:*   Observe an even number

(c) *C:*   Observe a number greater than 2

(d) *D:*   Observe an even number and a number greater than 2

6.10   Refer to Exercise 6.9. Which of the events (*A*, *B*, and *C*) are independent? Mutually exclusive?

6.11   Consider the following outcomes for an experiment:

Outcome	1	2	3	4	5
Probability	.20	.25	.15	.10	.30

Let event *A* consist of outcomes 1, 3, and 5, and event *B* consist of outcomes 4 and 5.

(a) Find $P(A)$ and $P(B)$.

(b) Find P(both *A* and *B* occur).

(c) Find P(either *A* or *B* occurs).

6.12   Refer to Exercise 6.11. Does P(either *A* or *B* occurs) = $P(A) + P(B)$? Why or why not?

6.13   A large corporation has spent considerable time developing employee performance rating scales to evaluate an employee's job performance on a regular basis, so major adjustments can be made when needed and employees who should be considered for a "fast track" can be isolated. Keys to this latter determination are ratings on the employee's ability to perform job capabilities and on his or her formal training for the job.

Workload	Formal Training			
Capacity	None	Little	Some	Extensive
Low	.01	.02	.02	.04
Medium	.05	.06	.07	.10
High	.10	.15	.16	.12

The probabilities for being placed on a fast track are as indicated for the 12 categories in the table. The following three events (A, B, and C) are defined:

A: An employee works at the high-capacity level.
B: An employee falls into the highest (Extensive) formal training category.
C: An employee has little or no formal training and works below high capacity.

(a) Find $P(A)$, $P(B)$, and $P(C)$.
(b) Find $P(A|B)$, $P(A|\bar{B})$ and $P(\bar{B}|C)$.
(c) Find $P(A|B)$, $P(A|\ )$ and $P(\ |C)$.

6.14   A survey of a number of large corporations gave the following probability table for events related to the offering of a promotion involving a transfer.

Promotion/ transfer	Married		Unmarried	Total
	Two-career marriage	One-career marriage		
Rejected	.184	.0555	.0170	.2565
Accepted	.276	.3145	.1530	.7435
Total	.46	.37	.17	

Use the probabilities to answer the following questions:

(a) What is the probability that a professional (selected at random) would accept the promotion? Reject it?

(b) What is the probability that a professional (selected at random) is part of a two-career marriage? A one-career marriage?

6.15 Refer to Exercise 6.14. Describe the following events in terms of event relations and conditional probabilities and give the probabilities of these events.

(a) A randomly selected professional is unmarried and would accept a promotion involving a transfer.

(b) A randomly selected professional would accept a transfer given that the professional selected is unmarried.

(c) A randomly selected professional is married.

(d) A randomly selected professional would accept a promotion involving a transfer given that the professional is married.

6.6 RANDOM VARIABLES

The basic language of probability developed in this chapter deals with many different kinds of events. We are interested in calculating the probabilities associated with both quantitative and qualitative events. For example, we developed techniques that could be used to calculate the probability that a person selected at random for a Nielsen survey of television viewing habits would favor the ABC nightly news program (as opposed to that of CBS or NBC). These same techniques are also applicable to finding the probability that a person selected for the Nielsen survey watches television more than 30 hours per week.

These qualitative and quantitative events can be classified as events (or outcomes) associated with qualitative and quantitative variables. For example, in the Nielsen survey, responses to the question "Which evening television news program do you prefer: ABC, CBS, or NBC?" are observations on a qualitative variable, since the possible responses vary in kind but not in any numerical degree. Because we cannot predict with certainty what a particular person's response will be, the variable is classified as a **qualitative random variable.** Other examples of qualitative random variables that are commonly measured are political party affiliation, socioeconomic status, geographic location, and sex/race classification.

qualitative random variable

There are a finite (and typically quite small) number of possible outcomes associated with any qualitative variable. Using the methods of this chapter, it is possible to calculate the probabilities associated with these events.

quantitative random variable

Many times the events of interest in an experiment are quantitative outcomes associated with a **quantitative random variable,** since the possible responses vary in numerical magnitude. For example, in a Nielsen survey, responses to the question "How many hours a week do you watch television?" are observations on a quantitative random variable. Events of interest, such as viewing television more than 30 hours per week, are measured by this quantitative random variable. Other examples of quantitative random variables are the change in earnings per share of a stock over the next year, the increase in total sales over the next year, and the number of persons voting for the incumbent in an upcoming election. Again, the methods of this chapter can be applied to calculate the probability associated with any particular event.

There are major advantages to dealing with quantitative random variables. The numerical yardstick underlying a quantitative variable makes the mean and standard deviation (for instance) sensible. With qualitative random variables, there isn't much more to be said than has already been said. The methods of this chapter can be used to calculate the probabilities of various events, and that's about all. With quantitative random variables we can do much more: we can average the resulting quantities, find standard deviations, and assess probable errors, among other things. Hereafter, we use the phrase *random variable* to mean quantitative random variable; virtually all texts on probability theory use the phrase this way.

Most events of interest result in numerical observations or measurements. If a quantitative variable measured (or observed) in an experiment is denoted by the symbol x, we are interested in the values that x can assume. These values are called numerical outcomes. The number of students in a class of 50 who earn an A in their biology course is a numerical outcome. The percentage of registered voters who cast ballots in a given election is also a numerical outcome. The quantitative variable x is called a random variable because the value that x assumes in a given experiment is a chance or random outcome.

Random variables are classified as one of two types.

Definition 6.5

discrete random variable

When observations on a quantitative random variable can assume only a countable number of values, the variable is called a **discrete random variable.**

These are examples of discrete random variables:

1. The number of bushels of apples per acre for a given orchard this year
2. The number of accidents per year at an intersection
3. The number of voters in a sample favoring candidate Jones

Note that it is possible to count the number of values that each of these random variables can assume.

Definition 6.6

continuous random variable

When observations on a quantitative random variable can assume any one of the countless number of values in a line interval, the variable is called a **continuous random variable.**

For example, the daily maximum temperature in Rochester, New York, can assume any of the infinitely many values on a line interval. It could be 89.6, 89.799, or 89.7611114. Typical continuous random variables are temperature, pressure, height, weight, and distance.

The distinction between discrete and continuous random variables is pertinent when we are seeking the probabilities associated with specific values of a random variable. The need for the distinction will be apparent when we discuss probability distributions.

6.7   PROBABILITY DISTRIBUTIONS FOR DISCRETE RANDOM VARIABLES

As we know, it is necessary to learn the probability of observing a particular sample outcome in order to make an inference about the population from which the sample was drawn. To do this we need to know the probability associated with each value of the variable x. Viewed as relative frequencies, these probabilities generate a distribution of theoretical relative frequencies, called the probability distribution of x. Probability distributions differ for discrete and continuous variables but the interpretation is essentially the same for both.

probability distribution

The **probability distribution** for a discrete variable displays the probability $P(x)$ associated with each value of x. This display can be presented as a table, a graph, or a formula. To illustrate, consider the tossing of two coins, and let x be the number of heads observed. Then x can take the values 0, 1, or 2. From the data of Table 6.1, we can determine the approximate probability for each value of x, as given in Table 6.3.

We point out that the relative frequencies in the table are very close to the theoretical relative frequencies (probabilities), which can be shown to be .25, .50, and .25 (using the classical interpretation of probability). If we had tossed the coins 2,000,000 times instead of 2000, the relative frequencies for $x = 0$, 1, and 2 would most likely be indistinguishable from the theoretical probabilities.

TABLE 6.3   Empirical Sampling Results for x, the Number of Heads in 2000 Tosses of Two Coins

x	Frequency	Relative frequency
0	474	.237
1	998	.499
2	528	.264

The probability distribution for x, the number of heads in the toss of two coins, is shown in Table 6.4. It is presented graphically as a probability histogram in Figure 6.1.

TABLE 6.4	Probability Distribution for the Number of Heads When Two Coins Are Tossed

x	$P(x)$
0	.25
1	.50
2	.25

FIGURE 6.1   Probability Distribution for the Number of Heads When Two Coins Are Tossed

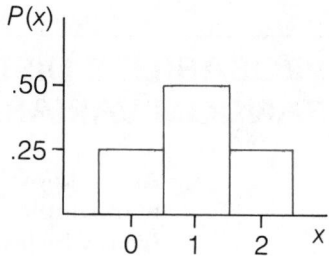

The probability distribution for this simple discrete random variable illustrates three important properties of discrete random variables.

Properties of Discrete Random Variables

1. The probability associated with every value of x lies between 0 and 1.
2. The sum of the probabilities for all values of x equals 1.
3. The probabilities for a discrete variable are additive. Hence the probability that x equals either 1 or 2 is equal to $P(1) + P(2)$.

The relevance of the probability distribution to statistical inference will be emphasized when we discuss the probability distribution for the binomial random variable.

6.8   A USEFUL DISCRETE RANDOM VARIABLE: THE BINOMIAL

Many populations of interest to business persons and scientists can be viewed as large sets of 0s and 1s. For example, consider the set of responses of all adults in the United States to the question, "Do you favor the development of nuclear energy?" If we disallow "no opinion," the responses will constitute a set of "yes" responses and "no" responses. If we assign a 1 to each "yes" and a 0 to each "no", the population will consist of a set of 0s and 1s and the sum of the 1s will equal the total number of persons favoring the development. The sum of the 1s divided by the number of adults in the United States will equal the proportion of people who favor the development.

Gallup and Harris polls are examples of the sampling of 0, 1 populations. People are surveyed and their opinions are recorded. Based on the sample responses, Gallup and Harris estimate the proportions of people in the population who favor some particular issue or possess some particular characteristic.

Similar surveys are conducted in the biological sciences, engineering, and business, but they may be called experiments rather than polls. For example, experiments are conducted to determine the effect of new drugs on small animals, such as rats or mice, before progressing to larger animals and, eventually, to human subjects. Many of these experiments bear a marked resemblance to a poll in that the experimenter records only whether or not the drug was effective. Thus if 300 rats are injected with a drug and 230 show a favorable response, the experimenter has conducted a "poll"—a poll of rat reaction to the drug, 230 "in favor" and 70 "opposed."

Similar "polls" are conducted by most manufacturers to determine the fraction of a product that is of good quality. Samples of industrial products are collected before shipment and each item in the sample is judged "defective" or "acceptable" according to criteria established by the company's quality control department. Based on the number of defectives in the sample, the company can decide whether the product is suitable for shipment. Note that this example, as well as those preceding, has the practical objective of making an inference about a population based on information contained in a sample.

The public opinion poll, the consumer preference poll, the drug-testing experiment, and the industrial sampling for defectives are all examples of a common, frequently conducted sampling situation known as a binomial experiment. The binomial experiment is conducted in all areas of science and business and only differs from one situation to another in the nature of objects being sampled (people, rats, electric light bulbs, oranges). Thus it is useful to define its characteristics. We can then apply our knowledge of this one kind of experiment to a variety of sampling experiments.

Definition 6.7

binomial experiment

A **binomial experiment** is one that possesses the following properties:

1. The experiment consists of n identical trials.

2. Each trial results in one of two outcomes. We will label one outcome a success, S, and the other a failure, F.

3. The probability of success on a single trial is equal to π, and π remains constant from trial to trial.

4. The trials are independent—that is, the outcome of one trial does not influence the outcome of any other trial.

5. The experimenter is interested in x, the number of successes observed during the n trials.

Now let us check some experiments to see whether or not they are binomial experiments.

Example 6.1

A survey is conducted to determine the proportion of adults in a certain locale who favor raising the legal drinking age from 19 to 21. A random sample of 300 adults is selected from the list of registered voters. They are interviewed and the number of those favoring the change is recorded. Is this a binomial experiment?

Solution   To answer this question, we will check each of the five characteristics of a binomial experiment to determine if they are satisfied.

1. Are there n identical trials? Yes ($n = 300$ interviews are conducted in an identical manner).

2. Does each trial result in one of two outcomes? Yes, each adult interviewed either favors or does not favor the change.

3. Is the probability of success the same from trial to trial? Yes, if we let success denote a person favoring the change, then, assuming the list of registered voters is large, the probability of selecting an adult favoring the change will remain (for all practical purposes) constant from trial to trial.

4. Are the trials independent? Yes, the outcome of one interview is unaffected by the results of the other interviews.

5. For this experiment the random variable of interest to the experimenter is x, the number of successes in the sample.

Since all five characteristics are present, the survey represents a binomial experiment.

Example 6.2

A biologist randomly selects 10 portions of water, each equal to .1 cubic centimeter in volume, from the local reservoir and counts the number of bacteria present in each portion. The biologist then totals the number of bacteria for the 10 portions to obtain an estimate of the number of bacteria per cubic centimeter present in the reservoir water. Is this a binomial experiment?

Solution   Check this experiment against the characteristics of a binomial experiment. This experiment consists of $n = 10$ trials, each resulting in the examination of a .1-cubic-centimeter portion of water. However, the second characteristic, that each trial results in only one of two outcomes, is not satisfied. In this experiment we observe the number of bacteria per portion, and this number can be any of the values 0, 1, 2, 3, . . . Thus this experiment is not a binomial experiment.

binomial random variable

The **binomial random variable** x is a discrete random variable that can assume the values $x = 0, 1, 2, . . . , n$. The binomial probability distribution, which gives the probabilities associated with each value of x, can be computed by using the following formula. The details of the origin of this formula are omitted from the text.

The Binomial Probability Distribution

$$P(x) = \frac{n!}{x!(n-x)!} \pi^x (1 - \pi)^{n-x} \qquad \text{for } x = 0, 1, 2, . . . , n$$

where n is the number of trials; π is the probability of a "success" on a single trial; $(1 - \pi)$ is the probability of a "failure" on a single trial; x is the number of successes in n trials; and

$$n! = n(n - 1)(n - 2) \cdots (3)(2)(1)$$

The probability of observing x successes in n trials can be calculated by substituting the values for x, n, π, and $1 - \pi$ into the formula for $P(x)$. As indicated above, the symbol $n!$, read "n factorial," means $n(n - 1)(n - 2) \cdots (3)(2)(1)$. For example, $5! = (5)(4)(3)(2)(1) = 120$. The quantity $0!$ is defined to be 1.

To see how the formula for the binomial probability distribution can be used to calculate the probability for a specific value of x, consider the following examples.

Example 6.3

An experiment consists of tossing a coin two times. If the probability of a head on a single toss is .5, compute the probability distribution for x, the number of heads in two tosses, using the binomial $P(x)$. Compare your results to those given in Table 6.4.

Solution   Using the formula

$$P(x) = \frac{n!}{x!(n - x)!}\, \pi^x(1 - \pi)^{n - x}$$

and substituting for $n = 2$, $\pi = .5$, $x = 0, 1, 2$, we obtain

$$P(0) = P(x = 0) = \frac{2!}{0!2!}\, (.5)^0(.5)^2 = .25$$

$$P(1) = P(x = 1) = \frac{2!}{1!1!}\, (.5)^1(.5)^1 = .50$$

$$P(2) = P(x = 2) = \frac{2!}{2!0!}\, (.5)^2(.5)^0 = .25$$

Note that the results are identical to those presented in Table 6.4.

Example 6.4

Suppose that a sample of households is randomly selected from all the households in the city in order to estimate the percentage in which the head of the household is unemployed. To illustrate the computation of a binomial probability, suppose that the unknown percentage is actually 10% and that a sample of $n = 5$ (we are selecting a small sample to make the calculation manageable) is selected from the population. What is the probability that all 5 heads of the households are employed?

Solution   We must carefully define which outcome we wish to call a success. For this example we will define a success as being employed. Then the probability of success when one person is selected from the population is $\pi = .9$ (because the proportion unemployed is .1). We wish to find the probability that $x = 5$ (all 5 are employed) in 5 trials.

$$P(x = 5) = \frac{5!}{5!(5 - 5)!}\, (.9)^5(.1)^0$$

$$= \frac{5!}{5!0!}\, (.9)^5(.1)^0$$

$$= (.9)^5 = .590$$

The binomial probability distribution for $n = 5$, $\pi = .9$ is shown in Figure 6.2. The probability of observing 5 employed in a sample of 5 is shaded in the figure.

FIGURE 6.2 The Binomial Probability Distribution for
$n = 5, \pi = .9$

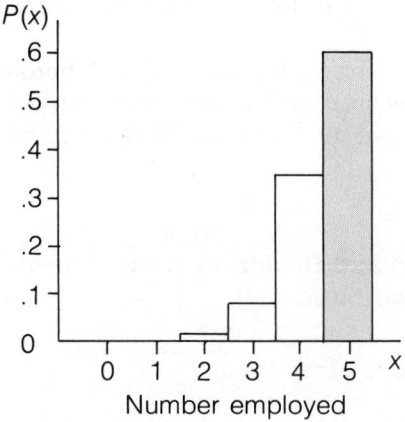

Number employed

Example 6.5

Refer to Example 6.4 and calculate the probability that exactly one person in the sample of 5 households is unemployed. What is the probability of one or less being unemployed?

Solution Since x is the number of employed in the sample of 5, one unemployed person would correspond to 4 employed ($x = 4$). Then

$$P(4) = \frac{5!}{4!(5 - 4)!} (.9)^4(.1)^1$$

$$= \frac{(5)(4)(3)(2)(1)}{(4)(3)(2)(1)(1)} (.9)^4(.1)$$

$$= 5(.9)^4(.1)$$

$$= .328$$

Thus the probability of selecting 4 employed heads of households in a sample of 5 is .328, or, roughly, one chance in three.

The outcome "one or less unemployed" is the same as the outcome "4 or 5 employed." Since x represents the number employed, we seek the probability that $x = 4$ or 5. Because the values associated with a random variable represent mutually exclusive events, the probabilities for discrete random variables are additive (this property of probabilities is discussed in Section 6.7). Thus we have

$$P(x = 4 \text{ or } 5) = P(4) + P(5)$$

$$= .328 + .590$$

$$= .918$$

That is, the probability that a random sample of 5 households will yield either 4 or 5 employed heads of households is .918. This high probability is consistent with our intuition: we could expect the number of employed in the sample to be large if 90% of all heads of households in the city are employed.

Like any relative frequency histogram, a binomial probability distribution possesses a mean μ and a standard deviation σ. Although we omit the derivations, we give the formulas for these parameters.

Mean and Standard Deviation of the Binomial Probability Distribution

$$\mu = n\pi \quad \text{and} \quad \sigma = \sqrt{n\pi(1 - \pi)}$$

where π is the probability of success in a given trial, and n is the number of trials in the binomial experiment.

Knowing π and the sample size, n, we can calculate μ and σ to locate the center and describe the variability for a particular binomial probability distribution. Thus we can quickly determine those values of x that are probable and those that are improbable.

Example 6.6

Calculate the mean and standard deviation for a binomial probability distribution with $\pi = .5$ and $n = 20$. The probability distribution for the number of successes is shown in Figure 6.3.

Solution Substituting into the formulas, we obtain

$$\mu = n\pi = 20(.5) = 10$$
$$\sigma = \sqrt{n\pi(1 - \pi)} = \sqrt{(20)(.5)(.5)} = \sqrt{5} = 2.24$$

Note that $x = 0$ is more than 4σ away from the mean $\mu = 10$. If we apply the Empirical Rule to this mound-shaped distribution, we see it is highly improbable that in 20 trials we would observe such a small value of x if π really is equal to .5.

FIGURE 6.3 Binomial Probability Distribution for x When
              $n = 20$ and $\pi = .5$

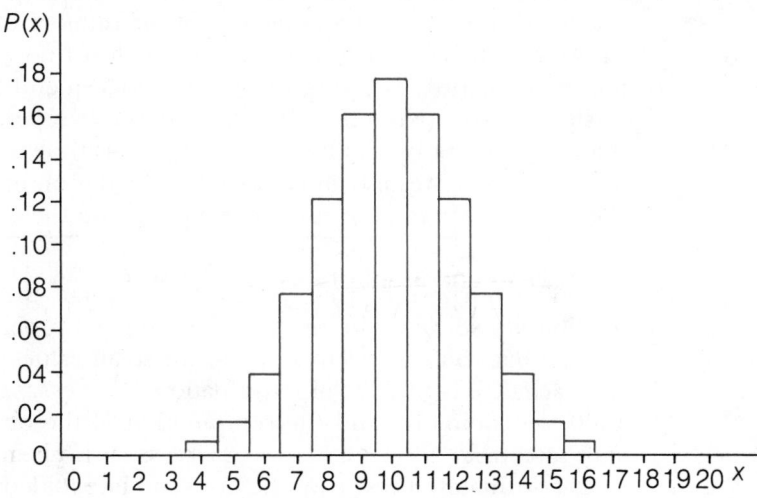

Example 6.7

A poll shows that 516 of 1218 voters favor the reelection of a particular political candidate. Do you think that the candidate will win?

Solution To win the election, the candidate will need at least 50% of the votes. Let us see whether $x = 516$ is too small a value of x to imply a value of π (the proportion of voters favoring the candidate) equal to .5 or larger. If $\pi = .5$,

$$\mu = n\pi = (1218)(.5) = 609$$
$$\sigma = \sqrt{n\pi(1 - \pi)} = \sqrt{(1218)(.5)(.5)}$$
$$= \sqrt{304.5} = 17.45$$

and $3\sigma = 52.35$.

FIGURE 6.4 Location of the Observed Value of x (x = 516)
              Relative to μ

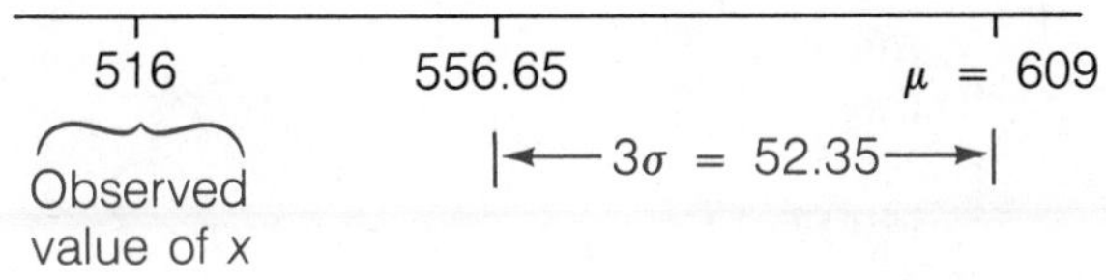

516	556.65	$\mu = 609$

Observed
value of x

$\vert\!\leftarrow\!\!-3\sigma = 52.35\!-\!\!\rightarrow\!\vert$

You can see from Figure 6.4 that $x = 516$ is more than 3σ, or 52.35, away from $\mu = 609$. In fact, if you wish to check, you will see that $x = 516$ is more than 5σ away from $\mu = 609$, the value of μ if π were really equal to 5. Thus it appears that the number of voters in the sample who favor the candidate is much too small if the candidate does, in fact, possess a majority favoring reelection. Consequently, we conclude that he or she will lose. (Note that this conclusion is based on the assumption that the set of voters from which the sample was drawn is the same as the set who will vote. We also must assume that the opinions of the voters will not change between the time of sampling and the date of the election.)

The purpose of this section is to present the binomial probability distribution so that you can see how binomial probabilities are calculated and so that you can calculate them for small values of n, if you so desire. In practice, n is usually large (in national surveys, sample sizes as large as 1500 are common), and the computation of the binomial probabilities is very tedious.

Fortunately, these computations can be avoided. Table 1 of Appendix 3 gives individual binomial probabilities $P(x)$ for various values of n, x, and π. For example, from Table 1 $P(x = 2)$ for $n = 10$ and $\pi = .4$ is found by reading down the x column on the left for $\pi = .4$. The correct probability is .1209. For $\pi > .50$, you must refer to the x column on the right. So for $n = 10$, $\pi = .6$, $P(x = 2)$ we locate the $\pi = .6$ column and read up the x column on the right. The correct value is .0106. For most purposes exact values of the probabilities are unnecessary. We will present a simple procedure in Chapter 7 for obtaining approximate values to the probabilities we need in making inferences. We can also use some very rough procedures for evaluating probabilities by using the mean and standard deviation of the binomial random variable x along with the Empirical Rule.

EXERCISES

6.16 An appliance store has the following probabilities for x = number of major appliances sold on a given day.

x	$P(x)$
0	.100
1	.150
2	.250
3	.140

(continued)

x	$P(x)$
4	.090
5	.080
6	.060
7	.050
8	.040
9	.025
10	.015

(a) Construct a graph of $P(x)$.

(b) Find $P(x \le 2)$.

(c) Find $P(x \ge 7)$.

(d) Find $P(1 \le x \le 5)$.

6.17 The weekly demand for copies of a popular word processing program at a computer store has the probability distribution shown here.

x	$P(x)$
0	.06
1	.14
2	.16
3	.14
4	.12
5	.10
6	.08
7	.07
8	.06
9	.04
10	.03

(a) What is the probability that three or more copies will be demanded in a particular week?

(b) What is the probability that the demand will be for at least two but no more than six?

(c) If the store has eight copies of the program available at the beginning of each week, what is the probability the demand will exceed the supply in a given week?

6.18 Let x be a binomial random variable; compute $P(x)$ for each of the following situations.

(a) $n = 10, \pi = .2, x = 3$

(b) $n = 4, \pi = .4, x = 2$

(c) $n = 16, \pi = .7, x = 12$

6.19 Let x be a binomial random variable with $n = 8$ and $\pi = .4$. Find

    (a) $P(x \leq 4)$ (b) $P(x > 4)$

    (c) $P(x \leq 7)$ (d) $P(x > 6)$

6.20 Examine the accompanying newspaper clipping. Does this sampling appear to satisfy the characteristics of a binomial experiment?

Poll Finds Opposition to Phone Taps

New York—People surveyed in a recent poll indicated they are 81 to 13% against having their phones tapped without a court order.

The people in the survey, by 68 to 27%, were opposed to letting the government use a wiretap on citizens suspected of crimes, except with a court order.

The survey was conducted for 1495 households and also found the following results:

— The people surveyed are 80 to 12% against the use of any kind of electronic spying device without a court order.

— Citizens are 77 to 14% against allowing the government to open their mail without court orders.

— They oppose, by 80 to 12%, letting the telephone company disclose records of long distance phone calls, except by court order.

For each of the questions, a few of those in the survey had no responses.

6.21 A survey is conducted to estimate the percentage of pine trees in a forest that are infected by the pine shoot moth. A grid is placed over a map of the forest, dividing the area into 25-feet-by-25-feet square sections. One hundred of the squares are randomly selected and the number of infected trees is recorded for each square. Is this a binomial experiment?

6.22 A survey was conducted to investigate the attitudes of nurses working in Veterans Administration hospitals. A sample of 1000 nurses was contacted using a mailed questionnaire and the number favoring or opposing a particular issue was recorded. If we confine our attention to the nurses' responses to a single question, would this sampling represent a binomial experiment? As with most mail surveys, some of the nurses will not respond. What effect might nonresponses in the sample have on the estimate of the percentage of all Veterans Administration nurses who favor the particular proposition?

6.23 A random sample of 10 members was obtained to ascertain opinions concerning a new wage package proposal to a local union by union leaders. If we assume that $\pi = .6$ of all the members have disagreements with the wage package, compute the following probabilities.

    (a) All disagree

    (b) Exactly 6 disagree

(c) 6 or more disagree

(d) All agree

(Hint: Use Table 1 of Appendix 3, which gives binomial probabilities for various values of n, π, and x.)

6.24 Refer to Exercise 6.23.

(a) Compute the same probabilities for $\pi = .3$.

(b) Indicate how you would compute $P(x \leq 100)$ for $n = 1000$ and $\pi = .3$.

6.25 An experiment is conducted to test the effect of an anticoagulant drug on rats. A random sample of 4 rats is employed in the experiment. If the drug manufacturer claims that 80% of the rats will be favorably affected by the drug, what is the probability that none of the 4 experimental rats will be favorably affected? One of the 4? One or less?

6.26 A criminologist claims that the probability of "reform" for a first-offender embezzler is .9. Suppose that we define "reform" as meaning the person commits no criminal offenses within a 5-year period. Three paroled embezzlers were randomly selected from the prison records, and their behavioral histories were examined for the 5-year period following prison release. If the criminologist's claim is correct, what is the probability that all three were reformed? At least two?

6.27 **Class Exercise** Consider the following experiment: Toss three coins and observe the number of heads x. Repeat the experiment 100 times and construct a relative frequency table for x. Note that these frequencies give approximations to the exact probabilities that $x = 0, 1, 2,$ and 3. (Note: These probabilities can be shown to be 1/8, 3/8, 3/8, and 1/8, respectively.)

6.28 Refer to Exercise 6.27. Use the formula for the binomial probability distribution to show that $P(0) = 1/8$, $P(2) = 3/8$, and $P(3) = 1/8$.

6.29 Suppose you match coins with another person a total of 1000 times. What is the mean number of matches? The standard deviation? Calculate the interval $(\mu \pm 3\sigma)$. (Hint: The probability of a match in the toss of a single pair of coins is $\pi = .5$.)

6.9 PROBABILITY DISTRIBUTIONS FOR CONTINUOUS RANDOM VARIABLES

You will recall that a continuous random variable is one that can assume values associated with infinitely many points in a line interval. Without elaboration, we state that it is impossible to assign a small amount of probability to each value of x (as was done for a discrete random variable) and retain the property that the probabilities sum to 1. For example, suppose the Social Security Administration is examining the distribution of ages for those receiving disability benefits under social security. Although we usually display ages to the nearest year for adults, we could use decimals to be more specific. We might have ages of 65.3 years, 65.328 years, 65.3286417 years, and so on. Using this degree of accuracy the variable age could be considered continuous; clearly it would be difficult to assign a small amount of probability to each and every age while retaining the property that the sum of the probabilities equals 1.

We overcome this difficulty by reverting to the concept of the relative frequency histogram of Chapter 4, where we talked about the probability of x falling in a given interval. Recall that the relative frequency histogram for a population containing a large number of measurements will almost be a smooth curve because the number of class intervals can be made large and the width of the intervals can be decreased. Thus we envision a smooth curve that provides a model for the population relative frequency distribution generated by repeated observation of a continuous random variable. This will be similar to the curve shown in Figure 6.5.

FIGURE 6.5 Probability Distribution for a Continuous Random Variable

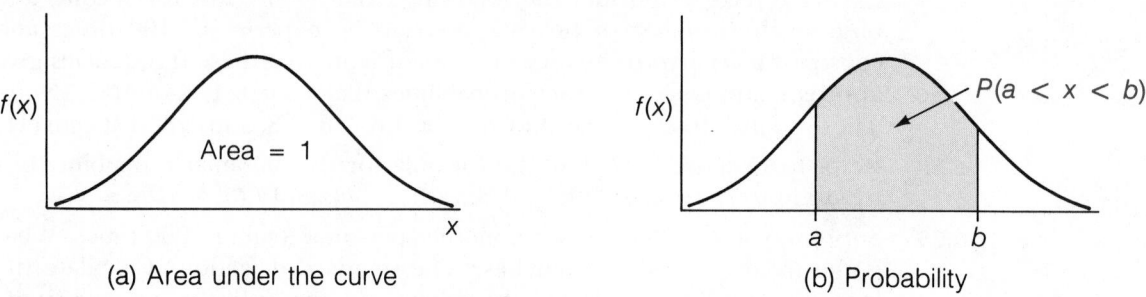

(a) Area under the curve   (b) Probability

Recall that the histogram relative frequencies are proportional to areas over the class intervals and that these areas possess a probabilistic interpretation. That is, if a measurement is randomly selected from the set, the probability that it will fall in an interval is proportional to the histogram area above the interval. Since a population is the whole (100%, or 1), we want the total area under the probability curve to equal 1. If we let the total area under the curve equal 1, then areas over intervals are exactly equal to the corresponding probabilities.

The graph for the probability distribution for a continuous random variable is shown in Figure 6.5. The ordinate (height of the curve) for a given value of x is denoted by the symbol $f(x)$. Many people are tempted to say that $f(x)$, like $P(x)$ for the binomial random variable, designates the probability associated with the continuous random variable x. But, as mentioned before, it is impossible to assign a positive probability to each of the infinitely many possible values of a continuous random variable. Thus all we will say is that $f(x)$ represents the height of the probability distribution for a given value of x.

The probability that a continuous random variable falls in an interval, say between two points a and b, follows directly from the probabilistic interpretation given to the area over an interval for the relative frequency histogram (Section 4.7) and is equal to the area under the curve over the interval a to b, as shown in Figure 6.5. This probability is written $P(a < x < b)$.

There are curves of many shapes that can be used to represent the population relative frequency distribution for measurements associated with a continuous random variable. Fortunately the areas under some of these curves have been tabulated and are ready for use. Thus if we know that student examination scores possess a particular probability distribution, as in Figure 6.6, and if areas under the curve have been tabulated, we could find the probability that a particular student will score more than 80%, by looking up the tabulated area (shaded).

FIGURE 6.6 Hypothetical Probability Distribution for Student Examination Scores

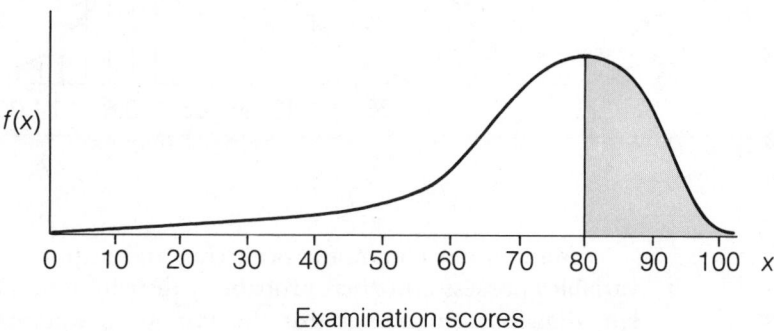

Examination scores

We will find that data collected on many continuous variables in nature possess mound-shaped frequency distributions and that many of these are nearly bell-shaped. A continuous variable (the normal) and its probability distribution (the bell-shaped normal curve) provide a good model for these types of data. The normally distributed variable also plays a very important role in statistical inference. We will study its bell-shaped probability distribution in detail in the next section.

6.10   A USEFUL CONTINUOUS RANDOM VARIABLE: THE NORMAL

normal curve

Many textbooks for introductory statistics note that the distribution of heights for males (or females) can be approximated by a smooth bell-shaped curve that is known as a **normal curve** or a normal distribution. Searching through the literature we found that Newman and White (1951) studied the physical characteristics of army personnel. In Figure 6.7 we present a histogram based on the study displaying the relative frequency distribution for the heights (in inches) of a sample of 24,404 United States Army males at the time of their release from the service. We also superimpose a normal curve over the histogram to show the close approximation obtained from the sample of servicemen.

FIGURE 6.7   Relative Frequency Histogram for the Heights of 24,404 Servicemen, with Normal Curve Superimposed

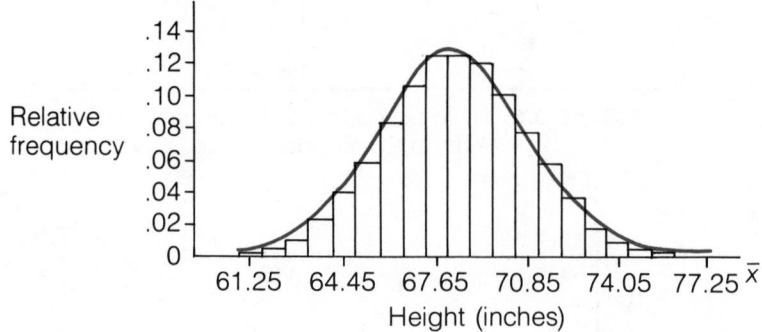

normal distribution

Many other commonly occurring and important continuous random variables possess a normal probability distribution. The **normal,** or Gaussian, **distribution** (named for the famous mathematician Karl Friedrich Gauss, 1777–1855) is a continuous bell-shaped curve as shown in Figure 6.8. The total area under the normal curve is equal to 1 and the probability that a normal random variable x assumes a value in a particular interval, say $a < x < b$, is the area under the curve that lies over the interval (see the shaded area).

There are infinitely many normal curves, one corresponding to each pair of values that you might assign to μ and σ. But all are symmetrical about the mean and are bell shaped, and for all such curves the areas within a specified number of standard deviations of the mean are identical. For example, the area within one standard deviation of the mean will

---

FIGURE 6.8   A Normal Probability Distribution

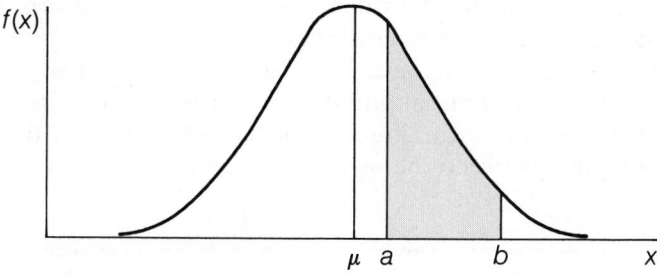

FIGURE 6.9   Characteristics of a Normal Probability
Distribution

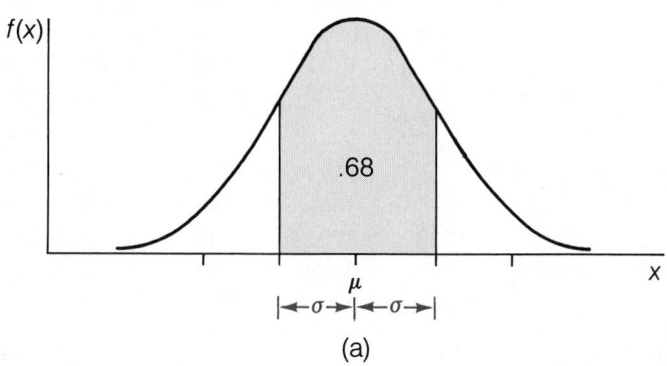

(a)

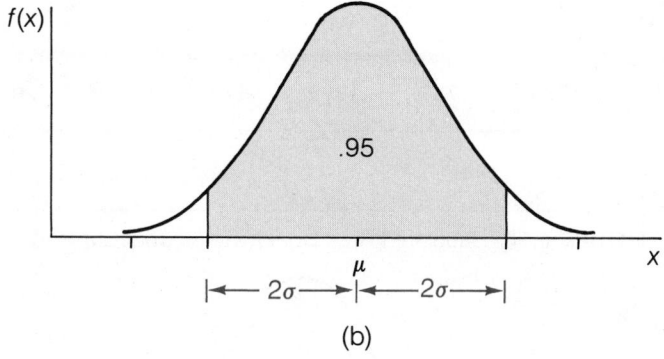

(b)

always (to two decimal places) equal .68. See Figure 6.9(a). Likewise, the area within two standard deviations of the mean will always (to two decimal places) equal .95. See Figure 6.9(b).

The probability that a normal random variable x assumes values in an interval, say $a < x < b$, can be obtained by using a **table of areas** under

table of areas

the normal curve. Table 2 in Appendix 3 gives areas under the normal curve between the mean and a point any number, say z, of standard deviations $(z\sigma)$ to the right of μ. A partial reproduction of this table is shown in Table 6.5. The tabulated area is shown in Figure 6.10.

In the table, areas to the left of the mean need not be tabulated because the normal curve is symmetric about the mean. Thus the area between the mean and a point 2σ to the right of the mean is the same as the area between the mean and a similar point 2σ to the left.

TABLE 6.5   Format of the Table of Normal Curve Areas, Table 2 in Appendix 3

z	.00	.01	.02	.03	.04	.05	.06	.07	.08	.09
0.0	.0000	.0040	.0080	.0120	.0160	.0199	.0239	.0279	.0319	.0359
0.1	.0398	.0438	.0478	.0517	.0557	.0596	.0636	.0675	.0714	.0753
0.2	.0793	.0832	.0871	.0910	.0948	.0987	.1026	.1064	.1103	.1141
0.3	.1179	.1217	.1255	.1293	.1331	.1368	.1406	.1443	.1480	.1517
0.4	.1554	.1591	.1628	.1664	.1700	.1736	.1772	.1808	.1844	.1879
.	.	.	.	.	.	.	.	.	.	.
.	.	.	.	.	.	.	.	.	.	.
1.0	.3413	.3438	.3461	.3485	.3508	.3531	.3554	.3577	.3599	.3621
1.1	.3643	.3665	.3686	.3708	.3729	.3749	.3770	.3790	.3810	.3830
1.2	.3849	.3869	.3888	.3907	.3925	.3944	.3962	.3980	.3997	.4015
.	.	.	.	.	.	.	.	.	.	.
.	.	.	.	.	.	.	.	.	.	.
1.6	.4452	.4463	.4474	.4484	.4495	.4505	.4515	.4525	.4535	.4545
.	.	.	.	.	.	.	.	.	.	.
.	.	.	.	.	.	.	.	.	.	.
2.0	.4772	.4778	.4783	.4788	.4793	.4798	.4803	.4808	.4812	.4817

FIGURE 6.10   Tabulated Area Under the Normal Curve

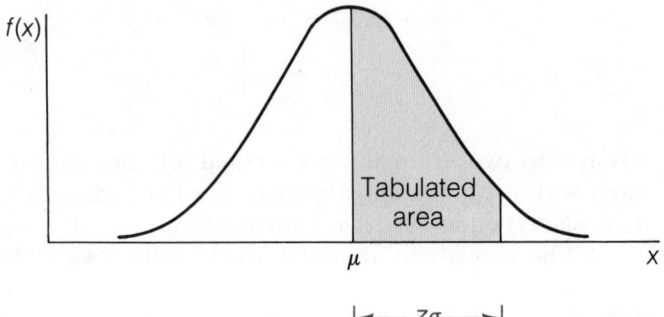

The number z of standard deviations is given to the nearest tenth in the left-hand column of Table 2, Appendix 3. Adjustments to take z to the nearest hundredth are given in the top row of the table. Entries in the table are the areas corresponding to particular values of z. For example, the area between the mean and a point $z = 2$ standard deviations to the right of the mean is shown in the second column of the table opposite $z = 2.0$. This area, shaded in Figure 6.11(a), is .4772. Likewise, the area between the mean and a point two standard deviations to the left of the mean, shown in Figure 6.11(b), is also .4772. Then the area within two standard deviations of the mean is $2(.4772) = .9544$. This explains the origin of the "approximately 95%" in the Empirical Rule.

FIGURE 6.11   Tabulated Area Corresponding to $z = 2$

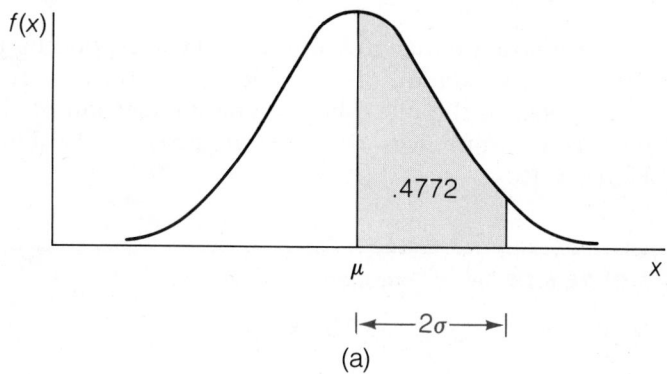

(a)

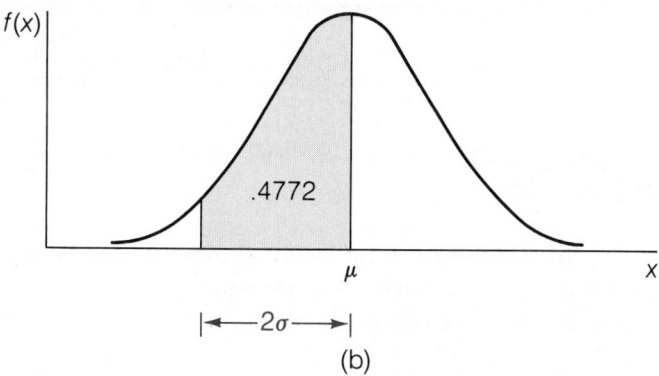

(b)

Similarly, the area between the mean and a point one standard deviation to the right of the mean (that is, $z = 1$) is .3413. The area within one standard deviation of the mean is .6828, or approximately 68%, as stated in the Empirical Rule. This area is shown in Figure 6.12.

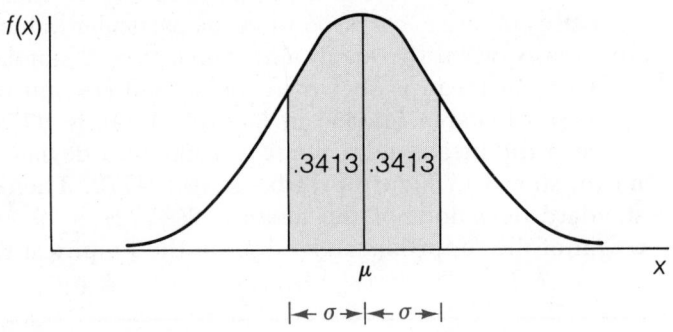

FIGURE 6.12 Area Within One Standard Deviation of the Mean

Suppose we wish to find the area corresponding to $z = 1.64$. Proceed down the left column of the table to the row $z = 1.6$ and across the top of the table to the .04 column. The intersection of the $z = 1.6$ row with the .04 column gives the desired area, .4495. This area is shown in Figure 6.13.

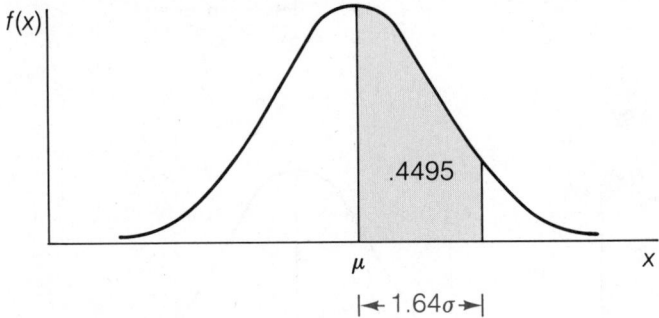

FIGURE 6.13 Area Corresponding to $z = 1.64$

To determine how many standard deviations a measurement x is from the mean μ, first we determine the distance between x and μ. Recall that this distance can be represented as

$$\text{distance} = x - \mu$$

Then the distance between x and μ can be converted into a number of standard deviations by dividing by σ, the standard deviation of x. This

z-score

standardized distance is often called a **z-score**.

$$z = \frac{\text{distance}}{\text{standard deviation}} = \frac{x - \mu}{\sigma}$$

standard normal distribution

standard normal random variable

The probability distribution for z, which has $\mu = 0$ and $\sigma = 1$, is called the **standard normal distribution** (see Figure 6.14), and the variable z is called a **standard normal random variable.** The area under the curve between $z = 0$ and a specified value of z, say z_0, has been tabulated in Table 2 of Appendix 3, and is shown in Figure 6.14. These tabulated areas can be used to find the area under any normal curve if you know the mean μ and the standard deviation σ.

FIGURE 6.14   Standard Normal Distribution

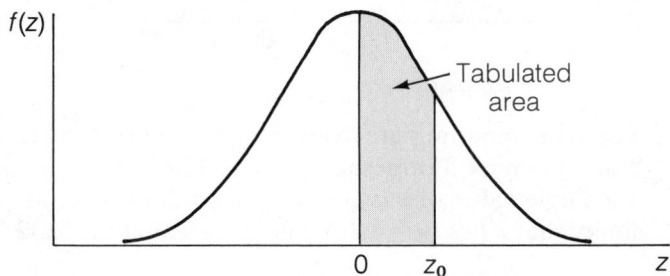

To calculate the area under a normal curve between the mean μ and a specified value x to the right of the mean, first we determine the number z of standard deviations x is from μ, using

$$z = \frac{x - \mu}{\sigma}$$

Then we refer to Table 2 in Appendix 3 and obtain the entry corresponding to the calculated value of z. This entry is the desired area (probability) under the curve between μ and the specified value of x.

We illustrate the use of the table of normal curve areas with a simple example and then proceed to more practical applications.

Example 6.8

Suppose that x is a normally distributed random variable with mean $\mu = 8$ and standard deviation $\sigma = 2$. Find the probability that x is in the interval from 8 to 11. That is, what fraction of the total area under the curve is between 8 and 11 (see the shaded portion of Figure 6.15)?

Solution   To determine the desired area, we compute the number of standard deviations that separate $x = 11$ from the mean $\mu = 8$:

$$z = \frac{x - \mu}{\sigma} = \frac{11 - 8}{2} = 1.5$$

FIGURE 6.15 Area Under the Curve Over the Interval from $\mu = 8$ to $x = 11$

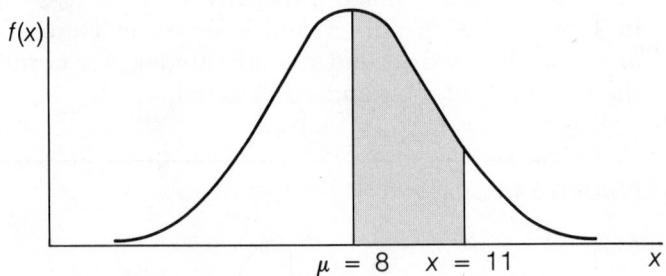

The corresponding area can then be determined from the entry in Table 2 in Appendix 3 opposite $z = 1.5$. The desired area is .4332. Therefore, if a single value of x is selected at random from its population, the probability that x lies between 8 and 11 is equal to .4332.

Example 6.9

The quantitative portion of a nationally administered achievement test is scaled so that the mean score is 500 and the standard deviation is 100.

(a) If we assume the distribution of scores is normal (bell shaped), what percentage of the students throughout the country should score between 500 and 682?

(b) What percentage should score between 340 and 682?

Solution   Consider Figure 6.16.

FIGURE 6.16   Area Between $x = 340$ and $x = 682$

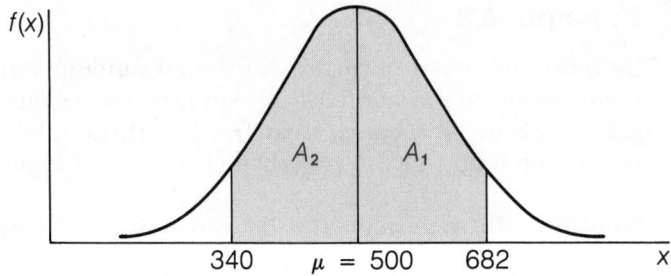

(a) To determine the percentage of the students that should score between 500 and 682, we must compute the area A_1 between $\mu = 500$ and $x = 682$:

$$z = \frac{x - \mu}{\sigma} = \frac{682 - 500}{100} = 1.82$$

The tabulated area for this value of z (Table 2, Appendix 3) is $A_1 = .4656$. Thus we expect 45.56% of the students to score between 500 and 682.

(b) To determine the percentage of students that should score between 340 and 682, note that the area between 340 and 682 is equal to the sum of A_1 and A_2 in Figure 6.16. To find A_2, we compute the number of standard deviations that separate $x = 340$ from $\mu = 500$. Hence

$$z = \frac{x - \mu}{\sigma} = \frac{340 - 500}{100} = -1.6$$

Negative values of z indicate a point to the left of the mean. The appropriate area can be found using Table 2 of Appendix 3 by ignoring the negative sign for z. This area is .4452. Thus we would expect $A_1 + A_2 = .4656 + .4452 = .9108$ or 91.08% of the students to score between 340 and 682 on the examination.

Example 6.10

Records maintained by the budget office of a particular state indicate that the amount of time elapsed between the submission of travel vouchers and the reimbursement of funds has an approximately normal distribution with a mean equal to 45 days and a standard deviation equal to 5 days.

(a) What is the probability that the elapsed time between submission and reimbursement will exceed 58 days for a travel expense report selected at random?

(b) If you had submitted a travel expense report and had still not been reimbursed after 58 days, what would you conclude?

Solution   A sketch of the desired area, A_2, is shown in Figure 6.17.

(a) To determine the area A_2, we must first compute a z score:

$$z = \frac{x - \mu}{\sigma} = \frac{58 - 45}{5} = 2.6$$

Thus the value 58 is 2.6 standard deviations above the mean $\mu = 45$. The area under a normal curve from the mean out to a point 2.6 standard

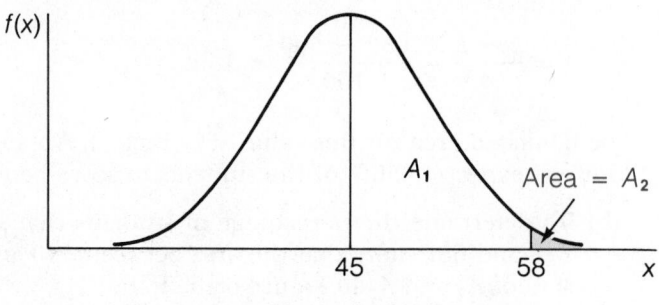

FIGURE 6.17   Probability That the Elapsed Time Will
Exceed 58 Days

deviations above the mean is, from Table 2 of Appendix 3, .4953. This area is indicated by A_1 in Figure 6.17. Since the total area under a normal curve to the right of the mean is .5, we can compute A_2 by subtracting $A_1 = .4953$ from .5. Thus

$$A_2 = .5 - .4953 = .0047$$

(b) Since the probability of having to wait more than 58 days is so small, .0047, we would conclude that something happened to the travel expense report—lost in the mail or misplaced.

In this section we have computed areas under a normal curve. To do this, we converted the distance between a measurement x and the mean μ into a number of standard deviations and then referred to Table 2 in Appendix 3. The resulting areas are equal to the probabilities that measurements will fall in particular intervals.

EXERCISES

6.30   Use Table 2 of Appendix 3 to find the area under the normal curve between these values:

(a) $z = 0$ and $z = 1.3$ (b) $z = 0$ and $z = -1.9$

6.31   Repeat Exercise 6.30 for these values:

(a) $z = 0$ and $z = .7$ (b) $z = 0$ and $z = -1.2$

6.32   Repeat Exercise 6.30 for these values:

(a) $z = 0$ and $z = 1.29$ (b) $z = 0$ and $z = -.77$

6.33 Repeat Exercise 6.30 for these values:

(a) $z = -.21$ and $z = 1.35$ (b) $z = .37$ and $z = 1.20$

6.34 Repeat Exercise 6.30 for these values:

(a) $z = 1.43$ and $z = 2.01$ (b) $z = -1.74$ and $z = -.75$

6.35 Find the probability that z is greater than 1.75.

6.36 Find the probability that z is less than 1.14.

6.37 Find a value for z, say z_0, such that $P(z > z_0) = .5$.

6.38 Find a value for z, say z_0, such that $P(z > z_0) = .025$.

6.39 Find a value for z, say z_0, such that $P(z > z_0) = .0089$.

6.40 Find a value for z, say z_0, such that $P(z > z_0) = .05$.

6.41 Find a value for z, say z_0, such that $P(-z_0 < z < z_0) = .95$.

6.42 Let x be a normal random variable with mean equal to 100 and standard deviation equal to 8. Find these probabilities:

(a) $P(x > 100)$ (b) $P(x > 110)$

(c) $P(x < 115)$ (d) $P(88 < x < 12)$

(e) $P(100 < x < 108)$

6.43 The College Boards, which are administered each year to many thousands of high school students, are scored in such a way as to yield a mean of 500 and a standard deviation of 100. These scores are close to being normally distributed. What percentage of the scores can be expected to be

(a) Greater than 600?   (b) Greater than 700?

(c) Less than 450?   (d) Between 450 and 600?

6.44 Let x be a normal random variable with $\mu = 20$ and $\sigma = 4$. Find the following probabilities.

(a) $Pr(x < 15)$ (b) $P(x > 25)$

(c) $P(x > 30)$ (d) $P(15 < x < 25)$

6.45 Sales figures (on a monthly basis) for a particular food industry tend to be normally distributed with mean of 150($000) and a standard deviation of 35($000). Compute the following probabilities.

(a) $P(x > 200)$ (b) $P(x > 220)$

(c) $P(x < 120)$ (d) $P(100 < x < 200)$

6.46 Refer to Exercise 6.43. An exclusive club wishes to invite those scoring in the top 10% on the College Boards to join.

(a) What score is required to be invited to join the club?

(b) What score separates the top 60% of the population from the bottom 40%? What do we call this value?

6.47 The mean for a normal distribution is 50 and the standard deviation is 10.

(a) What percentile is the value of 38? Choose the appropriate answer:

   88.49   38.49   49.99   0.01   11.51

(b) Which of the following is the z-score corresponding to the 67th percentile?

   1.00   0.95   0.44   2.25   None of these

6.48 The distribution of weights of a large group of high school boys is normal with $\mu = 120$ lb and $\sigma = 10$ lb. Which of the following is true?

(a) About 16% of the boys will be over 130 lb.

(b) Probably fewer than 2.5% of the boys will be below 100 lb.

(c) Half of the boys can be expected to weigh less than 120 lb.

(d) All of the above are true.

SUMMARY

In this chapter we presented some of the basic concepts of probability and the concept of a random variable. The probability distributions for two particularly important random variables, the binomial and the normal, were presented along with examples showing how they can be applied.

Probability, which measures our belief in the occurrence of a particular outcome of an experiment, is viewed conceptually as the relative frequency of occurrence of the outcome (event) when the experiment is repeated over and over again. Probability is important to statistics because the observation of a sample selected from a population is an experiment. Based on the probability of an observed sample, statisticians (and you) make inferences about the population from which the sample was selected. For example, if we theorize that only 5% of all telephone subscribers are delinquent in paying their telephone bills and then sample 10 bills and find that all are overdue, we would reject our theory and infer that the proportion of overdue bills in the population is larger than 5%. We reach this conclusion not because it is impossible to draw 10 unpaid bills in a sample of 10, assuming our 5% theory is correct, but because it is *highly improbable*.

Three probabilistic concepts play important roles in sampling and statistical inferences. These are the additive property of the probabilities for mutually exclusive events, conditional probability, and independent events. As noted in this chapter, most experiments result in numerical events obtained by the observation of random variables. These events, the values that a random variable may assume, are mutually exclusive. Therefore, to find the probability that a random variable when observed will assume one of a set of specific values, we sum the probabilities corresponding to these values.

The concept of independence plays an important role in statistics, especially as it relates to random sampling (to be discussed in Chapter 7), because the manner in which a measurement is selected (sampled) affects the probability of the occurrence of events of interest.

Random variables observed in real-life experiments or surveys can be discrete (usually data representing counts, such as the number of bacteria per cubic centimeter of water) or continuous (such as the length of time required to obtain service at a medical clinic). Two of the most important are the discrete binomial random variable and the continuous normal random variable. In addition to noting the characteristics of these two random variables and the situations to which they might apply, we gave their probability distributions. The probability distribution for the binomial random variable gives the probability associated with each of the values 0, 1, 2, . . . , n that the random variable can assume. To find the probability that a discrete random variable x will assume one of a set of these mutually exclusive numerical events (one event corresponding to each value), we simply sum the probabilities corresponding to the values. Thus, for example,

$$P(3 \leq x \leq 5) = P(3) + P(4) + P(5)$$

The probability distribution for the normal random variable is a smooth bell-shaped curve that is essentially the theoretical relative frequency distribution for a normally distributed population of measurements. For a continuous random variable x, the probability that x will assume a value in the interval $a \leq x \leq b$ is defined to be the area under the probability distribution curve over the interval $a \leq x \leq b$. And, since there is no area lying above a single-point, say $x = a$, it follows that $P(x = a) = 0$. However, in practical problems, we will be interested in probabilities of the form $x \leq a$, $x \geq a$, or $a \leq x \leq b$ rather than in the probability that x assumes some specific value, say $x = a$.

In the following chapters we will employ sample statistics to make inferences about population parameters. The probability distribution of a sample statistic, called its *sampling distribution*, plays a key role in selecting good statistics and in evaluating their reliability. Sampling distributions are the topic of Chapter 7.

KEY TERMS

experiment (*def*)
outcome
event
classical interpretation of probability

relative frequency concept
subjective probability
mutually exclusive events (*def*)
complement (*def*)

conditional probability

unconditional probability

independent events (*def*)

qualitative random variable

quantitative random variable

discrete random variable (*def*)

continuous random variable (*def*)

probability distribution

binomial experiment (*def*)

binomial random variable

normal curve

normal distribution

table of areas

z-score

standard normal distribution

standard normal random variable

KEY FORMULAS

1. For two independent events A and B

$$P(A|B) = P(A) \quad \text{and} \quad P(B|A) = P(B)$$

2. Binomial probability distribution

$$P(x) = \frac{n!}{x!(n-x)!} \pi^x (1 - \pi)^{n-x}$$

3. Mean and standard deviation of the binomial distribution

$$\mu = n\pi, \quad \sigma \sqrt{n\pi(1 - \pi)}$$

4. z-score

$$z = \frac{x - \mu}{\sigma}$$

SUPPLEMENTARY EXERCISES

6.49 **Class Exercise** Each student should perform the following experiment 20 times:

Toss 5 coins and observe the number of heads.

The possible values of x, the number of heads in 5 coin tosses, are 0, 1, 2, 3, 4, and 5. Each student should keep track of the number of times each outcome is observed and combine his or her results with those of the rest of the class to construct the following table.

x	Frequency	Relative frequency (approximate probability)	P(x)
0			.031
1			.156
2			.313
3			.313
4			.156
5			.031

Note that the exact value of the probability, $P(x)$, associated with each value of x is listed in the last column of the table. Your relative frequencies computed from this experiment should be approximately the same as the exact probabilities. Remember that the degree of accuracy increases as the number of repetitions of the experiment increases.

6.50 Refer to Exercise 6.49 and the observed relative frequencies. Approximate the probability of observing x equal to 0 or 1. What is the actual probability of observing 0 or 1 head in a toss of 5 coins?

6.51 Identify the following variables as being either discrete or continuous.

(a) The number of patient arrivals per hour at a medical clinic.

(b) The number of accidents at a given intersection for each year.

(c) The average amount of electricity (measured in kilowatt-hour units) consumed per household per month in New York City.

(d) The number of deaths per year attributed to lung cancer.

(e) The age of freshman United States senators when they take the oath of office.

(f) The gross national product for the United States per year.

6.52 Some parts of California are particularly earthquake-prone. Suppose that in one such area, 40% of all homeowners are insured against earthquake damage. Twelve homeowners are to be selected at random.

(a) What is the probability that exactly 3 of the 12 homeowners will have earthquake insurance? (Use formula for $P(x)$.)

(b) What is the probability that at least 7 of the 12 homeowners will have earthquake insurance? (Use Table 2, Appendix 3.)

(c) Determine the mean and standard deviation of the random variable x, which represents the number among the twelve who have earthquake insurance.

6.53 The mean contribution per person for a college alumni fund drive was $101 with a standard deviation of $32. Assuming that contributions follow a mound-shaped distribution, what percentage of the contributions

(a) were between $37 and $165?

(b) exceeded $165?

(c) were between $37 and $133?

(d) were less than $69?

6.54 The lifetime of a color TV picture tube is normally distributed with a mean of 7.2 years and a standard deviation of 2.3 years.

(a) What is the probability that a randomly chosen picture tube will last more than 10 years?

(b) If the manufacturer guarantees the picture tubes for 3 years, what is the chance that a randomly chosen picture tube will wear out before the guarantee is up?

(c) If the manufacturer is willing to replace only 2% of all picture tubes due to early failure, how should it change the guarantee?

6.55 A brand of water-softener salt comes in packages marked "net weight 40 lb." The company claims the bags contain an average of 40 lb of salt with standard deviation in weights of 1.5 lb. Furthermore it is known that the weights are normally distributed. What is the probability that the weight of a randomly selected bag will be 39 lb or less, if the company's claim is true?

6.56 A data processing company requires applicants for computer programming positions to take a test. The company policy is to reject all applicants whose score is below 50. The mean score for applicants has been 60 with a standard deviation of 10. The distribution of the applicants' scores is mound shaped. Approximately what percentage of the applicants have been rejected by this test?

6.57 The following table concerns the members of the local chapter of the Building Trades Union. Use the data to answer the following questions. (As usual, carry out all your calculations and your answers to at least four places.)

	Carpenter	Plumber	Bricklayer	Total
Employed	81	48	34	163
Unemployed	56	42	49	147
Total	137	90	83	310

(a) Find each of these probabilities.

1. P(Employed or Bricklayer).

2. P(Carpenter : Employed).

(b) Are the two events Unemployed and Plumber independent? Carefully explain your answer.

6.58 An accounting office has six incoming telephone lines. The probability distribution of busy lines, x, is

x	$P(x)$
0	0.052
1	0.154
2	0.232
3	0.240
4	0.174
5	0.105
6	0.043

(a) Is x a discrete or continuous random variable?

(b) What is the probability of at least 4 lines being busy?

(c) What is the probability of between 2 and 4, inclusive, being busy?

(d) At least one?

6.59 The table shown below gives the votes cast in a recent election for an increase in the tax levy for the school district comprising the towns of Bazetta, Delightful, and Shihola.

	Bazetta	**Delightful**	**Shihola**	**Total**
For the levy	192	84	381	657
Against the levy	448	940	145	1533
Total	640	1024	526	2190

Consider the following events.

B: The voter lives in Bazetta.

D: The voter lives in Delightful.

S: The voter lives in Shihola.

F: The voter voted for the levy.

A: The voter voted against the levy.

(a) Find each of the following probabilities:
$P(S)$
$P(D|A)$
$P(D \text{ or } A)$

(b) Are the events F and B *independent*? Carefully explain your answer.

6.60 Define each of the underlined terms.
The events A and B are <u>independent</u>.
The events A and B are <u>mutually exclusive</u>.

6.61 In the mining town of Wounded Brook, Colorado, many of the mines have been closed because of new environmental regulations. Not surprisingly, 80% of the people in the town are against the new regulations. What is the probability of randomly selecting five people from Wounded Brook and discovering that three of the five are *in favor* of the regulations? Show your formulas and computations!

6.62   If a person responds to a set of five true-false test questions on a random basis,

(a) What is the probability of getting exactly two questions correct? (Choose the correct answer.)

.0312   .3125   .0500   .625

(b) What is the probability of getting three or more questions correct? (Choose the correct answer.)

.0312   .1875   .5000   .6875

6.63   Suppose a telephone survey is to be done in the Atlanta area. Previous telephone surveys indicate that 60% of the time the phone will be answered by a woman. Suppose that we make three calls initially. What is the probability that:

(a) A woman will answer for each call?

(b) A woman will answer exactly one call?

(c) A woman will not answer any of the three calls?

6.64   A man's blood test shows a cholesterol count of 310 milligrams per deciliter. If the doctor says that the average cholesterol count for men of a similar age is 200 mg/dl and that only about 2½% of the male population will have a value higher than 310,

(a) Determine the standard deviation assuming the counts are normally distributed.

(b) What chance is there of finding a male with a cholesterol count less than 100?

6.65   Pay rates for hourly employees in a particular industry are assumed to be normally distributed with a mean of $7.00 and a standard deviation of $1.50.

(a) What is the percentage of employees with hourly pay rates between $7 and $10?

(b) What is the probability of finding an employee with a rate greater than $8.50?

6.66   The average weight of newborn, term babies at a local hospital is 7.5 lb with a standard deviation of 1.6 lb. Assuming that the distribution of weights is approximately normal, find the probability that a newborn weighs

(a) Less than 6 lb.

(b) More than 8 lb.

(c) Between 6 and 8 lb.

6.67   The weather forecaster has been the object of many jokes due to errors in forecasting. Of course, these errors are not completely avoidable. After obtaining information (measurements) on many different variables such as wind direction, wind velocity, and barometric pressure from local sources and satellite communications, the forecaster must interpret these data and supply an inference (weather forecast). How effectively weather forecasters have employed statistics to prepare their forecasts is open to question, but it is clear that probability has become an integral part of a weather forecast. We've all heard weather forecasts that state, "There is a 50 percent chance of rain this morning, decreasing to a 30 percent chance this afternoon and evening." Give your interpretation of this statement.

Area Forecast

Partly cloudy to occasionally cloudy through tomorrow with a chance of thundershowers mainly during the afternoons and evenings. Low today and tomorrow will be near 80. High will be near 90. Winds will be southwest to west 5 to 15 mph, stronger and gusty near showers. Rain probability is 50 percent today and 30 percent tonight.

6.68   Read the accompanying news clipping. Explain why this might or might not be a binomial experiment. What information, missing in the article, is needed to conclude firmly that the survey is a binomial experiment?

Study of Divided Families Shows Positive Attitudes

Chicago — A study of divorced mothers and their children has revealed some positive attitudes among members of divided families. Perhaps a broken home is not the psychological disaster for family members that society has suspected.

The study, involving 20 mothers with one or more children between the ages of 6 and 18, was conducted to determine the basic concerns of divorced mothers and their children. There were 20 mothers and 35 children involved in the study.

All the women were working full time. Most of them had made plans toward bettering their earning power. The women had been divorced from 3 months to 15 years. The educational level of the women in the study was high, compared to the national average: 12 years to 18 years of education.

A key aim of the study was to determine the feelings of the women and their children about their acceptance in society.

Eight-six percent of the children felt that at school they were treated the same as children whose parents were married. Children aged 10 through 12 especially preferred that teachers and friends be told about the home situation. They wanted news of the divorce not to come as a surprise to others or to be a source of embarrassment for them.

In general, the children were doing well in school and even excelled in some areas.

Although the trend among most of the women was to socialize mainly with single persons, 80% of them felt accepted in their neighbor-hoods. Half of them said they felt accepted at church.

Among the children, 91% indicated they were treated no differently at Sunday school. Ninety percent of the sample were active church members.

Most of the women, 85%, said that after their divorces their attitudes toward divorce had shifted from negative to positive. The same proportion saw advantages for their children, in terms of understanding life and people, as a result of the divorce.

6.69 Suppose you are the personnel manager for a manufacturing concern and are responsible for safety procedures in your plant. Records are maintained on the number of accidents on a daily basis and these are totaled by the month. Explain why these data are or are not measurements on a binomial random variable.

6.79 A recent survey suggests that Americans anticipate a reduction in living standards, and that a steadily increasing consumption no longer may be as important as it was in the past. Suppose that a poll of 2000 people indicated 1373 in favor of forcing a reduction in the size of American automobiles by legislative means. Would you expect to observe as many as 1373 in favor of this proposition if, in fact, the general public was split 50–50 on the issue? Why?

6.71 An experiment was conducted to test for the presence or absence of fungus on tobacco plants. Four hundred plants were observed to have been infected by the fungus.

(a) Does this appear to be a binomial experiment? Explain why it might or might not satisfy the characteristics of a binomial experiment.

(b) Suppose the characteristics of a binomial experiment are satisfied. What interpretation can you give to π?

(c) Previous experience suggests that the fungus affects 50% of a planting of tobacco seedlings. What is the mean value of x, the number of plants infected by the fungus? The standard deviation of x? If π really equals .5, is it probable that the observed number of infected plants could be as large as (or larger than) $x = 242$? Explain.

6.72 Answer the following questions for the survey discussed in the accompanying news article.

(a) Does this appear to be a binomial experiment?

(b) Explain why it might or might not satisfy the five characteristics of a binomial experiment.

Alcoholism Reported Up in Army

Large numbers of young American soldiers are becoming alcoholics. This parallels the increase of alcohol use among young civilians, researchers report.

In a study of 1873 Army men randomly selected from bases of the United States, nearly 2 out of every

5 soldiers were found to be either actual alcoholics, borderline alcoholics, or potential alcoholics.

The study showed that the largest percentage of problem drinkers were under age 20 and had ranks below sergeant.

6.73 Experience has shown that a lie detector will show a positive reading (indicate a lie) 10% of the time when a person is telling the truth and 95% of the time when a person is lying. Suppose that a sample of five suspects is subjected to a lie detector test regarding a recent one-person crime. What is the probability of observing no positive reading if all suspects plead innocent and are telling the truth?

6.74 A large stadium utilizes floodlights to illuminate the field. If the supplier of these lights claims that the time to failure is approximately normal with mean 40 hours and standard deviation 4 hours, answer the following questions:

(a) What is the probability that a randomly selected floodlight will burn for at least 30 hours?

(b) If the stadium buys 1500 floodlights, how many would you expect to last at least 30 hours?

(c) What might you conclude if only 1400 lights lasted at least 30 hours?

6.75 Define a z score in terms of x, a normally distributed random variable with mean μ and standard deviation σ.

6.76 Using Table 2 in Appendix 3, calculate the area under the normal curve between the following:

(a) $z = 0$ and $z = 1.5$ (b) $z = 0$ and $z = 1.8$

6.77 Repeat Exercise 6.76 for the following:

(a) $z = 0$ and $z = 2.5$ (b) $z = -1.5$ and $z = 0$.

6.78 Repeat Exercise 6.76 for the following:

(a) $z = -.08$ and $z = 0$ (b) $z = -0.8$ and $z = 0.8$

6.79 Repeat Exercise 6.76 for the following:

(a) $z = -1.96$ and $z = 1.96$ (b) $z = -2.58$ and $z = 2.58$

6.80 Repeat Exercise 6.76 for the following:

(a) $z = -0.12$ and $z = 1.8$ (b) $z = 1.65$ and $z = 2.0$

6.81 Find the value of z such that 30% of the area lies to its right. (Note: This is the 70th percentile of the standard normal distribution.)

6.82   Find the value of z such that 5% of the area lies to its right.

6.83   Find the value of z such that 2.5% of the area lies to its right.

6.84   A normally distributed variable x possesses a mean and standard deviation equal to 7 and 2, respectively. Find the z value corresponding to $x = 6$.

6.85   Refer to Exercise 6.84. Find the value of z corresponding to $x = 8.5$.

6.86   Refer to Exercise 6.84. Find the probability that x lies in the interval 6 to 8.5.

$ 6.87   Over a long period of time in a large multinational corporation, 10% of all sales trainees were rated as outstanding, 75% as excellent/good, 10% as satisfactory, and 5% as unsatisfactory. Find the following probabilities for a sample of ten trainees selected at random.

   (a) The probability that two are rated as outstanding.

   (b) The probability that two or more are rated as outstanding.

   (c) The probability that eight of the ten are rated either outstanding or excellent/good.

   (d) The probability that none of the trainees is rated as unsatisfactory.

6.88   A new technique, balloon angioplasty, is being widely used to open clogged heart valves and vessels. The balloon is inserted via a catheter and is inflated, opening the vessel; thus, no surgery is required. Left untreated, 50% of the people with heart valve disease die within about two years. If experience with this technique suggests that approximately 70% live for more than two years, would the next five patients of the patients treated with balloon angioplasty at a hospital constitute a binomial experiment with $n = 5$, $\pi = .70$? Why or why not?

6.89   A prescription drug firm claims that only 12% of all new drugs shown to be effective in animal tests ever make it through a clinical testing program and onto the market. If a firm has fifteen new compounds that have shown effectiveness in animal tests, find the following probabilities:

   (a) None reach the market.

   (b) One or more reach the market.

   (c) Two or more reach the market.

6.90   Does Exercise 6.54 satisfy the properties of a binomial experiment? Why or why not?

6.91   A survey was conducted to investigate the attitudes of nurses working in Veterans Administration hospitals. A random sample of 1000 nurses was contacted using a mailed questionnaire, and the number favoring or opposing a particular issue was recorded. If we confine our attention to the nurses' responses to a single question, would this sampling represent a binomial experiment? As with most mail surveys, some of the nurses will not respond. What effect might nonresponders in the sample have on the estimate of the percentage of all Veterans Administration nurses who favor the particular proposition?

6.92 The dollar sales per salesperson for a large company averaged $60,000 per year, with a standard deviation of $7,000. What fraction of the salespeople might be expected to sell less than $50,000 per year?

6.93 The length of time to complete a standard achievement test possesses a mean of 58 minutes, with a standard deviation of 9.5 minutes. If the professor wants to time the exam so that it will be completed by 90% of the students, how long an examination period must be scheduled? (Hint: Using Table 2 of Appendix 3, we can find the z-value corresponding to an area of .4. Then from the formula

$$z = \frac{x - \mu}{\sigma}$$

we can solve for the required value of x. See Figure 6.18.

FIGURE 6.18 Probability Distribution of the Length of Time to Complete Achievement Test (Exercise 6.93.)

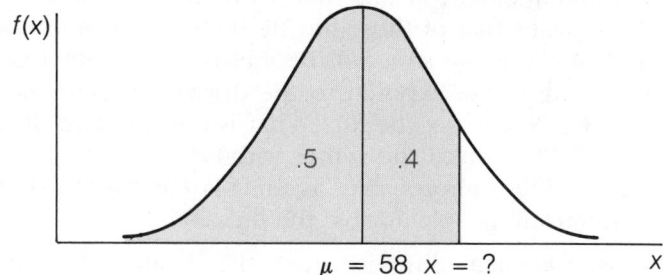

6.94 Refer to Exercise 6.93. Find the 80th percentile for the length of time to complete the test. (Note: The 80th percentile here means that 80% of the students would have completion times less than or equal to this value.) Find the lower quartile.

6.95 Give the mean and standard deviation of a binomial random variable with $n = 30$ and $\pi = .2$.

6.96 There is one disadvantage of z scores: they are difficult to explain to one not well versed in statistics. A college professor, in a stroke of genius that turned out to be highly misguided, once decided to report the results of an exam as z-scores. He was quickly besieged by anxious students who did not understand what a z-score of 0 meant. The professor knows you are taking a course in statistics so he asks you to explain what a z-score of 0 actually represents in terms of this exam. Give a *clear* but *short* explanation.

6.97 Weights of college students on a large campus are assumed to be normally distributed with $\mu = 150$. In this experiment 97.5% of the students weighed less than 210 pounds.

(a) What is the standard deviation of the normal distribution?

(b) If four students are randomly selected, what is the probability that all students will weigh less than 180 pounds?

(c) What percentage of the students weigh less than 105 pounds?

6.98 The probability of surviving a rare disease of the nervous system is one in twenty. A new drug has been developed to combat the disease and is tried on three patients. One out of the three survives. To some experimenters, this survival rate (33%) may appear to be a tremendous improvement over the old established rate of 5%. But remember, the 33% survival rate occurred only in a sample of three, not in a large number of cases.

If the survival rate, using the new drug, is still only 5% (i.e., the drug is worthless), what is the probability of observing one or more survivals in a sample of three? Examining this probability, what would you be inclined to say (as a matter of intuition) about the effectiveness of the new drug? Explain your reasoning.

6.99 An immunologist claims that a flu shot is 80% effective against the flu. This means that of those having the shot, 80% will be immune to the flu and for 20% the shot will be ineffective. If 8000 people receive the flu shot and all are exposed to the disease, what is the expected number x who would not get the flu? What is the standard deviation of x? Suppose that 6200 survived the winter without contracting the flu. Is this value of x ($x = 6200$) improbable, assuming that the flu shot is 80% effective in immunizing people against the flu?

6.100 Class Exercise Suppose that 50% of all apartment dwellings in a city exhibit one or more violations of the fire code. If 10 apartments are selected at random, what is the probability that 7 or more will be found to be in violation of the fire code?

Although we can solve this problem using the formula of Section 6.8, we can easily obtain an approximate answer (and thereby verify our formula) by a coin-tossing experiment. Tossing a single coin once is analogous to selecting a single apartment in the population described above. A head could correspond to an apartment in violation of the fire code, a tail to one that is in conformance with the code. Flipping the coin 10 times (or tossing 10 coins once) would be equivalent to sampling 10 apartments from the population.

Select 10 coins, mix thoroughly, toss, and record the number x of heads (apartments in violation of the code) in the sample. Record whether x is 7 or larger. Repeat this process $n = 100$ times and count the number of times x is 7 or larger. This will give you the number of times n_A your event of interest was observed. Then calculate an approximate value for the probability of observing 7 or more apartments in violation of the code in a sample of 10, using

$$P(A) = P(7 \text{ or more}) \approx \frac{n_A}{n} = \frac{n_A}{100}$$

This approximation may be poor because $n = 100$ repetitions of the experiment is not large enough to acquire an accurate approximation to $P(A)$.

To obtain a more accurate approximation to $P(A)$, combine the data collected by all the members of your class and use these data to approximate $P(A)$. This value x should be close to the exact value of $P(A)$, which is .172. (You can compute the exact value by using the binomial formula $P(x)$ for $x = 7, 8, 9,$ and 10.)

To see how the experimentally obtained approximations for $P(A)$ vary, collect the approximations from each member of your class and describe them by using a relative frequency histogram. Notice how they cluster about the exact value $P(A) = .172$. Although we only have one approximation based on the larger value of n for the combined data, you can see that this value will tend to fall closer to .172 than the approximations based on $n = 100$ repetitions of the experiment.

6.101 The following relative frequencies were generated from a survey of the employees of a large insurance company in the midwest.

Table of Hair Color by Eye Color of Employees

Hair color	Eye color					Total
	Black	**Blue**	**Brown**	**Green**	**Hazel**	
Black	.0106	.0053	.0582	.0053	.0053	.0847
Blonde	.0000	.1429	.0053	.0476	.0265	.2223
Brown	.0000	.1746	.2751	.1005	.1164	.6666
Red	.0000	.0052	.0159	.0053	.0000	.0264
Total	.0106	.3280	.3545	.1587	.1482	1.0000

Assuming that the relative frequencies in the above table are population probabilities, determine the following.

(a) Determine the probability that an employee is either green-eyed or hazel-eyed.

(b) Determine the probability that an employee is either blue-eyed or black-haired or both.

(c) Given that an employee is brown-haired, what is the probability that the employee has brown eyes?

(d) Illustrate whether the events "green-eyed" and "brown-haired" are independent or not independent?

6.102 A survey has shown that .30 of the customers of a Rockford, Illinois bank consider accuracy of the bank to be "good." If a sample of ten customers of the bank is randomly selected, what is the probability that two or fewer will consider the accuracy of the bank to be "good?"

6.103 If the mean and standard deviation of the distribution of prices for new textbooks in a student bookstore are $26.80 and $4.14, respectively, and the mean and standard deviation for used textbooks are $17.80 and $2.80, respectively, which would be the more unusual price, $16.00 for a new book or $12.50 for a used book? Assume that the bookstore prices are approximately normally distributed. Determine what percentage of the prices of new texts fall in the range from $20.00 to $30.00.

6.104 Getting from here to there could be a problem for some University of Kentucky students, according to a news release. Out of 2735 students entering the introductory level geography courses, 793 could not locate Lexington, Kentucky, the home of the University of Kentucky, on a map of Kentucky. What is the probability that a random sample of 5 students entering introductory geography would show 0 who couldn't locate Lexington on a map of Kentucky?

EXERCISES FROM THE DATABASE

6.105 Refer to the HAM-D total scores for the four treatment groups in the clinical trial data base in Appendix 1. If we assume the mean for the combined scores is 14.2 and the standard deviation is 6.3, determine z-scores for total scores of 5, 16, and 24.

6.106 Refer to Exercises 6.105. What total scores correspond to z-scores of -1, 0, and 2, respectively? (Hint: Use the formula for z and solve for x.)

6.107 Refer to the clinical trial data base and Exercise 6.105. If the combined HAM-D total scores are assumed to be normally distributed, determine the probability of having a total score higher than 24. Less than 10.

CHAPTER 7

SAMPLING DISTRIBUTIONS

7.1 Introduction ▪ **7.2** Random sampling ▪ **7.3** The sampling distribution for $\bar{x}$ and the central limit theorem ▪ **7.4** Normal approximation to the binomial ▪ **7.5** Sampling distribution for $\hat{\pi}$ ▪ **7.6** Using computers ▪ Summary ▪ Key terms ▪ Key formulas ▪ Supplementary exercises

---

7.1 INTRODUCTION

As noted in earlier chapters, a statistic is a numerical descriptive measure of the sample, whereas a parameter is a numerical descriptive measure of the population. We will use statistics to make inferences about population parameters—particularly to *estimate* the value of a population parameter or to make a *decision* about its value. As you will subsequently learn, we

sampling distribution

will need to know the probability distribution, called a **sampling distribution,** for the statistic in order to evaluate the reliability of inferences. For example, we will want to know the probability that a statistic will give an estimate that will fall close to (say within some specified distance of) the actual value of the population parameter. Or, if we base a decision about a population parameter on the observed value of a sample statistic, we will want to know the probability that the sample statistic has led us to an incorrect decision.

This chapter is about sampling and sampling distributions. We will define the simplest and most common method of sampling and will give the sampling distributions for some important statistics computed from this type of sample. As we proceed through this chapter, you will see how probability and probability distributions (Chapter 6) play key roles in statistical inference.

7.2 RANDOM SAMPLING

The Environmental Protection Agency (EPA) announced that nine areas have been tentatively designated as Air Quality Maintenance Areas. This designation means that local officials in the affected areas must develop a ten-year master plan to improve the air quality.

Upon receiving the EPA announcement, an engineer for one of the areas stated that he planned to fight the EPA designation because the sampling stations used were ill placed and did not provide accurate readings of the air quality. Further investigation showed that at least one of the stations was located near a dusty clay road. This alone could have accounted for the poor air quality of samples at that location. After acquiring additional information to substantiate the fact that the sampling was biased and tended to show a much higher rate of pollution than actually existed, the engineer sent an appeal through proper channels, and the air quality maintenance designation for the area was lifted by the EPA.

This example illustrates the important role that sampling plays in statistical inferences. Particularly, it shows that statistical inferences depend on how the sample is selected.

Random sampling is the most common type of sampling. It is defined as follows:

Definition 7.1

random sample of one measurement

A **random sample of one measurement** from a population of N measurements is one in which each of the N measurements has an equal probability of being selected.

The notion of random sampling can be extended to the selection of more than one measurement, say n measurements. For purposes of illustration we will take a population that contains a very small number of measurements, say $N = 4$, the 4 measurements being 1, 4, 3, and 2. Further let's assume that we wish to select a random sample of $n = 2$ from the 4. How many different and distinct samples could we select? The six possible samples are listed in Table 7.1. A random sample of $n = 2$ measurements taken from the population of $N = 4$ measurements is one in

TABLE 7.1   Six Possible Samples from a Population of Four Measurements

Possible samples	Measurements in sample
1	1, 2
2	1, 3
3	1, 4
4	2, 3
5	2, 4
6	3, 4

which each of the six different samples listed in Table 7.1 has an equal chance of being selected.

Definition 7.2

random sample of *n* measurements

A **random sample of *n* measurements** from a population is one in which every different subset of size *n* from the population has an equal probability of being selected.

It is unlikely that we would draw a sample that satisfies exactly our definition of a random sample, but we can achieve a very close approximation to random sampling in many situations. When a population is finite, we can use a **table of random numbers,** such as Table 11 in Appendix 3, to select (approximately) a random sample. A reproduction of Table 11 is shown in Table 7.2.

table of random numbers

TABLE 7.2 *Format of the Random Number Table (Table 11 in Appendix 3)*

Line	Column									
	1	*2*	*3*	*4*	*5*	*6*	*7*	*8*	*9*	*10*
1	75029	50152	25648	02523	84300	83093	39852	91276	88988	12439
2	73741	30492	19280	41255	74008	72750	70420	67769	72837	27098
3	07049	98408	27011	76385	15212	03806	85928	81312	14514	55277
4	01033	08705	42934	79257	89138	21506	26797	67223	62165	67981
5	48399	78564	35787	07647	23794	73938	29477	11420	03228	16586
6	70459	73480	06740	79124	14078	72352	07410	93292	93057	18715
7	74770	80185	08181	27417	90866	98444	72870	51219	51481	47916
8	24167	13753	65011	66288	12633	79199	61497	56186	83643	96184
9	24316	80240	62592	53393	57028	61626	56508	84407	97873	27571
10	84565	59254	94435	33322	50014	00180	50954	04099	66005	59141
11	60794	32497	47830	94509	36576	68874	84062	84503	50454	42199
12	99104	14833	97062	48867	19645	78069	91602	46991	57523	22219
13	15604	93654	21487	86036	22827	62637	70378	58539	17827	80108
14	20204	00253	19678	15789	17628	63667	23348	67083	92361	50413
15	71233	73676	00958	42662	47344	00104	74530	46238	06655	23791
16	82846	82954	52107	66054	27358	69664	71760	03577	75622	21536
17	48613	97858	49627	17036	55574	80116	80533	62146	48083	29177
18	42313	91287	66900	79817	76803	42462	63542	99089	22655	44130
19	60879	68102	60700	51281	61386	06782	88214	68246	15552	79093
20	34593	95713	62942	16236	30933	39470	58423	95304	46017	18364
21	96033	10917	01205	08978	43021	77321	76736	64527	96534	98457
22	21932	45476	75464	43497	81807	99369	59945	65349	52588	27386
23	91019	99635	78638	75114	42943	81629	03283	85036	80666	18675
24	86053	48238	14952	55565	98821	92843	67663	70387	13356	46650
25	59700	38346	92770	11506	34101	01051	99390	86884	26788	78768

When you look at this table, concentrate on the digits 0, 1, 2, 3, . . . , 9 rather than on the groups of digits that form larger numbers. Proceed across a row of the table, down a column, or in any other path that you desire, and accumulate a large number of digits. If you were to calculate the proportions of 0s, 1s, 2s, . . . , 9s in this set, you would find that the proportions are near .1. In other words, if you let your finger fall on any digit in the table, the probability that it will hit a particular digit, say a 7, is (with a high degree of accuracy) near .1. For any *pair* of digits selected from the table—the numbers 00, 01, 02, . . . , 99—the probability of each is approximately equal to .01. Similarly for any group of three digits—the numbers 000, 001, 002, . . . , 999—the probability of each is approximately equal to .001.

We now illustrate how to use Table 7.2 to select a random sample of n from a population containing N measurements.

Example 7.1

Select a random sample of $n = 5$ measurements from a population that contains $N = 2000$ measurements.

Solution   Number the 2000 population measurements in sequence, starting with 0000 and ending with 1999. To identify which $n = 5$ of these $N = 2000$ measurements should appear in the random sample, we randomly select a starting point in Table 7.2 and select $n = 5$ four-digit numbers (we only need four-digit numbers because the largest number used to identify the population measurements is 1999). For example, suppose that we decide to start with the first four digits of the random number appearing in the first row and third column of Table 7.2 and to proceed down the column, *discarding numbers larger than 1999*. We obtain 1928, 0674, 0818, 1967, and 0095. This tells us that the five measurements that are numbered 1928, 674, 818, 1967, and 95 should appear in our random sample.

It is relatively easy to select a random sample from a population when all the population measurements are available and can be numbered for purposes of identification. But when the sample measurements are generated by experimentation, such as measuring the acidity (pH) of a solution in a chemistry laboratory, the population (pH readings for the solution) is conceptual. That is, for this situation the population consists of the very large number of pH measurements that we could make on the solution if we were able to repeat the process *ad infinitum*. A sample of $n = 5$ pH readings would be obtained by thoroughly mixing the solution, drawing five specimens from the container and measuring the pH of each. But considering the infinitely large number of different solution specimens that could be selected from the container (with one pH reading corresponding to each specimen), there is no guarantee that our method of drawing the specimens will give every different sample of five pH readings an equal probability of selection. But with thorough mixing of the solution, we would expect the sampling procedure to very closely approximate random sampling.

conceptual population

Other types of laboratory measurements, selected from **conceptual populations**—populations that we can conceptualize but are not within our grasp—yield samples that are not so easily visualized as random samples. For example, if we measure the blood pressure of five experimental rats treated with a new drug, we obtain five measurements from the conceptual population of measurements for all rats that might be treated with the drug. Whether or not the sample of $n = 5$ measurements is a random sample from the conceptual population is a matter for conjecture.

So what do we do in practice when we wish to draw a random sample? We do the best that we can. If the population is finite and is available, we use a table of random numbers to select the sample. But, even with the use of a table of random numbers, careful planning and a certain amount of ingenuity are required to have even a decent chance to approximate random sampling. This is especially true when the population from which the sample is to be drawn involves people. People can be difficult to work with; they have a tendency to discard mail questionnaires and refuse to participate in personal interviews. So unless we're careful, the data we obtain may be full of biases having unknown effects on the inferences we're trying to make.

When the data arise as the result of experimentation, we take every possible precaution to avoid biasing the sample data. Particularly, we try to select the data so that the observation of any one measurement will not influence the observation of any others. We hope that independent selection of the sample measurements will give every subset of n measurements an equal probability of selection. Most important, if we have doubts that the sampling procedure satisfies the requirements of random sampling, we make note of this fact when we report particular inferences derived from the sample. Then persons reading the conclusions of a statistical analysis of experimental data or a sample survey will be aware of the conditional nature of the conclusions.

EXERCISES

7.1 Define what is meant by a random sample. Is it possible to draw a truly random sample? Why?

7.2 Suppose that you wish to randomly sample the opinions of $n = 10$ persons from a population of 800. Use Table 11 of Appendix 3 to identify the persons to appear in your sample.

7.3 Suppose that you wish to sample the opinions of the homeowners in a community regarding the desirability of increasing local expenditures to improve the quality of the public schools. You randomly select the households by using a table of random numbers, and you discard any households in your sample for which the homeowner is not at home when visited by the interviewer. Do you think this process is likely to approximate random sampling? Explain.

7.4 Use Table 11 of Appendix 3 to identify the measurements to be included in a random sample of $n = 10$ from a population containing $N = 1000$ measurements.

7.5 A psychologist is interested in studying women who are in the process of obtaining a divorce to determine whether there are significant attitudinal changes after the divorce has been finalized. Existing records show that there are 798 couples who are in the process of filing for a divorce in a certain county. If a sample of 25 women is desired, use Table 11 of Appendix 3 to determine which women should be interviewed. (Begin in column 2, row 1 and proceed down.)

7.6 Refer to Exercise 7.5. As is the case in most surveys, not all persons chosen for a study will agree to participate. Suppose that 5 of the 25 women selected refuse to be interviewed. Determine 5 more women to be included in the study.

7.3 THE SAMPLING DISTRIBUTION FOR $\bar{x}$ AND THE CENTRAL LIMIT THEOREM

We discussed several different measures of central tendency and variability in Chapter 5, and distinguished between numerical descriptive measures of a population (parameters) and numerical descriptive measures of a sample (statistics). Thus μ and σ are parameters, whereas $\bar{x}$ and s are statistics.

The numerical value that a sample statistic will have cannot be predicted exactly in advance. Even if we knew that a population mean μ was 216.37 and that the population standard deviation σ was 32.90 (even if we knew the complete population distribution), we could not say that the sample mean $\bar{x}$ would be exactly equal to 216.37. A sample statistic is a random variable; it is subject to random variation because it is based on a random sample of measurements selected from the population of interest. And, like any other random variable, a sample statistic has a probability distribution. We call the probability distribution of a sample statistic the sampling distribution of that statistic. Stated differently, the sampling distribution of a statistic is the population of all values for that statistic.

**sampling
distribution
of $\bar{x}$**

The actual mathematic derivation of sampling distributions is one of the basic problems of mathematical statistics. We will illustrate how the **sampling distribution of $\bar{x}$** can be obtained for a simplified population.

Example 7.2

The sample mean $\bar{x}$ is to be calculated from a random sample of size 2 taken from a population consisting of the five values ($2, $3, $4, $5, $6). Find the sampling distribution of $\bar{x}$, based on a sample of size 2.

Solution One way to find the sampling distribution is by counting. There are ten possible samples of two items from the five items. These are shown here:

Possible samples of size 2	Value of $\bar{x}$
2,3	2.5
2,4	3
2,5	3.5
2,6	4
3,4	3.5
3,5	4
3,6	4.5
4,5	4.5
4,6	5
5,6	5.5

Assuming each sample of size 2 is equally likely, it follows that the sampling distribution for $\bar{x}$ based on $n = 2$ observations selected from this population is as indicated here:

$\bar{x}$	$P(\bar{x})$
2.5	1/10
3	1/10
3.5	2/10
4	2/10
4.5	2/10
5	1/10
5.5	1/10

This sampling distribution is shown as a graph in Figure 7.1.

FIGURE 7.1   *Sampling Distribution for x̄, Example 7.2*

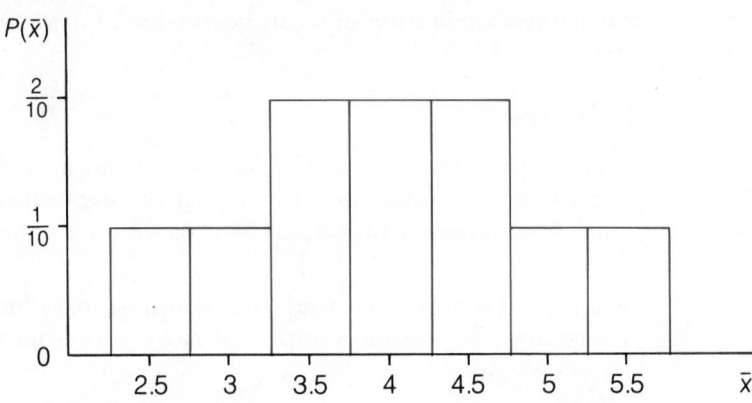

Interpretations for Sampling Distributions

Sampling distributions can be interpreted in several different ways. One way makes use of the classical interpretation of probability. Imagine listing all possible samples that can be drawn from a population; the probability that a sample statistic has a particular value (say $\bar{x} = 3.5$) is the proportion of all possible samples that yield that value. In Example 7.2, $P(\bar{x} = 3.5) = 2/10$ corresponds to the fact that 2 of the 10 possible samples have a sample mean of 3.5.

A second interpretation uses the long-run relative frequency approach. Imagine taking repeated samples of a fixed size from a population and calculating the value of the sample statistic for each sample. In the long run, the relative frequencies for the possible values of the sample statistic will approach the corresponding sampling distribution probabilities. For example, if one would take a large number of samples from the population of Example 7.2 and compute the sample mean for each sample, approximately 20% will have $\bar{x} = 3.5$.

In practice, though, a sample is taken only once, and only one value of the sample statistic is calculated. A sampling distribution is not something you see in practice; rather it is a theoretical concept. However, if we know what would have happened had we repeated the process over and over again, we can make inferences based on the results of a single sample.

Explanation for Why Some Sampling Distributions Are Normal

Quite a few sample statistics will have a sampling distribution that is normal (bell shaped). A very plausible explanation for this is offered by the

Central Limit Theorem. We will illustrate its applicability by way of example, and then state it formally.

Central Limit Theorem

The **Central Limit Theorem,** one of the most important theorems in statistics, provides information on the sampling distribution of $\bar{x}$. It states that the sampling distribution of sums, or means, based on repeated random samples of measurements from a population will be approximately bell shaped. This idea can best be illustrated with an example: Recall that a die is a cube with the faces of the cube showing from 1 to 6 dots. Thus there are six possible values that could appear face up when a die is rolled (tossed). We can simulate a population of outcomes by throwing a die a large number of times. The resulting relative frequency histogram for x (the number of dots appearing face up) would be as indicated in Figure 7.2.

FIGURE 7.2   Relative Frequency Distribution for a Population of Die Tosses

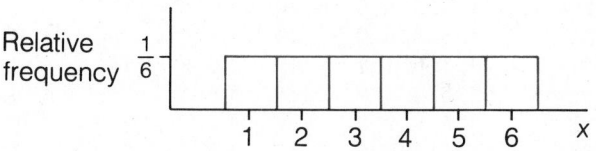

Suppose we now draw samples of five measurements ($n = 5$) from this distribution by tossing the die five times and recording the value of x observed each time. (See Table 7.3; note that the values of x observed for the first sample of five measurements are 3, 5, 1, 3, 2, respectively.) We repeat this process 100 times to obtain 100 samples of five measurements. The sum Σx and sample mean $\bar{x}$ are shown for each sample in columns three and seven, and four and eight, respectively.

TABLE 7.3   The Sums and Means for 100 Samples of 5 Die Tosses

Sample number	Sample measurements	Σx	$\bar{x}$	Sample number	Sample measurements	Σx	$\bar{x}$
1	3, 5, 1, 3, 2	14	2.8	8	3, 5, 5, 5, 5	23	4.6
2	3, 1, 1, 4, 6	15	3.0	9	6, 5, 5, 1, 6	23	4.6
3	1, 3, 1, 6, 1	12	2.4	10	5, 1, 6, 1, 6	19	3.8
4	4, 5, 3, 3, 2	17	3.4	11	1, 1, 1, 5, 3	11	2.2
5	3, 1, 3, 5, 2	14	2.8	12	3, 4, 2, 4, 4	17	3.4
6	2, 4, 4, 2, 4	16	3.2	13	2, 6, 1, 5, 4	18	3.6
7	4, 2, 5, 5, 3	19	3.8	14	6, 3, 4, 2, 5	20	4.0

(continued)

TABLE 7.3   The Sums and Means for 100 Samples of 5 Die Tosses

Sample number	Sample measurements	Σx	$\bar{x}$	Sample number	Sample measurements	Σx	$\bar{x}$
15	2, 6, 2, 1, 5	16	3.2	58	5, 5, 4, 3, 2	19	3.8
16	1, 5, 1, 2, 5	14	2.8	59	5, 4, 4, 6, 3	22	4.4
17	3, 5, 1, 1, 2	12	2.4	60	3, 2, 5, 3, 1	14	2.8
18	3, 2, 4, 3, 5	17	3.4	61	2, 1, 4, 1, 3	11	2.2
19	5, 1, 6, 3, 1	16	3.2	62	4, 1, 1, 5, 2	13	2.6
20	1, 6, 4, 4, 1	16	3.2	63	2, 3, 1, 2, 3	11	2.2
21	6, 4, 2, 3, 5	20	4.0	64	2, 3, 3, 2, 6	16	3.2
22	1, 3, 5, 4, 1	14	2.8	65	4, 3, 5, 2, 6	20	4.0
23	2, 6, 5, 2, 6	21	4.2	66	3, 1, 3, 3, 4	14	2.8
24	3, 5, 1, 3, 5	17	3.4	67	4, 6, 1, 3, 6	20	4.0
25	5, 2, 4, 4, 3	18	3.6	68	2, 4, 6, 6, 3	21	4.2
26	6, 1, 1, 1, 6	15	3.0	69	4, 1, 6, 5, 5	21	4.2
27	1, 4, 1, 2, 6	14	2.8	70	6, 6, 6, 4, 5	27	5.4
28	3, 1, 2, 1, 5	12	2.4	71	2, 2, 5, 6, 3	18	3.6
29	1, 5, 5, 4, 5	20	4.0	72	6, 6, 6, 1, 6	25	5.0
30	4, 5, 3, 5, 2	19	3.8	73	4, 4, 4, 3, 1	16	3.2
31	4, 1, 6, 1, 1	13	2.6	74	4, 4, 5, 4, 2	19	3.8
32	3, 6, 4, 1, 2	16	3.2	75	4, 5, 4, 1, 4	18	3.6
33	3, 5, 5, 2, 2	17	3.4	76	5, 3, 2, 3, 4	17	3.4
34	1, 1, 5, 6, 3	16	3.2	77	1, 3, 3, 1, 5	13	2.6
35	2, 6, 1, 6, 2	17	3.4	78	4, 1, 5, 5, 3	18	3.6
36	2, 4, 3, 1, 3	13	2.6	79	4, 5, 6, 5, 4	24	4.8
37	1, 5, 1, 5, 2	14	2.8	80	1, 5, 3, 4, 2	15	3.0
38	6, 6, 5, 3, 3	23	4.6	81	4, 3, 4, 6, 3	20	4.0
39	3, 3, 5, 2, 1	14	2.8	82	5, 4, 2, 1, 6	18	3.6
40	2, 6, 6, 6, 5	25	5.0	83	1, 3, 2, 2, 5	13	2.6
41	5, 5, 2, 3, 4	19	3.8	84	5, 4, 1, 4, 6	20	4.0
42	6, 4, 1, 6, 2	19	3.8	85	2, 4, 2, 5, 5	18	3.6
43	2, 5, 3, 1, 4	15	3.0	86	1, 6, 3, 1, 6	17	3.4
44	4, 2, 3, 2, 1	12	2.4	87	2, 2, 4, 3, 2	13	2.6
45	4, 4, 5, 4, 4	21	4.2	88	4, 4, 5, 4, 4	21	4.2
46	5, 4, 5, 5, 4	23	4.6	89	2, 5, 4, 3, 4	18	3.6
47	6, 6, 6, 2, 1	21	4.2	90	5, 1, 6, 4, 3	19	3.8
48	2, 1, 5, 5, 4	17	3.4	91	5, 2, 5, 6, 3	21	4.2
49	6, 4, 3, 1, 5	19	3.8	92	6, 4, 1, 2, 1	14	2.8
50	4, 4, 4, 4, 4	20	4.0	93	6, 3, 1, 5, 2	17	3.4
51	2, 3, 5, 3, 2	15	3.0	94	1, 3, 6, 4, 2	16	3.2
52	1, 1, 1, 2, 4	9	1.8	95	6, 1, 4, 2, 2	15	3.0
53	2, 6, 3, 4, 5	20	4.0	96	1, 1, 2, 3, 1	8	1.6
54	1, 2, 2, 1, 1	7	1.4	97	6, 2, 5, 1, 6	20	4.0
55	2, 4, 4, 6, 2	18	3.6	98	3, 1, 1, 4, 1	10	2.0
56	3, 2, 5, 4, 5	19	3.8	99	5, 2, 1, 6, 1	15	3.0
57	2, 4, 2, 4, 5	17	3.4	100	2, 4, 3, 4, 6	19	3.8

We can construct a relative frequency histogram for $\bar{x}$ (the mean of the five sample measurements) by first preparing a frequency table from the data in Table 7.3. The tabulation of relative frequencies is shown in Table 7.4, and the resulting relative frequency histogram is shown in Figure 7.3(b). According to the Central Limit Theorem, this relative frequency histogram of $\bar{x}$ should be approximately normal.

TABLE 7.4   Relative Frequency Table for 100 Values of $\bar{x}$

Class	Class boundaries	Frequency	Relative frequency
1	1.3–1.5	1	1/100
2	1.5–1.7	1	1/100
3	1.7–1.9	1	1/100
4	1.9–2.1	1	1/100
5	2.1–2.3	3	3/100
6	2.3–2.5	4	4/100
7	2.5–2.7	6	6/100
8	2.7–2.9	10	10/100
9	2.9–3.1	7	7/100
10	3.1–3.3	9	9/100
11	3.3–3.5	11	11/100
12	3.5–3.7	9	9/100
13	3.7–3.9	11	11/100
14	3.9–4.1	10	10/100
15	4.1–4.3	7	7/100
16	4.3–4.5	1	1/100
17	4.5–4.7	4	4/100
18	4.7–4.9	1	1/100
19	4.9–5.1	2	2/100
20	5.1–5.3	0	0
21	5.3–5.5	1	1/100

Note that although the relative frequencies for the individual values of x are equal and hence the relative frequency distribution is flat (see Figure 7.3a), the distribution of the sample means is mound shaped and even somewhat bell shaped (See Figure 7.3b). You can visually compare these two relative frequency distributions by observing Figure 7.3. The irregularities in the curve of Figure 7.3(b) are due to the small number of samples used to illustrate this concept. These irregularities would be less obvious if the sampling were conducted a large number of times. (Such extensive sampling is a time-consuming task to perform manually, but it can be done easily with the aid of a computer). The result would verify the Central Limit Theorem, which we now state as it applies to means.

FIGURE 7.3   Illustration of the Central Limit Theorem

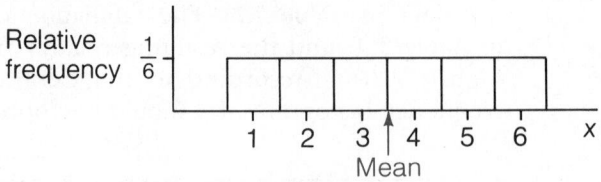

(a) Relative frequency distribution for x

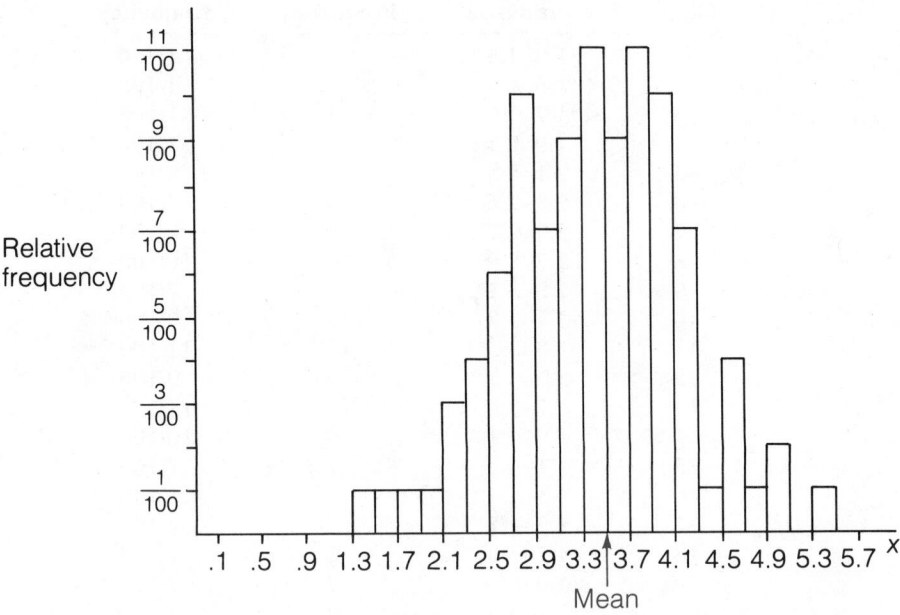

(b) Relative frequency distribution for the 100 sample means in Table 7.3

Theorem 7.1: The Central Limit Theorem, Applied to Means

If random samples containing a fixed number of n of measurements are repeatedly drawn from a population with finite mean μ and standard deviation σ, then if n is large, the sample means will have a distribution that is approximately normal (bell shaped), with mean $\mu_{\bar{x}} = \mu$ and standard deviation (called the **standard error of $\bar{x}$**) $\sigma_{\bar{x}} = \sigma/\sqrt{n}$.

standard error of $\bar{x}$

We hope the die-tossing data of Table 7.3 clarify the meaning of the Central Limit Theorem. Although proof is omitted, it can be shown that the mean and the standard deviation for x in the die-tossing distribution (Figure 7.3a) are $\mu = 3.50$ and $\sigma = 1.71$. For these values the Central Limit Theorem states that the sampling distribution for sample means based on $n = 5$ measurements possesses the same mean as the original population: $\mu_{\bar{x}} = \mu = 3.5$. The standard error is

$$\sigma_{\bar{x}} = \frac{\sigma}{\sqrt{n}} = \frac{1.71}{\sqrt{5}} = .76$$

Our experimental die-tossing sampling verifies these results. You can see that the mean of the distribution of $\bar{x}$ s, Figure 7.3(b), is approximately 3.5. In addition the range of the $\bar{x}$ is $(5.4 - 1.4) = 4.0$, so the standard error is approximately $(4.0)/4$, or 1.0. Hence the standard error of the distribution of 100 sample means, Figure 7.3(b), is near the value stated by the Central Limit Theorem, $\sigma/\sqrt{n} = .76$. As noted earlier, the relative frequency histogram of Figure 7.3(b) is approximately bell shaped. Thus the die-tossing sampling experiment provides us with practical verification of the Central Limit Theorem.

The Central Limit Theorem also applies to the distribution of sums of sample observations, and the die-tossing data can also be used to provide empirical evidence to support this version of the Central Limit Theorem. We leave it to you to construct a histogram of the sample sums Σx if you seek graphic evidence of the validity of the Central Limit Theorem as it applies to sums.

Theorem 7.2: The Central Limit Theorem, Applied to Sums

If random samples containing a fixed number n of measurements are repeatedly drawn from a population with finite mean μ and standard deviation σ, then if n is large, the sums of the sample measurements will have a distribution that is approximately normal (bell shaped), with mean equal to $n\mu$ and standard error equal to $\sigma\sqrt{n}$.

Broad Applicability of the Central Limit Theorem

Careful reading of the Central Limit Theorem suggests its broad applicability: it applies to the distribution of sample means drawn from any population with a finite mean μ and standard deviation σ. The resulting distribution of means will be approximately normal, with mean and standard error related not only to the mean and standard deviation of the population from which the samples are drawn but also to the sample size n. Note that the normal approximation to the distribution of sample means or sums becomes more and more accurate as the sample size n increases. Recall, however, that the approximation is quite good for an n as small as 5 in the die-tossing experiment.

The significance of the Central Limit Theorem is twofold:

1. It explains why many measurements have bell-shaped frequency distributions. For example, we might imagine that the test score for an individual on a national aptitude test is influenced by random factors or variables, such as the amount of sleep the individual had the night before, the length of time spent preparing for the exam, the individual's IQ, and so forth. If each of these factors in some way affects the final score, that score is the sum of random variables. The Central Limit Theorem may then help to explain why such scores are approximately normally distributed.

2. It is useful in statistical inference. Many estimators of population parameters used for purposes of statistical inference are sums of averages of the sample measurements. To illustrate, we will use the sample mean $\bar{x}$ to estimate a population mean μ. Where sums or averages are involved and the sample size n is large, the many estimates generated in repeated sampling can be expected to possess a bell-shaped normal distribution. Then we can use the properties of the normal distribution to describe the behavior of $\bar{x}$. (This application of the Central Limit Theorem is explained in Chapter 8.)

To summarize, the Central Limit Theorem tells us the nature of the sampling distribution of $\bar{x}$ when the sample size is large. This information is as follows:

Properties of the Sampling Distribution of Sample Mean $\bar{x}$

1. The sampling distribution of $\bar{x}$ is approximately normal for large sample sizes (n large).

2. The mean of the sampling distribution is equal to the population mean: $\mu_{\bar{x}} = \mu$.

3. The standard error of the sampling distribution is $\sigma_{\bar{x}} = \sigma/\sqrt{n}$, where σ is the standard deviation of the sampled population.

An obvious question is: how large should the sample size be in order for the Central Limit Theorem to hold? Numerous studies have been conducted over the years and the results of these studies show that, in general, the Central Limit Theorem will hold for $n > 30$. However, one should not apply the rule blindly. If the population is heavily skewed, the sampling distribution for $\bar{x}$ will still be skewed even for $n > 30$. On the other hand, if the population is symmetric, the Central Limit Theorem holds for $n < 30$. For example, the Central Limit Theorem worked very well for a sample size of $n = 5$ in the die-tossing experiment. So take a look at the sample data. If the sample histogram is heavily skewed, then probably the population will also be skewed. Consequently a value of n much larger than 30 may be required before the distribution of $\bar{x}$ is nearly normal.

Example 7.3

Reclaimed phosphate land in Polk County, Florida, has been found to emit a higher mean radiation level than other nonmining land in the county. Suppose that the radiation level for the reclaimed land has a distribution with mean $\mu = 5.0$ working levels (WL) and a standard deviation of .5 WL. Suppose that 20 houses built on reclaimed land are randomly selected and the radiation level measured in each. What is the probability (approximately) that the sample mean for the 20 houses exceeds 5.2 WL?

Solution   According to the Central Limit Theorem the sample mean of $n = 20$ randomly selected radiation level measurements will be approximately normally distributed with

$$\mu_{\bar{x}} = \mu = 5.0 \text{ WL}$$
$$\sigma_{\bar{x}} = \frac{\sigma}{\sqrt{n}} = \frac{.5}{\sqrt{20}} = .11$$

The distribution will be as shown in Figure 7.4
    The probability that $\bar{x}$ exceeds 5.2 WL is the shaded area shown in Figure 7.4. To find this area, we must determine how many standard errors ($\sigma_{\bar{x}} = .11$) the point 5.2 lies to the right of $\mu = 5.0$. This number is

$$z = \frac{\bar{x} - \mu}{\sigma_{\bar{x}}} = \frac{5.2 - 5.0}{.11} = 1.82$$

Turning to Table 2 of Appendix 3, we find that the area A corresponding to $z = 1.82$ is

$$A = .4656$$

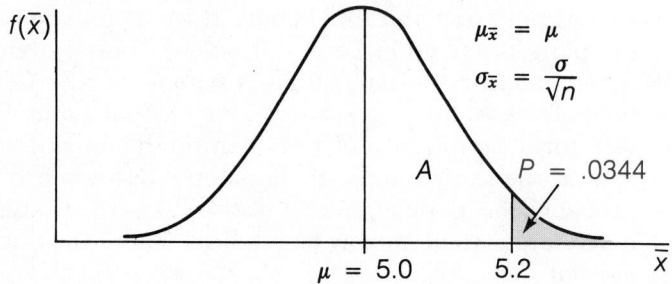

FIGURE 7.4 Sampling Distribution of $\bar{x}$ for Samples of $n = 20$ Randomly Selected from a Population with $\mu = 5.0$ and $\sigma = .5$

Since the total area to the right of the mean is equal to .5, the probability that $\bar{x}$ will exceed 5.2 WP is P, where

$$P = .5 - A = .5 - .4656 = .0344$$

EXERCISES

7.7   A population consists of the 5 measurements 2, 6, 8, 0, 1. Use the Central Limit Theorem to answer the following questions.

(a) How many different samples of size $n = 2$ can be drawn from the population? (Hint: List them.)

(b) What is the mean of the sampling distribution of $\bar{x}$? (Hint: Compute the mean for the population.)

(c) What is the standard error of the sampling distribution of $\bar{x}$ if $\sigma = 4.43$?

7.8   Complete the following sentence. As the sample size increases, the standard error of $\bar{x}$ _____.

7.9   A random sample of $n = 100$ measurements is obtained from a population with $\mu = 55$ and $\sigma = 20$. Describe the sampling distribution for $\bar{x}$, giving $\mu_{\bar{x}}$ and $\sigma_{\bar{x}}$.

7.10   A random sample of $n = 60$ measurements is obtained from a population with $\mu = 192$ and $\sigma = 43$. Describe the sampling distribution for $\bar{x}$.

7.11   Describe the sampling distribution for Σx in Exercise 7.10.

7.12   Refer to Table 7.3. Using the sample data from the 100 samples of $n = 5$ die tosses, construct a frequency table similar to Table 7.4 for Σx, the sum of the sample measurements. Use an interval width of 1 with a starting point of 6.5.

7.13   Using the frequency table of Exercise 7.12, construct a relative frequency histogram for Σx. The mean of the distribution of x, the number of dots appearing on the upper face, can be shown to be $\mu = 3.5$, and the size of each sample is $n = 5$. Thus by the Central Limit Theorem the relative frequency histogram should be mound shaped, with a mean approximately equal to $n(\mu) = 5(3.5) = 17.5$. Because the relative frequency histogram of Σx is based on only 100 samples, the approximation could be improved by using more repetitions of the experiment.

7.14   A random sample of $n = 25$ measurements is selected from a population with mean equal to 80 and standard deviation equal to 7.

(a) What is the probability that the sample mean will exceed 81?

(b) What is the probability that the sample mean will fall in the interval $79 \le \bar{x} \le 81$?

(c) What is the probability that the sample mean will be less than 78?

7.15   A random sample of 16 measurements is drawn from a population with a mean of 60 and a standard deviation of 5. Describe the sampling distribution of $\bar{x}$, the sample mean. Within what interval would you expect $\bar{x}$ to lie approximately 95% of the time?

7.16   Refer to Exercise 7.15. Describe the sampling distribution for the sample sum Σx_i. Is it unlikely (improbable) that Σx_i would be more than 70 units away from 960? Explain.

7.17   Psychomotor retardation scores for a large group of manic-depressive patients were found to be approximately normal with a mean of 930 and a standard deviation of 130.

(a) What fraction of the patients scored between 800 and 1100?

(b) Less than 800?

(c) Greater than 1200?

7.18   Refer to Exercise 7.17. Find the 90th percentile for the distribution of manic-depressive scores. [Hint: Solve for x in the expressions $z = (x - \mu)/\sigma$, where z is the number of standard deviations the 90th percentile lies above the mean μ.]

7.19   The oxygen content in water must exceed some minimum value in order to support aquatic life. Suppose that this value is approximately 6.0 parts per million (ppm). In one experiment $n = 5$ jars of water are randomly selected from a stream. Records from previous months suggest that the mean oxygen content is $\mu = 6.0$ ppm and the standard deviation is $\sigma = .7$ ppm.

(a) What is the probability that the sample mean exceeds 6.5 ppm?

(b) Suppose that the sample mean equals 7.0 ppm. Intuitively, what would you conclude about the mean oxygen content (μ) of the stream? Has the situation changed from previous months? (Note: In Chapter 8 we will answer this question by using a statistical decision procedure.)

7.4   NORMAL APPROXIMATION TO THE BINOMIAL

The Central Limit Theorem just discussed will enable us to calculate probabilities for a binomial random variable by approximating the binomial distribution with a normal curve and using normal curve areas as approximations to the desired probabilities. We have seen that probabilities associated with values of x can be computed for a binomial experiment for any values of n or π, but the task becomes more difficult when n gets large (Section 6.8). For example, suppose a sample of 1000 voters was polled to determine sentiment toward the consolidation of a city and county government. What would be the probability of observing 460 or fewer favoring consolidation if we assume that 50% of the entire population favor the change? Here we have a binomial experiment with $n = 1000$ and π, the probability of selecting a person favoring consolidation, equal to .5. To determine the probability of observing 460 or fewer favoring consolidation in the random sample of 1000 voters, we could compute $P(x)$, using the binomial formula for $x = 460, 459, \ldots, 0$. The desired probability would then be

$$P(x = 460) + P(x = 459) + \cdots + P(x = 0)$$

There would be 461 probabilities to calculate, with each one being somewhat difficult due to the factorials. For example, the probability of observing 460 favoring consolidation is

$$P(x = 460) = \frac{1000!}{460! \ 540!} (.5)^{460}(.5)^{540}$$

A similar calculation would be needed for all other values of x.

The normal distribution can be used in many situations to approximate the binomial probability distribution, and areas under the normal curve can be used to approximate the actual binomial probabilities. This process of approximating a binomial distribution with a normal distribution is called a **normal approximation to the binomial.** The normal dis-

normal approximation to the binomial

tribution that provides the best approximation to the binomial probability distribution has a mean and a standard deviation given by the following formulas.

Normal approximation to the Binomial Probability Distribution

$$\mu = n\pi \qquad \sigma = \sqrt{n\pi(1 - \pi)}$$

Note: This approximation can be used if

$$n \geq \frac{5}{\min(\pi,\ 1 - \pi)}$$

that is, if n is greater than or equal to 5 divided by the minimum of π and $1 - \pi$. Equivalently, the normal approximation can be used if $n\pi \geq 5$ and $n(1 - \pi) \geq 5$.

To illustrate how well the distribution of x can be approximated by a normal distribution, we show the binomial probability distribution for $n = 20$ and $\pi = .5$ in Figure 7.5, with the approximating normal curve superimposed. For this situation the mean and standard deviation are

$$\mu = n\pi = (20)(.5) = 10$$
$$\sigma = \sqrt{n\pi(1 - \pi)} = \sqrt{(20)(.5)(.5)} = 2.24$$

FIGURE 7.5 Comparison of a Binomial Probability Distribution, for $n = 20$ and $\pi = .5$, and the Approximating Normal Curve

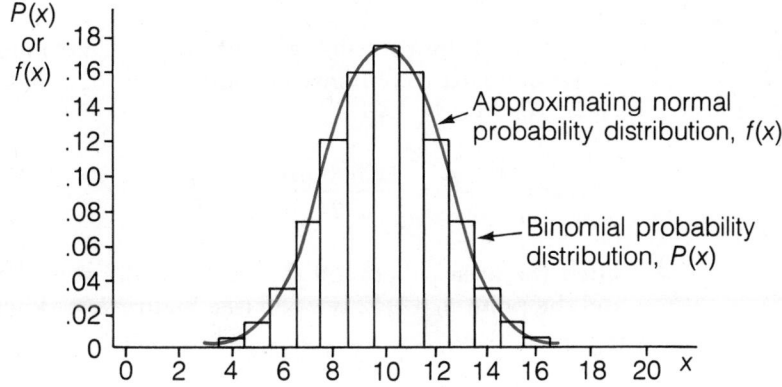

As we can see, this approximation is surprisingly good, considering the small value of n. However, when π is close to 0 or 1, we would need a much larger sample in order for the normal distribution to provide a good approximation to the binomial probability distribution.

Example 7.4

Use the normal approximation to the binomial probability distribution, Figure 7.5, to find the probability that $x \geq 13$.

Solution   The probability that we wish to approximate is the sum of the probability rectangles, Figure 7.5, corresponding to $x = 13, 14, \ldots, 20$. To approximate the area corresponding to these rectangles, we need to find the area under the normal curve to the right of $x = 12.5$; see Figure 7.6 (Notice that using only the area to the right of $x = 13$ would be incorrect, because then we would be omitting the area corresponding to the left half of the rectangle for $x = 13$.)

FIGURE 7.6   Required Area for Example 7.4

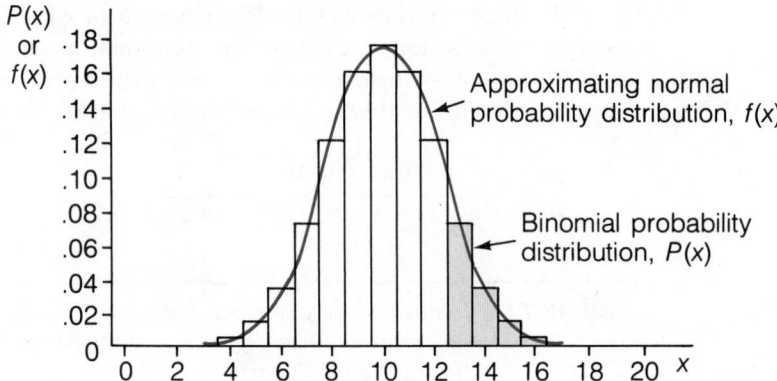

To find the probability P that $x \geq 13$, we must first determine how many standard deviations the point $x = 12.5$ lies to the right of $\mu = n\pi = 10$. This is

$$z = \frac{x - \mu}{\sigma} = \frac{12.5 - 10}{2.24} = 1.12$$

Then the area A between the mean of the standard normal distribution and the point $z = 1.12$ is .3686 (see Figure 7.7). Therefore, the probability is

$$P = .5 - A = .5 - .3686 = .1314$$

FIGURE 7.7   Normal Curve for Example 7.4

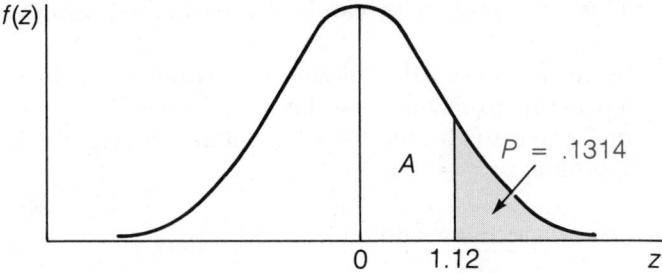

Example 7.5

Refer to the example discussed in the beginning of this section. Use the normal approximation to the binomial to compute the probability of observing 460 or fewer in a sample of 1000 favoring consolidation if we assume that 50% of the entire population favor the change.

Solution   The normal distribution used to approximate the binomial distribution will have

$$\mu = n\pi = 1000(.5) = 500$$
$$\sigma = \sqrt{n\pi(1 - \pi)} = \sqrt{1000(.5)(.5)} = 15.8$$

Note that

$$n \geq \frac{5}{.5} = 10$$

Hence we may use the normal approximation to the binomial. The desired probability is represented by the shaded area shown in Figure 7.8.

FIGURE 7.8   Approximating Normal Distribution for the Binomial Distribution of Example 7.5 with $\mu = 500$ and $\sigma = 15.8$

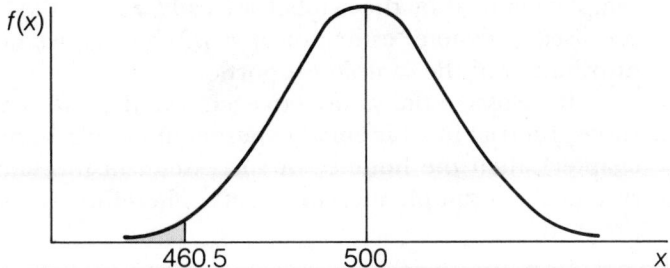

As noted earlier, we could compute

This same probability can be approximated using the area under the nor-

using the binomial probability distribution with $n = 1000$ and $\pi = .5$. This same probability can be approximated using the area under the normal curve to the left of $x = 460.5$ (see Figure 7.8). The z-score corresponding to $x = 460.5$ is

$$z = \frac{x - \mu}{\sigma} = \frac{460.5 - 500}{15.8} = -2.50$$

Referring to Table 2 of Appendix 3, we find that the area under the normal curve between 460.5 and 500 (i.e., for $z = -2.50$), is .4938. Thus the probability of observing 460 or fewer favoring consolidation is approximately $.5 - .4938 = .0062$.

Example 7.6

Refer to Example 7.5. Suppose that 460 of a sample of 1000 potential voters favor the consolidation of the city and county governments. Would you expect the consolidation issue to pass?

Solution   In Example 7.5 we computed $P(x \leq 460)$ to be approximately .0062 when $\pi = .5$ and $n = 1000$. Since this probability is so small, an observed value of x equal to 460 is contradictory to the assumption that $\pi = .5$ (or more) of the voters favor consolidation. Because of this contradiction, we conclude that the consolidation issue will not pass.

7.5   SAMPLING DISTRIBUTION FOR $\hat{\pi}$

Binomial populations are sampled in order to make inferences about π, the proportion of successes in the population. The most obvious statistic to select to make these inferences is the proportion of successes in the sample, denoted by the symbol $\hat{\pi}$ (read "π hat"; the "hat" over the symbol π is used to denote "estimator of π"). What can we say about the sampling distribution of the sample proportion $\hat{\pi}$?

   If we assign the value 0 to each trial that fails and the value 1 to each successful trial in a binomial experiment (as was suggested in the previous chapter), then the number of successes x in the sample of n trials is the sum of the n sample measurements. Therefore, it follows that the sample proportion

$$\hat{\pi} = \frac{x}{n} = \frac{\text{sum of the sample measurements}}{n}$$

sampling distribution of $\hat{\pi}$

is the sample mean and that the **sampling distribution of $\hat{\pi}$** will be approximately normally distributed when n is large (because of the Central Limit Theorem). It can be shown (proof omitted) that the mean and the standard deviation of this approximating normal distribution are

$$\mu_{\hat{\pi}} = \pi$$

$$\sigma_{\hat{\pi}} = \sqrt{\frac{\pi(1 - \pi)}{n}}$$

The sampling distribution for the sample proportion $\hat{\pi}$ is as shown in Figure 7.9.

FIGURE 7.9   The Sampling Distribution for $\hat{\pi}$

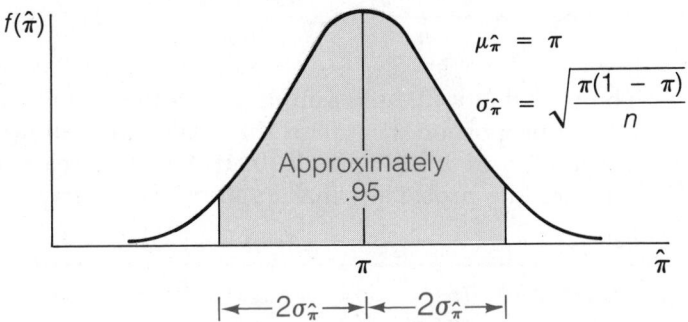

Properties of the Sampling Distribution of a Binomial Sample Proportion $\hat{\pi}$

1. The sampling distribution for $\hat{\pi}$ is approximately normal for large sample sizes (n large).
2. The mean $\mu_{\hat{\pi}}$ of the sampling distribution is equal to π, the population proportion of successes.
3. The standard error of the sampling distribution for $\hat{\pi}$ is equal to

$$\sigma_{\hat{\pi}} = \sqrt{\frac{\pi(1 - \pi)}{n}}$$

   In the next example we show how this information can be used in a practical application.

Example 7.7

Two thousand new automobile steering mechanisms were tested in order to estimate the proportion π of all the steering mechanisms that might be faulty. If the true proportion is $\pi = .03$, what is the probability that the sample proportion will be within .01 of π?

Solution The 2000 steering mechanisms can be viewed as a random sample. Thus the sampling distribution for $\hat{\pi}$ will appear as shown in Figure 7.10 with mean

$$\pi = .03$$

and standard error

$$\sigma_{\hat{\pi}} = \sqrt{\frac{\pi(1 - \pi)}{n}} = \sqrt{\frac{(.03)(.97)}{2000}} = .0038$$

The probability that the sample proportion $\hat{\pi}$ falls within .01 of the population proportion $\pi = .03$ is the shaded area shown under the sampling distribution of $\hat{\pi}$ in Figure 7.10. If A is the area over the interval .03 to .04, then the probability that $\hat{\pi}$ falls in the interval .02 to .04 is $2A$.

FIGURE 7.10   The Sampling Distribution for $\hat{\pi}$, $n = 2000$ and $\pi = .03$

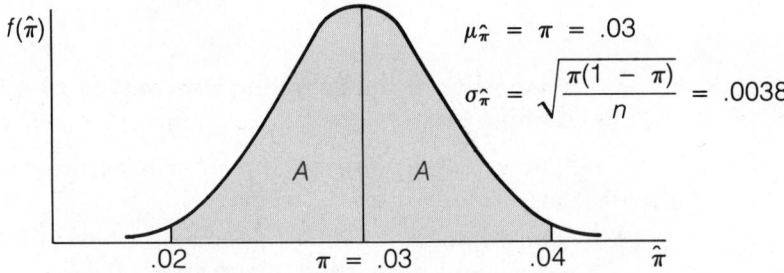

To find the number of standard deviations between $\pi = .03$ and a value of $\hat{\pi}$ equal to .04, we calculate the z-score:

$$z = \frac{\hat{\pi} - \pi}{\sigma_{\hat{\pi}}} = \frac{.04 - .03}{.0038} = 2.63$$

Since the area A to the right of the mean, corresponding to $z = 2.63$, is .4957, it follows that

$$P(.02 \le \hat{\pi} \le .04) = 2A = 2(.4957) = .9914$$

In other words, if the population proportion of steering mechanisms that are defective is $\pi = .03$, the proportion of steering mechanisms that malfunction in a sample of $n = 2000$ should be within .01 of π with a high probability—namely, .9914. Therefore, you can see that in this case the sample proportion will provide an accurate estimate of π.

EXERCISES

7.20 A random sample of $n = 1000$ measurements is obtained from a binomial population with $\pi = .7$. Describe the sampling distribution for the sample proportion $\hat{\pi}$.

7.21 Refer to Exercise 7.20. Use the sampling distribution of $\hat{\pi}$ to find the probability that $\hat{\pi}$ lies in the interval from .6 to .8.

7.22 Suppose that you select a sample of $n = 10$ from a binomial population with $\pi = .5$.

    (a) Use the binomial probability distribution to calculate the probability that x falls in the interval $3 \le x \le 5$.

    (b) Use the normal approximation to the binomial probability distribution to find the approximate probability that $3 \le x \le 5$. Compare this result with your answer to part (a).

$ 7.23 An insurance company states that 10% of all fire insurance claims are fraudulent. Suppose that the company is correct and that it receives 100 claims.

    (a) What is the probability that more than 12 will be fraudulent?

    (a) What is the probability that 15 or more will be fraudulent?

$ 7.24 Refer to Exercise 7.23.

    (a) What assumptions must be made in order that your answers be valid?

    (b) Suppose that only one claim in the sample was fraudulent. Would you have any doubts about the company's statement? Explain.

7.25 Suppose that you wish to conduct a poll to estimate the proportion of adult Americans who favor less governmental control of private business. Further, suppose that this unknown proportion is really $\pi = .5$. If you draw a random sample of $n = 1000$ adults from the United States population, what is the probability that the sample proportion will differ from the population proportion ($\pi = .5$) by more than .02?

7.26 If 2% of all babies in the United States are born with one or more congenital malformations and if 3 million babies are born in a given year,

(a) Determine the mean and standard deviation of the number of malformed infants born in that year.

(b) Determine the probability of observing 50,000 or fewer babies with malformations in that year.

7.27 Refer to Exercise 7.26.

(a) Describe the sampling distribution for $\hat{\pi}$, the proportion of malformed infants observed in the sample.

(b) What is the probability of observing 3% or more sampled infants with malformations? 4% or more?

7.6 USING COMPUTERS

The Minitab software system was used to simulate the sampling distribution for $\bar{x}$ based on 50 samples of $n = 5$ measurements drawn from a normal population with $\mu = 20$ and $\sigma = 4$. As you can see, we had to use the statements RANDOM, NORMAL, and RMEAN (row mean) to generate each sample of $n = 5$ measurements and compute the sample mean. Numerical descriptive measures of the sampling distribution and a histogram are shown in the output. Note how the mean of the 50 sample means (20.06) is very close to the theoretical value, $\mu = 20$; similarly, the standard deviation of the $\bar{x}$s, called the standard error of the mean, is 1.71, which is close to the theoretical value, $\sigma/\sqrt{n} = 1.79$. These two values would come closer and closer to the theoretical values with more and more samples of size $n = 5$.

The Minitab software system can be used for such simulations as the one illustrated here. Another example is shown in Exercise 7.65. Further details about simulations of other distributions are discussed in *Minitab Handbook* (1985).

The Minitab program for generating a random sample of $n = 50$ observations from a normal distribution with $\mu = 20$ and $\sigma = 4$ is:

```
MTB > RANDOM 50 C1-C5;
SUBC> NORMAL MU=20 SIGMA=4.
MTB > RMEAN C1-C5 INTO C6
MTB > PRINT C1-C6

ROW        C1        C2        C3        C4        C5        C6

  1    17.7675   19.8461   26.3163   16.9780   20.0225   20.1861
  2    21.7446   20.3663   23.4076   25.9735   23.9176   23.0819
```

3	15.6995	23.2978	22.9142	26.8342	18.0461	21.3583
4	22.1651	21.0854	21.7027	25.2676	11.7937	20.4029
5	20.6046	18.7233	21.8044	23.0735	15.2243	19.8860
6	20.6167	19.4629	23.9151	10.2321	14.6927	17.7839
7	20.0495	18.6123	19.6686	16.1272	23.8354	19.6586
8	22.8409	23.7688	19.2506	18.2076	19.9236	20.7983
9	20.1568	17.1087	22.7679	22.2152	21.2758	20.7049
10	18.2649	16.9100	13.4511	17.8793	19.2428	17.1496
11	23.2222	21.9275	26.0291	14.2883	19.5703	21.0075
12	18.8884	29.9938	20.0412	12.3697	21.2671	20.5121
13	20.8680	12.0775	18.8344	17.6789	15.4499	16.9817
14	17.2733	14.4175	23.1749	18.1204	16.4413	17.8855
15	18.0425	16.2597	15.5968	18.2761	8.5550	15.3461
16	18.2190	20.2101	28.3832	14.2130	24.0248	21.0100
17	24.5287	26.9656	18.7591	23.2235	17.4729	22.1899
18	20.3078	20.9958	21.1693	26.0732	13.9906	20.5073
19	19.9555	23.5640	19.3437	23.6595	11.9906	19.7027
20	18.5220	15.9421	26.3148	21.3247	24.5102	21.3228
21	21.1549	18.2365	16.6500	18.9231	13.7666	17.7462
22	20.0756	17.8309	24.6306	22.1602	18.9316	20.7258
23	20.1765	25.5803	25.2298	12.1687	30.1985	22.6708
24	23.7893	16.7882	20.7636	17.6004	23.3660	20.4615
25	17.3698	25.3004	19.4000	24.3134	24.7438	22.2255
26	15.6699	18.1690	25.9357	17.0216	16.5050	18.6722
27	25.9732	20.3175	18.5377	16.7085	18.4339	19.9942
28	21.7629	21.4541	20.8077	21.5750	19.9204	21.1040
29	17.7795	15.3596	15.9085	22.2014	19.1648	18.0827
30	19.2793	17.8108	14.4859	18.1489	19.7862	17.9022
31	19.6149	22.3171	29.6498	19.9469	17.6377	21.8333
32	14.1724	17.7072	18.6619	18.8721	21.6695	18.2166
33	22.0559	26.1348	21.3086	20.0742	17.2110	21.3569
34	27.1653	19.4272	20.7409	24.6679	23.6852	23.1373
35	21.4947	13.0971	23.1049	21.4793	24.9358	20.8224
36	21.7922	23.6901	13.4055	20.1015	20.1107	19.8200
37	18.9085	18.5206	24.9700	24.4045	19.4918	21.2591
38	14.7031	22.9635	18.8457	25.4396	16.7987	19.7501
39	27.4691	8.8521	21.6883	19.0457	22.4229	19.8956
40	17.2028	19.8334	21.9562	19.4756	18.0777	19.3091
41	26.2857	19.9121	18.6084	17.7153	22.1454	20.9334
42	17.5304	15.5321	18.3837	19.2592	23.2736	18.7958
43	18.3001	20.1210	21.2264	18.7826	23.8089	20.4478
44	18.7209	16.7202	18.9731	26.2228	22.2727	20.5820
45	16.7666	12.9603	24.9859	17.4112	17.1398	17.8528
46	23.5268	23.7971	19.2165	25.4780	9.8634	20.3763
47	18.7708	17.1806	20.3488	21.9186	22.5578	20.1553
48	20.4666	20.2224	18.1312	20.8058	14.9315	18.9115
49	22.3293	20.3925	18.1794	19.4384	14.2383	18.9156
50	20.4841	25.9499	23.2712	28.6011	19.6735	23.5960

```
MTB > DESCRIBE C6

            N     MEAN   MEDIAN   TRMEAN    STDEV   SEMEAN
C6         50   20.061   20.390   20.085    1.709    0.242

          MIN      MAX       Q1       Q3
C6     15.346   23.596   18.883   21.034

MTB > HISTOGRAM C6

Histogram of C6   N = 50
```

```
    Midpoint    Count
          15        1   *
          16        0
          17        2   **
          18        7   *******
          19        5   *****
          20       13   *************
          21       15   ***************
          22        3   ***
          23        3   ***
          24        1   *

MTB > STOP
```

SUMMARY

In this chapter we introduced random sampling and sampling distributions, which are two important background topics for developing statistical inference.

In order to develop the methods for making statistical inferences presented in Chapters 8–17, we assume the sample data were drawn "at random" from the population of interest. So the data-gathering step in making sense of data must involve random sampling. We discussed ways to accomplish random sampling through the use of a table of random numbers.

The second major topic, the sampling distribution of a statistic, was discussed in some detail. Problems in statistical inference involve the selection of a random sample of measurements and the computation of sample statistics (such as $\bar{x}$, $\hat{\pi}$, etc.) from the sample data. Although we will *not* repeat this sampling/computation process over and over again, if we know the sampling distribution of the sample statistic, we know what would have happened if we had repeated the sampling/computation process many more times.

This knowledge will be essential for making statistical inferences and, since many of the sample statistics used in statistical inferences are sums or averages of sample measurements (obtained in a random sample), the Central Limit Theorem helps explain why their sampling distributions (for sufficiently large sample sizes) will be normally distributed.

Chapters 6 and 7 have provided the background material for analyzing data and drawing statistical inferences. We will now discuss the methods for making these inferences relative to specific population parameters.

KEY TERMS

sampling distribution
random sample of one measurement (*def*)
random sample of *n* measurements (*def*)

table of random numbers
conceptual population
sampling distribution of $\bar{x}$
Central Limit Theorem
standard error of $\bar{x}$
normal approximation to the binomial
sampling distribution of $\hat{\pi}$

KEY FORMULAS

1. Sampling distribution for $\bar{x}$

Mean: μ

Standard error: $\sigma_{\bar{x}} = \dfrac{\sigma}{\sqrt{n}}$

2. Sampling distribution for Σx

Mean: $n\mu$
Standard error: $\sqrt{n}\,\sigma$

3. Normal approximation to the binomial

Mean: $n\pi$
Standard deviation: $\sqrt{n\pi(1 - \pi)}$
provided $n \geq \dfrac{5}{\min(\pi,\, 1 - \pi)}$

4. Sampling distribution for $\hat{\pi}$

Mean: π

Standard error: $\sigma_{\hat{\pi}} = \sqrt{\dfrac{\pi(1 - \pi)}{n}}$

SUPPLEMENTARY EXERCISES

7.28 A random sample of $n = 25$ measurements is selected from a population with mean $\mu = 3$ and standard deviation $\sigma = 1$.

(a) Find the approximate probability that $\bar{x} \geq 3.1$.

(b) Find the approximate probability that $2.8 \leq \bar{x} \leq 3.2$.

(c) Find the approximate probability that $\Sigma x \geq 80$.

7.29 A marketing research firm believes that approximately 25% of all persons mailed a "sweepstakes" offer will respond. If a preliminary mailing of 5000 is conducted in a region,

(a) What is the probability that 1000 or fewer will respond?

(b) What is the probability that 3000 or more will respond?

7.30 Let x be the IQ of any college student. It is believed that x has a normal distribution with a mean of 107 and a standard deviation of 15. If a sample of 25 college students is selected at random, find the probability that the *sample mean* is:

(a) Greater than 110.

(b) Between 104 and 113.

(c) Below 102.

7.31 (a) Simulate 100 times the experiment of rolling a pair of dice and finding the sum of the two numbers. Get a histogram of the sum.

(b) Simulate 1000 times the experiment of rolling a pair of dice and finding the sum of the two numbers. Get a histogram of the sum.

(c) Describe and explain the differences between the histograms in parts (a) and (b).

7.32 Suppose, in the past year, that 40% of all cars sold by a dealership were small cars. Assume that the current population of car buyers has not changed in its preference for small cars.

(a) If 8 people enter the dealership to look at cars, what is the probability that exactly 5 of them will prefer small cars?

(b) If 800 people enter the dealership in a month, what are the mean and standard deviation of the number that will prefer small cars?

7.33 At Central Library, the mean number of books checked out in a day is 320 and the standard deviation is 75. If we choose 30 days at random from the year, what is the approximate probability that the average number of books checked out in these 30 days is between 335 and 350?

7.34 Medical studies indicate that the lead level in a child's body is dangerous if it is over .30 mg/mL in a blood sample. Suppose that, in New York City, the proportion of children with lead level above the danger point is .13. Take a random sample of 700 children in New York City.

(a) What is the chance that the proportion of children in the sample with lead levels above the danger point is higher than .16?

(b) What is the chance that the proportion of children in the sample with lead levels above the danger point is within .02 of the true population proportion?

7.35 At a large bank the amounts of money in personal savings accounts have a mean of $289.56 and a standard deviation of $124.

(a) Assuming that these amounts are approximately normally distributed, find the percentage of accounts that have a balance in excess of $250.

(b) If 120 of these accounts are selected randomly, what is the probability that the sample mean will be in excess of $250?

7.36 In the town of Centerville, there are 48,500 adult residents. Of those residents, 31,040 are registered to vote. If we were to take a random sample of 200 adults (with replacement), what is the approximate probability that more than 135 of them are registered to vote?

7.37 Circle exactly one number in each part.

(a) As the sample size n increases, the standard deviation of the sampling distribution:

(1) increases

(2) stays the same

(3) decreases

(4) not enough information to say for sure

(b) As the sample size n increases, the mean of the sampling distribution:

(1) increases

(2) stays the same

(3) decreases

(4) not enough information to say for sure

(c) As the sample size n increases, the sampling distribution:

(1) looks more and more like the distribution from which the samples were drawn.

(2) looks more and more like a normal distribution.

(3) becomes more and more tightly clustered about its mean.

(4) both (2) and (3) above.

(5) none of the above.

7.38 Let us define an English word as being "short" if it contains five or fewer letters; otherwise, the word is "long." Suppose 40% of all English words are short. Find the probability that 200 randomly selected words contain at most 85 short words.

7.39 The values 1, 3, 7, 9 constitute a population with $\mu = 5$, $\sigma = 3.65$.

(a) Construct a frequency distribution for the population.

(b) Determine the sampling distribution of $\bar{x}$ based on $n = 2$.

(c) Graph the sampling distribution

(d) Determine the mean and standard error of the sampling distribution based on part (b). Does this agree with what the Central Limit Theorem states?

💲 7.40   Last year a company initiated a program to compensate its employees for unused sick days, paying each employee a bonus of one-half the usual wage earned for each unused sick day. The question that naturally arises is "Did this policy motivate employees to use fewer allotted sick days?" Before last year, employees averaged 7 sick days per year, with a standard deviation of 2. Assuming these parameters did not change last year, find the approximate probability that the sample mean number of sick days used by 100 employees chosen at random was less than or equal to 6.4 last year.

👫 7.41   A student taking a sampling course is required to conduct a survey of 200 fellow students to "estimate" the proportion of students from the entire study body favoring a semester system as compared to a school year based on quarters.

(a) How might the student select a random sample of 200 students from the student body? (Use your university as an example.)

(b) Why does the student need a random sample? Why can't he/she take a group of friends and use their opinions to ascertain the proportion of the entire student body favoring a semester system?

💲 7.42   The daily shrinkage due to theft in the inventory of a men's department store possesses a probability distribution with mean equal to $320 and standard deviation equal to $80. The store reports the shrinkage as a total T for a 4-week period. Describe the probability distribution for the shrinkage for the 4-week period and justify your conclusions.

💲 7.43   Refer to Exercise 7.42. Suppose that the mean daily shrinkage μ is unknown to you and that you randomly sample $n = 50$ days to estimate its value. What is the probability that the sample mean $\bar{x}$ will deviate from μ by more than $20?

⚙ 7.44   Due to temperature variations, the expansion or contraction of a gas pipeline per 1000 feet is normally distributed with mean $\mu = 1$ in. and standard deviation $\sigma = .5$ in. If the pipeline is 50,000 ft. long, what is the probability that the pipe might expand more than 55 in.?

💲 7.45   From returns in previous years it has been found that approximately 70% of the tax returns in a given income category are incorrectly filed. A spot check of 5000 returns drawn at random shows that only $x = 2600$ have been filed incorrectly. Assuming that 70% of the returns will be incorrect this year also, find the mean and standard deviation of the random variable x and use this information to describe its variability in repeated sampling.

💲 7.46   Judging from the results obtained in Exercise 7.45, would you anticipate approximately 70% of the returns this year to be incorrectly filed? Explain.

7.47   Federal resources have been tentatively approved for funding the construction of an out-patient clinic. In order for the designers to present plans for a facility that would handle patient-load requirements while still staying within a limited budget, a study of patient demand was made. From studying a similar facility in the area, it was found that the distribution of the number of patients requiring hospitalization during a week could be approximated by a normal distribution with a mean of 125 and a standard deviation of 32.

    (a) Use the Empirical Rule to describe the distribution of x, the number of patients requesting service in a week.

    (b) If the facility was built with a 160-patient capacity, what fraction of the weeks might the clinic be unable to handle the demand?

7.48   Refer to Exercise 7.47. What size facility should be built so that the probability of the patient load exceeding the clinic capacity is .05? .01?

7.49   The distribution of the milkfat percentages for Holstein cattle in a particular state during the 1960s was approximately normal with a mean of 3.7 and a standard deviation of .3.

    (a) What percentage of the Holsteins had a milkfat percentage less than 3?

    (b) Greater than 4.5%?

7.50   Refer to Exercise 7.49.

    (a) Find the limits within which 90% of the milkfat percentages fell.

    (b) Compute the 95th percentile for the distribution of milkfat percentages.

7.51   Refer to Exercise 7.49. Suppose a random sample of $n = 25$ Holsteins is selected from the population of Holstein cattle in the state.

    (a) Describe the distribution of $\bar{x}$, the mean milkfat percentage for the sample of 25 cattle.

    (b) Compare the distribution of $\bar{x}$ in part (a) to a distribution of $\bar{x}$ from a sample of 100 Holsteins.

    (c) What is the probability that the sample mean milkfat percentage would exceed 4 in part (a)?

7.52   A manufacturer claims that 95% of the components that it is supplying for a new jet transport meet a specified rigid standard of performance. Suppose 400 of the components are put on test. Find the probability of observing 30 or more that do not meet the standard of performance, assuming the manufacturer's claim is correct.

7.53   Refer to Exercise 7.52. Suppose that the test of 400 components is performed and that 30 of the components fall below the rigid standard of performance. What might you conclude about the manufacturer's claim?

7.54   An airline has found over the past several years that 10% of the persons making reservations on a particular flight will not show up at flight time. On a given day, records indicate that the flight is fully booked at 300, with

35 more people waiting on standby. What is the probability that all 35 people on standby will have a seat available at flight time?

7.55 Data collected over a long period of time indicate that a particular birth defect occurs in one of every 1000 live births. Data collected from a medical center in a particular section of the country found 10 children with the birth defect from the total of 20,000 birth records examined. If we assume the 20,000 records examined represent a random sample of birth records, what is the probability of observing 10 or fewer children with birth defects in the sample?

7.56 Research on sales displays in grocery stores has shown that 70% of all people who pick up a particular item will purchase it. As part of a class project in a small community, a random sample of 300 shoppers is observed to see how many handle and then purchase the sale item on display at the front of the grocery store. Assuming that 70% of all shoppers who handle the item will eventually purchase it, what is the probability of observing fewer than 200 who handle and then buy the item in the sample of 300 shoppers?

7.57 If you randomly sample $n = 400$ observations from a binomial population with parameter $\pi = .3$, what is the probability that the sample proportion will differ from $\pi = .3$ by more than .02?

7.58 A supplier of a particular semiconductor claims that no more than .5% are defective. If you randomly sample and test 10,000 of these semiconductors and find .9% defective, what would you conclude about the manufacturer's claim? Explain your reasoning. (Note: In later chapters we will give a statistical procedure for making this decision.)

7.59 **Class Exercise** Each 3-digit number shown in a random number table appears with a probability near .001. That is, they possess a probability distribution as shown in Figure 7.11.

FIGURE 7.11. A Probability Distribution for Three-Digit Random Numbers Selected from a Random Number Table

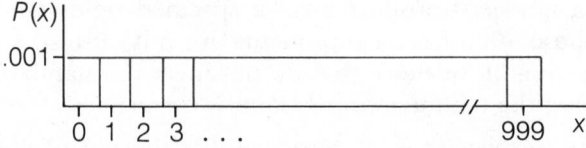

Simulate random sampling from a population possessing the probability distribution of Figure 7.11 by selecting $n = 3$-digit random numbers from Table 11 of Appendix 3 and calculate the sample range.

Range = difference between the largest and the smallest measurements in the sample

Repeat this process 200 times to obtain 200 sample ranges, and construct the relative frequency histogram. This histogram will be similar to the sampling distribution for the sample range based on samples of $n = 3$ measurements selected from a population possessing a relative frequency histogram similar to the one in Figure 7.11. You could obtain a better approximation to this sampling distribution by increasing the number of samples used to construct your histogram—say, 1000 samples of $n = 3$ measurements each rather than 200.

As noted in Chapter 5, the range is a good measure of data variation for small samples. In fact, for small samples it is easier to compute and it is almost as good an estimator of the population standard deviation σ as is the sample standard deviation s. For this reason the range is often used to measure data variation in industrial quality control, where a manufacturing process is monitored by taking frequent small samples of one or more measures of product quality over time (for example, one sample every hour). The sampling distribution that you have derived would be useful in evaluating the properties of the sample range for quality-control measurements on a random variable that possess the probability distribution shown in Figure 7.11.

7.60 If 65% of all of the married women who work do so primarily for the money, what is the probability that a random sample of 100 married women who work will show 68 or more who work primarily for the money? What proportion of the time will a sample of 100 show exactly 65 who work primarily for the money?

7.61 Today's new fathers are much more willing to help with the family chores than their counterparts in the past, according to sociologists. As an example, studies have shown that 80% of the new fathers will change diapers for their children. If this is correct, what is the probability that in a sample of 1000, there are 700 or less who will change diapers for their children?

7.62 The average daily temperature in January in Seattle, Washington, is 44.4 degrees with a standard deviation of 1.8 degrees. If a random sample of 30 January days is drawn, what is the probability that the sample mean will be greater than 44.4 degrees? Greater than 45 degrees?

7.63 The playing time of cuts on phonograph records of popular recording artists averages 180 secs with standard deviation of 48 secs. A random sample of 40 selections from various albums showed a mean time of 206 secs. Is this a likely outcome if the true mean is 180 secs? Take a random sample of 30 selections from records you or your friends own and see how they compare with the overall average of 180 secs.

7.64 A survey taken by the *New York Times* in 1984 showed that 90% of those surveyed thought that there would be better government if there were more females in Congress. If .90 is the true population proportion, what is the probability that a sample of 400 would show a proportion less than or equal to .80?

7.65 Use a computer program to simulate the sampling distribution for $\bar{x}$ based on 40 samples of size $n = 16$ drawn from a normal population with $\mu = 60$ and $\sigma = 5$. A portion of a Minitab program is shown here.

```
MTB > RANDOM 40 C1-C16;
SUBC> NORMAL MU=60 SIGMA=5.
MTB > RMEAN C1-C16 INTO C17
MTB > PRINT C17

C17
    58.9095    57.8750    58.6716    60.2102    60.0189    62.6226    59.3248
    60.7206    59.6111    60.2043    58.3992    60.2296    58.5047    58.8738
    57.8458    59.3341    59.4595    61.2888    58.6576    59.5844    60.0695
    57.9392    60.0738    60.3633    59.1295    60.9662    60.2626    60.2545
    59.6890    59.4249    62.4426    59.2217    60.4746    59.6225    59.1339
    62.3339    62.0966    57.5982    58.5845    58.7122

MTB > MEAN C17

   MEAN    =        59.718

MTB > STOP
```

PART 5

STEP THREE: METHODS FOR ANALYZING DATA

CHAPTER 8

INFERENCES ABOUT μ

8.1 Introduction ▪ **8.2** Estimation of μ ▪ **8.3** Elements of a statistical test ▪ **8.4** A statistical test about μ ▪ **8.5** The level of significance of a statistical test ▪ **8.6** Inferences about μ, with σ unknown ▪ **8.7** Assumptions underlying the analysis methods of this chapter ▪ **8.8** Using computers ▪ Summary ▪ Key terms ▪ Key formulas ▪ Supplementary exercises ▪ Exercises from the data base

8.1 INTRODUCTION

Inference, specifically decision making and prediction, is centuries old and plays a very important role in our lives. Each of us is faced daily with personal decisions and situations that require predictions concerning the future. The government is concerned with predicting the flow of gold to Europe. A stockbroker wants to know how the stock market will behave. A metallurgist would like to use the results of an experiment to determine whether a new type of steel is more resistant to temperature changes than another. A veterinarian investigates the effectiveness of a new product for treating worms in cattle. The inferences that these individuals make should be based on relevant facts, which we call observations, or data.

In many practical situations the relevant facts are abundant, seemingly inconsistent, and, in many respects, overwhelming. As a result, a careful decision or prediction is often little better than an outright guess. You need only refer to the "Market Views" section of the *Wall Street Journal* to observe the diversity of expert opinion concerning future stock market behavior. Similarly, a visual analysis of data by scientists and engineers often yields conflicting opinions regarding conclusions to be drawn from an experiment. See Figure 8.1.

Many individuals tend to feel that their own built-in inference-making equipment is quite good. But experience suggests that most people are incapable of utilizing large amounts of data, mentally weighing each bit of relevant information, and arriving at a good inference. (You may test your own inference-making ability by using the exercises in

FIGURE 8.1 *Given the Data Shown Here for the Years 1900 Through 1940, Do You Think That Sociologists, Statisticians, and Agricultural Experts Would Have Predicted the Decreases in the Farm Population Shown for 1950, 1960, 1970, and 1980?*

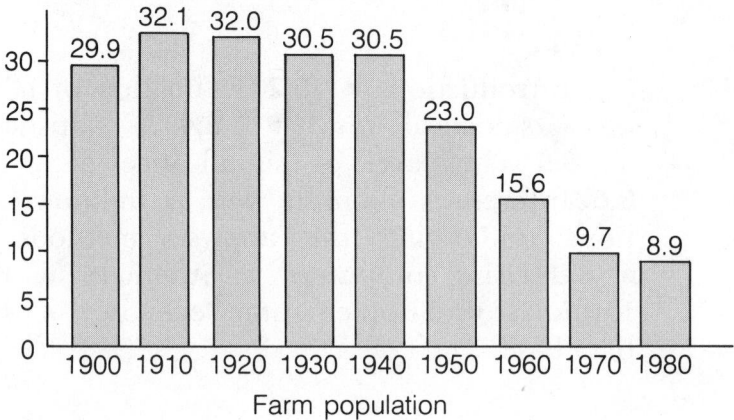

Chapters 8 through 11. Scan the data and make an inference before you use the appropriate statistical procedure. Then compare the results.) The statistician, rather than relying upon his or her own intuition, uses statistical results to aid in making inferences. Although we have touched upon some of the notions involved in statistical inference in preceding chapters, we will now collect our ideas in a presentation of some of the basic ideas involved in statistical inference.

The objective of statistics is to make inferences about a population based on information contained in a sample. Populations are characterized by numerical descriptive measures called *parameters*. Typical population parameters are the mean μ, the standard deviation σ, the area under the probability distribution to the right (or left) of some value of the random variable, or the area between two values of the variable. Most practical inferential problems you will encounter can be phrased to imply an inference about one or more parameters of a population. For example, in an experiment in which we wish to predict the average amount of money paid to welfare recipients in a given year, the population of interest is the set of all yearly welfare payments, and we are interested in estimating the value of the population mean μ.

Methods for making inferences about parameters fall into one of two categories. Either we will **estimate** (predict) the value of the population parameter of interest or we will **test a hypothesis** about the value of the parameter. These two methods of statistical inference—estimation and hypothesis testing—involve different procedures, and, more important, they answer two different questions about the parameter. In estimating a pop-

estimation
hypothesis
testing

ulation parameter we are answering the question, "What is the value of the population parameter?" In testing a hypothesis we are answering the question, "Is the parameter value equal to this specific value?"

Consider a study in which an investigator is interested in examining the effectiveness of a drug product in reducing anxiety levels of anxious patients. A screening procedure is employed to identify a group of anxious patients. After the patients are admitted into the study, each one's anxiety level is measured on a rating scale immediately before he or she receives the first dose of the drug and then at the end of one week of drug therapy. These sample data can be used to make inferences about the population from which the sample was drawn either by estimation or by a statistical test:

Estimation: Information from the sample can be used to estimate (or predict) the mean decrease in anxiety ratings for the set of all anxious patients who may conceivably be treated with the drug.

Statistical test: Information from the sample can be used to determine whether the population mean decrease in anxiety ratings is greater than zero.

Notice that the inference related to estimation is aimed at answering the question, "What is the mean decrease in anxiety ratings for the population?" In contrast, the statistical test attempts to answer the question, "Is the mean drop in anxiety ratings greater than zero?"

We will consider estimation of a population mean μ and a statistical test about μ.

EXERCISES

8.1 A researcher is interested in estimating the percentage of registered voters in her state who have voted in at least one election over the past two years.

(a) Identify the population of interest to the researcher.

(b) How might you select a sample of voters to gather this information?

8.2 Refer to Exercise 8.1. Is the researcher faced with a problem related to estimation or testing a hypothesis? What is the parameter of interest?

 8.3 A manufacturer claims that the average lifetime of a particular fuse is 1500 hours. Information from a sample of 35 fuses shows that the average lifetime is 1380 hours. What can be said about the manufacturer's claim?

(a) Identify the population of interest to us.

(b) Would an answer to the question posed involve estimation or testing a hypothesis?

8.4 Refer to Exercise 8.3. How might you select a sample of fuses from the manufacturer to test the claim?

8.2   ESTIMATION OF μ

Estimation procedures can be classified into two categories: point estimation and interval estimation. For example, we might use the sample data to specify a single number—say 15—as an estimate of the mean decrease in anxiety ratings. Alternatively, we might estimate that the mean decrease in anxiety ratings is in some interval—say from 12 to 18. See Figure 8.2.

FIGURE 8.2   Point Estimate and Interval Estimate

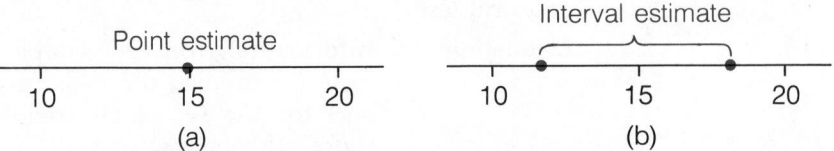

The sample mean $\bar{x}$ is a logical point estimate of the population mean μ. We can also use the point estimate $\bar{x}$ to form an interval estimate for the population mean μ. From previous work we know that when n is 30 or more, the sampling distribution for $\bar{x}$ will be approximately normal with mean μ and standard error $\sigma_{\bar{x}}$. Thus the interval ($\mu \pm 2\sigma_{\bar{x}}$), or, more precisely, ($\mu \pm 1.96\sigma_{\bar{x}}$), includes 95% of the $\bar{x}$s in repeated sampling. See Figure 8.3.

Consider the interval ($\bar{x} \pm 1.96\sigma_{\bar{x}}$). Any time $\bar{x}$ lies in the interval ($\mu \pm 1.96\sigma_{\bar{x}}$), the interval ($\mu \pm 1.96\sigma_{\bar{x}}$) will contain (or capture) the parameter μ (see Figure 8.4), and this will occur with probability .95. The interval ($\mu \pm 1.96\sigma_{\bar{x}}$) represents an interval estimate of μ.

We evaluate the goodness of an interval estimation procedure by examining the fraction of times in repeated sampling that the intervals contain the parameter being estimated. This fraction, called the **confidence coefficient,** is .95 when using the formula ($\bar{x} \pm 1.96\sigma_{\bar{x}}$). That is, 95% of the time in repeated sampling, intervals calculated by using the formula ($\bar{x} \pm 1.96\sigma_{\bar{x}}$) will contain the population mean μ.

confidence coefficient

FIGURE 8.3   Sampling Distribution for $\bar{x}$

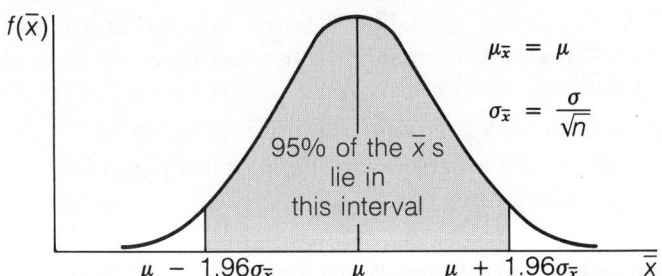

FIGURE 8.4 When the Observed Value of $\bar{x}$ Is in the Interval $(\mu \pm 1.96\sigma_{\bar{x}})$, the Interval $(\bar{x} \pm 1.96\sigma_{\bar{x}})$ Contains the Parameter μ

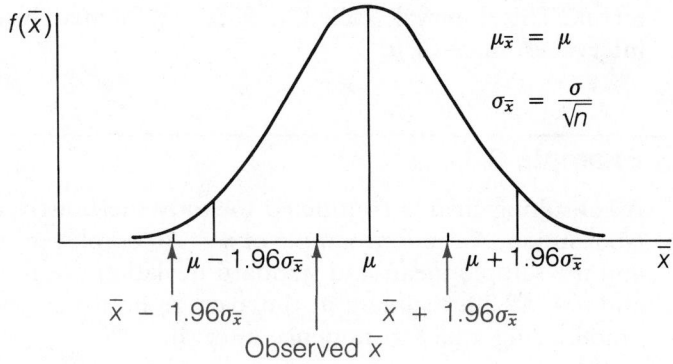

This situation is illustrated in Figure 8.5. Twenty different samples were drawn from a population with mean μ and standard deviation σ. For each sample an interval estimate was computed by using the formula $(\bar{x} \pm 1.96\sigma_{\bar{x}})$. Note that although the intervals bob about, most of them

FIGURE 8.5 Twenty Interval Estimates Computed by Using $(\bar{x} \pm 1.96\sigma_{\bar{x}})$

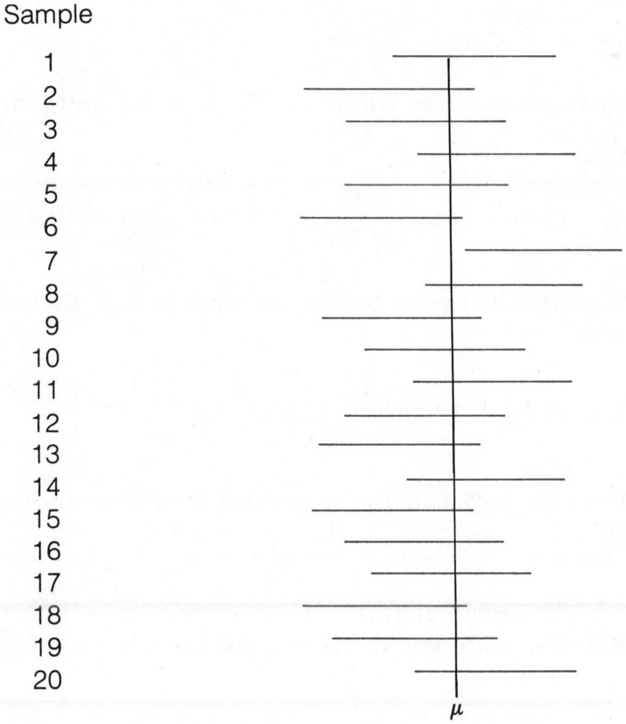

Sample

1
2
3
4
5
6
7
8
9
10
11
12
13
14
15
16
17
18
19
20

μ

intersect the vertical line and hence contain μ. In fact, if we repeated this process over and over again, 95% of the intervals formed would contain μ.

In a given experimental situation we will calculate only one such interval. This interval, called a **95% confidence interval,** represents an interval estimate of μ.

95% confidence interval

<hr/>

Example 8.1

A consulting firm is contracted to study the hourly wages in a particular labor union. A random sample of $n = 70$ employee records are examined, and the sample mean and standard deviation are found to be $\bar{x} = \$13.00$ and $s = \$2.20$. Estimate μ, the average hourly wage for the entire labor union, using a 95% confidence interval.

Solution   The appropriate 95% confidence interval is computed by using the formula $(\bar{x} \pm 1.96\sigma_{\bar{x}})$. Although σ is unknown, we can substitute the sample standard deviation s for σ in the formula for $\sigma_{\bar{x}}$ when the sample size is reasonably large; the $n > 30$ criterion mentioned in the previous chapter will suffice for mound-shaped distributions. A better way to handle this issue will be discussed in Section 8.6.

lower confidence limit

The lower point for the confidence interval, called **lower confidence limit,** is

$$\bar{x} - 1.96\sigma_{\bar{x}}$$

where $\sigma_{\bar{x}} = \sigma/\sqrt{n}$. Substituting s for σ, we find that the lower confidence limit is

$$13.00 - 1.96\left(\frac{2.20}{\sqrt{70}}\right) = 13.00 - .515 = 12.485$$

upper confidence limit

Similarly, the **upper confidence limit** is $(\bar{x} \pm 1.96\sigma_{\bar{x}})$, and in this example its value is

$$13.00 + 1.96\left(\frac{2.20}{\sqrt{70}}\right) = 13.515$$

Then the 95% confidence interval for the mean hourly wage is \$12.49 to \$13.52.

<hr/>

The general formula for a confidence interval for μ is given by $(\bar{x} \pm z\sigma_{\bar{x}})$. Different values of z are used depending on the desired degree of confidence.

Confidence Interval for μ; σ Known

$$\bar{x} \pm z\sigma_{\bar{x}}$$

where $\sigma_{\bar{x}} = \sigma/\sqrt{n}$. Note: The values of z for a 90%, a 95%, or a 99% confidence interval for μ are 1.645, 1.96, or 2.58, respectively. We assume that σ is known or that the sample size is large enough to replace σ with s.

Example 8.2

Refer to Example 8.1

(a) Construct a 90% and a 99% confidence interval for μ using the same sample data. Compare your findings to the 95% confidence interval of Example 8.1.

(b) Determine a 95% confidence interval for μ based on $n = 200$ rather than $n = 70$. Compare this interval to the one obtained in Example 8.1.

Solution

(a) The 90% confidence interval for μ is of the form $\bar{x} \pm 1.645\sigma_{\bar{x}}$; the corresponding 99% confidence interval is $\bar{x} \pm 2.58\sigma_{\bar{x}}$. By using $\bar{x} = \$13.00$ and replacing σ with s in the formula for $\sigma_{\bar{x}}$, we obtain:

$$90\% \text{ confidence interval:} \quad 13.00 \pm 1.645\left(\frac{2.20}{\sqrt{70}}\right)$$

$$\text{or} \quad 13.00 \pm 0.433$$

that is, 12.567 to 13.433.

$$99\% \text{ confidence interval:} \quad 13.00 \pm 2.58\left(\frac{2.20}{\sqrt{70}}\right)$$

$$\text{or} \quad 13.00 \pm .678$$

that is, 12.322 to 13.678.

This part of Example 8.2 illustrates the relation between the width of the confidence interval and the degree of confidence for the interval; the width of the confidence interval increases (decreases) as the degree of confidence increases (decreases).

(b) The 95% confidence interval for μ is of the form $\bar{x} \pm 1.96\sigma_{\bar{x}}$. If the sample size was $n = 200$ rather than $n = 70$, as it was for Example 8.1, the corresponding 95% confidence interval for μ would be

$$13.00 \pm 1.96\left(\frac{2.20}{\sqrt{200}}\right)$$

or $13.00 \pm .305$

that is, 12.695 to 13.305.

Note that this interval is narrower than the one in Example 8.1 for $n = 70$. This illustrates that, for a given confidence level, as the sample size increases (decreases), the width of the confidence interval decreases (increases).

EXERCISES

8.5 If a random sample of 100 observations yields $\bar{x} = 35.00$ and $s = 7.5$, construct a 90% confidence interval for μ, the population mean.

8.6 Refer to Exercise 8.5. Use the same data to compare a 95% and a 99% confidence interval for μ. Compare the 90%, 95%, and 99% confidence intervals for these data.

8.7 In a psychological study of depth perception 42 students were asked to judge the distance between two stationary objects. For each student's response the difference between the actual distance and the perceived distance (ignoring sign) was calculated. The mean and the standard deviation of the sample differences (in feet) were $\bar{x} = 6.5$ and $s = 2.6$. Use the sample data to estimate the population mean difference, using a 95% confidence interval.

8.8 Refer to Exercise 8.7. Use the same sample data to compute a 99% confidence interval for μ. Compare your result to that for Exercise 8.7.

8.9 In a study of the effects of a diuretic, 30 healthy adult males were given a single dose of the drug and were closely monitored to determine their urinary output over the next 24 hours. Construct a 95% confidence interval for μ if the mean urinary output (in milliliters) for the sample was 3300, with $s = 500$.

8.10 Refer to Exercise 8.9. Assuming s remains approximately the same, determine the width of a 95% confidence interval based on samples of size 30, 60, 90, and 120. Discuss the effect of sample size on the width of a confidence interval.

8.11 A problem of interest to the United States, other governments, and world councils concerned with the critical shortage of food throughout the world is finding a method to estimate the total amount of grain crops that will be produced throughout the world in a particular year.

One method of predicting total crop yields is based on satellite photographs of the earth's surface. Because a scanning device will read the total acreage of a particular type of grain with error, it will be necessary to have the device read many equal-sized plots of a particular planting in order to calibrate the reading on the scanner with the actual acreage. Satellite photographs of one hundred 50-acre plots of wheat were read by the scanner and gave a sample average and standard deviation

$$\bar{x} = 3.27 \quad s = .23$$

Find a 95% confidence interval for the mean scanner reading for the population of all 50-acre plots of wheat. Explain the meaning of this interval.

8.12 Another agricultural problem concerns the production of protein, an important component of human and animal diets. While it is common knowledge that grains and legumes contain high amounts of protein, it is not as well known that certain grasses provide a good source of protein. For example, Bermuda grass contains approximately 20% protein by weight. In a study to verify these results one hundred 1-pound samples were analyzed for protein content. The mean and standard deviation of the sample were

$$\bar{x} = .18 \text{ lb} \quad s = .08 \text{ lb}$$

Estimate the mean protein content per pound for the Bermuda grass from which this sample was selected. Use a 95% confidence interval. Explain the meaning of this interval.

8.13 The caffeine content (in milligrams) was examined for a random sample of 50 cups of black coffee dispensed by a new machine. The mean and standard deviation were 110 mg and 7.1 mg, respectively. Use these data to construct a 98% confidence interval for μ, the mean caffeine content for cups dispensed by the machine.

8.14 A random sample of the year-end statements of 22 small businesses (under $500,000 in sales) in a city shows the mean gross profit margin to be 5.2% (of sales) with a standard deviation of 3.3%. Use these data to place a 90% confidence interval for μ. Assume $\sigma \approx 3.3$.

8.15 Recent data from a national survey of 1350 women indicated that the average woman goes to a hair salon once every 5 weeks and spends about $26.40. With a standard deviation of $12.00, use these data to construct a 99% confidence interval for μ.

8.16 A social worker is interested in estimating the average length of time spent outside of prison for first offenders who later commit a second crime and are sent to prison again. A random sample of $n = 150$ prison records in the county courthouse indicates that the average length of prison-free life between first and second offenses is 3.2 years, with a standard deviation of 1.1 years. Use the sample information to estimate μ, the mean prison-free life between first and second offenses for all prisoners on record in the county courthouse. Construct a 95% confidence interval for μ. Assume that σ can be replaced by s.

8.17 Refer to Exercise 8.14. What impact would a doubling of the sample size have on the confidence interval?

8.3 ELEMENTS OF A STATISTICAL TEST

The second kind of inferential procedure is the statistical test, in which we attempt to answer a specific question about the value of the unknown population parameter. In the next two sections, we will discuss the elements of a statistical test and illustrate the procedure for running a test about μ when σ is known.

research
hypothesis

null hypothesis

alternative
hypothesis

A statistical test of hypothesis can be likened to a court trial. We begin with a **research hypothesis,** something that we wish to support. For example, in the court trial the research hypothesis is that the defendant is guilty. The prosecuting attorney attempts to support this hypothesis by showing that its antithesis—that the defendant is innocent—is false. This latter hypothesis, called a **null hypothesis,** is the crux of a statistical test. If we can collect evidence to show that the null hypothesis is false, we can conclude that the research hypothesis (the **alternative hypothesis**) is true.

How does the statistician, or the court, decide which hypothesis is true: the null hypothesis or the alternative hypothesis? In both cases evidence is collected; the evidence for the statistician is the information contained in a sample selected from the population. This evidence is then weighed (or considered), so that a decision can be made. In the court trial a jury functions as a decision maker, weighing the evidence to reach a decision. In a statistical test of a hypothesis we utilize a **test statistic,** some quantity computed from the sample measurements, to assist us in reaching a decision. We use this quantity in the following way: If the test statistic takes a value that is contradictory to the null hypothesis, we reject the null hypothesis and conclude that the alternative is true. Similarly, in a court trial, if the evidence presented to a jury is highly contradictory to the hypothesis of innocence (null hypothesis), the jury rejects the null hypothesis and concludes that the defendant is guilty (i.e., that the alternative hypothesis is true).

test statistic

A statistical test is made up of four parts: a null hypothesis, an alternative (research) hypothesis, a test statistic, and a rejection region. These four elements are defined as follows:

The Four Elements of a Statistical Test

1. A **null hypothesis:** a hypothesis about a population parameter. We sometimes designate a null hypothesis by the symbol H_0.

2. An **alternative (research) hypothesis:** a hypothesis we will accept if the null hypothesis is rejected. We sometimes designate an alternative hypothesis by the symbol H_a.

3. A **test statistic:** a quantity computed from the sample data.

4. A **rejection region:** a set of values for the test statistic that are contradictory to the null hypothesis and imply its rejection.

To summarize, we test a hypothesis in much the same manner as a court tries a defendant. We begin by specifying the null hypothesis and the alternative hypothesis. Then a random sample is drawn from the population of interest and the value of a test statistic is computed from the sample values. The decision to accept or reject the null hypothesis depends on the computed value of the test statistic.

How do we decide which values of the test statistic imply rejection of the null hypothesis and which do not? The answer is that we consider the set of all values that the test statistic could possibly assume. Then we divide the set into two regions: one corresponding to a **rejection region** and one to an **acceptance region.** (How this division is made will be explained subsequently.)

rejection region
acceptance
region

This situation is shown symbolically in Figure 8.6. A point on the horizontal line corresponds to a possible value of the test statistic. We symbolically divide these values with a vertical line to obtain two sets, one corresponding to rejection and the other to acceptance of the null hypothesis. If the computed value of the test statistic falls in the rejection region, the null hypothesis is rejected (and the alternative hypothesis is supported).

FIGURE 8.6   All Possible Values of a Test Statistic

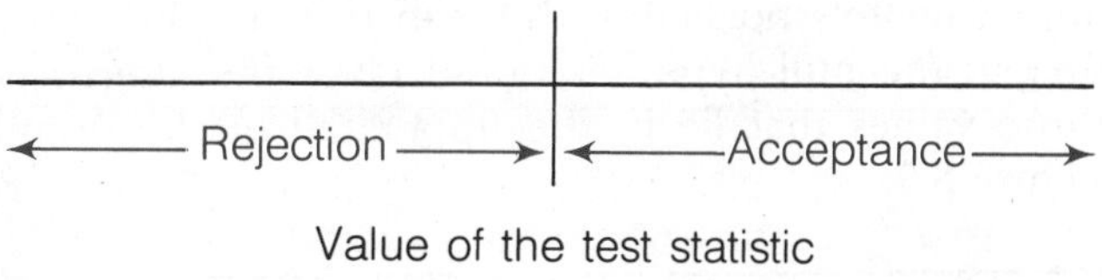

←——— Rejection ——→|←———Acceptance——→

Value of the test statistic

If the observed value of the test statistic falls in the acceptance region rather than the rejection region, we do not reject the null hypothesis. However, this does not mean that we automatically accept the null hypothesis. More will be said on this later.

8.4   A STATISTICAL TEST ABOUT μ

Consider now the problem of conducting a statistical test to answer the question: "Is the population mean μ equal to μ_0 (a specified value)?" For example, in studying the properties of a new antihypertensive drug product, a pharmaceutical company would examine the drop in diastolic blood pressure following a single fixed dose of the antihypertensive product. In particular, it may be important to show that the average drop in standing diastolic blood pressure two hours after administration of a single fixed dose is greater than 10 mm Hg, the average drop for a previously studied, weak antihypertensive product. Thus for this example the research hypothesis of interest to the pharmaceutical company is $\mu > 10$. The null hypothesis for this example could be written as $\mu \leq 10$, but the crucial value of μ is the "boundary" value $\mu = 10$. If the test statistic (computed from the sample data) contradicts the null hypothesis value $\mu = 10$ and supports the research hypothesis $\mu > 10$, the test statistic would also contradict other null hypothesis values of $\mu < 10$. It is for this reason that we take the null hypothesis to be the single value $\mu = 10$.

Once the alternative and null hypotheses have been stated, a random sample of hypertensive patients (e.g., patients having a standing diastolic blood pressure between 105 mm Hg and 130 mm Hg) would be screened and entered into the trial. The standing diastolic blood pressure would be recorded for each patient immediately prior to administration of the single fixed dose of medication and again two hours afterward. The response of interest is the drop in standing diastolic blood pressure at two hours post-medication.

The decision to accept or reject the null hypothesis in favor of the alternative hypothesis would be based on the value of a test statistic computed from the sample data. A logical test statistic for a test of hypothesis related to μ is $\bar{x}$, the sample mean drop in standing diastolic blood pressure.

If we choose $\bar{x}$ as the test statistic, and if we assume that the null hypothesis is true, we know that the sampling distribution of $\bar{x}$ will be approximately normal, with mean $\mu = 10$. Values of $\bar{x}$ that are contradictory to the null hypothesis (and favor the alternative hypothesis) will be those values that lie in the upper tail of the sampling distribution. See Figure 8.7.

FIGURE 8.7   Assuming H_0 Is True, Contradictory Values of $\bar{x}$ Are in the Upper Tail

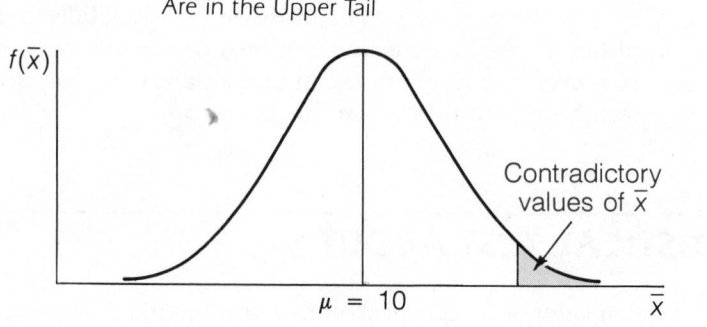

The contradictory values of $\bar{x}$ form a rejection region for our statistical test. If the observed value of $\bar{x}$ falls in the rejection region of Figure 8.7, we will reject the null hypothesis ($\mu = 10$) in favor of the alternative hypothesis ($\mu > 10$). Note that we are attempting to support the alternative hypothesis by contradicting the null hypothesis. If the observed value of $\bar{x}$ falls in the acceptance region, we are unable to reject the null hypothesis.

Before we give a precise location for the acceptance and rejection regions of Figure 8.7, we should consider the two types of errors that can be made while performing a statistical test of a hypothesis. As with any two-way decision, such as a jury decision in a court trial, we can make an error by incorrectly rejecting the null hypothesis (convicting an innocent defendant) or by incorrectly accepting the null hypothesis (acquitting a guilty defendant). These errors are called type I and type II errors, respectively (see Table 8.1).

TABLE 8.1 Decisions and Corresponding Errors

	Null hypothesis	
Decision	*False*	*True*
Reject H_0	Correct	Type I error
Accept H_0	Type II error	Correct

Definition 8.1

type I error

A **type I error** is committed if the null hypothesis is accepted when it is true. The **probability** of a type I error is denoted by the Greek letter α (**alpha**).

Definition 8.2

type II error

A **type II error** is committed if the null hypothesis is accepted when it is false. The **probability** of a type II error is denoted by the Greek letter β (**beta**).

The probability α of making a type I error is the probability of rejecting the null hypothesis when it is true. The probability β of a type II error is the probability of accepting the null hypothesis when it is false. Ideally we would like both α and β to be zero, but this is impossible. As an alternative we might try to make both α and β small. Although it is usually much more difficult to determine β than α, there is a relationship between these two probabilities. For a given sample size, α and β are inversely related; that is, as one goes up, the other goes down, as shown in Figure 8.8.

FIGURE 8.8   Relationship Between α and β

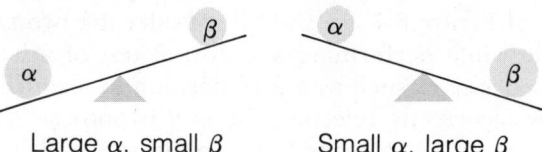

Large α, small β Small α, large β

As we will show subsequently, the experimenter must specify a tolerable value for α prior to running the statistical test. Thus he or she may choose α to be .01, .05, and so on. Specification of a value for α locates the rejection region. Determination of β is more complicated and is beyond the scope of this text. So in this text, if the observed value of the test statistic does not fall in the rejection region, we will withhold judgment by specifying that there is insufficient evidence to reject the null hypothesis, thus eliminating the possibility of committing a type II error.

Let us now see how the choice of α locates the rejection region. Returning to our investigation involving the antihypertensive drug product, we would reject the null hypothesis ($\mu = 10$) for large values of the sample mean $\bar{x}$. Suppose that we decide to take a random sample of $n = 45$ hypertensive patients and found the average drop in standing diastolic blood pressure 2 hours after administration of a single dose of the product to be $\bar{x} = 15.3$ and the standard deviation to be $s = 5.1$. Can we reject H_0: $\mu = 10$ and conclude that $\mu > 10$?

Before answering this question we must specify α. If we are willing to take the risk that one time in 40 we will incorrectly reject the null hypothesis, then $\alpha = 1/40 = .025$. An appropriate rejection can be specified for this (or any other) value of α by referring to the sampling distribution of $\bar{x}$. If the null hypothesis is true, then $\bar{x}$ is normally distributed, with mean $\mu_{\bar{x}} = \mu = 10$ and standard deviation $\sigma_{\bar{x}} \approx s/\sqrt{n} \approx 5.1/\sqrt{45} = .76$. In Figure 8.9 the shaded area corresponds to α, and so we must locate the rejection region in order for an area $\alpha = .025$ to lie in the upper tail (above $\mu = 10$). From our knowledge of the normal curve, we know that the rejection region is located at a distance of 1.96 standard errors ($1.96\sigma_{\bar{x}}$) above the mean $\mu = 10$. If the observed value of $\bar{x}$ is more than 1.96 standard errors above $\mu = 10$, we reject the null hypothesis; otherwise there is insufficient evidence to reject H_0.

FIGURE 8.9   Rejection Region for the Antihypertensive Drug
Example When $\alpha = .025$

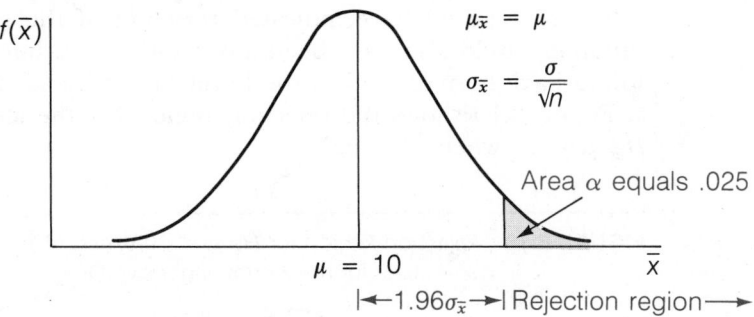

Example 8.3

Set up all parts of the statistical test for the antihypertensive drug example. Use the sample data ($\bar{x} = 15.3$, $s = 5.1$, $n = 45$) to reach a decision
when $\alpha = .025$.

Solution   The four parts of the statistical test are as follows:

Null hypothesis: $\mu = 10$

Alt. hypothesis: $\mu > 10$

Test statistic: $\bar{x}$

Rejection region: For $\alpha = .025$ reject H_0: $\mu = 10$ if $\bar{x}$ is more than
1.96 standard errors above $\mu = 10$

To determine the number of standard errors $\bar{x} = 15.3$ lies above
$\mu = 10$, we compute a z-score, using the formula

$$z = \frac{\bar{x} - \mu}{\sigma_{\bar{x}}}$$

Substituting $\bar{x} = 15.3$, $\mu = 10$, and $\sigma_{\bar{x}} = .76$, we obtain

$$z = \frac{15.3 - 10}{.76} = 6.97$$

Since the observed value of $\bar{x}$ is 6.97 standard errors above the hypothesized mean ($\mu = 10$), we reject H_0: $\mu = 10$ and conclude that the mean
drop in standing diastolic blood pressure is greater than 10 mm Hg at the
two-hour, postdrug time period.

one-tailed test

      The statistical test we conducted in Example 8.3 is called a **one-tailed test,** because the rejection region is located in only one tail of the sampling distribution for $\bar{x}$. If our alternative hypothesis had been H_a: $\mu < 10$, small values of $\bar{x}$ would have indicated rejection of the null hypothesis. This situation would also have been a one-tailed test, but the rejection region would have been located in the lower tail of the sampling distribution of $\bar{x}$. Figure 8.10 shows the rejection region for the alternative hypothesis: H_a: $\mu < 10$ when $\alpha = .025$.

FIGURE 8.10   Rejection Region for H_a: $\mu < 10$ When
$\alpha = .025$; for the Antihypertensive Drug

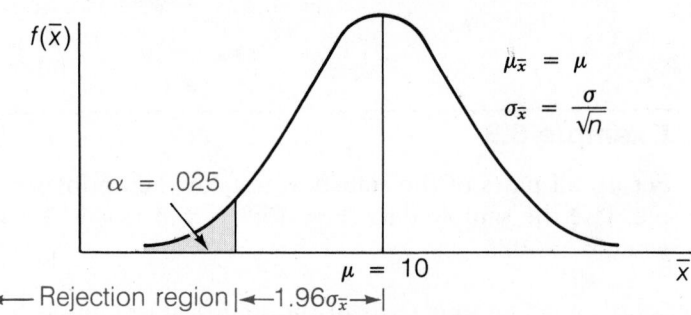

two-tailed test

      We can also formulate a **two-tailed test** when, for example, the alternative hypothesis is of the form H_a: $\mu \neq 10$. Here we would be interested in detecting whether μ is larger or smaller than $\mu = 10$, the value specified in the null hypothesis. For a two-tailed test we locate the rejection region in both tails of the sampling distribution of $\bar{x}$. The rejection region for a two-tailed test of H_a: $\mu \neq 10$ when the probability of a type I error is set at $\alpha = .05$ is shown in Figure 8.11.

FIGURE 8.11   Two-Tailed Rejection Region for H_a: $\mu \neq 10$
When $\alpha = .05$; for the Antihypertensive
Drug Example

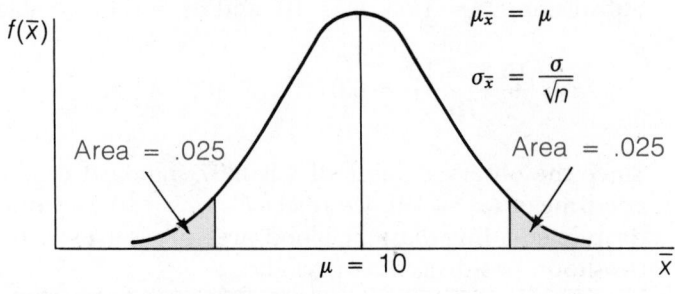

Example 8.4

A study was conducted to determine whether the average amount of money expended per household per week for food in a particular community differed from the national average ($98.00 per week). A random sample of $n = 50$ households in the community gave a mean and a standard deviation of $100.00 and $10.28, respectively. Do these data provide sufficient evidence to indicate that the mean weekly household expenditure for the community is different from the national average? Use $\alpha = .05$ and assume that σ can be replaced by s.

Solution: The four parts of the statistical test for this example are as follows:

Null hypothesis: $\mu = 98$, where μ is the average amount of money expended per household per week for food in this community

Alt. hypothesis: $\mu \neq 98$

Test statistic: $\bar{x}$

Rejection region: Reject H_0: $\mu = 98$ if $\bar{x}$ is more than 1.96 standard errors from $\mu = 98$. (See Figure 8.12)

FIGURE 8.12   Rejection Region for Example 8.4

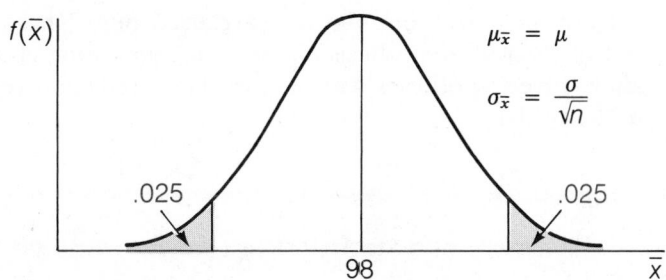

←—Rejection region|←—$1.96\sigma_{\bar{x}}$—→|←—$1.96\sigma_{\bar{x}}$—→| Rejection region—→

For our example, $\bar{x} = 100$ and $\sigma_{\bar{x}} = \sigma/\sqrt{n}$. Substituting s for σ and 50 for n, we obtain

$$\sigma_{\bar{x}} = \frac{10.29}{\sqrt{50}} = \frac{10.28}{7.07} = 1.45$$

The z-score corresponding to $\bar{x} = 100$ is

$$z = \frac{\bar{x} - \mu}{\sigma_{\bar{x}}} = \frac{100 - 98}{1.45} = 1.38$$

Since $\bar{x} = 100$ is 1.38 standard errors above $\mu = 98$, we have insufficient evidence to reject H_0 and hence are unable to conclude that the average food expenditure per week in this community differs from \$98.

Now that we know how to conduct a statistical test about μ, we can simplify the mechanics of the test by making the test statistic z (the z-score) rather than $\bar{x}$. For example, if H_0 and H_a are as shown,

Null hypothesis: $\mu = \mu_0$ (where μ_0 is some specific value)

Alt. hypothesis: $\mu > \mu_0$

Test statistic: $z = \dfrac{\bar{x} - \mu_0}{\sigma_{\bar{x}}}$

When H_a is $\mu > \mu_0$ and $\alpha = .025$, we reject the null hypothesis if the computed value of z is greater than 1.96—that is, if $\bar{x}$ is more than 1.96 standard errors above the hypothesized mean μ_0. Similarly, for $H_0: \mu = \mu_0$, $H_a: \mu \neq \mu_0$, and $\alpha = .05$, we reject the null hypothesis if the computed value of z is greater than 1.96 or less than -1.96. Stated more simply, we reject H_0 if $|z| > 1.96$.

The statistical test for a population mean can now be summarized. For $H_0: \mu = \mu_0$, three different alternative hypotheses (two one-sided alternatives and one two-sided alternative) are shown. The corresponding rejection regions are shown also for $\alpha = .05$ and $\alpha = .01$. Although theoretically the experimenter may choose any value for α, values of .05 and .01 are the most common and are used here. In conducting a statistical test of the null hypothesis $H_0: \mu = \mu_0$, you must choose one of the three alternative hypotheses with its associated rejection region and set α equal to .05 or .01.

Summary of a Statistical Test for μ; σ Known

Null hypothesis: $\mu = \mu_0$ (μ_0 is specified)

Alt. hypothesis: For a one-tailed test:

1. $\mu > \mu_0$
2. $\mu < \mu_0$

For a two-tailed test:

3. $\mu \neq \mu_0$

Test statistic: $z = \dfrac{\bar{x} - \mu_0}{\sigma_{\bar{x}}}$ where $\sigma_{\bar{x}} = \dfrac{\sigma}{\sqrt{n}}$

Rejection region: For $\alpha = .05$ (or .01) and for a one-tailed test:

1. Reject H_0 if $z > 1.645$ (or 2.33)

2. Reject H_0 if $z < -1.645$ (or -2.33)

For $\alpha = .05$ (or .01) and for a two-tailed test:

3. Reject H_0 if $|z| > 1.96$ (or 2.58)

For the time being, if σ is unknown but $n \geq 30$, you may replace σ by s and proceed with the test. A more detailed discussion of the σ unknown case is presented in Section 8.6.

EXERCISES

8.18 The mean and the standard deviation of a random sample of $n = 50$ measurements are $\bar{x} = 63.7$ and $s = 14.2$. Conduct a statistical test of H_0: $\mu = 68$ against the alternative H_a: $\mu < 68$, using $\alpha = .05$.

8.19 Refer to Exercise 8.18. Would your conclusion be different if you had selected $\alpha = .01$? Explain.

8.20 To evaluate the success of a one-year experimental program designed to increase the mathematical achievement of underprivileged high school seniors, the mathematics scores for a sample of $n = 100$ underprivileged seniors were obtained for comparison with the previous year's statewide average of 525 for underprivileged seniors. You wish to examine whether there has been an increase in the mean achievement level over last year's statewide average. Discuss whether you would use a one-tailed or a two-tailed test. Set up all parts of the statistical test for μ, using $\alpha = .05$.

8.21 Refer to Exercise 8.20. Suppose you wish to examine whether the mean achievement has changed (up or down) over the past year. Would you use a two-tailed test? Explain. Set up all parts of the statistical test for μ, using $\alpha = .01$.

8.22 To study the effectiveness of a weight-reducing agent, a clinical trial was conducted in which 35 overweight males were placed on a fixed diet. After a two-week period each male was weighed and then given a supply of the weight-reducing agent. The diet was to be maintained and, in addition, a single dose of the weight reducing agent was to be taken each day. At the end of the next two-week period, weights were again obtained. Set up all parts of the statistical test for the alternative hypothesis that μ, the average weight loss, is greater than 0. Why is a one-tailed test appropriate? Use $\alpha = .05$.

8.23 Refer to Exercise 8.22.

(a) The average weight loss for the second two-week period was $\bar{x} = 10.3$ pounds, and the standard deviation was $s = 4.6$. Perform a statistical test and draw conclusions. Use $\alpha = .05$.

(b) Based on the results for part (a), can you conclude that the weight-reducing agent is effective? Explain.

8.24 Transportation, getting people to their destination and home again, is a national problem. One aspect of this problem currently being studied by the Federal Highway Administration is how to successfully merge automobiles entering at high speed with congested interstate traffic. To study this problem, an automobile merging system was installed on the entrance to I-75 at Tampa, Florida. Through the use of a series of display lights, a driver is told whether or not he is traveling at an appropriate speed to merge successfully into the existing traffic on the highway. Prior to the installation of the system, investigators measured the stress levels of many drivers merging onto the highway during the 4 to 6 PM rush hour period. Similar testing on a random sample of 50 drivers was conducted after the merging system was installed.

For the purposes of illustration, suppose that the average stress level prior to the installation of the system was 8.2 (measured on a 10-point scale). Set up appropriate null and alternative hypotheses to test the research hypothesis that the average stress level for drivers under the merging system is less than that observed prior to the installation of the system. Is this a one- or two-tailed test?

8.25 Refer to Exercise 8.24. Suppose the sample mean and standard deviation for the 50 drivers tested using the merging system were, respectively, 7.6 and 1.8. Use these data to test the alternative hypothesis of Exercise 8.24. Use $\alpha = .05$.

8.26 Tooth decay generally develops first on those teeth that have irregular shapes (typically molars). The most susceptible surfaces on these teeth are the chewing surfaces. Usually the enamel on these surfaces contains tiny pockets that tend to hold food particles. Bacteria begin to eat the food particles to create an environment in which the tooth surface will decay.

Of particular importance in the decay rate of teeth, in addition to the natural hardness of the teeth, is the form of the food eaten by the individual. Some forms of carbohydrates are particularly detrimental to dental health. Many studies have been conducted to verify these findings, and we can imagine how the study might have been run. A random sample of 60 adults was obtained from a given locale. Each person was examined and then maintained a diet supplemented with a sugar solution at all meals. At the end of a one-year period the average number of newly decayed teeth for the group was .70, and the standard deviation was .4. Do these data present sufficient evidence to indicate that the mean number of newly decayed teeth for people whose diet includes a sugar solution is greater than .30, a rate that had been shown to apply to a person whose diet did not contain the sugar solution supplement? Why would a two-tailed test be inappropriate? Use $\alpha = .05$.

8.5 THE LEVEL OF SIGNIFICANCE OF A STATISTICAL TEST

In the previous sections we presented an introduction to hypothesis testing that is rather traditional: we defined the parts of a statistical test along with the two types of errors and their associated probabilities, α and β. In recent years many scientists and other users of statistics have objected to this decision-based approach to hypothesis testing. They argue that, rather than conducting a statistical test with a preset value of α, we should specify the alternative and null hypotheses, collect the sample data, and determine the weight of the evidence for rejecting the null hypothesis. This weight, given in terms of a probability, is called the **level of significance** (or *p*-value) of the statistical test. The following example illustrates the calculation of a level of significance.

level of significance

Example 8.5

Consider the previous experimental situation related to the testing of an antihypertensive drug product (Example 8.3). Rather than specifying a preset value for α, determine the level of significance for a test of H_0: $\mu = 10$ against H_a: $\mu > 10$ if $\bar{x} = 11.8$ and $\sigma_{\bar{x}} = 76$.

Solution   From the sample data given here, the computed z-score is

$$z = \frac{\bar{x} - \mu_0}{\sigma_{\bar{x}}} = \frac{11.8 - 10}{.76} = 2.37$$

The level of significance for this test is the probability of observing a value of z greater than 2.37. This probability can be found by referring to Table 2 in Appendix 3. The level of significance for this test is $.5 - .4911 = .0089$ (see Figure 8.13).

FIGURE 8.13   Level of Significance for Example 8.5

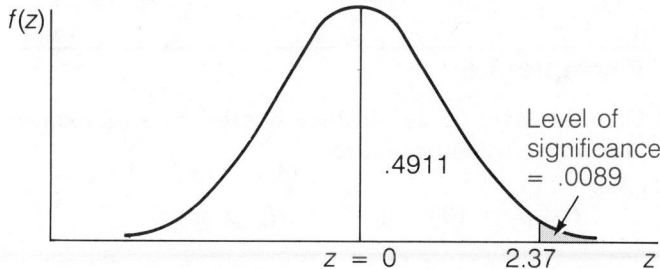

As shown in Example 8.5, the level of significance represents the probability of observing a sample outcome more contradictory to H_0 than the observed sample result if, in fact, H_0 is true. The smaller the value of this probability, the heavier is the weight of the sample evidence for rejecting H_0. For example, a statistical test with a level of significance of .01 has more evidence for the rejection of H_0 than another statistical test with a level of significance of .20.

If the null and alternative hypotheses in Example 8.5 had been

$$H_0: \mu = 10 \quad \text{and} \quad H_a: \mu < 10$$

and the computed value of z had been -2.37, the level of significance would still have been .0089 (see Figure 8.14).

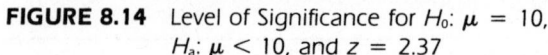

FIGURE 8.14 Level of Significance for H_0: $\mu = 10$, H_a: $\mu < 10$, and $z = 2.37$

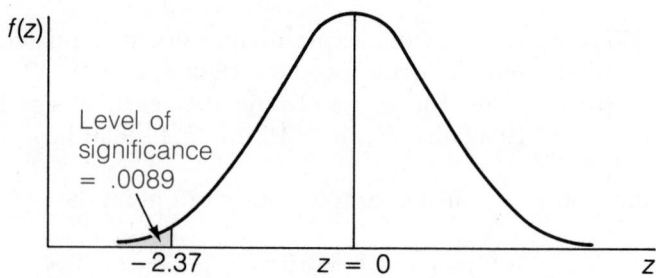

The level of significance for a two-tailed test must take into account that contradictory values are located in both tails of the distribution. For example, when testing H_0: $\mu = \mu_0$ against the alternative hypothesis H_a: $\mu \neq \mu_0$, if the computed value of z is 2.33, the level of significance is determined by finding the probability of observing a z-value greater than 2.33 or less than -2.33. This value is .0198. We illustrate such a calculation in the next example.

Example 8.6

Give the level of significance for the data of Example 8.5 if the null and alternative hypotheses are

$$H_0: \mu = 10 \quad \text{and} \quad H_a: \mu \neq 10$$

and the computed value of z is 2.37.

Solution   Values of z that are at least as contradictory to H_0 as the observed z (assuming H_0 is true) are those that are too large (i.e., $z > 2.37$) or too small (i.e., $z < -2.37$). Since the probability of observing a value of z greater than 2.37 or less than -2.37 is .0178, the level of significance for this test is .0178.

There is much to be said in favor of using the level of significance to summarize the results for any statistical test. Rather than reaching a decision directly, the statistician (or person performing the statistical test) can present the experimenter with the weight of evidence for rejecting the null hypothesis. The experimenter can then draw his or her own conclusions. Many professional journals have followed this approach by reporting the results of a statistical test in terms of its level of significance. Thus we might read that a particular test was significant at the .05 level or perhaps the .01 level.

One word of warning should be voiced. The .05 level of significance has become a magic level, and many seem to feel that a particular null hypothesis should not be rejected unless the test achieves a p-value of .05 or less. This, of course, has resulted in part from attitudes held by people who for many years have used the decision-based approach with α preset at .05. Keep this in mind when reading journal articles or reporting the results of statistical tests. After all, statistical significance at a particular level does not dictate practical significance; after determining the level of significance of a test, the experimenter should always consider the practical significance of the finding.

Throughout the text we will conduct statistical tests both from the decision-based approach and from the standpoint of reporting levels of significance, to familiarize you with both avenues of thought.

EXERCISES

8.27   Sample data for a statistical test of H_0: $\mu = 40$ yielded a z-score of 1.86.

(a) Determine the level of significance for a test of H_a: $\mu > 40$.

(b) Determine the level of significance for a test of H_a: $\mu \neq 40$.

8.28   A random sample of 36 cigarettes of a certain brand was tested for nicotine content. The sample mean and standard deviation (in milligrams) are, respectively, 15.1 and 3.8. Give the level of significance of the statistical test of H_0: $\mu = 14$ (the claimed nicotine content) against the alternative hypothesis H_a: $\mu > 14$.

8.29   A psychological experiment was conducted to investigate the length of time (time delay) between the administration of a stimulus and the observation of a specified reaction. A random sample of 36 persons was subjected to the stimulus and observed for the time delay. The sample mean and standard deviation were 2.2 and .57 seconds, respectively. Test the null hypothesis that the mean time delay for the hypothetical set of all persons who may be subjected to the stimulus is $\mu = 1.6$ against the alternative hypothesis that the mean time delay differs from 1.6. Use $\alpha = .05$.

INFERENCES ABOUT μ, WITH σ UNKNOWN

The estimation and test procedures about μ presented earlier in this chapter were based on the assumption that the population standard deviation was known or that we had enough observations so that s could be substituted for σ. In this section we will present a statistical test and confidence interval for μ that can be applied when σ is unknown no matter what the sample size, provided the sample measurements are drawn from a population that is symmetrical. More discussion on this requirement will be presented later. Here is an example of this test procedure.

Example 8.7

An experiment was performed to determine if the use of pictures would facilitate or impede a child's ability to learn the meanings of words. A random sample of 10 kindergarten children was assigned to a class that used pictures to assist in learning. Each child was tested after the experimental period and the number of words correctly identified from a total of 20 words chosen for the test was recorded. This test was repeated for five successive days, and the score assigned to each student was the average of the five separate tests. These data are shown below.

12.0   13.6   15.2   14.4   17.8   8.2   9.6   16.0   12.2   18.8

We want to compare the achievement of this experimental group of children with the performance of others who have not used pictures to aid (or impede) their ability to learn the meaning of words. Fortunately, sample test data have been collected over a long period of time for a very large group of children who have not employed pictures in their learning process. The mean for their tests was found to be 17.1.

Suppose we regard 17.1 to be the true mean achievement for children who have learned without the use of pictures. Do these data provide sufficient evidence to indicate that the use of pictures either improves or impedes word learning? That is, do the data provide sufficient evidence to show that the mean μ for the test population differs from 17.1? Note that we cannot use the z-test of Section 8.4 because the population standard deviation σ is unknown and the sample size is too small (less than 30) to substitute s for σ. Consequently, we will defer the solution of this example until we explain how to test hypotheses about a population mean when σ is unknown.

Problems similar to Example 8.7 were encountered by experimenters early in the twentieth century, when the only available test statistic was

$$z = \frac{\bar{x} - \mu}{\sigma/\sqrt{n}}$$

What statistical test did they use? Faced with unknown σ and small samples, early experimenters computed the sample standard deviation s to approximate σ and substituted it into the formula for z. Although aware that their procedure might be invalid, they took the only course open to them—they used the z-test of Section 8.4.

However, many of these experimenters were quite concerned that their statistical tests, based on z, were leading them to incorrect decisions. One of these experimenters, W. S. Gosset, translated his concern into action. Gosset was a chemist for the Guinness Breweries, and, as you might suspect, he was only provided with small samples for use in tests of quality.

Gosset was faced with the problem of finding the sampling distribution for

$$t = \frac{\bar{x} - \mu}{s/\sqrt{n}}$$

a statistic similar to z but with s substituted for σ. Gosset found that this test statistic, which he called t, possessed a probability distribution similar in appearance to the normal distribution but with a much wider spread. In fact, the smaller the sample size n, the greater was the spread in the probability distribution for t.

A z-distribution (normal distribution with $\mu = 0$ and $\sigma = 1$) and a t-distribution based on a sample of six measurements are shown in Figure 8.15. Note that t is more "spread out" than z.

Gosset published his work on the t-distribution in 1908; because of company policy, he used the pen name Student. His publication included the exact form of the distribution as well as the tail-end values of t, which are helpful in locating rejection regions. Gosset's statistic has many other applications in statistical decision making and has achieved a position of major importance in the field of statistics. His unique choice of a pen

Student's t   name has caused the statistic to be called **Student's t**.

FIGURE 8.15 *A z-Distribution (Standard Normal Distribution) and a t-Distribution Based on* $n = 6$ *Measurements*

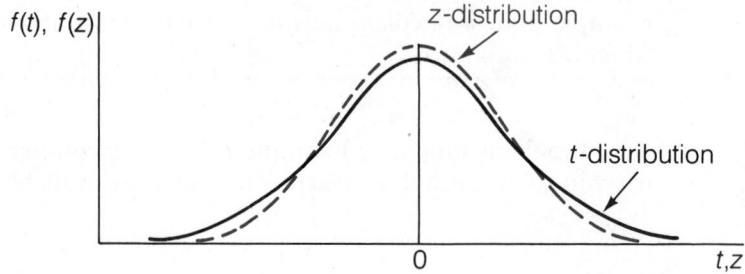

The *t*-distribution possesses the following characteristics.

Characteristics of the t-Distribution

1. Like *z*, it is mound shaped and symmetrical about 0 (see Figure 8.15).

2. It is more variable than *z* since both $\bar{x}$ and *s* change for each sample that is drawn from a population,

3. There are many *t*-distributions. We determine a particular one by specifying a quantity called the *degrees of freedom*, which is directly related to the sample size.

4. As the degrees of freedom (or, equivalently, the sample size *n*) become large, the *t*-distribution becomes almost a standard normal (*z*) distribution. This change is reasonable because *s* provides a better estimate of σ as *n* increases.

degrees of freedom

The phrase "degrees of freedom" sounds a bit mysterious. The formal definition requires advanced mathematics beyond the scope of this text. On a more informal, nontechnical basis, **degrees of freedom** represent pieces of information in the sample that are used for estimating the population standard deviation σ. The sample standard deviation *s*, which we use to estimate σ, is based on the deviations $x - \bar{x}$. Because $\Sigma(x - \bar{x}) = 0$ always, we can determine $n - 1$ of the deviations, but the last one (*n*th one) is fixed to make $\Sigma(x - \bar{x}) = 0$. So, in a sample of *n* measurements, there are $n - 1$ pieces of information (degrees of freedom) for estimating σ.

The values of t used to locate the rejection region for a statistical test are presented in Table 3 of Appendix 3. Since the t-distribution is symmetrical about $t = 0$, we give only right-tail values. A value in the left tail is simply the negative of the corresponding right-tail value. An entry in the table specifies a value of t, say t_a, such that an area a lies to its right (see Figure 8.16). For example, $t_{.05}$ is the value of t such that an area equal to .05 lies to its right.

FIGURE 8.16   Use of the t-table

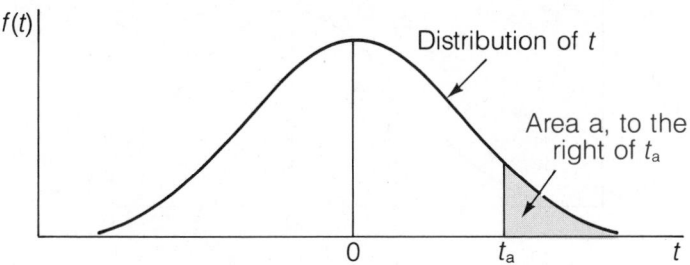

The format of Table 3 of Appendix 3 is shown in Table 8.2. Columns corresponding to various values of a are shown at the top of the table for a = .100, .05, .025, .010, and .005. The symbol df, denoting degrees of freedom, is shown in the first column of the table.

TABLE 8.2   Format of the t-table, Table 3 in Appendix 3

df	a = .100	a = .05	a = .025	a = .010	a = .005
1	3.078	6.314	12.706	31.821	63.657
2	1.886	2.920	4.303	6.965	9.925
.	.	.	.	.	.
.	.	.	.	.	.
.	.	.	.	.	.
9	1.383	1.833	2.262	2.821	3.250

For the test statistic

$$t = \frac{\bar{x} - \mu}{s/\sqrt{n}}$$

the degrees of freedom will always be one less than the sample size, that is, df = $(n - 1)$.

The numbers recorded in the table give the values of t_a. For example, suppose that we have a sample of $n = 10$ measurements and wish to use a one-tailed t-test, rejecting in the upper tail of the t-distribution with a probability of a type I error of $\alpha = .05$. Then we would want to find the value of t_a corresponding to an area a $= .05$. (See Figure 8.17.)

FIGURE 8.17   Rejection Region for a One-Tailed t-Test for
$n = 10$ and $a = .05$

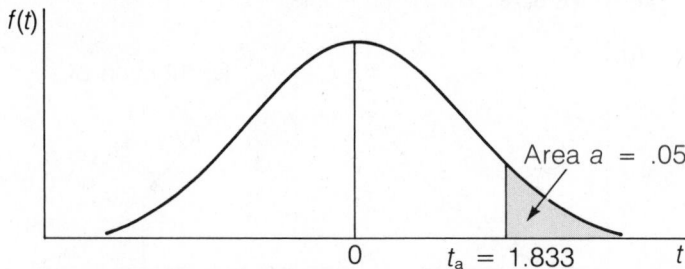

To find the value of t corresponding to a $= .05$ and $n = 10$, look at Table 3 in Appendix 3. Select the column corresponding to a $= .05$ and proceed down the column to df $= (n - 1) = 9$. The value of t_a is 1.833. That value is shown in Figure 8.17.

You will recall that we deferred the solution of Example 8.7 until we had a statistical test about a population mean when σ is unknown. Now let us return to testing the hypothesis that the mean number of words correctly identified in a posttest for children learning to identify words using illustrative pictures differs from 17.1, the average for children learning without such pictures. Use $\alpha = .05$.

Solution to Example 8.7   The appropriate null and alternative hypotheses are

H_0: $\mu = 17.1$ (the population mean for the plan with pictures is 17.1; i.e., it is the same as the population mean for the plan with no pictures)

H_a: $\mu \neq 17.1$

The test statistic is

$$t = \frac{\bar{x} - \mu_0}{s/\sqrt{n}}$$

where $\bar{x}$ and s are the sample mean and standard deviation, respectively, and μ_0 is the hypothesized value of the population mean (in this case, 17.1).

Our next step is to calculate $\bar{x}$ and s for the data.

$$\Sigma x = 12.0 + 13.6 + \cdots + 18.8 = 137.8$$
$$\Sigma x^2 = (12.0)^2 + (13.6)^2 + \cdots + (18.8)^2 = 2001.88$$

Hence

$$\bar{x} = \frac{\Sigma x}{n} = \frac{137.8}{10} = 13.78$$

$$s^2 = \frac{\Sigma x^2 - (\Sigma x)^2/n}{n-1} = \frac{2001.88 - (137.8)^2/10}{9} = 11.444$$

and

$$s = \sqrt{11.444} = 3.383$$

The test statistic is

$$t = \frac{\bar{x} - \mu_0}{s/\sqrt{n}}$$

where μ_0 is the hypothesized value of μ, that is, $\mu_0 = 17.1$. Substituting, we obtain

$$t = \frac{13.78 - 17.1}{3.383/\sqrt{10}} = -3.10$$

To locate the rejection region for our test, we must specify the degrees of freedom. The sample size is $n = 10$ and hence df $= 9$. Then, for a two-tailed test with $\alpha = .05$, we can locate the rejection region from Table 3 in Appendix 3 for df $= 9$ and a $= .025$. The table value of t is 2.62 (see Figure 8.18). Since the calculated value of t does fall in the re-

FIGURE 8.18   Rejection Region for Testing the Hypothesis $\mu = 17.1$ Against the Alternative Hypothesis $\mu \neq 17.1$ for df $= 9$ and a $= .025$

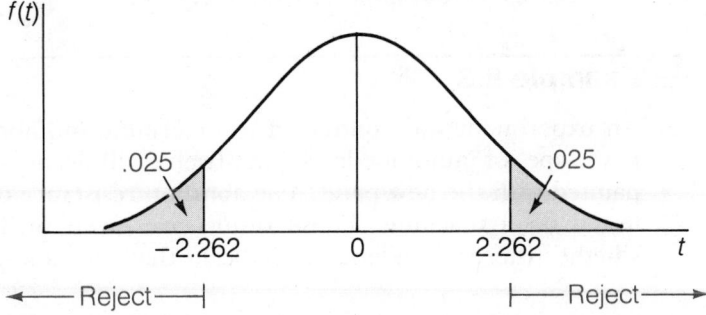

jection region, we reject H_0 and conclude that the mean number of correctly identified words for the hypothetical population of children under the experimental program is different from the mean for all children not involved in the new program. Because the computed value of t falls in the lower tail of the t-distribution, it appears that the average is less than 17.1. Indeed, the use of pictures appears to be interfering with a child's ability to learn words.

We summarize the parts of the t-test next. The only difference between the z-test given in Section 8.4 and the test given here is that t replaces z; this test (rather than the z-test) should be used when σ is unknown.

Statistical Test About μ, with σ Unknown

Null hypothesis: $\mu = \mu_0$ (μ_0 is specified)

Alt. hypothesis: For a one-tailed test:

1. $\mu > \mu_0$
2. $\mu < \mu_0$

For a two-tailed test:

3. $\mu \neq \mu_0$

Test statistic: $t = \dfrac{\bar{x} - \mu_0}{s/\sqrt{n}}$

Rejection region: For a specified value of α, df $= (n - 1)$, and for a one-tailed test:

1. Reject H_0 if $t > t_a$, where a $= \alpha$
2. Reject H_0 if $t < -t_a$, where a $= \alpha$

For a specified value of α, df $= n - 1$, and for a two-tailed test:

3. Reject H_0 if $|t| > t_a$, where a $= \alpha/2$

Let us look at another example.

Example 8.8

An experiment was conducted to determine the abrasion resistance of a new type of automobile paint. Twelve different strips of metal were painted with the new paint. The abrasion resistance of each strip was then tested on a machine. These results are given in Table 8.3. Determine whether there is evidence to indicate that the new paint has a mean re-

TABLE 8.3 *Abrasion Resistance of Paint*

Strip	Abrasion resistance	Strip	Abrasion resistance
1	2.1	7	3.2
2	4.3	8	4.0
3	4.2	9	2.9
4	3.3	10	2.5
5	3.7	11	2.3
6	2.8	12	3.1

sistance to abrasion greater than 2.9, the mean for the paint that is now being used. Use $\alpha = .05$.

Solution   The four parts of the *t*-test are as follows:

Null hypothesis: $\mu = 2.9$, where μ is the mean of the hypothetical population of abrasion resistances for the new paint.

Alt. hypothesis: $\mu > 2.9$

Test statistic: $$t = \frac{\bar{x} - 2.9}{s/\sqrt{n}}$$

Rejection region: For a one-tailed test, with $\alpha = .05$ and $n = 12$, we look up the *t*-value for a $= .05$ and df $= 11$. From Table 3 of Appendix 3 this value is 1.796. So we will reject H_0 if $t > 1.796$.

Before computing t, we must first obtain $\bar{x}$ and s for the sample data. We find that

$$\Sigma x = 2.1 + 4.3 + \cdots + 3.1 = 38.40$$
$$\Sigma x^2 = (2.1)^2 + (4.3)^2 + \cdots + (3.1)^2 = 128.76$$

Hence

$$\bar{x} = \frac{\Sigma x}{n} = \frac{38.4}{12} = 3.2$$

$$s^2 = \frac{\Sigma x^2 - (\Sigma x)^2/n}{n - 1} = \frac{128.76 - (38.4)^2/12}{11}$$

$$= \frac{128.76 - 122.88}{11} = .535$$

and

$$s = \sqrt{.535} = .731$$

Substituting into the test statistic, we have

$$t = \frac{3.2 - 2.9}{.731/\sqrt{12}} = \frac{(.3)(3.464)}{.731} = 1.42$$

Since the computed value of t does not fall in the rejection region, we have insufficient evidence to indicate that the new paint is more resistant to abrasion.

What assumptions are required when using a t-test? The exact probability distribution of the quantity

$$\frac{\bar{x} - \mu}{s/\sqrt{n}}$$

depends on the form of the distribution from which the sample observations x were obtained. Only if these observations come from a normal distribution with mean μ and standard deviation σ will $(\bar{x} - \mu)/(s/\sqrt{n})$ possess a t-distribution based on $(n - 1)$ degrees of freedom. Gosset's results, and hence the tail values in Table 3 of Appendix 3, are based on this assumption (that the original sample was selected from a population that possesses a normal probability distribution).

In addition to being able to run a statistical test about μ when σ is unknown, we can construct a confidence interval using t. The confidence interval for μ when σ is unknown is identical to the "σ known" confidence interval, with z replaced by t and σ replaced by s.

The only restriction that we have in using this confidence interval is that again we must assume the sample was drawn from a population that is approximately normal.

General Confidence Interval for μ: σ Unknown

$$\bar{x} \pm \frac{ts}{\sqrt{n}}$$

The value of t corresponding to a 90%, a 95%, or a 99% confidence interval is found in Table 3 of Appendix 3, for df $= (n - 1)$ and a $= .05, .025,$ or $.005,$ respectively.

We illustrate the use of the confidence interval with an example.

Example 8.9

A pharmaceutical firm has been conducting highly restricted studies on small groups of people to determine the effectiveness of a measles vaccine. The following measurements are readings on the antibody strength of five individuals injected with the vaccine:

$$1.2 \quad 3.0 \quad 2.5 \quad 2.4 \quad 1.9$$

Use the sample data to estimate μ, the population mean antibody strength for individuals vaccinated with the new drug. Use a 95% confidence interval.

Solution   Assuming the data were drawn from a normal population, we can proceed with an analysis. For this example we have

$$\Sigma x = 11.00 \quad \text{and} \quad \Sigma x^2 = 26.06$$

Thus

$$\bar{x} = \frac{11}{5} = 2.2$$

$$s^2 = \frac{\Sigma x^2 - (\Sigma x)^2/n}{n - 1} = \frac{26.06 - (11)^2/5}{4}$$

$$= \frac{26.06 - 24.20}{4} = .465$$

$$s = \sqrt{.465} = .682$$

The general form of the confidence interval is

$$\bar{x} \pm \frac{ts}{\sqrt{n}}$$

From Table 3 in Appendix 3, for a = .025 and df = $(n - 1)$ = 4, we find that t = 2.776. Substituting into the formula, we obtain a lower and an upper confidence limit of

$$2.2 - \frac{2.776(.682)}{\sqrt{5}} = 2.2 - .847 = 1.353$$

$$2.2 + \frac{2.776(.682)}{\sqrt{5}} = 2.2 + .847 = 3.047$$

The 95% confidence interval is then 1.353 to 3.047. We note that this interval either does or does not enclose the true mean μ. However, in repeated sampling, 95% of the intervals calculated in this way will enclose μ. Hence we are 95% confident that the true mean antibody strength for the measles vaccine is in the interval 1.353 to 3.047.

EXERCISES

8.30 Why is the z-test of Section 8.4 inappropriate for testing H_0: $\mu = \mu_0$ when $n < 30$?

8.31 Set up the rejection region based on t for H_0: $\mu = \mu_0$ when $\alpha = .05$ and for the following conditions:

(a) H_a: $\mu < \mu_0$, $n = 15$

(b) H_a: $\mu \neq \mu_0$, $n = 23$

(c) H_a: $\mu > \mu_0$, $n = 6$

8.32 Repeat Exercise 8.31 with $\alpha = .01$.

8.33 The sample data for a t-test of H_0: $\mu = 15$ and H_a: $\mu > 15$ are $\bar{x} = 16.2$, $s = 3.1$, and $n = 18$. Use $\alpha = .05$ to draw your conclusions.

8.34 The voter turnout in upper-middle-class areas of a city has averaged 650 for every 1000 registered voters in past years. This year a random sample of five precincts shows a turnout of 635, 655, 640, 643, and 620 per 1000 registered voters. Do these data indicate an overall average of less than 650? Use $\alpha = .05$.

8.35 The sample mean and standard deviation from a random sample of $n = 5$ measurements were $\bar{x} = 38.6$ and $s = 5.7$. Construct a 95% confidence interval for μ.

8.36 Organic chemists often purify organic compounds by a method known as fractional crystallization. An experimenter desired to prepare and purify 4.85 grams of aniline. Ten 4.85-gram quantities of aniline were individually prepared and purified to acetanilide. The following dry yields were recorded:

$$3.85 \quad 3.90 \quad 3.72 \quad 3.85 \quad 4.01$$
$$3.88 \quad 3.62 \quad 3.80 \quad 3.36 \quad 3.83$$

Estimate the mean number of grams of acetanilide that could be recovered from an initial amount of 4.85 grams of aniline. Use a 95% confidence interval. (You can verify that the sample standard deviation is .181.)

8.37 A random sample of ten students in a fourth-grade reading class were thoroughly tested to determine reading speed and reading comprehension. Based on a fixed-length standardized test-reading passage, the following speeds (in minutes) and comprehension scores (based on a 100-point scale) were obtained.

Student	1	2	3	4	5	6	7	8	9	10
Reading speed	5	7	15	12	8	7	10	11	13	9
Reading comprehension	60	76	96	100	81	75	85	88	98	83

(a) Use the reading speed data to place a 95% confidence interval on μ, the average speed for all fourth-grade students in the large school from which the sample was drawn.

(b) Interpret the interval estimate in part (a).

(c) How would your inference change by using a 98% confidence interval?

8.38 Refer to Exercise 8.37. Using the reading comprehension data, test the research hypothesis that the mean for all fourth graders on the standardized examination is greater than 80, the statewide average for comparable students the previous year. Give the level of significance for your test. Interpret your findings.

8.39 Refer to Exercise 8.38.

(a) Set up all parts for a statistical test of the research hypothesis that the mean score for all fourth graders is different from 80, which was the statewide average the previous year.

(b) Give the level of significance for this test.

8.40 Refer to Exercise 8.37. Use a computer program to construct a 90% confidence interval for the mean total reading score (speed plus comprehension). The following minitab program is shown for illustrative purposes:

```
MTB > READ C1 C2
DATA>   5    60
DATA>   7    76
DATA>  15    96
DATA>  12   100
DATA>   8    81
DATA>   7    75
DATA>  10    85
DATA>  11    88
DATA>  13    98
DATA>   9    83
DATA> END
      10 ROWS READ

MTB > LET C3=C1+C2
MTB > TINTERVAL 90 PERCENT C3

              N     MEAN   STDEV   SE MEAN   90.0 PERCENT C.I.
C3           10    93.90   15.14     4.79   ( 85.12,  102.68)
```

```
MTB > PRINT C1-C3

  ROW     C1      C2      C3

    1      5      60      65
    2      7      76      83
    3     15      96     111
    4     12     100     112
    5      8      81      89
    6      7      75      82
    7     10      85      95
    8     11      88      99
    9     13      98     111
   10      9      83      92

MTB > STOP
```

8.41 The amount of sewage and industrial pollutants dumped into a body of water affects the health of the water by reducing the amount of dissolved oxygen available for aquatic life. Suppose that weekly readings are taken from the same location in a river over a two-month period. Use the summary data from the computer printout given below to conduct a statistical test of the research hypothesis that the mean dissolved oxygen content is less than 5.0 parts per million, a level some scientists think is marginal for supplying enough dissolved oxygen for fish.

5.100000000	5.300000000	
4.900000000	5.200000000	
5.600000000	8.000000000	sample size
4.200000000	4.950000000	$\bar{x}$
4.800000000	.2028571428	s^2
4.500000000	.4503966505	s

8.7 ASSUMPTIONS UNDERLYING THE ANALYSIS METHODS OF THIS CHAPTER

Now that we have looked at the first methods for analyzing data (dealing in particular with methods and inferences concerning a population mean μ), what assumptions must be made in order to use these methods and draw these inferences?

Any statistical method for analyzing data involves assumptions. Some assumptions are general and apply to many different methods for analyzing data; others are specific to a particular method. What assumptions apply to the methods of this chapter?

First, the methods of this chapter apply only to data obtained from a random sample. For example, the plus-or-minus factor that is part of a confidence interval is an allowance for random error, not an allowance for biases inherent in the method of data collection. So we see how important the first step is—collecting the data—to the inferences (sense) that can be derived from the sample data. If the data have been collected in a haphazard way, the confidence interval (or test) computed from these data will likely be in error, simply because of the biases in the data collection. There are no known ways to compensate for the biases inherent in badly chosen samples.

If the data collection stage has been handled well and you have indeed obtained a random sample, there are two assumptions that must be considered for the analysis methods of this chapter—*independence* of the measurements within a sample and *normality* of the underlying population from which the sample measurements were drawn.

All the methods described in this chapter assume the measurements in a sample are independent of each other. Not all random-sampling methods yield independent observations. For example, suppose that a real-estate assessor chooses 22 city blocks of homes to evaluate from the tax lists of a city, and then assesses the market value of all homes in each block. Assuming that the assessor does, in fact, choose the blocks randomly, there is no systematic bias in favor of low-value homes or high-value homes. But there is a dependence problem. Given the well-established tendency of high-value homes to cluster together (and also of low-value homes to occur in bunches), if one home in the sample has higher than average values, so do adjacent homes. The assessment may involve, say, 300 homes, but the method does not give 300 separate, independent measurements of home values. In fact, the data arising from the assessor's evaluations would be more appropriately evaluated by the cluster-sampling methods described in Chapter 3.

The most common problem with the assumption of independence occurs in time-series data, data collected in a well-defined chronological order. Suppose, for example, that we measure the dollar volume of back orders for a particular manufacturer on twenty consecutive Friday afternoons. It is reasonable to suppose that a high back-order volume on one Friday is likely to be followed by high back-order volumes on succeeding Fridays, and the same for low volumes. The standard error formulas that we use in confidence intervals depend heavily on the assumption of independence of observations. When there is dependence, the standard error formulas may underestimate the actual uncertainty in an estimate. Even for most dependencies, the degree of underestimation may be serious.

Beyond the assumption of independence, methods for analyzing data concerning a population mean μ involve an assumption that the underlying population is normally distributed. In practice no population is exactly normal.

How restrictive is this assumption? In practice, provided the population of measurements is symmetric, the results based on a t-distribution will hold. This property of the t-distribution and the common occurrence of mound-shaped distributions in practice make the Student's t an invaluable tool for use in statistical inference.

How does one know what to do in practice? The answer is quite simple—look at the data. If a plot of the data (such as a histogram) suggests any gross skewness, there is a good chance that the population is skewed, and hence the *t*-method should not be used. In situations such as this, you should consult a professional statistician for an alternative method.

8.8 USING COMPUTERS

A Minitab program for computing a 90% and a 95% confidence interval for μ is shown here for the data of Table 8.3. (Note that we obtain a wider confidence interval when a higher degree of confidence is required.) These same data were also used to illustrate how one could run a *t*-test about μ. The statement "TEST OF MU = 2.90 VS MU G.T. 2.90" indicates that the alternative hypothesis is H_a: $\mu > 2.9$ (as it was in Example 7.8). Simple modification of the code in the program allows one to obtain any desired confidence interval and to conduct a statistical test about μ using the desired one- or two-tailed test (see Ryan, Joyner, and Ryan, *Minitab Handbook*, 1985. Additional examples of Minitab output are shown in Exercises 8.69 and 8.75.

```
MTB > SET INTO C1
DATA> 2.1  4.3  4.2  3.3  3.7  2.8
DATA> 3.2  4.0  2.9  2.5  2.3  3.1
DATA> END

MTB > PRINT C1

C1
   2.1    4.3    4.2    3.3    3.7    2.8    3.2    4.0    2.9    2.5    2.3
   3.1

MTB > TINTERVAL 90 PERCENT C1

               N      MEAN    STDEV   SE MEAN    90.0 PERCENT C.I.
C1            12     3.200    0.731    0.211    (  2.821,   3.579)

MTB > TINTERVAL 95 PERCENT C1

               N      MEAN    STDEV   SE MEAN    95.0 PERCENT C.I.
C1            12     3.200    0.731    0.211    (  2.735,   3.665)

MTB > TTEST OF MU = 2.90, ALTERNATIVE = +1 C1

TEST OF MU = 2.900 VS MU G.T. 2.900

               N      MEAN    STDEV   SE MEAN        T    P VALUE
C1            12     3.200    0.731    0.211     1.42      0.091

MTB > STOP
```

SUMMARY

In this chapter we introduced you to important concepts about making references (in the form of estimates or decisions) based on sample data. First we considered how to construct a confidence interval for μ when σ is known. The corresponding statistical test for μ when σ is known is composed of four parts: null hypothesis, alternative (or research) hypothesis, test statistic, and rejection region. It employs the technique of proof by contradiction.

To support the research hypothesis we gather information to contradict the null hypothesis H_0: $\mu = \mu_0$. As with any two-decision problem, there are two types of errors that can be committed, the rejection of H_0 when H_0 is true—a type I error—and the acceptance of H_0 when H_0 is false and some alternative is true—a type II error. The probabilities for these errors, designated by α and β, respectively, measure the goodness of the test procedure.

We also considered an alternative to the traditional decision-based approach for a statistical test for a hypothesis. Rather than relying on a preset level of α, we compute the weight of evidence for rejecting the null hypothesis. This weight, expressed in terms of a probability, is called the level of significance for the test. Most professional journals summarize the results of a statistical test using the level of significance.

The final topic we discussed concerned inferences about μ when σ is unknown (which is almost always the case). Through the use of the t-distribution we can construct both confidence intervals and a statistical test for μ. Since the t-values of the t-distribution approach the z-values of a normal distribution as the sample size n increases and since σ is almost never known, it is convenient to use t-results for all inferences about μ (large or small sample).

KEY TERMS

estimation
hypothesis testing
confidence coefficient
95% confidence interval
lower confidence limit
upper confidence limit
research hypothesis
alternative hypothesis
null hypothesis
test statistic

rejection region
acceptance region
type I error (*def*)
type II error (*def*)
two-tailed test
one-tailed test
level of significance
Student's *t*
degrees of freedom

KEY FORMULAS

1. General confidence interval for μ; σ known

$$\bar{x} \pm \frac{z\sigma}{\sqrt{n}}$$

2. General confidence interval for μ; σ unknown

$$\bar{x} \pm \frac{ts}{\sqrt{n}}$$

3. Statistical test for μ; σ known

$$H_0: \mu = \mu_0$$

Test statistic $z = \dfrac{\bar{x} - \mu_0}{\sigma_{\bar{x}}}$ where $\sigma_{\bar{x}} = \dfrac{\sigma}{\sqrt{n}}$

4. Statistical test for μ; σ unknown

$$H_0: \mu = \mu_0$$

Test statistic $t = \dfrac{\bar{x} - \mu_0}{s/\sqrt{n}}$

SUPPLEMENTARY EXERCISES

8.42 What are parameters and how do they differ from statistics?

8.43 Is there a need for estimation procedures in situations in which the sample constitutes the population of interest? For example, would there be a need to determine a confidence interval for an opinion survey involving all the United States senators? Why or why not?

8.44 A wine manufacturer sells a cabernet wine whose label asserts an alcohol content of 11%. Fifteen bottles of this wine are selected at random and analyzed for alcohol content, with a resulting mean of 10.2% and a standard deviation of 1.2%.

(a) Find a 95% confidence interval for the mean alcohol content.

(b) Based on your answer to part (a), do you think the label is correct? Explain.

8.45   A paint manufacturer wishes to validate its claim that a gallon of its paint covers 400 square feet, and it sets up a test based on a random sample of fifty 1-gallon cans of paint. The hypothesis to be tested is H_0: $\mu = 400$ vs H_a: $\mu > 400$; the significance level is $\alpha = .05$.

   (a) In words, what is the parameter of interest (μ)?

   (b) Give the rejection region (including critical value) for the test.

   (c) If the sample of 50 cans shows an average coverage of 412 square feet and a standard deviation of 38 square feet, what is the conclusion of the test?

   (d) Find the p-value of the test.

8.46   A study of the operation of a parking garage showed that, in the past, average parking time was 220 minutes. Recently the garage was remodeled and charges increased. The management wants to know if the changes have had any effect on mean parking time. Thus we wish to test H_0: $\mu = 220$ vs H_a: $\neq 220$, and we will use $\alpha = .05$. A random sample of 50 cars had an average parking time of 208 minutes and a standard deviation of 40 minutes.

   (a) Give the rejection region, including test-statistic formula and critical value.

   (b) Give the observed value of the test statistic.

   (c) Give your conclusion about changes in parking time.

   (d) Give the significance level (p-value) of the test.

8.47   A random sample of 35 city buses showed the mean number of passengers (per day, per bus) to be 225 with a standard deviation of 60 passengers.

   (a) Find a 95% confidence interval for the average number of passengers.

   (b) In words, describe the parameter of interest in this problem and give its value (if it is known).

   (c) In words, describe a sample statistic in this problem and give its value (if it is known).

8.48   An office manager wishes to estimate the mean time required to handle a customer complaint. A sample of 38 complaints shows a mean of 28.7 minutes and a standard deviation of 12 minutes.

   (a) Give a point estimate for the true mean time required to handle customer complaints.

   (b) Construct a 90% confidence interval estimate for the true mean time required to handle customer complaints.

8.49   The concentration of mercury in a lake has been measured many times. This population of measurements has an average of 1.20 mg/m^3 (milligrams per cubic meter) with a standard deviation of 0.30 mg/m^3. Following an accident at a smelter on the shore of the lake, nine more measurements were taken. These have an average mercury concentration of 1.45 mg/m^3. Report the level of significance of the evidence from this sample that the mean mercury concentration in the lake has increased.

8.50 Answer each question by circling T for true or F for false.

(a) T F   Given any particular random sample, if we form the 95% confidence interval for the sample mean there is a 95% chance that the population mean lies in this confidence interval.

(b) T F   If a large number of random samples are selected and we form the 95% confidence interval for each sample mean, the population mean will lie in about 95% of these confidence intervals.

(c) T F   If a sample size is larger than 30, there is a 95% chance that the sample mean equals the population mean.

(d) T F   If a very large number of random samples are selected, there is a 95% chance that one of the sample means is equal to the population mean.

(e) T F   The 95% confidence interval around a given sample mean is wider than the 90% confidence interval around that mean.

(f) T F   In order to prove that $\mu = \mu_0$ with type I error .05, we must select a sample and fail to reject the null hypothesis H_0: $\mu = \mu_0$ using $\alpha = .05$.

(g) T F   To find the critical value for a *two-tailed* test with type I error .04, we can look in Table 2, Appendix 3 for the z-score corresponding to the area .4800.

(h) T F   To find the critical value for a *one-tailed* test with type I error .02, we can look in Table 2, Appendix 3 for the z-score corresponding to the area .4800.

(i) T F   If we rejected the null hypothesis at the $\alpha = .05$ level, then we would also have rejected it at the $\alpha = .01$ level.

8.51 Modified True/False Questions. Circle T or F. If your answer is F, change the statement to make it true. Only change the *underlined* words.

(a) T F   A type I error is committed when we Fail to Reject the Null Hypothesis H_0 when H_0 is actually false.

(b) T F   If we make a type II error, we have missed detecting an event or effect when there actually was one.

(c) T F   The probability of making a type I error is equal to β.

(d) T F   If we increase the probability of making a type II error, we will increase the probability of making a type I error.

8.52 Over the years, projected due dates for expectant mothers have been notoriously bad. In a recent survey of 100 mothers it was found that the average number of days-to-birth beyond the projected due date was 9.2, with a standard deviation of 12.4. Use these data to find a 95% confidence interval for the mean number of days-to-birth beyond the due date.

8.53 Refer to Exercise 8.52. Use these data to find a 90% confidence interval for the mean number of days-to-birth beyond the due date.

8.54 A corporation maintains a large fleet of company cars for its salespeople. In order to determine the average number of miles driven per month by all salespeople, a random sample of 70 records was obtained. The mean and the standard deviation for the number of miles were 3250 and 420, respectively. Estimate μ, the average number of miles driven per month for all the salespeople within the corporation, using a 99% confidence interval.

8.55 The length of time to assemble an electronic fuse was measured for 50 assemblers. The mean and the standard deviation were 3.2 and .3 minutes, respectively. Give a 90% confidence interval for the mean length of time to assemble a fuse.

8.56 A group of 40 rats was selected for study. Each rat's heart rate was measured prior to receiving a single dose of the test preparation and then again two hours after administration of the drug. The sample mean drop in blood pressure from the predrug reading to the two-hour postdrug reading was 30.2, and the standard deviation was 10.0. Use these data to test the null hypothesis H_0: $\mu_{drop} = 0$ against the alternative H_a: $\mu_{drop} > 0$. Use $\alpha = .05$.

8.57 Refer to Exercise 8.56. Give the level of significance for the test.

8.58 Refer to Exercise 8.29. Give a 99% confidence interval for these data.

8.59 The diameter of extruded plastic pipe varies about a mean value that is controlled by a machine setting. A random sample of the diameters of 50 pieces of plastic pipe gave a mean and a standard deviation of 4.05 and .12 inches, respectively. Do the data present sufficient evidence to indicate that the mean diameter is different from 4 inches? Use $\alpha = .05$.

8.60 A hospital claims that the average length of patient confinement is five days. A study of the length of patient confinements for 36 people showed $\bar{x} = 6.2$ and $s = 5.2$. Do these data present sufficient evidence to contradict the hospital's claim? Use $\alpha = .05$.

8.61 The manufacturer of an automatic control claims that the device will maintain a mean room humidity of 80%. The humidity in a controlled room was recorded for a period of 30 days, and the mean and the standard deviation were found to be 78.3 and 2.9, respectively. Do the data present sufficient evidence to contradict the manufacturer's claim? Use $\alpha = .05$.

8.62 A buyer wishes to determine whether the mean sugar content per orange shipped from a particular grove is less than .027 pound. A random sample of 50 oranges produced a mean sugar content of .025 and a standard deviation of .003 pound. Do the data present sufficient evidence to indicate that the mean sugar content is less than .027 pound? Use $\alpha = .05$.

8.63 One method for solving the electrical power shortage makes use of floating nuclear power plants located a few miles offshore in the ocean. Because there is great concern about the possibility of a ship colliding with the floating (but anchored) power plant, navigation experts have stated that it would be desirable if the average number of ships per day passing within 10 miles of the proposed power site location were less than 7. To verify this hypothesis for the proposed site, a random sample of 60 days was used throughout the peak shipping months. For each day the number of ships passing within the 10-mile limit was recorded. The sample mean and standard deviation were 6.3 ships and 2 ships, respectively. Use these data to test the navigation experts' alternative hypothesis. Use $\alpha = .05$.

8.64 Administrative officials for a university are concerned that the freshman students taking advantage of off-campus housing facilities have significantly lower grade-point averages (GPA) than all freshmen at the school. After the fall quarter the all-freshman average GPA was 2.1 (on a 4-point system). Since it was not possible to isolate grades for all students living in off-campus housing by university records, a random sample of 81 off-campus freshmen was obtained by tracing students through their permanent home addresses. The sample mean and standard deviation were found to be 1.92 and .2, respectively. Do these data present sufficient evidence to indicate that the average GPA for all off-campus freshmen is lower than the all-freshman average? Use $\alpha = .05$.

8.65 Industrial waste and sewage dumped into our rivers and streams absorb oxygen and thereby reduce the amount of dissolved oxygen available for fish and other forms of aquatic life. One state agency requires a minimum of 5 parts per million (ppm) of dissolved oxygen in order that the oxygen content be sufficient to support aquatic life. During the low-water season (July), 30 specimens taken from a river at a specific location gave a sample mean and standard deviation of 4.9 and .2 ppm, respectively, of dissolved oxygen. Do the data provide sufficient evidence to indicate that the average dissolved oxygen content is less than 5 ppm? Use $\alpha = .05$.

8.66 In a standard dissolution test for tablets of a particular drug product, the manufacturer must obtain the dissolution rate for a batch of tablets prior to release of the batch. Suppose that the dissolution test consists of assays for 36 individual 25-mg tablets. For each test the tablet is suspended in an acid bath and then assayed after 30 minutes. The sample mean and standard deviation after 30 minutes are 19.8 and .42 mg, respectively. Use these data to test H_0: $\mu = 20$ (80% of the labeled amount in the tablets) against the alternative hypothesis H_a: $\mu < 20$. Use $\alpha = .05$.

8.67 Refer to Exercise 8.66. Give the level of significance for the test when the alternative hypothesis is H_a: $\mu \neq 20$.

8.68 Statistics has become a valuable tool for auditors, especially where large inventories are involved. It would be costly and time consuming for an auditor to inventory each item in a large operation. Thus the auditor frequently resorts to obtaining a random sample of items and using the sample results to check the validity of a company's financial statement. For example, a hospital financial statement claims an inventory that averages $300 per item. An auditor's random sample of 20 items yielded a mean and standard deviation of $160 and $90, respectively. Do the data contradict the hospital's claimed mean value per inventoried item and indicate that the average is less than $300? Use $\alpha = .05$.

8.69 Over the past five years the mean time for a warehouse to fill a buyer's order has been 25 minutes. Officials of the company believe that the length of time has increased recently, either due to a change in the work force or due to a change in customer purchasing policies. The processing time (in minutes) was recorded for a random sample of 15 orders processed over the past month.

28	25	27	31	10
26	30	15	55	12
24	32	28	42	38

Do the data present sufficient evidence to indicate that the mean time to fill an order has increased? Use the accompanying output to reach a conclusion based on $\alpha = .01$.

```
MTB > SET INTO C2
DATA> 28   25   27   31   10
DATA> 26   30   15   55   12
DATA> 24   32   28   42   38
DATA> END

MTB > PRINT C2

C2
    28     25     27     31     10     26     30     15     55     12     24     32     28
    42     38

MTB > TTEST MU =25, ALTERNATE = +1 C2

TEST OF MU = 25.00 VS MU G.T. 25.00

              N      MEAN     STDEV    SE MEAN        T    P VALUE
C2           15     28.20     11.44       2.95     1.08       0.15

MTB > STOP
```

8.70 Give the level of significance for the statistical test in Exercise 8.69

8.71 If a new process for mining copper is to be put into full-time operation, it must produce an average of more than 50 tons of ore per day. A five-day trial period gave the results shown in the accompanying table. Do these figures warrant putting the new process into full-time operation? Test by using $\alpha = .05$.

Day	1	2	3	4	5
Yield in tons	50	47	53	51	52

8.72 A test was conducted to determine the length of time required for a student to read a specified amount of material. All students were instructed to read at the maximum speed at which they could still comprehend the material. Sixteen students took the test, with the following results (in minutes):

25	18	27	29	20	19	25	24
32	21	24	19	23	28	31	22

Estimate the mean length of time required for all students to read the material, using a 95% confidence interval.

8.73 A random sample of eight students participated in a psychological test of depth perception. Two markers, one labeled A and the other B, were arranged a fixed distance apart at the far end of the laboratory. One by one the students were ushered into the room and asked to judge the distance between the two markers at the other end of the room. The sample data (in feet) were as follows:

2.1	2.2	2.6	2.3
1.8	2.3	2.4	2.5

Construct a 90% confidence interval for μ, the mean judged distance for all students for which the sample is representative.

8.74 The lifetimes (in years) of ten automobile batteries of a certain brand are

2.4   1.9   2.0   2.1   1.8   2.3   2.1   2.3   1.7   2.0

Estimate the mean lifetime, using a 95% confidence interval.

8.75 A drug antibiotic manufacturer randomly sampled twelve different locations in the fermentation vat to determine average potency for the batch of antibiotic being prepared. Readings were as follows:

8.9	9.0	9.1	8.9
9.1	9.0	9.0	9.0
8.9	8.8	9.1	8.8

Use the output shown here to estimate the mean potency for the batch, based on a 95% confidence interval. Interpret the interval.

```
MTB > SET INTO C1
DATA> 8.9   9.0   9.1   8.9
DATA> 9.1   9.0   9.0   9.0
DATA> 8.9   8.8   9.1   8.8
DATA> END

MTB > TINTERVAL 95 PERCENT C1

                N      MEAN    STDEV   SE MEAN    95.0 PERCENT C.I.
C1             12     8.9667   0.1073   0.0310   ( 8.8985,  9.0349)

MTB > STOP
```

8.76 In a statistical test about μ, the null hypothesis was rejected. Based on this conclusion, which of the following statements are true?

(a) The type I error was committed.

(b) The type II error was committed.

(c) The type I error could have been committed.

(d) The type II error could have been committed.

(e) It is impossible to have committed both type I and type II errors.

(f) It is impossible that neither type I nor type II error was committed.

(g) Whether any error was committed is not known, but if an error was made it was type I.

(h) Whether any error was committed is not known, but if an error was made it was type II.

8.77 Indicate whether the following statements are True or False.

T F   In a test of hypothesis, a test statistic is computed from the sample data.

T F   A statistical test of a hypothesis employs the technique of proof by contradiction. That is, we try to show that the alternative hypothesis is true by showing that the null hypothesis is false.

T F   The sample size n plays an important role in testing hypotheses because it measures the amount of data (and hence information) upon which we base a decision. If the data are quite variable and n is too small, it is unlikely that we will reject the null hypothesis even when the null hypothesis is false.

8.78 Complete the following statements (more than one word may be needed).

(a) If we take all possible samples (of a given sample size) from a population, then the distribution of sample means tends to be _____ and the mean of these sample means is equal _____.

(b) The larger the sample size, other things remaining equal, the _____ the confidence interval.

(c) The larger the confidence coefficient, other things remaining equal, the _____ the confidence interval.

(d) The statement "If random samples of a fixed size are drawn from any population (regardless of the form of the population distribution), as n becomes larger, the distribution of sample means approaches normality," is known as the _____.

(e) By failing to reject a null hypothesis that is false, one makes a _____ error.

8.79 Refer to Exercise 8.65. Six water specimens taken from the river at a specific location during the low-water season (July) gave readings of 4.9, 5.1, 4.9, 5.0, 5.0, and 4.7 ppm of dissolved oxygen. Do the data provide sufficient evidence to indicate that the dissolved oxygen content is less than the required 5 ppm? Test by using $\alpha = .05$.

8.80 Suppose that the tar content of cigarettes is normally distributed with a mean of 10 and a standard deviation of 2.4 mg. A new manufacturing process is developed for decreasing the tar content. A sample of 16 cigarettes produced by the new process yielded a mean of 8.8 mg. Use $\alpha = .05$.

(a) Do a test of hypothesis to determine if the new process has significantly *decreased* the tar content. Use the following outline.

Null hypothesis
Alternative hypothesis
Assumptions
Rejection region(s)
Test statistic and computations
Conclusion
  In statistical terms
  In plain English

(b) Based on your conclusion, could you have made
_____ A type I error?
_____ A type II error?
_____ Neither error?
_____ Both type I and type II errors?

8.81 Measurements of water intake, obtained from a sample of 17 rats that had been injected with a sodium chloride solution, produced a mean and standard deviation of 31.0 cm^3 and 6.2 cm^3, respectively. The historical average water intake for noninjected rats observed over a comparable period of time is 22.0 cm^3. Do the data indicate that injected rats drink more water than noninjected rats? Test the hypothesis by using $\alpha = .05$.

8.82 The defense in a criminal case charged that the jury rolls, from which jury members are selected, were not representative of the population at large in the county. Specifically, the defense contended that the average salary of wage earners selected as jury members was greater than the average wage-earner salary for the county (there was information available to lead one to believe this). The most recent census showed that the average salary in the county was $8,500. The average salary of the last 100 wage earners selected for jury duty at this court was $22,890 with a standard deviation of $7,670. What statistical test could be used to test the defense attorney's claim?

8.83 A random sample of 45 blue-eyed individuals participating in the Miss America Pageant during the years of 1974–1983 showed a mean amount won of $2,567 with standard deviation of $4,230. Estimate the mean amount won for all blue-eyed participants using a 90% confidence interval.

8.84 Refer to Exercise 8.83. For the same group of 45 blue-eyed participants, the mean years of schooling was 15.3 with standard deviation of 1.06 years. Estimate the mean years of schooling for blue-eyed participants using a 95% confidence interval.

8.85 A large supermarket chain sells longhorn cheese in one-pound packages. As a city inspector you weigh 36 randomly selected packages of cheese and note that the sample mean is 14.58 ounces with standard deviation of 1.44 ounces. Does the sample provide sufficient evidence to reject or not reject the hypothesis that the cheese packages weigh 16 ounces or more? Use $\alpha = .01$.

8.86 A financial consultant has asserted that firms in selected "targeted for growth" industries with price/earnings ratios in the lowest 25% for such firms will have a current ratio of 2.0 or less. Since your organization felt that the consultant may have understated the mean current ratio, a random sample of 48 firms in the "targeted for growth" industries was taken. The mean and standard deviation for the sample were 2.58 and 2.35, respectively. Does the sample information support or contradict the consultant's assertion?

8.87 The personnel department of a major retail chain with its central offices in Chicago was interested in the length of time its employees had been with the organization. A random sampling of 16 employees produced the following data, which show years with the company.

3.4	6.6	1.2	3.3
11.0	5.1	7.4	3.5
1.4	2.8	7.9	6.3
6.5	5.6	4.2	4.9

Estimate the mean length of time of employee longevity with the company, using a 95% confidence interval.

8.88 The caffeine content (in milligrams) of a random sample of 50 cups of black coffee dispensed by a new machine is measured. The mean and standard deviation are 100 mg and 7.1 mg, respectively. Construct a 98% confidence interval for the true (population) mean caffeine content per cup dispensed by the machine.

8.89 The machine in Exercise 8.88 is capable of dispensing 3000 cups per day. The caffeine content varies because of variation of the ground coffee beans and because of variation in brewing time.

(a) Is the study in Exercise 8.88 questionable because such a small fraction of the machine's output is analyzed?

(b) The 50 cups sampled are taken consecutively from the machine. Does this make the study questionable?

8.90 A random sample of the year-end financial statements of a sample of 22 small (under $500,000 in sales) retail businesses in a city show that the average net margin on sales is .0210 and the standard deviation is .0114. Find a 90% confidence interval for the mean net margin for all small retail businesses in the city.

8.91 The data in Exercise 8.90 indicate that the distribution of net margins has a peak at about .015, with some businesses having much larger margins but none having lower margins. What does this fact indicate about the claimed 90% confidence interval?

8.92 A dealer in recycled paper places empty trailers at various sites; these are gradually filled by individuals who bring in old newspapers and the like. The trailers are picked up (and replaced by empties) on several schedules. One such schedule involves pickup every second week. This schedule is desirable if the average amount of recycled paper is more than 1600 ft^3 per two-week period. The dealer's records for eighteen 2-week periods show the following volumes (in cubic feet) at a particular site:

1660	1820	1590	1440	1730	1680	1750	1720	1900
1570	1700	1900	1800	1770	2010	1580	1620	1690

Assume that these figures represent the results of a random sample. It can be shown that $\bar{x} = 1718.3$ and $s = 137.8$. Do the data support the research hypothesis that $\mu > 1600$, using $\alpha = .10$? Write out all parts of the hypothesis-testing procedure.

8.93 A federal regulatory agency is investigating an advertised claim that a certain device can increase the gasoline mileage of cars. Seven such devices are purchased and installed in cars belonging to the agency. Gasoline mileage for each of the cars under standard conditions is recorded before and after installation.

	Car						
	1	*2*	*3*	*4*	*5*	*6*	*7*
Mpg before	19.1	19.9	17.6	20.2	23.5	26.8	21.7
Mpg after	20.0	23.7	18.7	22.3	23.8	19.2	24.6
Change	.9	3.8	1.1	2.1	.3	−7.6	2.9

The mean change is .50 mpg, and the standard deviation is 3.77.

    (a) Formulate appropriate null and research hypotheses.

    (b) Is the advertised claim supported at $\alpha = .05$? Carry out the steps of a hypothesis test.

8.94 Use the data of Exercise 8.93 to construct a 90% confidence interval for the mean change. On the basis of this interval, can one reject the hypothesis of no mean change? (Note that the two-sided 90% confidence interval corresponds to a one-tailed $\alpha = .05$ test.)

8.95 Would you say that the agency of Exercise 8.93 has conclusively established that the device has no effect on the average mileage of cars? What does the width of the interval in Exercise 8.94 have to do with your answer?

8.96 A small manufacturer has a choice between using the postal service or a private shipper. As a test, ten destinations are chosen and packages shipped to each by both services. The delivery times, in days, are as follows:

	Destination									
	1	*2*	*3*	*4*	*5*	*6*	*7*	*8*	*9*	*10*
Postal service	3	4	5	4	8	9	7	10	9	9
Private shipper	2	2	3	5	4	6	9	6	7	6
Difference	1	2	2	−1	4	3	−2	4	2	3

    (a) Calculate the mean and standard deviation of the differences.

    (b) Test the null hypothesis of no mean difference in delivery times against the research hypothesis that the private shipper has a shorter average delivery time. Use $\alpha = .01$.

 8.97 The search for alternatives to oil as a major source of fuel and energy will inevitably bring about many environmental challenges. These challenges will require solutions to problems in such areas as strip mining and many others. Let us focus on one. If coal is considered a major source of fuel and energy, we will have to consider ways to keep large amounts of sulfur dioxide (SO_2) and particulates from getting into the air. This is especially important at large government and industrial operations. Here are several possibilities.

    (a) Build the smokestack extremely high.

    (b) Remove the SO_2 and particulates from the coal prior to combustion.

    (c) Remove the SO_2 from the gases after the coal is burned out but before the gases are released into the atmosphere. (This is accomplished by using a scrubber.)

Several scrubbers have been developed in recent years. Suppose that a new one has been constructed and is set for testing at a given power plant. Fifty samples are obtained at various times from gases emitted from the stack. The mean SO_2 emission is .13 lb per million Btu, with a standard deviation of .05 lb. Use the sample data to construct a statistical test of the null hypothesis H_0: $\mu = .145$, the average emission level for one of the more efficient scrubbers that has been developed. Choose an appropriate alternative hypothesis, with $\alpha = .05$.

8.98 Refer to Exercise 8.97. Rather than being interested in testing the research hypothesis that $\mu < .145$, the average emission level for one of the more efficient scrubbers, we may wish to estimate the mean emission level for the new scrubber. Use the sample data to construct a 99% confidence level for μ. Interpret your results.

8.99 As part of an overall evaluation of training methods, an experiment was conducted to determine the average capacity of healthy male army inductees. To do this each male in a random sample of 35 healthy army inductees exercised on a bicycle ergometer (a device for measuring work done by the muscles) under a fixed workload until he tired. Blood pressure, pulse rates, and other indicators were carefully monitored to ensure that no one's health was in danger. The exercise capacities (mean time, in minutes) for the 35 inductees were:

23	19	36	12	41	43	19
28	14	44	15	46	36	25
35	25	29	17	51	33	47
42	45	23	29	18	14	48
21	49	27	39	44	18	13

(a) Use these data to construct a 95% confidence interval for μ, the average exercise capacity for healthy male inductees. Interpret your findings.

(b) How would your interval change using a 99% confidence interval?

8.100 We all remember being told, "Your fever has subsided and your temperature has returned to normal." What do we mean by the word *normal?* Most people use the benchmark 98.6°F, but this does not apply to all people—only the "average" person. Without putting words into someone's mouth, we might define a person's normal temperature to be his or her average temperature when healthy. But even this definition is cloudy because there is variation in a person's temperature throughout the day. To determine a subject's normal temperature, we recorded it for a random sample of 30 days. On each day selected for inclusion in the sample, the temperature reading was made at 7 AM. The sample mean and standard deviation for these 30 readings were, respectively, 98.4 and .15. Assuming the subject was healthy on the days examined, use these data to estimate the person's 7 AM "normal" temperature using a 90% confidence interval.

8.101 Refer to the data of Exercise 8.99. Suppose that the random sample of 35 inductees was selected from a large group of new army personnel being subjected to a new (and hopefully improved) physical fitness program. Assume previous testing with several thousand personnel over the past several years has shown an average exercise capacity of 29 minutes. Run a statistical test for the research hypothesis that the average exercise capacity is improved for the new fitness program. Give the level of significance for the test. Interpret your findings.

8.102 Doctors have recommended that we try to keep our caffeine intake at 200 mg or less per day. With the following chart, a sample of 35 office workers were asked to record their caffeine intake for a 7-day period.

coffee (6 oz)	100–150 mg
tea (6 oz)	40–110 mg
cola (12 oz)	30 mg
chocolate cake	20–30 mg
cocoa (6 oz)	5–20 mg
milk chocolate (1 oz)	5–10 mg

After the 7-day period, the average daily intake was obtained for each worker. The sample mean and standard deviation of the daily averages were 560 and 160, respectively. Use these data to estimate μ, the average daily intake, using a 90% confidence interval.

EXERCISES FROM THE DATA BASE

8.103 Refer to the clinical trials data base in Appendix 1 to construct a 95% confidence interval for the mean HAM-D total score of treatment group C. How would this interval change for a 99% confidence interval?

8.104 Using the clinical trials data base, give a 90% confidence interval for the Hopkins Obrist cluster score of treatment A.

INFERENCES ABOUT $\mu_1 - \mu_2$

9.1 Introduction ∎ **9.2** The sampling distribution for the difference between two sample means ∎ **9.3** Estimation of $\mu_1 - \mu_2$ ∎ **9.4** A statistical test about $\mu_1 - \mu_2$ ∎ **9.5** Assumptions ∎ **9.6** Using computers ∎ Summary ∎ Key terms ∎ Key formulas ∎ Supplementary exercises ∎ Exercises from the data base

9.1 INTRODUCTION

In Chapters 4 and 5 we described a set of measurements by using graphical and numerical descriptive techniques. In Chapters 6 and 7 we dealt with probability, probability distributions, and sampling distributions, showing how to reason from a known population to a sample, and thus provided the mechanism for making an inference about a population. We presented the two methods for making inferences—estimation and testing hypotheses—in Chapter 8. These techniques were illustrated for practical inferential problems related to a population mean μ when σ is known and when it is unknown. Thus we have stated that the objective of statistics is to make inferences, have explained how inferences are made, and have given practical illustrations. Where do we go now? To more practical applications. We will now consider inferences about the difference between two population means.

Rarely do you read one of the popular news magazines or the Sunday edition of a leading newspaper without confronting articles about the comparison of two populations. For example, we read that the average size of factory orders in July rose 1.7% over that in June, car production for September is scheduled to drop 4.5% in comparison with August, the public school teachers of a certain state receive salaries less than the national average, and the percentage of people suffering from arteriosclerosis is higher for individuals with cadmium in their water supply than for those whose water lacks the element. All these examples involve the comparison of two populations based on information contained in samples selected from each.

How can you tell whether the observed differences in the previous examples are real or whether they are due to random variation? People unfamiliar with statistics frequently answer this question by saying, "But you can see the difference, can't you? There's no question about it!" They forget—or do not know—that the difference they observe is based on samples, and hence they confuse this observed difference in the sample means or percentages with the corresponding population difference.

There are numerous examples that emphasize this point. For instance, recall that the Federal Trade Commission has placed pressure on many companies to restrain them from using misleading advertising. And research articles published in some professional journals can be very misleading. Typical of this practice is the reporting of results in a medical journal comparing two products, I and II, which, for purposes of illustration, might be drugs used to treat obesity. Suppose the author of the article reports that the average weight loss for 20 patients on drug product I is 15.6 pounds while the corresponding average weight loss for the 20 patients on drug product II is 10.1. If their data and a valid description of the actual experiments are presented in condensed form to a practicing physician who is overburdened with patients and who has little time to read the hundreds of similar articles on new treatments, instruments, and drugs, what will he or she conclude?

The sample means for these two groups of patients are substantially different—15.6 versus 10.1—but because the sample means are based on relatively small samples, the sample estimates may vary considerably about the true population means. Hence the sample data may not provide sufficient evidence to indicate a difference in weight loss for patients treated with the two drug products. Drug I may be better than II, II may be better than I, or they may be equally effective.

In this chapter we will present techniques for comparing two population means.

9.2 THE SAMPLING DISTRIBUTION FOR THE DIFFERENCE BETWEEN TWO SAMPLE MEANS

In many sampling situations we will select independent random samples from two populations in order to compare the population means or proportions. The statistics used to make these inferences will, in many cases, be the difference between the corresponding sample statistics. For example, suppose we select independent random samples of n_1 observations from one population and n_2 observations from a second population. We will use the difference between the sample means, $(\bar{x}_1 - \bar{x}_2)$, to make an inference about the difference between the population means, $(\mu_1 - \mu_2)$.

The following theorem will help in finding the sampling distribution for the difference between sample statistics computed from independent random samples.

Theorem 9.1

If two independent random variables x_1 and x_2 are normally distributed with means and variances (μ_1, σ_1^2) and (μ_2, σ_2^2), respectively, the difference between the random variables will be normally distributed with mean equal to $(\mu_1 - \mu_2)$ and variance equal to $(\sigma_1^2 + \sigma_2^2)$.

Note: The sum $(x_1 + x_2)$ of the random variables will also be normally distributed with mean $(\mu_1 + \mu_2)$ and variance $(\sigma_1^2 + \sigma_2^2)$.

Theorem 9.1 can be applied directly to find the sampling distribution of the difference between two independent sample means or two independent sample proportions. The Central Limit Theorem (discussed in Chapter 7) implies that if independent samples of sizes n_1 and n_2 are selected from two populations 1 and 2, then, when n_1 and n_2 are large, the sampling distributions of $\bar{x}_1$ and $\bar{x}_2$ will be approximately normal, with means and variances $(\mu_1, \sigma_1^2/n_1)$ and $(\mu_2, \sigma_2^2/n_2)$, respectively. Consequently, since $\bar{x}_1$ and $\bar{x}_2$ are independent, normally distributed random variables, it follows from Theorem 9.1 that the sampling distribution for the difference in the sample means, $(\bar{x}_1 - \bar{x}_2)$, will be approximately normal, with a mean

$$\mu_{\bar{x}_1 - \bar{x}_2} = \mu_1 - \mu_2$$

and a variance

$$\sigma_{\bar{x}_1 - \bar{x}_2}^2 = \sigma_{\bar{x}_1}^2 + \sigma_{\bar{x}_2}^2 = \frac{\sigma_1^2}{n_1} + \frac{\sigma_2^2}{n_2}$$

and a standard error

$$\sigma_{\bar{x}_1 - \bar{x}_2} = \sqrt{\frac{\sigma_1^2}{n_1} + \frac{\sigma_2^2}{n_2}}$$

The sampling distribution of the difference between two independent, normally distributed sample means is shown in Figure 9.1.

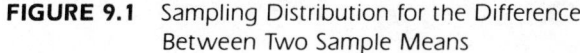

FIGURE 9.1   *Sampling Distribution for the Difference Between Two Sample Means*

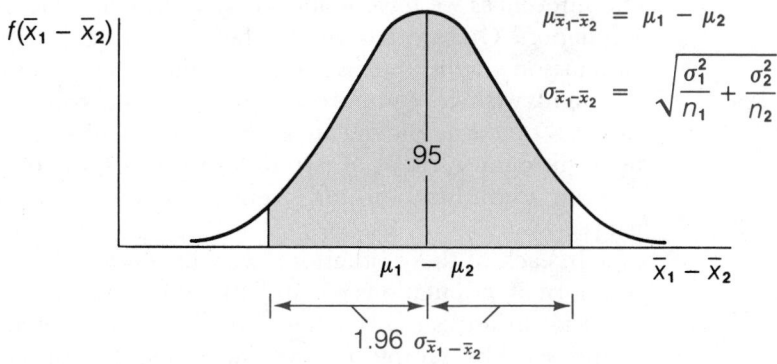

Properties of the Sampling Distribution for the Difference Between Two Sample Means, $(\bar{x}_1 - \bar{x}_2)$

1. The sampling distribution of $(\bar{x}_1 - \bar{x}_2)$ is approximately normal for large samples.

2. The mean of the sampling distribution, $\mu_{\bar{x}_1 - \bar{x}_2}$, is equal to the difference between the population means, $(\mu_1 - \mu_2)$.

3. The standard error of the sampling distribution is

$$\sigma_{\bar{x}_1 - \bar{x}_2} = \sqrt{\frac{\sigma_1^2}{n_1} + \frac{\sigma_2^2}{n_2}}$$

The sampling distribution for the difference between two sample means, $(\bar{x}_1 - \bar{x}_2)$, can be used to answer the same types of questions as were asked about the sampling distribution for $\bar{x}$ in Chapter 7. Since sample statistics are used to make inferences about corresponding population parameters, we can use the sampling distribution of a statistic to calculate the probability that the statistic will be within a specified distance of the population parameter. For example, we could use the sampling distribution of the difference in sample means to calculate the probability that $(\bar{x}_1 - \bar{x}_2)$ will be within a specified distance of the unknown difference in population means $(\mu_1 - \mu_2)$. Inferences (estimations or tests) about $(\mu_1 - \mu_2)$ will be discussed in succeeding sections of this chapter.

9.3 ESTIMATION OF $\mu_1 - \mu_2$

The inferences we have made so far have concerned the mean of a single population. Quite often we are faced with an inference that concerns a comparison of the means from two different populations. For example, we might wish to compare the mean corn crop yield for two different varieties of corn, the mean annual income for two ethnic groups, the mean nitrogen content of two different lakes, or the mean length of time between administration and eventual relief for two different antivertigo drugs.

In each of these situations, we will assume that we are sampling from two normal populations (1 and 2) with different means μ_1 and μ_2 but identical variances σ^2. We then draw independent random samples of size n_1 and n_2. The sample means are $\bar{x}_1$ and $\bar{x}_2$; the corresponding sample variances are s_1^2 and s_2^2, respectively. Using the data from the two samples, we would like to make a comparison between the population means μ_1 and μ_2. In particular, we will estimate and test a hypothesis concerning the difference $\mu_1 - \mu_2$.

point estimate of $\mu_1 - \mu_2$

A logical point estimate for the difference in population means is the sample difference $\bar{x}_1 - \bar{x}_2$. The standard error for the difference in sample means is more complicated than for a single sample mean, but the confidence interval has the same form: **point estimate** $\pm t$ (standard error). The value of t is determined from Table 3 of Appendix 3 for a given value of a and df $= n_1 + n_2 - 2$. The t-values for $a = .05$, $.025$, and $.005$ are used to construct 90%, 95%, and 99% confidence intervals, respectively.

s_p^2 (weighted average)

The quantity s_p in the confidence interval is the estimate of the standard deviation σ for the two populations and is formed by combining (pooling) information from the two samples. In fact, **s_p^2 is a weighted average** of the sample variances s_1^2 and s_2^2. For the special case in which the sample sizes are the same ($n_1 = n_2$), the formula for s_p^2 reduces to $s_p^2 = (s_1^2 + s_2^2)/2$, the mean of the two sample variances. The degrees of freedom for the confidence interval are a combination of the degrees of freedom for the two samples; that is, df $= (n_1 - 1) + (n_2 - 1) = n_1 + n_2 - 2$.

Confidence Interval for $\mu_1 - \mu_2$ Independent Samples

$$(\bar{x}_1 - \bar{x}_2) \pm t\, s_p \sqrt{\frac{1}{n_1} + \frac{1}{n_2}}$$

(continued)

where

$$s_p = \sqrt{\frac{(n_1 - 1)s_1^2 + (n_2 - 1)s_2^2}{n_1 + n_2 - 2}}$$

and

$$df = n_1 + n_2 - 2$$

We are assuming that the two populations from which we draw the samples are normal distributions with a common variance σ^2. If the confidence interval presented were valid only when these assumptions were met exactly, the estimation procedure would be of limited use. Fortunately, the confidence coefficient remains relatively stable if both distributions are mound shaped and the sample sizes are approximately equal.

Example 9.1

Company officials were concerned about the length of time a particular drug product retained its potency. A random sample, sample 1, of $n_1 = 10$ bottles of the product was drawn from the production line and analyzed for potency; a second sample, sample 2, of $n_2 = 10$ bottles was obtained and stored in a regulated environment for a period of one year.

The readings obtained from each sample are given in Table 9.1.

TABLE 9.1 Potency Reading for Two Samples

Sample 1		Sample 2	
10.2	10.6	9.8	9.7
10.5	10.7	9.6	9.5
10.3	10.2	10.1	9.6
10.8	10.0	10.2	9.8
9.8	10.6	10.1	9.9

Suppose we let μ_1 denote the mean potency for all bottles that might be sampled from the production line and μ_2 denote the mean potency for all bottles that may be retained for a period of one year. Estimate $\mu_1 - \mu_2$ by using a 95% confidence interval.

Solution   For this example, previous data suggest that it is safe to assume that the two populations are normal. Also, with equal sample sizes (and similar sample variances) it is safe to assume that the population variances are equal. The necessary sample calculations from the data of Table 9.1 are presented below.

Sample 1	Sample 2
$\sum\limits_j x_{1j} = 103.7$	$\sum\limits_j x_{2j} = 98.3$
$\sum\limits_j x_{1j}^2 = 1076.31$	$\sum\limits_j x_{2j}^2 = 966.81$

Then

$$\bar{x}_1 = \frac{103.7}{10} = 10.37 \qquad\qquad \bar{x}_2 = \frac{98.3}{10} = 9.83$$

$$s_1^2 = \frac{1}{9}\left[1076.31 - \frac{(103.7)^2}{10} \right] = .105 \qquad s_2^2 = \frac{1}{9}\left[966.81 - \frac{(98.3)^2}{10} \right] = .058$$

The estimate of the common standard deviation σ is

$$s_p = \sqrt{\frac{(n_1 - 1)s_1^2 + (n_2 - 1)s_2^2}{n_1 + n_2 - 2}} = \sqrt{\frac{9(.105) + 9(.058)}{18}}$$

which, for $n_1 = n_2 = 9$, reduces to

$$s_p = \sqrt{\frac{.105 + .058}{2}} = .285$$

The t-value based on df $= n_1 + n_2 - 2 = 18$ and a $= .025$ is 2.101. A 95% confidence interval for the difference in mean potencies is

$$(10.37 - 9.83) \pm 2.101(.285)\sqrt{\frac{1}{10} + \frac{1}{10}} \qquad \text{or} \qquad .54 \pm .268$$

We estimate that the difference in mean potencies, $\mu_1 - \mu_2$, lies in the interval .272 to .808. Since the possible values for $\mu_1 - \mu_2$ in this interval are all positive, the mean potency for bottles produced now (μ_1) is larger than the mean potency for bottles retained one year(μ_2).

Example 9.2

A study was conducted to determine whether persons in suburban district 1 have a different mean income from those in district 2. A random sample of 20 homeowners was taken in district 1. Although 20 homeowners were to be interviewed in district 2 also, one person refused to provide the information requested, even though the researcher promised to keep the interview confidential. So only 19 observations were obtained from district 2. The data, recorded in thousands of dollars, produced sample means and variances as shown in Table 9.2. Use these data to construct a 95% confidence interval for $(\mu_1 - \mu_2)$.

TABLE 9.2   Income Data for Example 9.2

	District 1	District 2
Sample size	20	19
Sample mean	18.27	16.78
Sample variance	8.74	6.58

Solution   Histograms plotted for the two samples suggest that the two populations are mound shaped (near-normal). Also the sample variances are very similar. The difference in the sample means is

$$\bar{x}_1 - \bar{x}_2 = 18.27 - 16.78 = 1.49$$

The estimate of the common standard deviation σ is

$$s_p = \sqrt{\frac{(n_1 - 1)s_1^2 + (n_2 - 1)s_2^2}{n_1 + n_2 - 2}}$$

$$= \sqrt{\frac{19(8.74) + 18(6.58)}{20 + 19 - 2}} = 2.77$$

The t-value for a = .025 and df = 20 + 19 − 2 = 37 is not listed, but taking the labeled value for the nearest df (df = 40) we have t = 2.021. A 95% confidence interval for the difference in mean incomes for the two districts is of the form

$$\bar{x}_1 - \bar{x}_2 \pm t\, s_p \sqrt{\frac{1}{n_1} + \frac{1}{n_2}}$$

Substituting into the formula we obtain

$$1.49 \pm 2.021(2.77) \sqrt{\frac{1}{20} + \frac{1}{19}}$$

or

$$1.49 \pm 1.79$$

Thus we estimate the difference in mean incomes to lie somewhere in the interval from $-.30$ to 3.28. If we multiply these limits by \$1,000, the confidence interval for the difference in mean incomes is $-$ \$300 to \$3,280. Since this interval includes both positive and negative values for $\mu_1 - \mu_2$, we are unable to determine whether the mean income for district 1 is larger or smaller than the mean income for district 2.

EXERCISES

 9.1 A study of attitude toward population control compared married women with unmarried women between the ages of 18 and 40. Note that a higher score is associated with a more positive attitude toward population control.

	n	$\bar{x}$	s^2
Unmarried	12	15	52
Married	12	10	56

Use these data to construct a 95% confidence interval for the difference in mean scores for the two groups. Explain (in words rather than numbers) what this interval means. What assumptions did you have to make?

 9.2 Refer to Exercise 9.1. Would a 99% confidence interval be narrower or wider than the 95% confidence interval constructed? Why?

9.3 The data shown below are the weight gains (in pounds) for babies from birth to six months of age. All babies weighed approximately the same at birth, but those in sample 1 were fed with a specific formula while those in sample 2 were breast fed. Use these data to construct a 90% confidence interval for the difference in mean weight gains for formula-fed and breast-fed babies.

Sample 1	Sample 2
5	9
7	10
8	8
9	11
6	7

9.4 A petroleum corporation was interested in running some preliminary tests to compare the performance of a new gasoline mixture to one currently on the market. Ten identical new automobiles were randomly assigned, five to gasoline A and five to gasoline B. Gasoline B contained a mileage additive, and gasoline A was regular gasoline. Each automobile was filled with 10 gallons of gasoline and driven over a test course until it stopped. The mileage was recorded for each in the table that follows:

Gasoline A	Gasoline B
282	284
279	285
280	286
278	277
275	283
$\bar{x}_1 = 278.80$	$\bar{x}_2 = 283.00$

Use these data to construct a 95% confidence interval for the difference in mean mileage for the two gasolines.

9.5 A sociologist gave a current events test to four blue-collar workers and four white-collar workers. The blue-collar workers made scores of 23, 18, 22, and 21; the white-collar workers made scores of 17, 22, 19, and 18. Estimate the difference in mean scores for blue-collar and white-collar workers using a 99% confidence interval.

9.4 A STATISTICAL TEST ABOUT $\mu_1 - \mu_2$

In the previous section we constructed a confidence interval for the difference between two population means based on independent random samples selected from each of the populations. Now let's run a statistical test about the difference between two population means. As with any test procedure, we begin by specifying the null and alternative hypotheses. If we let μ_1 and μ_2 denote the means for populations 1 and 2, respectively, the null and alternative hypotheses for a two-tailed test are

$$H_0: \mu_1 - \mu_2 = 0$$

and

$$H_a: \mu_1 - \mu_2 \neq 0$$

The test statistic is

$$t = \frac{\bar{x}_1 - \bar{x}_2}{s_p \sqrt{\dfrac{1}{n_1} + \dfrac{1}{n_2}}}$$

and the rejection region for the t-test is selected in the same manner as is the rejection region for a statistical test about μ (Section 8.4) except that df $= (n_1 - 1) + (n_2 - 1) = n_1 + n_2 - 2$. Thus for a two-tailed test with $\alpha = .05$, we look up the t-value in Table 3 of Appendix 3 with a $= .025$ and df $= n_1 + n_2 - 2$. We illustrate this two-sample t-test with an example.

Example 9.3

An experiment was conducted to investigate the effect of two diets on weight gain of 14-year-old children suffering from malnutrition. Ten children were subjected to diet 1 and nine to diet 2. The gains in weight over a nine-month period are shown in Table 9.3. Determine if the data indicate a difference between the mean gains in weight for children fed on the two diets. Use $\alpha = .05$.

TABLE 9.3 Data for Example 9.3

				Weight gain (in pounds)					
Diet I	14.0	12.5	10.2	9.8	10.5	11.2	15.0	22.0	13.0	9.6
Diet II	14.4	18.2	19.5	21.2	15.3	11.6	12.8	13.1	11.3	

Solution First we assume that weight gains for the two diets are normally distributed, with unknown means μ_1 and μ_2 and common unknown variance σ^2. The sample means and variances are given in Table 9.4.

TABLE 9.4 Means and Variances for the Data of Example 9.3

	Diet 1	Diet 2
Sample mean	12.78	15.27
Sample variance	13.88	12.81
Sample size	10	9

The estimate of the common population standard deviation is

$$s_p = \sqrt{\frac{(n_1 - 1)s_1^2 + (n_2 - 1)s_2^2}{n_1 + n_2 - 2}}$$

$$= \sqrt{\frac{9(13.88) + 8(12.81)}{10 + 9 - 2}} = \sqrt{\frac{124.92 + 102.48}{17}} = \sqrt{13.38}$$

and so

$$s_p = \sqrt{13.38} = 3.66$$

Let μ_1 and μ_2 be the means for the hypothetical populations of weight gains associated with diets 1 and 2, respectively; then the null and alternative hypotheses are

$$H_0: \mu_1 - \mu_2 = 0 \qquad H_a: \mu_1 - \mu_2 \neq 0$$

The value of the test statistic t for this test is

$$t = \frac{\bar{x}_1 - \bar{x}_2}{s_p\sqrt{\dfrac{1}{n_1} + \dfrac{1}{n_2}}} = \frac{12.78 - 15.27}{3.66\sqrt{\dfrac{1}{10} + \dfrac{1}{9}}} = \frac{-2.49}{3.66(.459)} = -1.48$$

The rejection region for $\alpha = .05$ utilizes a t-value corresponding to $a = .025$ and df $= (10 + 9 - 2) = 17$. From Table 3 of Appendix 3, this value is 2.110. Thus we reject the null hypothesis if the computed value of t is greater than 2.110 or less than -2.110. This rejection region is shown in Figure 9.2.

FIGURE 9.2 Two-Tailed Rejection Region for $\alpha = .05$ and df $= 17$

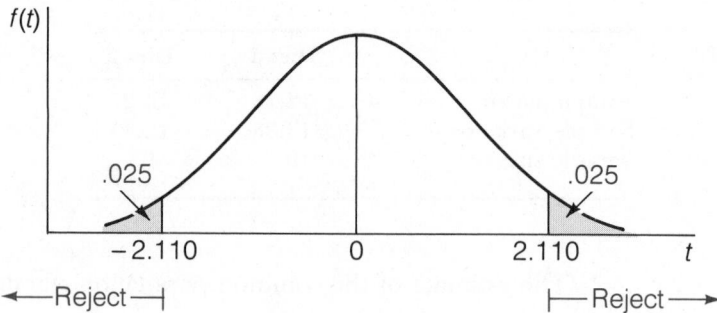

Noting that the computed value of t, $t = -1.48$, does not fall in the rejection region, we conclude that there is insufficient evidence to indicate a difference in the mean weight gains for the two diets.

Example 9.4

Refer to the data of Example 9.3 to determine the level of significance of the test.

Solution   Since Table 3 of Appendix 3 is not constructed in such a way as to obtain an exact p-value, we have to do the best we can to approximate the p-value. For these data the computed value of t is -1.48 and df $= 10 + 9 - 2 = 17$. Using the df $= 17$ row of Table 3, we see that a value of -1.48 lies between the a $= .10$ value (-1.333) and a $= .05$ value (-1.740). So the level of significance for this two-sided test is between $.10$ and $.20$; we write this as $.10 < p < .20$.

assumptions for
t-test

Before concluding our discussion of the **t-test**, we question whether the **assumptions** of normal populations and equal variances $(\sigma_1^2 = \sigma_2^2)$ must hold in order for the t-test to be valid. The test functions satisfactorily for populations possessing mound-shaped probability distributions. The assumption that $\sigma_1^2 = \sigma_2^2$ is more critical. But again, it does not seriously affect the properties of the test if n_1 and n_2 are approximately equal. Consequently, these assumptions are not too restrictive, giving the test wide applicability. More will be said about assumptions in Section 9.5.

A summary of the elements of the t-test for comparing two means is given next.

Summary of a Two-Sample t Test for Comparing Two Population Means

Null hypothesis: $\mu_1 - \mu_2 = 0$.

Alt. hypothesis: For a one-tailed test:

1. $\mu_1 - \mu_2 > 0$
2. $\mu_1 - \mu_2 < 0$

For a two-tailed test:

3. $\mu_1 - \mu_2 \neq 0$

Test statistic: $t = \dfrac{\bar{x}_1 - \bar{x}_2}{s_p\sqrt{\dfrac{1}{n_1} + \dfrac{1}{n_2}}}$ where

$$s_p = \sqrt{\dfrac{(n_1 - 1)s_1^2 + (n_2 - 1)s_2^2}{n_1 + n_2 - 2}}$$

Rejection region: For a specified value of α, for df $= (n_1 + n_2 - 2)$, and for a one-tailed test:

1. Reject H_0 if $t > t_a$,   where $a = \alpha$
2. Reject H_0 if $t < -t_a$,   where $a = \alpha$

For a specified value of α, for df $= (n_1 + n_2 - 2)$ and for a two tailed test:

3. Reject H_0 if $|t| > t_a$ where $a = \alpha/2$

EXERCISES

9.6 Set up the rejection regions for testing H_0: $\mu_1 - \mu_2 = 0$ for the following conditions:

(a) H_a: $\mu_1 - \mu_2 \neq 0$, $n_1 = 12$, $n_2 = 14$, and $\alpha = .05$

(b) H_a: $\mu_1 - \mu_2 > 0$, $n_1 = n_2 = 8$, and $\alpha = .01$

(c) H_a: $\mu_1 - \mu_2 < 0$, $n_1 = 6$, $n_2 = 4$, and $\alpha = .05$

(d) What assumptions must be made prior to applying a two-sample t-test?

9.7 Conduct a test of H_0: $\mu_1 - \mu_2 = 0$ against the alternative hypothesis H_a: $\mu_1 - \mu_2 < 0$ for the sample data shown here. Use $\alpha = .05$.

	Population	
	1	2
Sample size	16	13
Sample mean	71.5	79.8
Sample variance	68.35	70.26

9.8   Refer to the data of Exercise 9.7. Give the level of significance for your test.

9.9   In an effort to link cold environments with hypertension in humans, a preliminary experiment was conducted to investigate the effect of cold on hypertension in rats. Two random samples of six rats each were exposed to different environments. One sample of rats was held in a normal environment at 26°C. The other sample was held in a cold 5°C environment. Blood pressures and heart rates were measured for rats for both groups. The blood pressures for the twelve rats are shown in the accompanying table. Do the data provide sufficient evidence to indicate that rats exposed to a 5°C environment have a higher mean blood pressure than rats exposed to a 26°C environment? Test by using $\alpha = .05$.

	26°		5°
Rat	**Blood pressure**	**Rat**	**Blood pressure**
1	152	7	384
2	157	8	369
3	179	9	354
4	182	10	375
5	176	11	366
6	149	12	423

9.10   Refer to Exercise 8.65 (p. 290), in which we measured the dissolved oxygen content in river water to determine whether a stream possessed sufficient oxygen to support aquatic life. A pollution control inspector suspected that a riverside community was releasing semitreated sewage into a river and this, as a consequence, was changing the level of dissolved oxygen of the river. To check his theory, he drew five randomly selected specimens of river water at a location above the town and another five specimens below. The dissolved oxygen readings, in parts per million, are given in the accompanying table. Do the data provide sufficient evidence to indicate a difference in mean oxygen content between locations above and below the town? Use $\alpha = .05$.

Above town	4.8	5.2	5.0	4.9	5.1
Below town	5.0	4.7	4.9	4.8	4.9

9.11 The length of time to complete recovery was recorded for patients randomly assigned and subjected to two different surgical procedures. The data, in days, are given in the accompanying table. (Hint: Use the df row of Table 3 in Appendix 3 that is nearest to the degrees of freedom for these data.) Do the data present sufficient evidence to indicate a difference in mean recovery time for the two surgical procedures? Test by using $\alpha = .05$.

	Procedure 1	Procedure 2
Sample size	21	23
Sample mean	7.3	8.9
Sample variance	1.23	1.49

9.12 Refer to Exercise 9.11. Give the level of significance for your test.

9.13 Two different emission-control devices were being tested to determine the average amount of nitric oxide being emitted by an automobile over a one-hour period of time. Twenty cars of the same model and year were selected for the study. Ten cars were randomly selected and equipped with a type I emission-control device, and the remaining cars were equipped with type II devices. Each of the 20 cars was then monitored for a one-hour period to determine the amount of nitric oxide emitted. Use the following data to test the research hypothesis that the mean level of emission for type I devices (μ_1) is greater than the mean emission level for type II devices (μ_2). Use $\alpha = .01$.

Type I device		Type II device	
1.35	1.28	1.01	0.96
1.16	1.21	0.98	0.99
1.23	1.25	0.95	0.98
1.20	1.17	1.02	1.01
1.32	1.19	1.05	1.02

9.14 The costs of major surgery vary substantially from one state to another due to differences in hospital fees, doctors' fees, malpractice insurance cost, and rent. A study of hysterectomy costs was done in California and Montana. Based on a random sample of 20 patient records from each state, the following sample statistics were obtained:

	Sample mean	Sample standard deviation
California	$ 6,458	$520
Montana	$12,690	$305

Construct a 95% confidence interval for $\mu_1 - \mu_2$ (California minus Montana difference).

9.5 ASSUMPTIONS

The test procedures for comparing two population means in this chapter are based on several assumptions. The first and most critical one is that the two samples are independent. Practically, we mean that the two samples are drawn from two different populations and that the elements of one sample are unrelated to those of the second sample. If this assumption is not valid, then the t-methods of this section will likely be in error and other methods may be appropriate.

The second assumption we make is that the samples are drawn from normal populations. Fortunately, this assumption is less critical because, for modest-sized samples, the Central Limit Theorem applies and hence the sampling distributions for $\bar{x}_1$ and $\bar{x}_2$ will be approximately normal. With independent samples and the combined sample size $n_1 + n_2 > 30$, the t-methods of this section should be reasonably accurate even for a modest skewing in the two populations. A nonparametric alternative to the t-test for independent samples is presented in the next section; this alternative does not require normality.

The third and final assumption is that the two population variances σ_1^2 and σ_2^2 are equal. For now, just examine the sample variances to see that they are approximately equal; later we'll give a test for this assumption. Many efforts have been made to investigate the effect of deviations from the equal variance assumption on the t-methods for independent samples. The general conclusion is that for equal sample sizes, the population variances can differ by as much as a factor of 3 (for example, $\sigma_1^2 = 3\sigma_2^2$), and the t-methods will still apply. This is remarkable and provides a convincing argument to use equal sample sizes. When the sample sizes are different, the most serious case is when the smaller sample size is associated with the larger variance. In this situation and in others where the sample variance s_1^2 and s_2^2 suggest that $\sigma_1^2 \neq \sigma_2^2$, there is an approximate t-test using the test statistic

$$t' = \frac{\bar{x}_1 - \bar{x}_2}{\sqrt{\dfrac{s_1^2}{n_1} + \dfrac{s_2^2}{n_2}}}$$

Percentage points of a t-distribution with modified degrees of freedom can be used to set the rejection region for $H_0: \mu_1 - \mu_2 = D_0$ (Welch, 1938). This t-test is summarized here.

Approximate *t*-test for Independent Samples, Unequal Variance

H_0: $\mu_1 - \mu_2 = D_0$

H_a: 1. $\mu_1 - \mu_2 > D_0$
   2. $\mu_1 - \mu_2 < D_0$
   3. $\mu_1 - \mu_2 \neq D_0$

Test statistic: $t' = \dfrac{\bar{x}_1 - \bar{x}_2 - D_0}{\sqrt{\dfrac{s_1^2}{n_1} + \dfrac{s_2^2}{n_2}}}$

Rejection region: For a specified value of α,

   1. reject H_0 if $t' > t_\alpha$
   2. reject H_0 if $t' < t_\alpha$
   3. reject H_0 if $|t'| > t_{\alpha/2}$

where

$$ df = \frac{(n_1 - 1)(n_2 - 1)}{(n_2 - 1)c^2 + (1 - c)^2(n_1 - 1)} \qquad \text{where} \quad c = \frac{s_1^2/n_1}{\dfrac{s_1^2}{n_1} + \dfrac{s_2^2}{n_2}} $$

Note: If the computed value of df is not an integer, *round down* to the nearest integer.

The test based on the t'-statistic is sometimes referred to as the *separate-variance t-test* because we use the separate sample variances s_1^2 and s_2^2, rather than a pooled sample variance.

Example 9.5

An experiment was conducted to compare the mean number of tapeworms in the stomachs of sheep that had been treated for worms against the mean number of those that were untreated. Thirteen worm-infected sheep were randomly divided into two groups; seven were injected with the drug and the remaining six were left untreated. After a 6-month period, worm counts were recorded for all thirteen sheep. These data are shown here.

Drug-treated sheep	5	13	18	6	4	2	15
Untreated	40	54	26	63	21	37	

Test the research hypothesis $H_a: \mu_1 - \mu_2 < 0$ under the assumption that the two population variances are different. Use $\alpha = .05$.

Solution It is easy to verify that

$$\bar{x}_1 = 9.00 \qquad \bar{x}_2 = 40.17$$
$$s_1^2 = 38.67 \qquad s_2^2 = 258.17$$

Then the statistical test is set up as follows:

$$H_0: \mu_1 - \mu_2 = 0$$
$$H_a: \mu_1 - \mu_2 < 0$$

Test statistic:

$$t' = \frac{\bar{x}_1 - \bar{x}_2}{\sqrt{\dfrac{s_1^2}{n_1} + \dfrac{s_2^2}{n_2}}} = \frac{9 - 40.17}{\sqrt{\dfrac{38.67}{7} + \dfrac{258.17}{6}}} = -4.47$$

In order to compute the rejection region we need

$$c = \frac{s_1^2/n_1}{s_1^2/n_1 + s_2^2/n_2} = \frac{38.67/7}{38.67/7 + 258.17/6} = .114$$
$$c^2 = .013$$

and

$$\text{df} = \frac{(n_1 - 1)(n_2 - 1)}{(n_2 - 1)c^2 + (1 - c)^2(n_1 - 1)} = 6.283 \qquad \text{which is rounded to 6}$$

Rejection region: For $\alpha = .05$ and df = 6 reject H_0 if $t' < -1.943$.

Since $t' = -4.47$ is less than -1.943, we reject H_0 and conclude that μ_1, the mean worm count for treated sheep, is less than that for untreated sheep.

9.6 USING COMPUTERS

The weight gain data of Table 9.3 for Example 9.3 have been reworked to illustrate how Minitab can be used to compute a 95% confidence interval for $\mu_1 - \mu_2$ and to conduct a two-sample t-test about $\mu_1 - \mu_2$; the 3 Minitab TWOSAMPLE T procedure does both of these procedures. Compare the output shown here to the results obtained for Example 9.3. You may be a bit bewildered in noting that the degrees of freedom shown next to the p-value in the output are 16.9 rather than 17. For all practical purposes, you can ignore the df value printed out for Minitab. (The degrees of freedom shown in Minitab do correspond to the "unequal variance" test described briefly for the SAS output.)

Corresponding output from SAS shows how PROC TTEST can be used to conduct a two-sample t test about $\mu_1 - \mu_2$. The relevant t-test is shown in the "equal variance" row of the output. The equal variance row of the output corresponds to the test presented in this chapter. The "unequal" variance presents the results for an approximate t-test that is used when the assumption of equal variances is invalid. As a practical matter, in most cases conclusions drawn from the equal and unequal variance tests are identical. If the conclusions do not agree, the usual t-test of Section 9.4 is probably inappropriate because the assumption of equal variances is violated. More details about the unequal variance test are presented in Hildebrand and Ott (1987).

The Minitab and SAS programs illustrated in this section can be used for other data sets to make inferences about $\mu_1 - \mu_2$. Additional examples of Minitab output are shown in Exercises 9.33, 9.45, and 9.48.

Minitab Output

```
MTB > SET INTO C1
DATA> 14.0  12.5  10.2  9.8  10.5  11.2  15.0  22.0  13.0  9.6
DATA> END
MTB > SET INTO C2
DATA> 14.4  18.2  19.5  21.2  15.3  11.6  12.8  13.1  11.3
DATA> END

MTB > PRINT C1 C2

 ROW      C1      C2

   1     14.0    14.4
   2     12.5    18.2
   3     10.2    19.5
   4      9.8    21.2
   5     10.5    15.3
   6     11.2    11.6
   7     15.0    12.8
   8     22.0    13.1
   9     13.0    11.3
  10      9.6

MTB > TWOSAMPLE T FOR C1 VS C2;
SUBC> POOLED.

TWOSAMPLE T FOR C1 VS C2
        N      MEAN     STDEV   SE MEAN
C1     10     12.78      3.73      1.18
C2      9     15.27      3.58      1.19
```

(continued)

```
95 PCT CI FOR MU C1 - MU C2: (-6.032, 1.059)

TTEST MU C1 = MU C2 (VS NE): T= -1.48  P=0.16  DF=  17

POOLED STDEV =      3.66

MTB > STOP
```

SAS Output

```
OPTIONS LS=78 PS=60 NODATE NONUMBER;
DATA RAW;
  INPUT DIET $ WT_GAIN;
  LABEL WT_GAIN='WEIGHT GAIN';
  LIST;  CARDS;
  I    14.0
  I    12.5
  I    10.2
  I     9.8
  I    10.5
  I    11.2
  I    15.0
  I    22.0
  I    13.0
  I     9.6
  II   14.4
  II   18.2
  II   19.5
  II   21.2
  II   15.3
  II   11.6
  II   12.8
  II   13.1
  II   11.3
;
PROC PRINT N;
TITLE1 'WEIGHT GAIN USING TWO DIFFERENT DIETS';
TITLE2 'LISTING OF DATA';

PROC TTEST;
  CLASS DIET;
  VAR WT_GAIN;
TITLE2 'EXAMPLE OF PROC TTEST';
RUN;
```

SAS Output

```
        WEIGHT GAIN USING TWO DIFFERENT DIETS
                 LISTING OF DATA

            OBS     DIET    WT_GAIN

             1       I        14.0
             2       I        12.5
             3       I        10.2
             4       I         9.8
             5       I        10.5
             6       I        11.2
             7       I        15.0
             8       I        22.0
             9       I        13.0
            10       I         9.6
            11      II        14.4
            12      II        18.2
            13      II        19.5
            14      II        21.2
            15      II        15.3
            16      II        11.6
            17      II        12.8
            18      II        13.1
            19      II        11.3

                  N = 19
```

```
        WEIGHT GAIN USING TWO DIFFERENT DIETS
                EXAMPLE OF PROC TTEST

                  TTEST PROCEDURE

Variable: WT_GAIN      WEIGHT GAIN

DIET       N        Mean       Std Dev      Std Error      Minimum       Maximum
---------------------------------------------------------------------------------
I          10   12.78000000   3.72522930   1.17802094   9.60000000    22.00000000
II          9   15.26666667   3.57840747   1.19280249   11.30000000   21.20000000

Variances        T        DF      Prob>|T|
--------------------------------------------
Unequal       -1.4833    16.9      0.1564
Equal         -1.4800    17.0      0.1572

For HO: Variances are equal, F' = 1.08    DF = (9,8)    Prob>F' = 0.9205
```

SUMMARY

In this chapter we presented methods for estimating and testing hypotheses about the difference between two population means ($\mu_1 - \mu_2$). All the methods are similar to those of the corresponding t-methods for inferences about a population mean. The confidence intervals that we presented for μ (in Chapter 8) and for $\mu_1 - \mu_2$ in this chapter are of the form

point estimate $\pm$ (t-value) standard error

where the t-value comes from Table 3 of Appendix 3. Similarly the statistical tests for μ and $\mu_1 - \mu_2$ are composed of four parts (a null hypothesis, a research (or alternative) hypothesis, a test statistic, and a rejection region) and employ the technique of proof by contradiction. We try to verify the research hypothesis by gathering data to contradict the null hypothesis. The material in this and the preceding chapter lays the foundations for statistical inferences presented in the remainder of the text. Because of this, it will help to review the material periodically as new topics are introduced.

KEY TERMS

point estimate of $\mu_1 - \mu_2$
s_p^2 (weighted average)
assumptions for t-test

KEY FORMULAS

1. General confidence interval for $\mu_1 - \mu_2$

$$(\bar{x}_1 - \bar{x}_2) \pm t\, s_p \sqrt{\frac{1}{n_1} + \frac{1}{n_2}}$$

where $s_p = \sqrt{\dfrac{(n_1 - 1)s_1^2 + (n_2 - 1)s_2^2}{n_1 + n_2 - 2}}$

2. Statistical test for $\mu_1 - \mu_2$

Null hypothesis: $\mu_1 - \mu_2 = 0$

Test statistic: $t = \dfrac{\bar{x}_1 - \bar{x}_2}{s_p \sqrt{\dfrac{1}{n_1} + \dfrac{1}{n_2}}},$ df $= n_1 + n_2 - 2$

3. Approximate statistical test for $\mu_1 - \mu_2$, unequal variances

Null hypothesis: $\mu_1 - \mu_2 = 0$

Test statistic: $t' = \dfrac{\bar{x}_1 - \bar{x}_2}{\sqrt{\dfrac{s_1^2}{n_1} + \dfrac{s_2^2}{n_2}}}$

$\mathrm{df} = \dfrac{(n_1 - 1)(n_2 - 1)}{(n_2 - 1)c^2 + (1 - c)^2(n_1 - 1)}$ where $c = \dfrac{s_1^2/n_1}{s_1^2/n_1 + s_2^2/n_2}$

SUPPLEMENTARY EXERCISES

9.15 Two alloys, A and B, are used in the manufacture of steel bars. We wish to estimate the difference in load capacity of bars made of each alloy. A sample of nine bars of alloy A had a mean load capacity of 28.5 tons and a standard deviation of 2.5 tons, while a sample of thirteen bars of alloy B had an average load capacity of 23.2 tons with a standard deviation of 1.8 tons. Find a 90% confidence interval for the difference.

9.16 It is thought that exposure to ozone increases lung capacity. In order to investigate this possibility, a researcher exposed eight rats to ozone in the amount of 2 parts per million for a period of 30 days. The average lung capacity for these rats at the end of the 30 days was 9.4 mL, and the standard deviation was 0.8 mL. A control group of six rats did not have exposure to ozone and their lung capacity averaged 8.3 mL with a standard deviation of 0.7 mL.

(a) Is there sufficient evidence at a 5% significance level to support the original conjecture? Justify your answer statistically with specific numerical values.

(b) Give the p-value for the hypothesis test.

9.17 In a study of the possible factors that influence the frequency of birds being hit by aircraft (which, ironically, is viewed as a hazard to the aircraft), the noise level of various jets was measured just seconds after their wheels left the ground. The jets were either wide-bodied or narrow-bodied. Twenty-two wide-bodied jets had noise levels averaging 106.4 decibels (dB) and a standard deviation of 3.3 dB, while ten narrow-bodied jets had noise levels averaging 114.0 dB with a standard deviation of 2.0 dB.

Test whether the average noise levels in the two populations of jets are the same. Report the level of significance of the sample as evidence that the two types of jets have different noise levels.

9.18 A farmer was interested in determining which of two soil fumigants, A or B, is more effective in controlling the number of parasites in a particular agricultural crop. To compare the fumigants, four small fields were divided into equal areas: fumigant A was applied to one part and fumigant B to the other. Crop samples of equal size were taken from each of the eight plots and the numbers of parasites per square foot were counted. The data are in the following table. Do the data provide sufficient evidence to indicate a difference in the mean level of parasites for the two fumigants?

Field	A	B
1	15	9
2	5	3
3	8	6
4	8	4

9.19 A psychologist was interested in comparing the average length of time it takes individuals to complete two different psychological checklists. From a relatively homogeneous group of twenty individuals, ten were randomly assigned to list 1 and the other ten to list 2. The appropriate checklists were then administered, and the amount of time required to complete the task was recorded for each individual. These data are summarized here. Find a 95% confidence interval for $(\mu_1 - \mu_2)$, the difference in the mean completion times. What assumptions must you make?

List 1	List 2
$\bar{x}_1 = 54.3$ minutes	$\bar{x}_2 = 48.1$ minutes
$n_1 = 10$	$n_2 = 10$
$s_1^2 = 16.0$	$s_2^2 = 12.2$

9.20 Refer to the data of Exercise 9.19. Construct a 99% confidence interval for $(\mu_1 - \mu_2)$.

9.21 Use the data of Exercise 9.9 (p. 314) to construct a 90% confidence interval for $(\mu_1 - \mu_2)$, the difference in mean blood pressure in rats subjected to the two environments.

9.22 An experiment was conducted to compare the mean lengths of time required for the bodily absorption of two drugs, A and B. Ten people were randomly selected and assigned to each drug treatment. Each of the ten persons in the sample received an oral dosage of the assigned drug, and the length of time (in minutes) for the drug to reach a specified level in the blood was recorded. The means and variances for the two samples are given in the accompanying table. Find a 95% confidence interval for the difference in mean times for absorption.

	Drug A	**Drug B**
Sample mean	27.2	33.5
Sample variance	16.36	18.92

9.23 The accompanying computer output gives the drop in blood pressure for three groups of six rats from a strain of hypertensive rats. The six rats in the first group were treated with a low dose of an antihypertensive product, the second group with a higher dose of the same antihypertensive product, and the third group with an inert control. Note that the variability in blood pressure decreases, even for rats in the control group. Also note that negative values represent increases in blood pressure.

(a) Draw conclusions for a comparison of the mean drop for the high-dose group and the control group.

(b) Is there evidence to indicate a difference between the low- and high-dose groups? Explain.

```
DESCRIPTIVE STATISTICS

              FILE: Low-Dose Group
                   -51.00000
                    15.00000
                    48.00000
                    65.00000
                   -20.00000
                    75.00000


Low-dose Group

NUMBER:              6
  MEAN:         22.00000
STD DEV:        49.95198
```

```
DESCRIPTIVE STATISTICS

              FILE: High-Dose Group
                    69.00000
                    24.00000
                    63.00000
                    87.50000
                    77.50000
                    40.00000


High-Dose Group

NUMBER:              6
  MEAN:         60.16667
STD DEV:        23.86769
```

DESCRIPTIVE STATISTICS

FILE: Control Group
                        9.00000
                        12.00000
                        63.00000
                        77.50000
                       −7.50000
                        32.50000

Control Group

NUMBER: 6
  MEAN: 26.58333
STD DEV: 29.66381

UNPAIRED T TEST

	Low-Dose Group	High-Dose Group
	−51.00000	69.00000
	15.00000	24.00000
	48.00000	63.00000
	65.00000	87.50000
	−20.00000	77.50000
	75.00000	40.00000

NO. OF OBSERVATIONS	6.	6.
MEAN	22.00000	60.16667
STANDARD DEVIATION	49.95198	23.86769
STANDARD ERROR	20.39281	9.74394

RATIO OF MEANS (2ND/1ST)		2.73485
DIFFERENCE OF MEANS (2ND-1ST)		38.16667
STANDARD ERROR OF DIFFERENCE		22.60113
95% CONFIDENCE INTERVAL	−12.18865,	88.521991
FOR DIFFERENCE OF MEANS		
RATIO OF VARIANCES (2ND/1ST)		0.22831

T STATISTIC (EQUAL VARIANCES)	1.68871
DEGREES OF FREEDOM	10
PROBABILITY	0.12216

UNPAIRED T TEST

FILES

High-Dose Group	Control Group
69.00000	9.00000
24.00000	12.00000
63.00000	36.00000
87.50000	77.50000

(continued)

```
                    77.50000                              -7.50000
                    40.00000                              32.50000

        NO OF OBSERVATIONS                  6.            6.
        MEAN                             60.16667      26.58333
        STANDARD DEVIATION               23.86769      29.66381
        STANDARD ERROR                    9.74394      12.11020

        RATIO OF MEANS (2ND/1ST)                  0.44183
        DIFFERENCE OF MEANS (2ND-1ST)           -33.58333

        STANDARD ERROR OF DIFFERENCE             15.54353
        95% CONFIDENCE INTERVAL          -68.21432,    1.047661
            FOR DIFFERENCE OF MEANS
        RATIO OF VARIANCES (2ND/1ST)              1.54466

        T STATISTIC (EQUAL VARIANCES)            -2.16060
        DEGREES OF FREEDOM                       10
        PROBABILITY                               0.05605
```

9.24 Use the data of Exercise 9.23 to construct a 95% confidence interval for $(\mu_1 - \mu_3)$, the difference in population means for the low-dose group and the control group.

9.25 The elasticity of plastic can vary depending on the process by which the plastic is prepared. To compute the elasticity of plastic produced by two different processes, six samples from each process were analyzed for elasticity. These data are shown in the accompanying table. Do the data present sufficient evidence to indicate a difference in the mean elasticities for the two processes? Use $\alpha = .05$.

Process A	Process B
6.1	9.1
9.2	8.2
8.7	8.6
8.9	6.9
7.6	7.5
7.1	7.9
$\bar{x}_1 = 7.93$	$\bar{x}_2 = 8.03$
$s_1^2 = 1.46$	$s_2^2 = .61$

9.26 The purity of ore can vary greatly from one location to another. One determining factor, then, in choosing a site for mining would be the metal content of the ore. Two prospective locations were to be compared. Three ore samples were obtained from each location and analyzed to determine the metal content of the ore; see the accompanying table. Do the data provide sufficient evidence to indicate a difference in mean metal content for the two locations? Use $\alpha = .01$.

Location 1	50.1	49.6	51.2
Location 2	47.0	46.0	46.4

9.27 Refer to Exercise 9.26. Give the approximate level of significance for your test.

9.28 The amount of work accomplished on a construction job is frequently approximated by a visual estimate of the amount of material used per day. Six experienced men were employed to approximate the number of bricks used on two different jobs. Three were randomly assigned to job 1 and three to job 2. Each man, independent of the others, approximated the number of bricks used. The approximations (in thousands of bricks) are shown in the accompanying table. Assume that the men have been randomly selected from a very large set of experienced people. Thus μ_1 is the mean of the large set of approximations produced by people who visually estimate the number of bricks in job 1. Similarly, μ_2 is a corresponding mean of a large set of approximations that could be acquired for job 2. Do these data provide evidence to indicate that the mean number of bricks approximated for job 1 differs from the mean approximation for job 2? Use $\alpha = .05$.

Job 1	Job 2
107.2	103.2
108.1	105.9
105.7	104.1
$\bar{x}_1 = 107.00$	$\bar{x}_2 = 104.40$
$s_1^2 = 1.47$	$s_2^2 = 1.89$

9.29 Refer to Exercise 9.25. Estimate the difference in the mean elasticities of the two processes, using a 95% confidence interval.

9.30 Refer to Exercise 9.26. Estimate the difference in the mean metal content of the two locations, using a 90% confidence interval.

9.31 Refer to Exercise 9.28. Construct a 95% confidence interval for the difference in the mean estimates for the two jobs.

9.32 An experiment was conducted to investigate the effect of the drug Propranolol in reducing hypertension in rats. Two groups of rats were studied. One group received the drug, and the other group served as the control group. Hypertension was induced in the rats by exposure to a cold environment. The extent of the induced hypertension in a given rat was measured by monitoring its blood pressure. After six weeks of cold exposure the sample blood pressure data were summarized for the two groups; see the accompanying table. Use these data to determine whether there is evidence to indicate that rats treated with Propranolol have less hypertension, on the average, than those that are untreated. Use $\alpha = .05$.

	Group 1 (received Propranolol)	Group 2 (control)
Sample size	7	5
Sample mean	129.43	167.60
Sample variance	583.95	249.30

9.33 Refer to the data of Exercise 9.4 (p. 309). Use the Minitab output here to conduct a two-sample t-test for H_0: $\mu_1 - \mu_2 = 0$ versus H_a: $\mu_1 - \mu_2 \neq 0$. Give the p-value for your test and draw conclusions.

```
MTB > READ INTO C1 C2
DATA> 282    284
DATA> 279    285
DATA> 280    286
DATA> 278    277
DATA> 275    283
DATA> END

      5 ROWS READ
MTB > PRINT C1 C2

 ROW     C1    C2

   1     282   284
   2     279   285
   3     280   286
   4     278   277
   5     275   283

MTB > TWOSAMPLE T FOR C1 VS C2;
SUBC> POOLED.

TWOSAMPLE T FOR C1 VS C2
      N      MEAN    STDEV   SE MEAN
C1    5    278.80     2.59     1.16
C2    5    283.00     3.54     1.58

95 PCT CI FOR MU C1 - MU C2: (-8.720, 0.3200)

TTEST MU C1 = MU C2 (VS NE): T= -2.14  P=0.064  DF=  8

POOLED STDEV =         3.10

MTB > STOP
```

9.34 A processor of recycled aluminum cans is concerned about the levels of impurities (principally other metals) contained in lots from two sources. Laboratory analysis of sample lots yields the following data (in kilograms of impurities per hundred kilograms of product):

Source I: 3.8   3.5   4.1   2.5   3.6   4.3   2.1   2.9   3.2   3.7   2.8   2.7
        (mean = 3.267, standard deviation = .676)
Source II: 1.8   2.2   1.3   5.1   4.0   4.7   3.3   4.3   4.2   2.5   5.4   4.6
        (mean = 3.617, standard deviation = 1.365)

(a) Calculate the pooled variance and standard deviation.

(b) Calculate a 95% confidence interval for the difference in mean impurity levels.

(c) Can the processor conclude, using $\alpha = .05$, that there is a nonzero difference in means?

9.35 To compare the performance of microcomputer spreadsheet programs, teams of three students each choose whatever spreadsheet program they wish. Each team is given the same set of standard accounting and finance problems to solve. The time (in minutes) required for each team to solve the set of problems is recorded. The data shown here were obtained for the two most widely used programs; also displayed are the sample means, sample standard deviations, and sample sizes.

Program	Time										$\bar{x}$	s	n
A	39	57	42	53	41	44	71	56	49	63	51.50	10.46	10
B	43	38	35	45	40	28	50	54	37	29			
	36	27	52	33	31	30					38.00	8.67	16

(a) Calculate the pooled variance.

(b) Use this variance to find a 99% confidence interval for the difference of population means.

(c) According to this interval, can the null hypothesis of equal means be rejected at $\alpha = .01$?

9.36 Redo parts (b) and (c) of Exercise 9.35 using a separate-variance (t') method. Which method is more appropriate in this case? How critical is it to use one or the other?

9.37 Refer to Exercise 9.3 (p. 308). Suppose the experimenter felt that breast feeding might be better.

(a) State the null hypothesis in

(i) Statistical terms or symbols

(ii) Plain English

(b) State the alternative hypothesis.

(c) Locate the rejection region based on $\alpha = .05$.

(d) Compute the proper test statistic.

(e) Draw conclusions in

(i) Statistical terms

(ii) Plain English

9.38 Refer to Exercise 9.37. Suppose someone begins to question you because he had heard something about "statistical tests" that result in type I or type II errors. Based on your conclusion, could you have made

_____ A type I error?

_____ A type II error?

_____ Both kinds of errors?

_____ Neither error?

(Answer each by yes or no)

9.39 Educators compared scores of nursing degree students with scores of students from diploma and associate degree programs on a state licensing board examination. By random sampling procedures, the educators drew a sample of five from those completing the degree program, resulting in a mean score of 400 with a standard deviation of 15. A random sample of five drawn from the associate degree program had a mean of 370 with a standard deviation of 30. Can the licensing board conclude that the mean score of nursing students completing the degree program is higher than the mean score of those who complete the associate program? Base your answer on the results of a statistical test. Give the approximate p-value for your test.

9.40 An educator wants to compare the effects of two different teaching methods. Two classes of students are selected at random; class 1 receives method 1 and class 2 method 2. A comprehensive standard examination is administered to the two classes to determine the effectiveness of the two methods at the end of the test period. The relevant data are shown below.

	Class 1	**Class 2**
Sample size	$n_1 = 64$	$n_2 = 64$
Average test score	$\bar{x}_1 = 88$	$\bar{x}_2 = 80$
Sample variance	$s_1^2 = 56$	$s_2^2 = 56$

Determine a 95% confidence interval for the difference between the two population means on the basis of the difference between the two sample means. Would a 90% confidence interval be wider?

9.41 A study was conducted to see if food prices charged in a ghetto area are higher than those charged in a more affluent suburban area. Food prices were obtained from nine stores in each area and a food price index computed. The summary results for each area were as follows:

Ghetto area	Suburban area
$n_1 = 9$	$n_2 = 9$
$\bar{x}_1 = 11.1$	$\bar{x} = 10.5$
$s_2^2 = 2.5$	$s_2^2 = 1.5$

Conduct a statistical test of $H_0: \mu_1 - \mu_2 = 0$ versus $H_a: \mu_1 - \mu_2 > 0$. Show all steps in the test of hypothesis and state your conclusion in *non-statistical terms*. Use $\alpha = .05$. What type of error (type I or II) could you have made?

9.42 We are given the following data summarizing information on two independent samples taken from populations whose variances are known to be equal:

	n	$\bar{x}$	s^2
Sample 1	6	30	60
Sample 2	4	20	60

From these data, the following was computed:

$$t = \frac{30 - 20}{7.75\sqrt{\dfrac{1}{6} + \dfrac{1}{4}}} = 2$$

(a) Show, by computing it, how the 7.75 in the above computation was obtained.

(b) Suppose that the null hypothesis had been tested against a two-tailed alternative with $\alpha = .05$.

   (i) What would be the rejection region?

   (ii) Would the null hypothesis be rejected?

(c) Suppose the null hypothesis had been tested against the one-tailed alternative that the mean of population 1 was larger than the mean of population 2, with $\alpha = .05$.

   (i) What would be the rejection region for this test?

   (ii) Would the null hypothesis be rejected?

9.43 To test the research hypothesis that teacher expectation can improve student performance, two groups of 100 students were compared. Teachers of the experimental group were told that their students would show large IQ gains during the test semester, while teachers of the control group were told nothing. At the end of the semester, IQ change scores were calculated with the following results:

	Mean	Standard deviation	Sample size
Experimental	16.5	14.2	100
Control	7.0	13.1	100

(a) Test the null hypothesis of no effect on mean IQ change scores.

(b) State your conclusion in two ways:

(i) In statistical terms.

(ii) In nontechnical terms as you might explain it to an intelligent person who was not familiar with statistical terminology.

 **9.44** Those running for public office must now report the amount of money spent in each campaign. It has been reported that women candidates usually find it difficult to raise money and therefore spend less in their campaigns than men candidates. Suppose the accompanying data represent the campaign expenditures of a randomly selected group of men and women candidates who have just completed their campaigns for public office. Do the data support the claim that women candidates generally spend less in their campaigns for public office than men candidates?

(a) Would you use a one-tailed test or two-tailed test of hypothesis in this case? Why?

Cost of campaign (in thousands of dollars)		
	Women	Men
	138	134
	127	137
	134	135
	125	140
		130
		134
Sum	524	810
Mean	131	135

(b) State the null and alternative hypotheses in

(i) Statistical terms or symbols.

(ii) Plain English.

9.45 Refer to Exercise 9.44. Summary data for the two samples are shown in the accompanying Minitab output along with the results of a *t*-test for $\mu_1 - \mu_2$. What assumptions must we make in order to run a *t*-test? Which one (if any) could cause a problem for these data?

```
MTB > SET INTO C1
DATA> 138   127   134   125
DATA> END
MTB > SET INTO C2
DATA> 134   137   135   140   130   134
DATA> END
MTB > PRINT C1 C2

  ROW      C1      C2

    1     138     134
    2     127     137
    3     134     135
    4     125     140
    5             130
    6             134

MTB > TWOSAMPLE T FOR C1 VS C2;
SUBC> POOLED;
SUBC> ALTERNATIVE -1.

TWOSAMPLE T FOR C1 VS C2
         N      MEAN     STDEV    SE MEAN
C1   4       131.00      6.06       3.03
C2   6       135.00      3.35       1.37

95 PCT CI FOR MU C1 - MU C2: (-10.78, 2.782)

TTEST MU C1 = MU C2 (VS LT): T= -1.36   P=0.11   DF=  8

POOLED STDEV =          4.56

MTB > STOP
```

💲 9.46 Suppose you are the personnel manager for a company and you suspect a difference in the mean length of work time lost due to sickness for two types of employees: those who work at night versus those who work during the day. Particularly, you suspect that the mean time lost for the night shift exceeds the mean for the day shift. To check your theory, you randomly sample the records for ten employees for each shift category and record the number of days lost due to sickness within the past year. The data are shown next.

Night shift	Day shift
15	8
10	9
10	2
7	0
7	10
4	9
9	9
6	7
10	3
12	3

(a) Would you use a one-tailed test or two-tailed test in your test hypothesis? Why?

(b) What is the pooled estimate of σ?

(c) Conduct the statistical test and show all parts of the test.

    (i) Null hypothesis

    (ii) Alternative hypothesis

    (iii) Test statistic and computations

    (iv) Rejection region

    (v) Conclusion

        (a) In statistical terms

        (b) In plain English that nonstatisticians can understand

9.47 Refer to Exercise 9.46. Based on your decision could you have made

_____ A type I error

_____ A type II error

_____ Both type I and type II errors

_____ Neither error

(Answer each with yes or no)

9.48 Refer to the data of Exercise 9.9 (p. 314). Use the output shown here to determine a 95% confidence interval for $\mu_1 - \mu_2$.

```
MTB > READ INTO C3 C4
DATA> 152    384
DATA> 157    369
DATA> 179    354
DATA> 182    375
DATA> 176    366
DATA> 149    423
DATA> END
```

(continued)

```
        6 ROWS READ
MTB > PRINT C3 C4

   ROW     C3      C4

     1     152     384
     2     157     369
     3     179     354
     4     182     375
     5     176     366
     6     149     423

MTB > TWOSAMPLE T FOR C3 VS C4;
SUBC> POOLED;
SUBC> ALTERNATIVE -1.

TWOSAMPLE T FOR C3 VS C4
        N       MEAN      STDEV    SE MEAN
C3   6         165.8      14.8       6.03
C4   6         378.5      24.0       9.78

95 PCT CI FOR MU C3 - MU C4: (-238.3, -187.1)

TTEST MU C3 = MU C4 (VS LT): T= -18.51  P=0.0000  DF=  10

POOLED STDEV =          19.9

MTB > STOP
```

9.49 A study was carried out to determine whether nonworking wives from middle-class families have more voluntary association memberships than nonworking wives from working-class families. A random sample of housewives was obtained, and each was asked for information about her husband's occupation and her own memberships in voluntary associations. On the basis of their husbands' occupations, the women were divided into middle-class and working-class groups, and the mean number of voluntary association memberships was computed for each group.

For the 15 middle-class women, the mean number of memberships per woman was $\bar{x}_1 = 3.4$ with $s_1 = 2.5$. For the 15 working-class wives, $\bar{x}_2 = 2.2$ with $s_2 = 2.8$. Use these data to construct a 95% confidence interval for $\mu_1 - \mu_2$.

9.50 A regional IRS auditor ran a test on a sample of returns filed by March 15 to determine whether the average refund for taxpayers is larger this year than last year. Sample data are shown here for a random sample of 100 returns for each year.

	Last year	This year
Mean	320	410
Variance	300	350
Sample size	100	100

(a) In a test of hypothesis, would you use a one-tailed or two-tailed test? Why?

(b) What assumptions are required to conduct a t-test of H_0: $\mu_1 - \mu_2 = 0$? Do you think the assumptions hold, and why (or why not)?

9.51 Miss American Pageant officials maintain that their Pageant is not a beauty contest and that talent is more important than beauty when it comes to success in the Pageant. In an effort to evaluate the assertion, a random sample of 55 preliminary talent competition winners and a random sample of 53 preliminary swimsuit competition winners were taken to see if there was a significant difference in the mean amount won for the two groups. For the 55 preliminary talent competition winners the mean amount was $8,645 with standard deviation of $5,829; for the 53 preliminary swimsuit winners the mean amount won was $9,198 with standard deviation of $8,185. Compute a 95% confidence interval for the difference in the mean amount won by the two groups. Does your confidence interval confirm what the Pageant officials contend?

9.52 A visitor to the United States from France insisted that recordings made in Europe are likely to have selections with longer playing times than recordings made in the United States. In order to verify or contradict the contention a random sample of selections was taken from a group of records produced in France and Germany, and another random sample of selections was taken from American-produced records. The results of the samples were as given below.

	Foreign produced	American produced
Number in sample	14	14
Mean playing time in seconds	207.45	182.54
Standard deviation	41.43	37.32

Do the foreign-produced selections have longer mean playing times? Use $\alpha = .05$.

9.53 A major Federal agency located in Washington, D.C., regularly conducts classes in PL/1, a computer programming language used in the programs written within the agency. One week the course was taught by an individual associated with an outside consulting firm. The following week a similar course was taught by a member of the computer staff of the agency. The following results were achieved by the classes:

Taught by outsider	38	42	53	37	36	48	47	47	44
Taught by staff member	46	33	38	60	58	52	44	45	51

The values represent scores aggregated over the one-week course out of a potential maximum of 64. Do the data present sufficient evidence to indicate a difference in teaching effectiveness, assuming that the scores reflect teaching effectiveness? Use $\alpha = .05$.

💲 9.54   A firm has a generous but rather complicated policy concerning end-of-year bonuses for its lower-level managerial personnel. The policy's key factor is a subjective judgment of "contribution to corporate goals." A personnel officer took samples of 24 females and 36 male managers to see if there was any difference in bonuses, expressed as a percentage of yearly salary. The data are listed here:

Gender	Bonus percentage								
F	9.2	7.7	11.9	6.2	9.0	8.4	6.9	7.6	7.4
	8.0	9.9	6.7	8.4	9.3	9.1	8.7	9.2	9.1
	8.4	9.6	7.7	9.0	9.0	8.4			
M	10.4	8.9	11.7	12.0	8.7	9.4	9.8	9.0	9.2
	9.7	9.1	8.8	7.9	9.9	10.0	10.1	9.0	11.4
	8.7	9.6	9.2	9.7	8.9	9.2	9.4	9.7	8.9
	9.3	10.4	11.9	9.0	12.0	9.6	9.2	9.9	9.0

A computer program yielded the output shown here.

```
SAMPLE                    1        2

MEAN                   8.5333   9.6833

ST. DEV.               1.1890   1.0038

SAMPLE SIZE              24        36

SUM OF RANKS           48.10   1349.0

POOLED-VARIANCE T STATISTIC = -4.037
   2-TAILED P-VALUE = 0.0004
SEPARATE-VARIANCE T STATISTIC = -3.901
   2-TAILED P-VALUE = 0.0008
   APPROX DF = 43

RANK SUM Z STATISTIC = -3.787
   2-TAILED P-VALUE = 0.0002
```

(a) Identify the value of the pooled-variance t-statistic (the usual t-test based on the equal variance assumption).

(b) Identify the value of the t'-statistic.

(c) Use both statistics to test the research hypothesis of unequal means at $\alpha = .05$ and at $\alpha = .01$. Does the conclusion depend on which statistic is used?

9.55 Two possible methods for retrofitting jet engines to reduce noise are being considered. Identical planes are fitted with two systems. Noise-recording devices are installed directly under the flight path of a major airport. Each time one of the planes lands at the airport, a noise level is recorded. The data are analyzed by a computer software package (SAS). The relevant output is:

VARIABLE: DBREAD

SYSTEM	M	MEAN	STD DEV	STD ERROR	MINIMUM	MAXIMUM	VARIANCES	T	DF	PROB > |T|
H	42	100.90476190	2.99438111	0.46204304	95.00000000	110.00000000	UNEQUAL	4.4491	21.5	0.0002
R	20	92.50000000	8.19178022	1.83173774	79.00000000	111.00000000	EQUAL	5.9126	60.0	0.0001

(a) Locate the t-statistic.

(b) Locate the t'-statistic.

(c) Can the research hypothesis of unequal means be supported using $\alpha = .01$? Does it matter which statistic is used?

EXERCISES FROM THE DATA BASE

9.56 Refer to the clinical trials data base in Appendix 1. Use the HAM-D total score data to conduct a statistical test of H_0: $\mu_D - \mu_A = 0$ vs H_a: $\mu_D - \mu_A > 0$; that is, we want to know whether the placebo group (D) has a higher (worse) mean total depression score at the end of the study than the group receiving treatment A. Use $\alpha = .05$. What are your conclusions?

9.57 Refer to Exercise 9.56 and repeat this same comparison with the placebo group for treatment B, and then for treatment C. Give the p-value for each of these tests. Which of the three treatment groups (A, B, or C) appears to have the lowest mean HAM-D total score?

9.58 Use the clinical trials data base to construct a 95% confidence interval for $\mu_D - \mu_A$ based on the HAM-D anxiety score data. What can you conclude about $\mu_D - \mu_A$ based on this interval?

9.59 Refer to the clinical trials data base. Compare the mean ages for treatment groups B and D using a two-sided statistical test. Set up all parts of the test using $\alpha = .05$; draw a conclusion. Why might it be important to have patients with similar ages in the different treatment groups when studying the effects of several drug products on the treatment of depression?

9.60 Refer to Exercise 9.59. What other variables should be comparable among the treatment groups in order to draw conclusions about the effectiveness of the drug products for treating depression?

INFERENCES ABOUT π
AND $\pi_1 - \pi_2$

10.1 Introduction ▪ **10.2** Estimation of π ▪ **10.3** A statistical test about π ▪ **10.4** Comparing π_1 and π_2 ▪ Summary ▪ Key terms ▪ Key formulas ▪ Supplementary exercises ▪ Exercises from the data base

10.1 INTRODUCTION

In Chapters 8 and 9 we presented estimation and test procedures for making inferences about a population mean μ or the difference between two population means $\mu_1 - \mu_2$. This chapter will be devoted to making inferences about a binomial proportion π and the difference between two binomial proportions $\pi_1 - \pi_2$.

10.2 ESTIMATION OF π

In Chapter 6 we noted that public opinion polls, consumer preference polls, certain drug testing experiments, and some quality-control surveys provide examples of the binomial experiment. If x is a **binomial random variable,** then the probability distribution for x is

binomial random variable

$$P(x) = \frac{n!}{x!(n-x)!}\, \pi^{x}(1-\pi)^{n-x}$$

where

n = the number of trials
π = the probability of "success" on a single trial
(π is also the proportion of successes in the population)
x = the number of successes in n trials

How do we estimate π, the unknown proportion of successes in the population? The point estimate we will use is the one that you might have chosen, the proportion of successes observed in the sample. If x represents the number of successes observed in the sample, then the sample proportion of successes denoted by $\hat{\pi}$ (read π-hat) is

$$\hat{\pi} = \frac{x}{n}$$

sampling distribution for $\hat{\pi}$

When we use the Central Limit Theorem (Chapter 7), we note that as n becomes large, the binomial random variable x possesses a mound-shaped probability distribution that approaches a normal curve. We also note that the **sampling distribution for $\hat{\pi}$** will also be approximately normal when n is large, and has mean and standard error as shown here and in Figure 10.1.

Mean and Standard Error for the Sampling Distribution of $\hat{\pi}$

Mean: $\mu_{\hat{\pi}} = \pi$

Standard error: $\sigma_{\hat{\pi}} = \sqrt{\dfrac{\pi(1 - \pi)}{n}}$

FIGURE 10.1 Sampling Distribution of $\hat{\pi}$

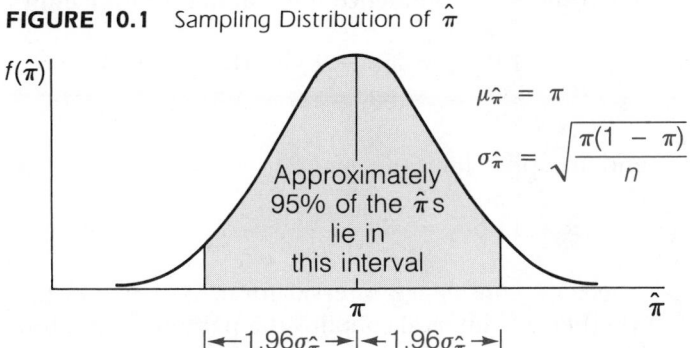

A 95% confidence interval for estimating the population proportion π has the same form as a 95% confidence interval for a population mean μ (see Section 8.2). We know from the sampling distribution of $\hat{\pi}$ that $\mu_{\hat{\pi}} = \pi$ and that the interval ($\pi \pm 1.96\sigma_{\hat{\pi}}$) includes 95% of the sample proportions $\hat{\pi}$ found in repeated sampling. Any time $\hat{\pi}$ lies in the interval

($\pi \pm 1.96\sigma_{\hat{\pi}}$) the interval ($\hat{\pi} \pm 1.96\sigma_{\hat{\pi}}$) will contain π. This occurs with a probability equal to .95. This gives us confidence (measured by the confidence coefficient .95) that in a practical situation, the interval calculated from a single sample when using the formula ($\hat{\pi} \pm 1.96\sigma_{\hat{\pi}}$) will enclose the parameter π. The interval estimate is called a 95% confidence interval.

Example 10.1

A random sample of 1000 working-class people in Great Britain was interviewed to determine each person's political party affiliation. If 680 identified with the major left-of-center party, use a 95% confidence interval to estimate the true proportion π of Great Britain's working class that identified with the left-of-center party.

Solution   The point estimate of π is

$$\hat{\pi} = \frac{x}{n} = \frac{680}{1000} = .68$$

Substituting $\hat{\pi} = .68$ into the formula for $\sigma_{\hat{\pi}}$, we find

$$\sigma_{\hat{\pi}} = \sqrt{\frac{.68(.32)}{1000}} = \sqrt{.000218} = .015$$

Then the 95% confidence interval has a lower limit of

$$\hat{\pi} - 1.96\sigma_{\hat{\pi}} = .68 - 1.96(.015)$$
$$= .68 - .03 = .65$$

and an upper limit of

$$\hat{\pi} + 1.96\sigma_{\hat{\pi}} = .68 + .03 = .71$$

The 95% confidence interval for π is then .65 to .71. We do not know whether this interval contains the parameter π. However, since 95% of the intervals ($\hat{\pi} \pm 1.96\sigma_{\hat{\pi}}$) contain π in repeated sampling, we are confident that the interval .65 to .71 includes the true proportion of Great Britain's working class that identify with the left-of-center political party.

In Section 8.2 we indicated that it is possible to construct a general confidence interval for μ (when σ is known) by using the formula ($\bar{x} \pm z\sigma_{\bar{x}}$). The corresponding general formula for estimating π is as follows:

Large-Sample Confidence Interval for π

$$\hat{\pi} \pm z\sigma_{\hat{\pi}}$$

where

$$\sigma_{\hat{\pi}} = \sqrt{\frac{\pi(1 - \pi)}{n}}$$

Note: The z-values corresponding to a 90%, a 95%, or a 99% confidence interval are, respectively, 1.645, 1.96, or 2.58. This confidence interval is valid only if $n\hat{\pi}$ and $n(1 - \hat{\pi})$ are both 5 or greater. Also, since π is unknown, we substitute $\hat{\pi}$ for π in the formula for $\sigma_{\hat{\pi}}$.

The confidence interval for π is based on a normal approximation to a binomial, which is appropriate provided n is sufficiently large. The rule we've specified is that both $n\pi$ and $n(1 - \pi)$ should be at least 5, but since π is the unknown parameter, we will require that $n\hat{\pi}$ and $n(1 - \hat{\pi})$ be at least 5. When the sample size is too small and violates this rule, the confidence interval usually will be too wide to be of any use. For example, with $n = 20$ and $\hat{\pi} = .2$, the rule is not satisfied, since $n\hat{\pi} = 4$. The 95% confidence interval based on these data would be $.025 < \pi < .375$, which is practically useless. Very few product managers would be willing to launch a new product if the expected increase in market share was between .025 and .375.

Keep in mind, however, that a sample size that is sufficiently large to satisfy the rule does not guarantee that the interval will be informative. It only judges the adequacy of the normal approximation to the binomial—the basis for the confidence level.

Example 10.2

A random sample of 100 seniors from a large university was selected to estimate the proportion of the graduating seniors going on to graduate school. If the sample produced 15 students who planned to attend graduate school, use these data to construct a 95% confidence interval for π, the proportion of seniors from the university who plan to attend graduate school.

Solution   The point estimate for π is

$$\hat{\pi} = \frac{x}{n} = \frac{15}{100} = .15$$

Since $n\hat{\pi} = 15$ and $n(1 - \hat{\pi}) = 85$ are both greater than 10, we can use the confidence interval formula $\hat{\pi} \pm z\sigma_{\hat{\pi}}$. Substituting into this formula we obtain the interval

$$\hat{\pi} \pm 1.96\sigma_{\hat{\pi}} = .15 \pm 1.96\sqrt{\frac{.15(.85)}{100}}$$
$$= .15 \pm .070$$

or .08 to .22. That is, we are 95% confident that the actual proportion of graduating seniors who plan to attend graduate school lies in the interval from .08 to .22.

EXERCISES

10.1   Refer to Example 10.1. Construct a 90% confidence interval for π and compare this result to the corresponding 95% confidence interval.

$ 10.2   A firm wishes to determine the proportion of new washing machines that require servicing sometime during their first six months of operation. A random sample of $n = 50$ sales records is examined. The records disclose that $x = 13$ of the machines have been serviced within six months of the date of purchase. Use this information to construct a 95% confidence interval for the proportion of all machines requiring servicing within six months.

$ 10.3   Use the data of Exercise 10.2 to construct a 99% confidence interval for π. Compare your results to those of Exercise 10.1.

▼ 10.4   It is not uncommon for patients to complain of certain side effects that accompany the use of a drug product. For example, certain muscle relaxants can cause drowsiness in some individuals and not in others. In a sample of 150 users of a given product, 38 complained of a particular side effect. Estimate the population proportion who would experience this side effect. How might your estimate differ if information about the side effect was solicited rather than volunteered?

$ 10.5   A federal highway study was undertaken to evaluate the effectiveness of a new commuter bus system for relieving highway congestion and reducing gasoline consumption. One measure of the effectiveness of the system is the proportion of people working in the destination areas that make use of the bus system. A random sample of $n = 1200$ people working in these areas was selected, using census information. Of the 1200 people interviewed 80 used the system. Construct a 95% confidence interval for the proportion of workers using the system.

10.6 Refer to Example 10.2. What would the 95% confidence interval be for a sample of $n = 1000$ students when $x = 150$? Comment on the widths of the intervals for $n = 100$ and $n = 1000$.

$ 10.7 A survey was conducted to investigate the proportion of registered nurses in a particular state that are actively employed. A random sample of 400 nurses selected from the state registry showed 274 actively employed. Find a 95% confidence interval for the proportion of registered nurses actively employed.

👫 10.8 A survey conducted to determine the proportion of college students favoring "more than equal" job-rights opportunities for women (to off-set past injustices) showed 258 of a random sample of 1000 favoring the proposal. Estimate the proportion of all college students favoring the proposal, using a 95% confidence interval.

🏛 10.9 The results of a recent national survey are presented in the news report shown here.

(a) Does the article provide all the information you need to be certain that the survey is valid? Explain.

(b) Can you compute a 95% confidence interval for the proportion of people who feel that the administration lacks a coherent foreign policy? Explain.

55% of People Disenchanted with U.S. Foreign Policy in Washington

Washington—According to a recent national survey, 55% of the adults who were questioned indicated that the administration lacks a coherent foreign policy.

Following the meeting of NATO ministers last month, a survey of adults in many regions of the United States was conducted. More than half of those questioned reported a lack of confidence in the depth and the extensiveness of the administration's foreign policy. This finding was not confined to one socioeconomic demographic group but rather was uniform throughout all subgroups examined.

The survey also showed that 63% of the people think we should deal with human rights issues, national defense, and arms reduction more effectively. One interesting finding of the survey was that only 20% of the people feel that we've made enough of an effort to negotiate an arms reduction with the Soviets.

🏛 10.10 Experts have predicted that approximately one-in-twelve tractor-trailer units will be involved in an accident this year. One of the reasons for this is that one-in-three tractor-trailer units has an imminently hazardous mechanical condition, probably related to the braking systems on the vehicle.

A survey of 50 tractor-trailer units passing through a weighing station confirmed that 19 had a potentially serious braking system problem.

(a) Do the binomial assumptions hold?

(b) Can a normal approximation to the binomial be applied here to get a confidence interval for π?

(c) Give a 95% confidence interval for π using these data. Is the interval informative? What could be done to decrease the width of the interval, assuming $\hat{\pi}$ remained the same?

10.11 In a study of self-medication practices, a random sample of 1230 adults completed a survey. Some of the medical conditions that were self-treated are shown here:

Medical condition	Home remedy	% Responding
Sore throat—not related to a cold	Salt water or baking soda mouthwash	30
Burns—other than sunburn	Cold water/butter	28
Overindulgence in alcohol	Homebrew	25
Overweight	Diet	22
Pain associated with injury	Hot or cold compresses	21

Summarize the results of this part of the survey using a 95% confidence interval for each medical condition.

10.12 Many individuals over the age of 40 develop an intolerance for milk and milk-based products. A dairy has developed a line of lactose-free products that are more tolerable to such individuals. To assess the potential market for these products, the dairy commissioned a market research study of individuals over 40 in its sales area. A random sample of 250 individuals showed that 86 of them suffer from milk intolerance. Calculate a 90% confidence interval for the population proportion that suffers milk intolerance based on the sample results.

10.13 Shortly before April 15 of the previous year, a team of sociologists conducted a survey to study their theory that tax cheaters tend to allay their guilt by holding certain beliefs. A total of 500 adults were interviewed and asked under what situations they think cheating on an income tax return is justified. The responses include:

56% agree that "other people don't report all their income."
50% agree that "the government is often careless with tax dollars."
46% agree that "cheating can be overlooked if one is generally law abiding."

Assuming the data are a simple random sample of the population of tax-payers (or tax-nonpayers), calculate 95% confidence intervals for the population proportion that agrees with each statement.

10.3   A STATISTICAL TEST ABOUT π

A statistical test about the binomial parameter π is very similar to the statistical test about μ (σ known) presented in Chapter 8. These results are summarized with three different alternative hypotheses along with their corresponding rejection region. Recall that only one alternative is chosen for a particular problem.

Summary of a Large-Sample Statistical Test About π

Null hypothesis: $\pi = \pi_0$ (π_0 is specified).

Alt. hypothesis: For a one-tailed test:

       1. $\pi > \pi_0$

       2. $\pi < \pi_0$

      For a two-tailed test:

       3. $\pi \neq \pi_0$

Test statistic: $z = \dfrac{\hat{\pi} - \pi_0}{\sigma_{\hat{\pi}}}$ where $\sigma_{\hat{\pi}} = \sqrt{\dfrac{\pi_0(1 - \pi_0)}{n}}$

Rejection region: For $\alpha = .05$ (or .01) and for a one-tailed test:

       1. Reject H_0 of $z > 1.645$ (or 2.33).

       2. Reject H_0 of $z < -1.645$ (or -2.33).

      For $\alpha = .05$ (or .01) and for a two-tailed test:

       3. Reject H_0 of $|z| > 1.96$ (or 2.58).

Note: This test is valid when $n\pi_0$ and $n(1 - \pi_0)$ are both 5 or more.

Example 10.3

According to recent marketing research reports, 12% of potential customers (businesses) favor a given brand of word-processing equipment. An extensive advertising and promotional campaign for this brand is conducted over a broad market area. At the end of the campaign, a sample of 300 potential new customers is polled to determine if the customer favors the advertised equipment over other competing systems. If π denotes the proportion of all potential customers that favors the specified brand of word processor, then it would be desirable from the company's standpoint to increase π throughout the campaign.

(a) Set up all parts of a statistical test about π to determine if the campaign was successful. Use $\alpha = .05$.

(b) Suppose 45 of the 300 sampled customers favor the advertised brand of word-processing equipment. Conduct a statistical test for these data.

Solution

(a) The statistical test is as follows:

Null hypothesis: $\pi = .12$

Alt. hypothesis: $\pi > .12$ (i.e., the campaign was successful)

Test statistic: $z = \dfrac{\hat{\pi} - .12}{\sigma_{\hat{\pi}}}$ where $\sigma_{\hat{\pi}} = \sqrt{\dfrac{(.12)(.88)}{300}} = .0188$

Rejection region: For $\alpha = .05$ and for a one-tailed test with H_a: $\pi > .12$, we will reject H_0 if $z > 1.645$

(b) For 45 "successes" from the sample of 300, the sample proportion is $\hat{\pi} = 45/300 = .15$. Substituting into z, we have

$$z = \frac{.15 - .12}{.0188} = 1.60$$

Since the computed value of z does not exceed 1.645, we have insufficient evidence to show that the advertising campaign was successful.

The z-test for π, like the confidence interval for π based on z, depends on the adequacy of the normal approximation to the binomial. When can you use the z-test for π? Generally speaking, you should view the results of a z-test for π skeptically if either $n\pi_0$ or $n(1 - \pi_0)$ is 2 or less. If both $n\pi_0$ and $n(1 - \pi_0)$ are at least 5, the z-test should be accurate. But for the same sample size n, z-tests based on more extreme values of π_0 are less accurate than are those based on values of π_0 closer to .5. For example, for $n = 5000$, a test of H_0: $\pi = .001$ (for which $n\pi_0 = 5$) would be much more suspect than would be a test of H_0: $\pi = .01$ (for which $n\pi_0 = 50$).

10.14   A professor wishes to determine whether a student's guessing ability on a true-false test differs from the results that could be obtained by flipping a coin to answer each question. An examination is composed of 200 questions. State the alternative (research) and null hypotheses for this statistical test.

10.15 Refer to Exercise 10.14. If the student answers $x = 110$ of the questions correctly, conduct the statistical test indicated in Exercise 10.14. Use $\alpha = .05$.

10.16 For $n = 400$, $x = 180$, and $\alpha = .05$, perform the calculations necessary to test $H_0: \pi = .4$ versus $H_a: \pi \neq .4$.

10.17 In a random sample of $n = 1000$ voters, $x = 560$ favored the passage of a controversial tax issue. Let π denote the proportion of all registered voters who favor the passage of the tax issue. Use these data to test $H_0: \pi = .5$ against the alternative $H_a: \pi > .5$. Use $\alpha = .05$.

10.18 The accounting department of a large manufacturing firm is concerned about the number of errors that are detected during routine audits. In a sample of 10,000 single-digit entries, the auditor detects 20 errors. Is there evidence to indicate that the error rate is greater than 1 in 1000? Use $\alpha = .05$.

10.19 Airlines have kept accurate records over the past few years concerning the number of persons who purchased tickets for a flight but then did not show up at the scheduled departure time. The no-show problem has led to the practice of overbooking flights. Suppose that the Civil Aeronautics Board runs a check of this practice by sampling 500 different flights to research the number of times people have been denied a seat on a scheduled flight after having purchased a ticket in advance. Suppose 40 of these sampled flights had one or more persons denied a seat. Use these data to test the alternative hypothesis that an overbooking policy has led to seat denial more than 5% of the time (i.e., $H_a: \pi > .05$). Use $\alpha = .05$.

10.20 A leakage test has been used to determine whether a large shipment of chemicals supplied in 16-ounce polyvinyl containers should be accepted from the supplier. A shipment of containers is not acceptable if the proportion of defective containers is greater than .10. Set up all parts of a statistical test for determining whether a shipment is unacceptable. Draw a conclusion if $n = 100$ containers are inspected and 12 are defective. Use $\alpha = .05$.

10.21 Refer to Exercise 10.20. If 18 containers were defective, what would your conclusion have been?

10.22 The benign mucosal cyst is the most common lesion of a pair of sinuses in the upper jawbone. In a random sample of 800 males, 35 persons were observed to have a benign mucosal cyst.

   (a) Would it be appropriate to use a normal approximation in conducting a statistical test of the null hypothesis $H_0: \pi = .096$ (the highest incidence in previous studies among males)? Explain.

   (b) Conduct a statistical test of the research hypothesis $H_a: \pi < .096$. Use $\alpha = .05$.

10.4 COMPARING π_1 AND π_2

Many practical problems require the comparison of two binomial parameters. For example, we might wish to compare the proportions of housewives who utilize prenatal health services before and after a campaign to publicize the services, or the proportions of households in two states that are entirely supported by welfare, or the proportions of voters favoring the Democratic candidate in a suburban and a rural area.

We assume that independent random samples are drawn from two binomial populations with unknown parameters π_1 and π_2. We further assume that the samples contain n_1 and n_2 observations, respectively. If x_1 represents the number of successes in the n_1 trials for sample 1 and x_2 the number of successes in n_2 trials for sample 2, then the sample proportions

$$\hat{\pi}_1 = \frac{x_1}{n_1} \quad \text{and} \quad \hat{\pi}_2 = \frac{x_2}{n_2}$$

are point estimates of π_1 and π_2, respectively. These results are summarized in Table 10.1.

TABLE 10.1   Sampling from Two Binomial Populations

	Population	
	1	*2*
Population proportion	π_1	π_2
Sample size	n_1	n_2
Number of successes	x_1	x_2
Sample proportion	$\hat{\pi}_1 = x_1/n_1$	$\hat{\pi}_2 = x_2/n_2$

sampling distribution for $\hat{\pi}_1 - \hat{\pi}_2$ Theorem 9.1 (p. 302) enables us to find the **sampling distribution for the difference in sample proportions, $\hat{\pi}_1 - \hat{\pi}_2$,** computed from independent random samples from two binomial populations. From the normal approximation to the binomial discussed in Chapter 7, we know that for large samples $\hat{\pi}_1$ and $\hat{\pi}_2$ will be approximately normally distributed with means and standard errors as shown here.

	Mean	**Standard error**
Sampling distribution for $\hat{\pi}_1$	π_1	$\sqrt{\dfrac{\pi_1(1 - \pi_1)}{n_1}}$
Sampling distribution for $\hat{\pi}_2$	π_2	$\sqrt{\dfrac{\pi_2(1 - \pi_2)}{n_2}}$

It follows from Theorem 9.1 that the sampling distribution for $\hat{\pi}_1 - \hat{\pi}_2$ will be approximately normal with mean and standard error, respectively, of

$$\mu_{\hat{\pi}_1 - \hat{\pi}_2} = \pi_1 - \pi_2$$

and

$$\sigma_{\hat{\pi}_1 - \hat{\pi}_2} = \sqrt{\frac{\pi_1(1 - \pi_1)}{n_1} + \frac{\pi_2(1 - \pi_2)}{n_2}}$$

The sampling distribution for $\hat{\pi}_1 - \hat{\pi}_2$ is shown in Figure 10.2.

FIGURE 10.2   Sampling Distribution of $(\hat{\pi}_1 - \hat{\pi}_2)$

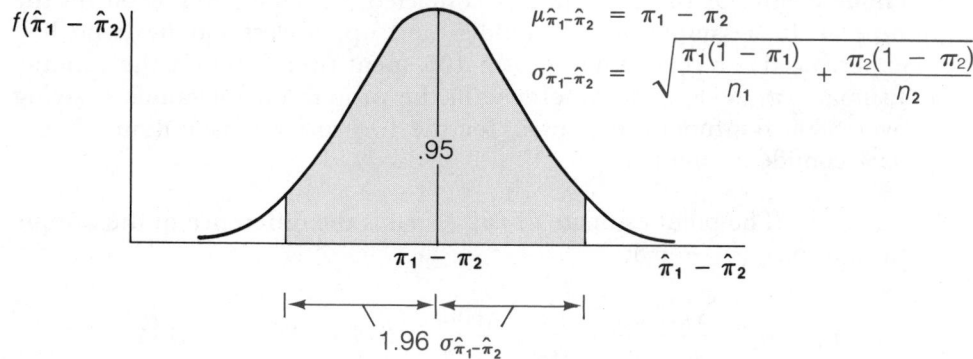

Knowing that the distribution of $\hat{\pi}_1 - \hat{\pi}_2$ is approximately normal for large samples allows us to construct a confidence interval for $\pi_1 - \pi_2$ in the familiar form

$$\hat{\pi}_1 - \hat{\pi}_2 \pm z\, \sigma_{\hat{\pi}_1 - \hat{\pi}_2}$$

Large-Sample Confidence Interval for $(\pi_1 - \pi_2)$

$$\hat{\pi}_1 - \hat{\pi}_2 \pm z\, \sigma_{\hat{\pi}_2 - \hat{\pi}_2}$$

where

$$\sigma_{\hat{\pi}_2 - \hat{\pi}_2} = \sqrt{\frac{\pi_1(1 - \pi_1)}{n} + \frac{\pi_2(1 - \pi_2)}{n}}$$

(continued)

Large-Sample Confidence Interval for $(\pi_1 - \pi_2)$ (continued)

Substitute $z = 1.645$, 1.96, or 2.58 for a 90%, a 95%, or a 99% confidence interval, respectively. Note: Use $\hat{\pi}_1$ and $\hat{\pi}_2$ for the unknown parameters π_1 and π_2 in the formula for $\sigma_{\hat{\pi}_1 - \hat{\pi}_2}$. This confidence interval is appropriate if $n\hat{\pi}$ and $n(1 - \hat{\pi})$ are greater than 5 for both samples.

Example 10.4

In a survey to analyze the cost of funeral expenditures for various social classes, a random sample of 162 families from the lower and working classes was interviewed to determine the funeral expenses for a recent family death. Of the 162 families contacted, 61 spent over $800 on the funeral. In a sample of 189 middle- and upper-class families who had experienced a recent family death, 106 spent over $800 on the funeral. Estimate $(\pi_1 - \pi_2)$, the difference in the proportions of families paying over $800 for funeral expenses, for the two social classifications. Use a 95% confidence interval.

The point estimate of $(\pi_1 - \pi_2)$ is the difference in the sample proportions, $(\hat{\pi}_1 - \hat{\pi}_2)$:

$$\hat{\pi}_1 - \hat{\pi}_2 = \frac{x_1}{n_2} - \frac{x_2}{n_2} = \frac{61}{162} - \frac{106}{189} = .377 - .561 = -.184$$

The standard error for $\hat{\pi}_1 - \hat{\pi}_2$ is

$$\sigma_{\hat{\pi}_1 - \hat{\pi}_2} = \sqrt{\frac{\pi_1(1 - \pi_1)}{n_1} + \frac{\pi_2(1 - \pi_2)}{n_2}}$$

$$\approx \sqrt{\frac{.377(.623)}{162} + \frac{(.561)(.439)}{189}} = .052$$

A 95% confidence interval for $\pi_1 - \pi_2$ has $z = 1.96$. Substituting into the confidence interval formula we have

$$\hat{\pi}_1 - \hat{\pi}_2 \pm z\,\sigma_{\hat{\pi}_1 - \hat{\pi}_2} = -.184 \pm 1.96(.052)$$

or $-.184 \pm .102$. That is, we are 95% confident that the difference in the proportions of families paying more than $800 per funeral for the working class (π_1) and the middle/upper class (π_2) lies in the interval $-.286$ to $-.082$. Because the confidence interval for $\pi_1 - \pi_2$ involves only negative numbers, the data suggest that π_2 is greater than π_1.

We can readily formulate a statistical test for the equality of two binomial parameters. The null hypothesis is $H_0: \pi_1 - \pi_2 = 0$. The test statistic has the familiar form

$$z = \frac{\text{point estimate} - \text{hypothesized value}}{\text{standard error}} = \frac{(\hat{\pi}_1 - \hat{\pi}_2) - 0}{\sigma_{\hat{\pi}_1 - \hat{\pi}_2}}$$

The standard error $\sigma_{\hat{\pi}_1 - \hat{\pi}_2}$ is slightly different from what we used in the formula for a confidence interval. When H_0 is true, the two population proportions are the same (i.e., $\pi_1 = \pi_2$); we'll call this common value π. Then the standard error formula becomes

$$\sigma_{\hat{\pi}_2 - \hat{\pi}_2} = \sqrt{\frac{\pi_1(1 - \pi_1)}{n_1} + \frac{\pi_2(1 - \pi_2)}{n_2}}$$

$$= \sqrt{\frac{\pi(1 - \pi)}{n1} + \frac{\pi(1 - \pi)}{n2}} = \sqrt{\pi(1 - \pi)\left(\frac{1}{n1} + \frac{1}{n2}\right)}$$

The best estimate of π, the proportion of successes common to both populations, is

$$\hat{\pi} = \frac{\text{total number of successes}}{\text{total number of trials}} = \frac{x_1 + x_2}{n_1 + n_2}$$

We summarize the test procedure next.

Large-Sample Statistical Test for Comparing Two Binomial Proportions

Null hypothesis: $\pi_1 - \pi_2 = 0$

Alt. hypothesis: For a one-tailed test:

    1. $\pi_1 - \pi_2 > 0$

    2. $\pi_1 - \pi_2 < 0$

    For a two-tailed test:

    3. $\pi_1 - \pi_2 \neq 0$

Test statistic:

$$z = \frac{\hat{\pi}_1 - \hat{\pi}_2}{\sigma_{\hat{\pi}_1 - \hat{\pi}_2}}$$

where $\sigma_{\hat{\pi}_1 - \hat{\pi}_2} = \sqrt{\pi(1 - \pi)\left(\frac{1}{n_1} + \frac{1}{n_2}\right)}$

and π is approximated by

$$\hat{\pi} = \frac{x_1 + x_2}{n_1 + n_2}$$

(continued)

Large-Sample Statistical Test for Comparing Two Binomial Proportions (continued)

Rejection region: For $\alpha = .05$ (or .01) and for a one-tailed test:

    1. Reject H_0 if $z > 1.645$ (or 2.33).

    2. Reject H_0 if $z < -1.645$ (or -2.33).

For $\alpha = .05$ (or .01) and for a two-tailed test:

    3. Reject H_0 if $|z| > 1.96$ (or 2.58).

Note: $n\hat{\pi}$ and $n(1 - \hat{\pi})$ must be greater than or equal to 5 for both populations.

Example 10.5

A class of 120 ninth-grade algebra students was randomly divided into two groups of equal size. Group 1 used a programmed learning text and had no formal lectures; group 2 was given formal lectures by a teacher. At the conclusion of a four-month period a comprehensive test was given to both groups to determine the proportion of students in each group who obtained a score of 85 (out of 100) or better. The results are given in Table 10.2.

TABLE 10.2   Data for Example 10.5

Group 1	Group 2
$n_1 = 60$	$n_2 = 60$
$x_1 = 41$	$x_2 = 24$

Let π_1 represent the population proportion of students taught with a programmed text who would achieve a score of 85 or more on the comprehensive test. Similarly, let π_2 represent the population proportion of students taught by formal lectures who would score 85 or more on the test. Test the research hypothesis that the two population proportions π_1 and π_2 are different. Use $\alpha = .01$.

Solution   For all practical purposes, sampling from each population satisfies the requirements of a binomial experiment. The four parts of the statistical test are as follows:

      Null hypothesis: $\pi_1 - \pi_2 = 0$

      Alt. hypothesis: $\pi_1 - \pi_2 \neq 0$

Test statistic:
$$z = \frac{\hat{\pi}_1 - \hat{\pi}_2}{\sqrt{\pi(1-\pi)\left(\dfrac{1}{n_1} + \dfrac{1}{n_2}\right)}}$$

Rejection region: For $\alpha = .01$ and a two-tailed test, we will reject H_0 if $|z| > 2.58$ (see Figure 10.3).

FIGURE 10.3 Rejection Region for Example 10.5

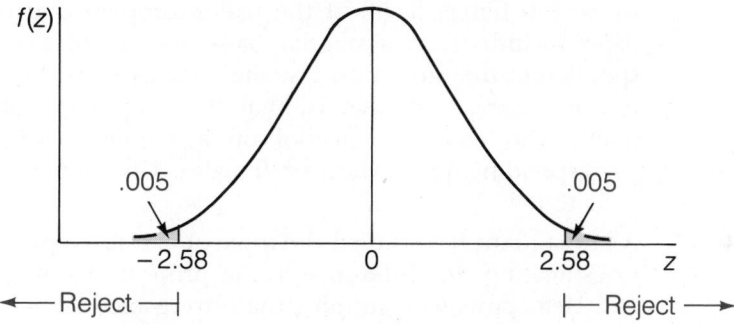

From the sample data we have

$$\hat{\pi}_1 = \frac{41}{60} = .68 \qquad \hat{\pi}_2 = \frac{24}{60} = .40$$

$$\hat{\pi} = \frac{41 + 24}{60 + 60} = .54$$

Substituting into the test statistic, we obtain

$$z = \frac{.68 - .40}{\sqrt{(.54)(.46)\left(\dfrac{1}{60} + \dfrac{1}{60}\right)}} = 3.08$$

Since the computed value of z exceeds 2.58, we reject H_0 and conclude that the two proportions are different. In fact, since $\hat{\pi}_1 = .68$ is greater than $\hat{\pi}_2 = .40$, we may conclude that the programmed teaching technique produces a higher proportion of students scoring 85 or more.

EXERCISES

10.23 A random sample of $n_1 = 1000$ measurements was obtained from a binomial population with $\pi_1 = .4$. Another random sample (independent of the first sample) of $n_2 = 1000$ measurements was selected from a second binomial population with $\pi_2 = .2$. Describe the sampling distribution of $(\hat{\pi}_1 - \hat{\pi}_2)$.

10.24 In a study to compare two binomial proportions, $n_1 = 50$, $n_2 = 40$, $x_1 = 18$, and $x_2 = 10$. Use these data to construct a 99% confidence interval for the difference between the corresponding population proportions.

10.25 Refer to Exercise 10.24. Use the same data to test $H_0: \pi_1 - \pi_2 = 0$ against the alternative hypothesis $H_a: \pi_1 - \pi_2 > 0$. Use $\alpha = .01$.

10.26 In a recent survey of county high school students ($n_1 = 100$ males and $n_2 = 100$ females), 58 of the males sampled said they consume alcohol (beer included) on a regular basis and 46 of the females sampled responded in the same way. Use these data to test the alternative hypothesis $H_a: \pi_1 - \pi_2 > 0$; that is, that the proportion of county high school males who consume alcohol on a regular basis is greater than the corresponding proportion of females. Give the level of significance for your test.

10.27 A law student has studied the topic of executive privilege and is interested in estimating the difference in the proportions of registered Republicans and Democrats who support the unrestricted right of executive privilege. To do this, she acquired independent random samples of 200 Republicans and 200 Democrats and found 46 Republicans and 37 Democrats in favor of the unrestricted right of executive privilege. Construct a 95% confidence interval for the difference in proportions for Democrats and Republicans.

10.28 In a comparison of the incidence of tumor potential in two strains of rats, 100 rats (50 males, 50 females) were selected from each of two strains and were examined for a period of one year. All the rats were approximately the same age and were housed and fed under comparable conditions. Use the accompanying one-year sample data to construct a 95% confidence interval for the difference in the proportions of rats exhibiting tumor potential for the two strains.

One-Year Results	Strain A	Strain B
Sample size	100	100
Number exhibiting tumor potential	25	15

10.29 A survey is conducted to determine whether a difference exists between the proportion of married persons and the proportion of single persons in the 20–29 age group who smoke. A sample of 200 persons from each group is polled, and 64 married persons and 80 single persons are found to smoke. Do the data provide sufficient evidence to indicate a difference in the proportions of smokers for the two populations? Use $\alpha = .05$.

☥ 10.30   Sixty of 87 couples sampled from the United States prefer a certain contraceptive technique; 40 out of 100 couples interviewed from a European country prefer the same technique. Estimate the difference in the population proportions, using a 95% confidence interval.

☥ 10.31   Refer to Exercise 10.30. Test the research hypothesis that couples from the United States and from the European country differ in their preference for the contraceptive technique. Use $\alpha = .01$.

💲 10.32   A retail computer dealer is trying to decide between two methods for servicing customers' equipment. The first method emphasizes preventive maintenance; the second emphasizes quick response to problems. Samples of customers are served by each of the two methods; of course, each customer is served by only one method. After six months, it is found that 171 of 200 customers served by the first method are very satisfied with the service, as compared to 153 of 200 customers served by the second method.

(a) Test the research hypothesis that the population proportions are different. Use $\alpha = .10$. State your conclusion carefully.

(b) State a p-value for the test in part (a).

💲 10.33   The media-selection manager for an advertising agency inserts the same advertisement for a client bank in two magazines. The ads are similarly placed in each magazine. One month later, a market research study finds that 226 of 473 readers of the first magazine are aware of the banking services offered in the ad, as are 165 of 439 readers of the second magazine (readers of both magazines are excluded).

(a) Calculate a 95% confidence interval for the difference of proportions of readers who are aware of the advertised services.

(b) Are the sample sizes adequate to use the normal approximation?

(c) Does the confidence interval indicate that there is a statistically significant difference using $\alpha = .05$?

10.34   Using the data of Exercise 10.33, perform a formal test of the null hypothesis of equal populations. Use $\alpha = .05$. How important is it whether or not the pooled proportion is used in the standard error?

SUMMARY

In this chapter we extended our study of statistical inferences beyond inferences about a single mean or two means. Using the normal approximation to the binomial for large samples, we can now formulate a confidence interval for π or $\pi_1 - \pi_2$ of the form

point estimate $\pm z$(standard error)

Similarly, the statistical test for π or $\pi_1 - \pi_2$ has a common format with the general test statistic

$$z = \frac{\text{point estimate} - \text{hypothesized value } (H_0)}{\text{standard error}}$$

KEY TERMS

binomial random variable
sampling distribution for $\hat{\pi}$
sampling distribution for $\hat{\pi}_1 - \hat{\pi}_2$

KEY FORMULAS

1. General confidence interval for π

$$\hat{\pi} \pm z\,\sigma_{\hat{\pi}} \quad \text{where} \quad \sigma_{\hat{\pi}} = \sqrt{\frac{\pi(1 - \pi)}{n}}$$

2. Statistical test about π

$H_0: \pi = \pi_0$ (π_0 is specified)

Test statistic: $z = \dfrac{\hat{\pi} - \pi_0}{\sigma_{\hat{\pi}}} \quad \text{where} \quad \sigma_{\hat{\pi}} = \sqrt{\dfrac{\pi_0(1 - \pi_0)}{n}}$

3. General confidence interval for $\pi_1 - \pi_2$

$$\hat{\pi}_1 - \hat{\pi}_2 \pm z\,\sigma_{\hat{\pi}_1 - \hat{\pi}_2}$$

$$\text{where} \quad \sigma_{\hat{\pi}_1 - \hat{\pi}_2} = \sqrt{\frac{\pi_1(1 - \pi_1)}{n_1} + \frac{\pi_2(1 - \pi_2)}{n_2}}$$

4. Statistical test about $\pi_1 - \pi_2$

$H_0: \pi_1 - \pi_2 = 0$

Test statistic:

$$z = \frac{\hat{\pi}_1 - \hat{\pi}_2}{\sigma_{\hat{\pi}_1 - \hat{\pi}_2}}$$

where $\quad \sigma_{\hat{\pi}_1 - \hat{\pi}_2} = \sqrt{\pi(1 - \pi)\left(\dfrac{1}{n_1} + \dfrac{1}{n_2}\right)}$

and $\quad\quad\quad \hat{\pi} = \dfrac{x_1 + x_2}{}$

SUPPLEMENTARY EXERCISES

10.35 A national poll was conducted recently to determine the proportion of adults who attend church in a typical week. If 1200 adults were interviewed and 580 indicated they attend church in a typical week, use these data to construct a 95% confidence interval for π. Explain the meaning of the interval.

10.36 It has been asserted that 40% of the persons belonging to a certain race have blood type O. Results of a recent study of 100 persons of this race show that 35 have type O blood.

(a) Carry out the appropriate test of hypothesis using a two-tailed test with $\alpha = .05$. Show all steps!

(b) Carefully state your conclusion:

(i) In statistical terms.

(ii) In plain English as it would be explained to someone not familiar with statistical language.

10.37 A museum purchased a $1 million painting, expecting that it would attract 75% of its visitors. To study this hypothesis, 50 people were chosen randomly as they entered the museum and followed until they left. Of these 50, it was found that 34 looked at the painting. Does this contradict the original hypothesis? Show the basis for your answer.

10.38 There is an effort each year to simplify income tax forms. However, errors continue to be a real problem. Recently, articles have appeared in the paper pointing out some of the common errors. Suppose a random sample of 250 tax forms filed by March 1 of this year showed that 55 of those examined had one or more errors.

(a) Compute a 95% confidence interval for the proportion of all income tax forms filed by March 1 that would have one or more errors.

(b) Suppose you had used the 90% confidence coefficient. Other things remaining equal, would the resulting confidence interval be larger or smaller than the interval in part (a)?

10.39 A radio commentator who has asked his listeners to write to him expressing their opinions on a subject on which he has definite views, finds that

90 percent of the letters received favor his view. Subsequently, he claims that at least 90 percent of the voting population agrees with him. A newspaper editor decides to test this claim and takes a random sample of 100 voters. Of these, 79 agree with the commentator and 21 disagree. Do the data contradict the radio commentator's claim? Use $\alpha = .05$. (Show the basis for your decision.)

10.40 In a physical fitness test, 240 out of 400 (60%) high school students could run a mile in less than 7 minutes. If the 400 students were a random sample, construct a 90% confidence interval for the proportion of all high school students that can run a mile in less than 7 minutes.

10.41 During a water shortage in Florida, restaurants were asked not to serve water unless a customer requested it. For the first three months in one restaurant, 45% of all customers requested water. Then the restaurant management put out signs on the tables explaining that it actually requires 24 ounces of water to serve an 8-ounce glass of water (due to washing, ice, etc.). After placement of the signs, a random sample of 150 customers had only 54 requesting water. Is there enough evidence at the 5% significance level to say that the signs reduced demand for water? Justify your answer.

10.42 It is thought that a large percentage of children between the ages of 3 and 7 do not know their home telephone numbers. In a study using a random sample of 500 children, it was found that 340 of them did not know their phone number. Estimate the true proportion of children between ages 3 and 7 who do not know their home phone number using a 95% confidence interval.

10.43 The Nielsen rating is a number that is supposed to show the percentage of all households that have their TV sets tuned to a particular show. Let us assume that 1000 selected families (Nielsen actually uses 1200 families but we will assume 1000 for ease of calculations) are an accurate cross section of all the households that have TV sets.

(a) Suppose a recent Nielsen rating for Monday Night Football was 40%. Construct a 95% confidence interval for π, the proportion of all households that watch Monday Night Football.

(b) Is the network justified in concluding with 95% confidence that the true population proportion, π, may be as high as 45%? Give the basis for your answer.

10.44 Selection of juries in a certain county is alleged to give inadequate representation to minority races in the county; minority races make up 24% of the population of the county. Would a jury panel of 100 members containing 16 from the minority groups give convincing evidence of discrimination?

(a) In answering the question would you use a one-tailed or two-tailed test of hypothesis? Explain your reason(s) for choosing one over the other.

(b) Carry out the proper test of hypothesis showing all steps and state your conclusion in

    (i) Statistical terms

and

    (ii) Nonstatistical terms so that a reasonably intelligent person will understand your results.

10.45 A social worker has reason to believe that fewer than 25% of the couples in a certain area have ever used any form of birth control. In order to see whether this is a reasonable assumption, the social worker selected a random sample of 120 couples from the area, and asked the question: "Have you ever used any form of birth control?" Of the 120 couples in the sample, 21 said they had used some method of birth control.

(a) The social worker posed the following null and alternative hypotheses.

$$H_0: \pi = .25$$
$$H_a: \pi <' .25$$

Would you agree with these hypotheses? If not, how would you change them?

(b) The rejection region based on a z-test for π was given as $z > 1.96$ or $z < -1.96$ for $\alpha = .05$. Do you agree? If not, then make the necessary correction(s) and explain why you differ.

(c) The computed z-score was given as

$$z = \frac{.175 - .25}{\sqrt{\dfrac{(.25)(.75)}{120}}} = \frac{-.075}{.0395} = -1.90$$

Do you agree? If not, give the corrected value of z.

10.46 Refer to Exercise 10.45. The social worker concluded that she could not reject the null hypothesis; therefore her belief that fewer than 25% of the couples have ever used any form of birth control was not supported by the data. Do you agree with this conclusion? If not, give the proper conclusion and explain why you differ.

10.47 Independent random samples of 800 Republicans and 800 Democrats showed 40% and 32%, respectively, in favor of the death sentence for major crimes. Do these data provide sufficient evidence to indicate a difference in the population proportions favoring the death sentence? Use $\alpha = .01$.

10.48 For the accompanying news article, comment on the appropriateness of the sample and the inferences that are drawn. Is there additional information that would have been helpful for you in drawing your own inferences?

Study of Divided Families Shows Positive Attitudes

Chicago—A study of divorced mothers and their children has revealed some positive attitudes among members of divided families. Perhaps a broken home is not the psychological disaster for family members that society has long suspected.

The study, involving 20 mothers with one or more children between the ages of 6 and 18, was conducted to determine the basic concerns of divorced mothers and their children. There were 20 mothers and 35 children involved in the study.

All the women were working full time. Most of them had made plans toward bettering their earning power. The women had been divorced from 3 months to 15 years. The educational level of the women in the study was high, compared to the national average: 12 years to 18 years of education.

A key aim of the study was to determine the feelings of the women and their children about their acceptance in society.

Eighty-six percent of the children felt that at school they were treated the same as children whose parents were married. Children aged 10 through 12 especially preferred that teachers and friends be told about the home situation. They wanted news of the divorce not to come as a surprise to others or to be a source of embarrassment for them.

In general, the children were doing well in school and even excelled in some areas.

Although the trend among most of the women was to socialize mainly with single persons, 80% of them felt accepted in their neighborhoods. Half of them said they felt accepted at church.

Among the children, 91% indicated they were treated no differently at Sunday school. Ninety percent of the sample were active church members.

Most of the women, 85%, said that after their divorces their attitudes toward divorce had shifted from negative to positive. The same proportion saw advantages for their children, in terms of understanding life and people, as a result of the divorce.

10.49 On a question concerning library hours on a college campus, 42 of 50 men interviewed favored the proposal to increase the hours of operation, and 75 of 80 women favored the proposal. Is there evidence of a difference in the proportions of all men and women favoring the proposal? Give the level of significance for your test.

10.50 The advent of teacher evaluations by students has seemed to change the criteria on which students are judged. While this might be arguable, it is a fact that at many universities the percentages of students awarded As and Bs have increased substantially in the past five years. Administrators who are concerned about this trend have taken a dim view of the excessively high percentages of As and Bs. The percentages of As and Bs awarded by two college philosophy professors were duly noted by the dean. Professor 1 achieved a rate of 53% as opposed to 40% for professor 2, based upon 200 and 180 students, respectively. Do the data indicate a difference in the rates of awarding As and Bs by the two philosophy professors? Use $\alpha = .05$.

10.51 An experiment was conducted to compare two different rations of feed for chicks. From a total of 200 chicks, 100 were randomly assigned to group 1 and fed ration A. The other 100 chicks were assigned to group 2 and fed ration B. One study of interest was a comparison of the proportions of chicks that show no major weight gain while being fed the two rations. Assuming that all factors other than the rations were the same for the two groups, use the accompanying sample data to conduct a statistical test to determine whether the proportions of chicks with no major weight gain are different for the two groups. Use $\alpha = .05$.

	Ration A	Ration B
Sample size	100	100
Number that show no major weight gain	14	8

10.52 A marketing research firm is interested in comparing the proportions of potential buyers of a soon-to-be-released product for chain stores and for independent stores. Use the accompanying sample data to construct a 95% confidence interval for the difference in proportions between the two types of stores.

	Chain stores	Independent stores
Sample size	110	650
Number of potential buyers	26	275

10.53 The traffic congestion in cities has become a major cause for concern not only from an environmental standpoint but also from an engineering viewpoint. The Federal Highway Administration in conjunction with the Urban Mass Transit Authority contracted with the University of Florida Transportation Center to test and evaluate the effectiveness of an exclusive bus lane constructed on a 10-mile (north-south) section of 7th Avenue in Miami and a parallel section of Interstate 95. Initially a parking facility and bus terminal were built at the Golden Glades exit to provide commuters with ready access to an inexpensive, fast form of transportation to their destinations in downtown Miami. Each bus was equipped with a device that enabled the driver to preempt traffic signals along the 10-mile stretch of highway. Thus a driver could flash ahead to a traffic signal if the bus was going to arrive at the intersection during the red phase of the traffic signal cycle and the light would either remain green or switch as quickly as possible to allow uninterrupted bus travel. One major task of the University of Florida Transportation Center was to compare the following:

(a) The effect of various preempting strategies (how much leeway was given to the individual bus driver in controlling the green phase of the signal) on the average travel time for buses from the terminal to each of the four destination areas.

(b) The effect of these preempting strategies on congestion at intersections along the 10-mile section of 7th Avenue.

In addition to these considerations, a major objective involved a comparison of the auto occupancies throughout various phases of the 3½ year project. For example, it was important to compare the proportion of cars with two or more persons through each of the test phases in order to determine if there were shifts in commuter car usage as a result of or in conjunction with changes in bus usage.

Suppose that initial results were as follows: During phase 0 (which ran for approximately three months) the buses were not given a preempting option at the intersections. Phase 1 allowed local preemption of a traffic signal by a bus but no other signal coordination. Of 10,000 cars observed during the sample period in phase 0, 10% had two or more persons in a car. Similarly, for 10,000 cars sampled during phase 1, 15% had two or more persons in a car. Do these data present sufficient evidence to indicate whether there is a difference in the proportion of cars carrying two or more persons for the two phases? Use $\alpha = .05$.

10.54 A pharmaceutical company compared two frequently used antibiotics, Erythromycin and Tetracycline, in a bacterial sensitivity study. Bacterial cultures from a random sampling of patients who had received injections of one of the two drugs were analyzed to ascertain antibacterial activity against specific bacteria. A summary of the survey data is given in the accompanying table for Erythromycin and Tetracycline for the *Streptococcus pyogenes* bacteria. Use these sample data to test whether there is evidence to indicate a difference in the proportions of patient cultures showing antibacterial activity against the *Streptococcus pyogenes* bacteria for the two test-drug products. Use $\alpha = .05$.

	Erythromycin	Tetracycline
Number of patient cultures analyzed	528	481
Number of patient cultures showing antibacterial activity	515	394

10.55 The accompanying article alludes to a study of the effects of orange juice in reducing symptoms of respiratory infections. Examine this article and comment on the published statistics contained in it. Be sure to comment on the adequacy of the sample, the inferences that are drawn, and additional material that might have been included in the article to help you reach a conclusion.

Researchers Study Effects of Orange Juice

Gainesville, Fla.—A quart of orange juice, taken every day, has significant effects in reducing the symptoms of respiratory infections caused by the rubella virus, report medical scientists from the University of Florida.

The Florida study involved 55 human volunteers. Many of those drinking orange juice produced antibodies earlier than those not drinking orange juice. The researchers said that the early antibody production may or may not be the means by which the symptoms were reduced. However, the appearance of this antibody production, they pointed out, may indicate that citrus has a localized effect in stimulating the immune system to fight respiratory infection.

Many of the volunteers who drank fresh frozen orange juice did not experience any sore throat or runny nose. Other volunteers, forbidden to drink orange juice or eat citrus fruit, did show these mild symptoms. Prior to the study all participants were free of infection and had normal antibody levels.

The volunteers were subjected to a weakened, nontransmissible form of rubella virus. The rubella virus was chosen because it can induce either respiratory or systemic infec-tions, depending on how the virus is administered. Introduced in the nasal passages, the virus causes respiratory infection; introduced under the skin, the virus causes systemic infection.

Half of the 55 volunteers were given the rubella virus by nose drops; the other half received the same virus by injection. Each group was again divided. Half of them were on a quart of orange juice a day; their controls were on regular diets with no orange juice or citrus fruit. Each volunteer was followed for 21 days after the virus was administered.

In the groups who received the virus by nose drops and were not allowed to drink orange juice, 77% showed symptoms of respiratory infection. Among the counterparts who received the same type of virus and drank orange juice, only 27% had respiratory symptoms.

"This represents a significant reduction in symptoms," one researcher said.

Volunteers in the study who received the virus by injection showed no difference in symptoms, whether they were drinking orange juice or not.

10.56 Pharmacologists sometimes use a series of tests in the screening of compounds for potential use in the mental health area. Those compounds that pass the screening test are then studied in more detail. One such test in a screening process is concerned with the proportion of rats that exhibit increased self-cleaning activity (called pernicious preening) after being given a single dose of a compound. In one study 27 of 40 rats exhibited signs of pernicious preening. Use these data to construct a 95% confidence interval for π, the proportion of rats that exhibit pernicious preening after receiving a single dose of the compound.

10.57   An accurate estimate of the percentage unemployed (unemployment rate) in a given area is an essential measure of social well-being. A random sample of 1000 adults in Alachua County, Florida, showed that 47 were unemployed. Estimate the proportion unemployed in the county using a 90% confidence interval.

10.58   The accompanying news article appeared recently in the press. You will note that a survey found that 80% of those interviewed (2500 adults) had adopted the new social values.

Self-Culture

Washington—A recent news survey involving approximately 2500 adults reveals some fascinating changes in the mood of the country. Questions related to one's values, attitudes, and lifestyles were posed. The results show that clearly the Puritan ethic, characterized by self-denial, conformity, hard work, and upward mobility, has given way to a new set of values: self-expression, self-gratification, and self-fulfillment. In the survey 80% of those interviewed indicated that they have adopted the new value system to some extent. This is in contrast to results from a similar survey done in the mid-1970s, when only 60% said they had adopted the new social values in some way.

(a) If this survey is to be of value to an unbiased observer, the sample proportion, .80, should be an estimate of the proportion of people all over the country who have adopted the new value system. If this is the case, what assumptions must be made concerning the sample of 2500 adults?

(b) Construct a 95% confidence interval for π, the proportion of adults that have adopted the new value system.

10.59   **Class Exercise**   If a balanced coin is flipped a very large number of times, the proportion of outcomes resulting in heads will (for all practical purposes) equal ½.

(a) To illustrate the notion of sampling error, flip a coin 25 times and estimate the probability, π, of tossing a head. Calculate a 95% confidence interval for π. Since you know $\pi = $ ½, then

$$\sigma_{\hat{\pi}} = \sqrt{\frac{\pi(1 - \pi)}{n}} = \sqrt{\frac{(1/2)\,(1/2)}{25}} = .1$$

(b) Have 20 different students repeat the process to obtain several estimates. Note how their estimates vary by about ½.

(c) Combine all your sample results into one sample to obtain a more accurate estimate of π. Calculate a 95% confidence interval. The width of this confidence interval will be smaller than for the samples of 25 because it is based on a larger number of tosses.

10.60 A survey of cancer patients treated in ten regional hospitals shows that of 874 patients treated for cancer, 257 have survived for a period of five years or longer. Estimate the proportion of cancer patients treated at one of these regional hospitals who will survive for a period of at least five years, using a 95% confidence interval. In order for this inference to be valid, what assumptions must be made regarding the sample data? Are you concerned about this being a random sample?

10.61 A random sample of insurance records of $n = 500$ physicians, selected from the files of an insurance company, shows that 10% of the physicians have been involved in one or more lawsuits. Estimate the proportion of all physicians covered by the insurance company who have been involved in lawsuits using a 90% confidence interval.

10.62 The accompanying news report gives the results of a recent national survey, which indicate that a frighteningly large number of teenagers either have driven while drunk or have been passengers in cars with drivers who had been drinking heavily.

Study of Teenage Drinkers Completed

Washington—A recent study conducted by the National Highway Traffic Safety Administration revealed that an "alarming . . . frightening" number of teenagers have either driven while drunk or have been passengers in cars with heavy-drinking drivers.

The study found that neither scare tactics nor legal threats discourage teenage drinking and driving. The legal consequences of being stopped by police are not considered as serious by teenagers. They also do not consider death or a crippling injury a likely consequence, according to the study.

The study indicated other aspects of teenage drinking:

- Many teenagers admit they have frequently driven when they are "pretty drunk."
- One-fourth say they have driven once or twice when they knew themselves to be too drunk to drive.
- Another one-fourth have driven three or more times when drunk.
- Of the teenagers interviewed, 32% said they were passengers at least once a month in a car operated by a heavily drinking driver.

The agency's spokesman said that its future activities would be directed toward encouraging social pressures against drinking and driving.

(a) Are you given enough information in the article to be certain that the survey is valid? Explain.

(b) Compute a 95% confidence interval for the proportion of teenagers who say they have driven once or twice when they knew themselves to be too drunk to drive. Explain.

(c) Might there be a tendency for some teenagers to understate the number of times they had been driving while under the influence of alcohol? Explain.

(d) How might the report of this survey be improved to allow the reader to draw his or her own conclusions?

10.63 Two hundred voters were randomly selected and 110 were found to favor candidate A. Estimate the fraction of voters favoring candidate A, using a 95% confidence interval.

10.64 Response to an advertising display was measured by counting the number of prospects who purchased out of the total number who were exposed to the display. If 330 purchased out of a total of 870 exposed, estimate the probability of purchase, using a 95% confidence interval.

10.65 In a random sample of 150 students on a college campus, 72 students were in favor of an increase in their activities fee to help fund a proposed coliseum. Estimate π, the proportion of the entire student body in favor of the fee increase, using a 95% confidence interval.

10.66 The state highway patrol wishes to estimate the proportion of automobiles with badly worn tires that use a particular turnpike. In a spot check (random sample) of 300 cars, 58 cars were observed to have bad tires. Estimate π, the proportion of the automobiles on the turnpike with badly worn tires, using a 90% confidence interval.

10.67 A random sample of 400 fuses was selected from a day's production and tested; 40 were found to be defective. Estimate the proportion defective in the day's production, using a 99% confidence interval.

10.68 Americans (as well as the people of most highly industrialized nations) have had to live with the realities of large national debts for several years, and people are still concerned with the problem. In a recent survey of 1540 American adults 18 years of age or older, 830 indicated that they felt the size of the national debt was the most important problem facing the nation today. Use those data to construct a 95% confidence interval for π, the proportion of adults who think the size of the national debt is the major problem facing our nation.

10.69 Class Exercise Conduct a sample survey to determine the attitude of your student body on a major political, social, or campus issue. Use the methods of Section 7.2, p. 209, to select a large random sample of n students (say $n = 400$ or more) from the student directory. Since each of the n students in the sample must be contacted by telephone to determine his or her attitude concerning the question, this chore should be divided so that each student (or a small group of students) is responsible for contacting a fixed number, say 25, of the total.

When the data have been collected, pool the n student responses and construct a 95% confidence interval for the proportion of students in the student body who favor the issue.

To observe sampling variation, let each student (or team) use the 25 students whom he or she was to contact to estimate the proportion in favor of the issue. Collect the estimates from each team and construct a histogram of the estimates. Locate the estimate based on all n students on the graph. Notice how the estimates based on the samples containing 25 students tend to vary and that the large-sample estimate (based on all n students) falls near the center of the histogram.

10.70 The manager of a discount store was interested in determining how its customers felt about selected facets of its operation. One of the questions of concern was whether or not customers knew the slogan used in its television commercials. In a properly conducted survey, 176 out of 240 in the "50 years and over" age group knew the slogan; 110 out of 225 in the "under 30 years" age group knew the slogan. Estimate the difference in the population proportions for the two groups using a 99% confidence interval.

10.71 Suppose that we wish to test the hypothesis that a voter population is equally distributed in its preferences between two candidates, A and B, against the alternative hypothesis that one of the candidates is preferred to the other. A random sample of 100 voters is selected, and the number x favoring B is recorded. Specify the appropriate null hypothesis. Construct a test of the null hypothesis, being certain to identify each of the four parts of the test. Set the rejection region so that α is approximately equal to .05.

10.72 Refer to Exercise 10.71. Suppose that we observe 61 voters ($x = 61$) favoring B. What would you conclude? When using this test, what is the probability that you will incorrectly reject the null hypothesis?

10.73 Refer to Exercise 10.72. Give the level of significance for the test.

10.74 A random group of 300 homemakers was interviewed to determine the preference for one of two types of fabric softeners. Brand A was favored by 135 homemakers; the others favored B. Do these data provide sufficient evidence to indicate a difference in preference for the two fabric softeners? Test by using $\alpha = .05$.

10.75 To determine consistency in evaluating student behavior, two evaluators were presented with a random group of 200 students for examination. Each student was examined by both of the evaluators. The evaluators agreed on 133 of the evaluations. Does this indicate that their agreement is due to reasons other than pure chance? Give the level of significance for your test.

10.76 A white mouse is running a maze that has two doors of equal size at one end. One of the doors has a piece of cheese behind it, the other does not.

If the mouse does not "learn" to choose the door with the cheese, he should choose either door with probability equal to ½. In 90 trials the mouse chooses the door with the cheese 62 times. Can we conclude that the mouse has "learned" where the cheese is? Use $\alpha = .05$.

10.77 A manufacturer claims that, at most, 5% of the goods it produces are defective. Out of 200 items randomly selected from production, 14 are found to be defective. Is there enough evidence to indicate that more than 5% are defective? Use $\alpha = .01$.

10.78 A data processing department claimed that in converting data from handwritten pages to a computer file, no more than .1% of the data entries (keystrokes) were in error. For a large study a sample of $n = 20,000$ keystrokes was checked against the handwritten copy; $x = 50$ errors were found. Conduct a statistical test of the data processing department's claim. Use $\alpha = .05$.

10.79 Refer to Exercise 10.78. Give the level of significance for the statistical test.

10.80 **Class Exercise** Select an issue that is of concern to your university, your college, or your local community. For example, at the time of this writing, our county commission is holding hearings preliminary to a decision to authorize the construction of a large shopping mall adjacent to a new elementary school. Some witnesses appear before the television camera and strongly support the mall; others denounce it. Evidence suggests that elected officials, such as our county commission, place great weight on the proportion of speakers who favor or oppose the proposed construction (all other things being equal). Indeed, we frequently observe similar behavior on the part of our congressmen who quote counts of letters supporting or opposing some issue, thereby implying public support or opposition.

Do the speakers who appear before a county commission (or the writers of letters to a senator) represent a random sample of an elected official's constituency? Why might the majority of these speakers (or letter writers) favor one side of an issue when the constituency they are supposed to represent favor the other side?

Conduct a survey of public opinion to test the theory that the public supports (or opposes) the issue. Obtain a listing of the public and select a random sample (see Section 7.2, p. 209, on how to draw a random sample) of 400 people from this group. Then test the hypothesis that the proportion π of people favoring the issue is equal to .5 against the alternative that π is greater than .5. Collect the data and draw your conclusions. Does the public favor the issue? Why are the conclusions based on your sample survey more valid than an inference based on the proportion of speakers (or letters to a senator) favoring the issue?

10.81 A psychiatrist contended that 25% or more of the individuals living in the United States experience feelings of guilt at least once very day. To evaluate the claim, you have taken a random sample of 400 persons and 64 agreed that they experience such feelings at least once every day. Can we reject the claim of the psychiatrist based on $\alpha = .05$?

10.82 Using a random sample of 50 printing and publishing firms, a group of financial consultants concluded that 31 of the firms in the sample had little danger of impending bankruptcy. At the 95% level of confidence, estimate the proportion of all firms in the printing and publishing industry that have little danger of bankruptcy.

10.83 In an analysis concerned with the banking habits of families, a random sample of 200 families was taken. In the sample 60 families indicated that they had accounts in more than one financial institution. Estimate the proportion of all families that have accounts in more than one financial institution, using a 90% confidence interval.

10.84 In the survey discussed in Exercise 10.83, one question was concerned with "Why did you choose your primary banking institution?" Fifty-seven percent of those in the survey marked the answer "Because it is the closest bank." Estimate the proportion of all families that choose their primary bank because it is the closest bank, using a 90% confidence interval.

10.85 Two random samples were drawn from students enrolled in various classes on a college campus. Only classes in which students were free to select their own seats were considered, and only classes of conventional size were included. One sample of students was taken from the first two rows of seats and the other sample consisted of students sitting in the back two rows of seats. Of the 200 selected from the first two rows, 122 were identified as those who participate regularly in classroom discussions. Of the 200 selected from the back two rows, 84 participated regularly in classroom discussions. Should one conclude that there is a difference in classroom participation between those sitting in the first two rows and those sitting in the back two rows? Use $\alpha = .05$.

10.86 A survey was taken to assess attitudes about the safety of toys. In response to the question, "Are toys safer now than 5 years ago?" 276 out of 400 females in the survey responded "Yes." Of the 280 males in the surveyed group, 152 responded "Yes." Compute a 95% confidence interval for the difference in population proportions for females and males. Considering the range of your interval, is it likely that there is no difference between population proportions for the two groups?

10.87 The director of a savings and loan association believes that the decision to take out a joint savings account with her organization is made by the husband more often than by the wife. You are asked to test this hypothesis. Data from a random sample of 200 joint account holders show the husband made the decision to open a savings account in 134 of the cases. What should you conclude about the hypothesis, based on $\alpha = .05$?

10.88 In a comparison of the incidence of tumor potential in two strains of rats, 100 rats (50 males, 50 females) were selected from each of two strains and were examined for a period of one year. All the rats were approximately the same age and were housed and fed under comparable conditions. Use the accompanying one-year sample data to construct a 95% confidence interval for the difference in the proportions of rats exhibiting tumor potential for the two strains.

One-Year Results	Strain A	Strain B
Sample size	100	100
Number exhibiting tumor potential	25	15

10.89 There may be a remedy for male pattern baldness—at least that's what millions of males hope, since the FDA approved Upjohn's Minoxidil for such a use. Minoxidil was investigated in a large, 27-center study where patients were randomly asigned to receive Minoxidil or an identical-appearing "placebo." Ignoring the center-to-center variation, suppose the preliminary results were as follows:

	Sample size	% with new hair growth
Minoxidil group	310	32
Placebo	309	20

(a) Use these data to test $H_0: \pi_1 - \pi_2 = 0$ versus $H_a: \pi_1 - \pi_2 \neq 0$. Give the p-value for your test.

(b) If you were working for the FDA, what additional information might you want to examine in this study?

10.90 Television research to date suggests that "zipping" and "zapping" of commercials by VCR viewers is uncommon. Zipping is the use of the VCR remote control to fast-forward past commercials; zapping is the use of the remote control to change channels when commercials appear. Based on a random sample of 2000 users of VCRs, 66% said that they did not skip the commercials. Obviously, with more than 30 million households with VCRs, widespread zipping and zapping would have a tremendous impact on the rates charged for TV advertising. Use the data to construct a 95% confidence interval for π, the proportion of VCR users that do not skip commercials.

10.91 Is cocaine deadlier than heroin? A study reported in the *Journal of the American Medical Association* found that rats with unlimited access to cocaine had poorer health, had more behavior disturbances, and died at a higher rate than did a corresponding group of rats given unlimited access to heroin. The death rates after 30 days on the study were as follows:

	% dead at 30 days
Cocaine group	90
Heroin group	36

(a) Suppose that 100 rats were used in each group. Conduct a test of $H_0: \pi_1 - \pi_2 = 0$ versus $H_a: \pi_1 - \pi_2 > 0$. Give the p-value for your test.

(b) What implications are there for human use of the two drugs?

10.92 Two researchers at Johns Hopkins University have studied the use of drug products in the elderly. Patients in a recent study were asked the extent to which physicians counseled them with regard to their drug therapies. The researchers found the following:

—25.4% of the patients said their physicians did not explain what the drug was supposed to do

—91.6% indicated they were not told how the drug might "bother" them

—47.1% indicated their physicians did not ask how the drug "helped" or "bothered" them after therapy was started

—87.7% indicated the drug was not changed after discussion on how the therapy was helping or bothering them.

(a) Assume that 500 patients were interviewed in this study. Summarize each of these results using a 95% confidence interval.

(b) Do you have any comments about the validity of any of these results?

EXERCISES FROM THE DATA BASE

10.93 Refer to the clinical trial data base in Appendix 1. Determine the proportions of patients in each treatment group with a therapeutic rating of either moderate or marked. Use these data to test the following hypotheses:

$$H_0: \pi_A - \pi_D = 0$$
$$H_0: \pi_B - \pi_D = 0$$
$$H_0: \pi_C - \pi_D = 0$$

For each null hypothesis, use a one-tailed, upper-tailed test, and give the corresponding p-value.

10.94 For the clinical trial data base determine the proportions of patients that experienced no adverse effects in the four treatment groups. Construct 95% confidence intervals for $\pi_A - \pi_D$, $\pi_B - \pi_D$, and $\pi_C - \pi_D$.

INFERENCES ABOUT σ^2 AND σ_1^2/σ_2^2

11.1 Introduction ■ **11.2** Estimation and tests for σ^2 ■ **11.3** Estimation and tests for comparing σ_1^2 and σ_2^2 ■ Summary ■ Key terms ■ Key formulas ■ Supplementary exercises ■ Exercises from the data base

11.1 INTRODUCTION

variability

When most people think of statistical inference, they think of inferences concerning population means. However, the population parameter that answers an experimenter's practical questions will vary from one situation to another, and sometimes the **variability** of a population is more important than its mean. For example, the producer of a drug product is certainly concerned with controlling the mean potency of tablets but must also worry about the variation in potency from one tablet to another. Excessive potency or an underdose could be very harmful to a patient. Hence the manufacturer would like to produce tablets with the desired mean potency and with as little variation in potency (as measured by σ or σ^2) as possible.

Inferential problems about a population variance are similar to those for a population mean. We can estimate or test hypotheses about a single population variance or compare two variances.

11.2 ESTIMATION AND TESTS FOR σ^2

The sample variance

s^2

$$s^2 = \frac{\Sigma(x - \bar{x})^2}{n - 1}$$

point estimate for σ^2

chi-square distribution

can be used for inferences concerning a population variance σ^2. For a random sample of n measurements drawn from a normal population with mean μ and variance σ^2, s^2 provides a **point estimate for σ^2.** In addition, the sampling distribution for the quantity $(n - 1)s^2/\sigma^2$ is a **chi-square distribution** with df $= n - 1$. The formula for a chi-square distribution will be omitted here, but the properties of the distribution are listed below.

Properties of a Chi-Square Distribution

1. Unlike z and t, values in a chi-square distribution are all positive.

2. A chi-square distribution, unlike the normal and t-distributions, is not symmetrical.

3. There are many chi-square distributions and each one has a different shape. We can specify a particular one by designating the degrees of freedom associated with the sample variance s^2. This quantity is df $= n - 1$.

4. Upper- and lower-tail values of a chi-square distribution are listed in Table 4 of Appendix 3.

From Table 4 of Appendix 3 we can determine the value of chi-square that has an area a to its right for a given chi-square distribution. For example, the value 18.3070 has an area a $= .05$ to its right for a chi-square distribution with degrees of freedom df $= 10$. We find this by locating the df $= 10$ row and the a $= .05$ column.

Knowledge of the properties of a chi-square distribution has given rise to confidence intervals for σ^2 and σ. You will observe that the confidence interval for σ^2 is not in the form used for μ, $\mu_1 - \mu_2$, π, and $\pi_1 - \pi_2$, namely, the point estimate $\pm z$ (or t) standard error. Rather, because chi-square distributions are not symmetrical, we will write confidence intervals for σ^2 in the form:

lower confidence limit $< \sigma^2 <$ upper confidence limit

General Confidence Interval for σ^2 (or σ)

$$\frac{(n - 1)s^2}{\chi_U^2} < \sigma^2 < \frac{(n - 1)s^2}{\chi_L^2}$$

where χ_U^2 is the upper-tail value and χ_L^2 the lower-tail value for a chi-square distribution with df $= n - 1$ (see Figure 11.1). We can deter-

(continued)

> **General Confidence Interval for σ^2 (or σ)** (continued)
>
> mine χ_U^2 and χ_L^2 for a specific value of df using Table 4 of Appendix 3.
>
> For example, for a 95% confidence interval we find the χ_U^2 value for the given df in the column headed by a = .025; the corresponding χ_L^2 value comes from the a = .975 column (i.e., 1 − .025) for the same degrees of freedom. Upper-tail and lower-tail values for 90% confidence intervals come from the .05 and .95 columns of Table 4, Appendix 3; corresponding tail values for 99% confidence intervals come from the .005 and .995 columns.
>
> Note: The confidence interval for σ is found by taking square roots throughout.

FIGURE 11.1   Upper-Tail and Lower-Tail Values of Chi-Square; df = 6

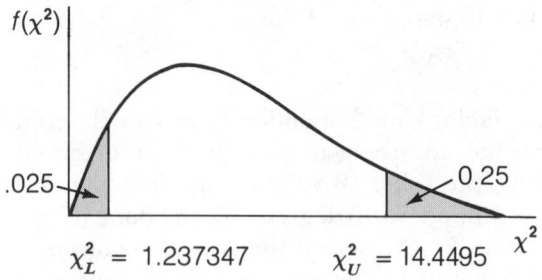

$\chi_L^2 = 1.237347$ $\chi_U^2 = 14.4495$

Example 11.1

The variability in milk production for a 305-day lactation period was observed for a random sample of 15 Holstein cows. Use the milk-yield data in Table 11.1 to estimate σ^2, the population variance of milk yields, using a 95% confidence interval.

TABLE 11.1   Milk production data (in 1000 lb) for Example 11.1

12.928	13.812	11.036
12.120	14.358	9.248
14.972	8.998	9.980
14.044	10.620	11.990
14.788	14.744	14.786

Solution For these data we find

$$\Sigma x = 188.424 \qquad \Sigma x^2 = 2431.470$$

Substituting into the shortcut formula for s^2 (Chapter 5), we have

$$s^2 = \frac{1}{n-1}\left[\Sigma x^2 - \frac{(\Sigma x)^2}{n}\right] = \frac{1}{14}\left[2431.470 - \frac{(188.424)^2}{15}\right] = 4.612$$

The confidence coefficient for our example is $1 - \alpha = .95$. The upper-tail chi-square value can be obtained from Table 4, Appendix 3, for df $= n - 1 = 14$ and a $= .025$. Similarly, the lower-tail chi-square value is obtained with a $= .975$. Thus

$$\chi_U^2 = 26.1190 \qquad \chi_L^2 = 5.62872$$

The 95% confidence interval is then

$$\frac{14(4.612)}{26.1190} < \sigma^2 < \frac{14(4.1612)}{5.62872}$$

or

$$2.472 < \sigma^2 < 11.471$$

In addition to estimating a population variance, we can construct a statistical test of the null hypothesis that σ^2 equals a specified value, σ^2. This test procedure is summarized here.

Statistical Test for σ^2 (or σ)

Null hypothesis: $\sigma^2 = \sigma_0^2$ (σ_0^2 is specified)

Alt. hypothesis: For a one-tailed test:

   1. $\sigma^2 > \sigma_0^2$

   2. $\sigma^2 < \sigma_0^2$

   For a two-tailed test:

   3. $\sigma^2 \neq \sigma_0^2$

Test statistic: $\chi^2 = \dfrac{(n-1)s^2}{\sigma_0^2}$

(continued)

> **Statistical Test for σ^2 (or σ)** (continued)
>
> Rejection region: For a specified value of α,
>
>   reject H_0 if χ^2 is greater than χ_U^2, the upper-tail value for a $= \alpha$ and df $= n - 1$;
>
>   reject H_0 if χ^2 is less than χ_L^2, the lower-tail value for a $= 1 - \alpha$ and df $= n - 1$;
>
>   reject H_0 if χ^2 is greater than χ_U^2, based on a $= \alpha/2$ and df $= n - 1$, or less than χ_L^2, based on a $= 1 - \alpha/2$ and df $= n - 1$.

Example 11.2

A manufacturer of a specific pesticide useful in the control of household bugs claims that his product retains most of its potency for a period of at least six months. More specifically, he claims that the drop in potency from zero to six months will vary in the interval from 0 to 8% To test the manufacturer's claim, a consumer group obtained a random sample of 20 containers of pesticide from the manufacturer. Each can was tested for potency and then stored for a period of six months at room temperature. After the storage period, each can was again tested for potency. The drop in potency was recorded for each can and the sample variance for the drops in potencies was computed to be $s^2 = 6.2$. Use these data to determine if there is sufficient evidence to indicate that the population of potency drops has more variability than that claimed by the manufacturer. Use $\alpha = .05$.

Solution   The manufacturer has claimed that the population of potency reductions has a range of 8%. Dividing the range by 4, we obtain an approximate population standard deviation of $\sigma = 2\%$ (or $\sigma^2 = 4$).

The appropriate null and alternative hypotheses are

H_0: $\sigma^2 = 4$ (i.e., the manufacturer's claim is correct)

H_a: $\sigma^2 > 4$ (i.e., there is more variability than claimed by the manufacturer)

Using the computed sample variance based on 20 observations, the test statistic and rejection region are as follows:

Test statistic: $\chi^2 = \dfrac{(n-1)s^2}{\sigma_0^2} = \dfrac{19(6.2)}{4} = 29.45$

Rejection region: For $\alpha = .05$, we will reject H_0 if the computed value of chi-square is greater than 30.1435, the figure obtained from Table 4 of Appendix 3 for $a = .05$ and df = 19.

Conclusion: Since the computed value of chi-square, 29.45, is less than the tabled value 30.1435, there is insufficient evidence to reject the manufacturer's claim, based on $\alpha = .05$. However, the consumer group is not prepared to accept H_0: $\sigma^2 = 4$. Rather, since $s^2 = 6.2$ and the p-value of the test is $.05 < p < .10$, it would be wise to do additional testing with a larger sample size before reaching a definite conclusion.

---

normality assumption

The inferences we discussed about σ^2 and σ using chi-square distributions are based on the assumption that the underlying population from which the sample measurements are drawn is normal. This **normality assumption** is very important, much more important than it was for the z- and t-methods of previous chapters. We were fortunate in dealing with t-tests about means, because the Central Limit Theorem had wide applicability. With tests about variances (or standard deviations) there is no comparable theorem that helps "assure" normality. So always plot the data. If there is a suggestion of non-normality, p-values obtained from the chi-square test or the confidence coefficient for an interval estimate for σ^2 (or σ) could be in serious error. If this happens, seek out a professional statistician to find an alternative solution to your problem.

EXERCISES

11.1 There are several alternatives for alleviating sulfur dioxide (SO_2) pollution from smokestacks in operations using coal as a major source of fuel for energy. One is to employ a scrubber that cleans the gases after the coal has been burned but before the gases are released into the atmosphere. In 50 samples of gases emitted from a stack equipped with a new scrubber, the sample mean and standard deviation SO_2 emission were .13 and .05 ppm Btu, respectively. Use these data to construct a 95% confidence interval for σ^2.

11.2 Refer to Exercise 11.1. Suppose that testing on the leading commercial scrubber on a comparable stack has indicated an SO_2 emission standard deviation of .50 ppm Btu. Is there sufficient evidence that the new scrubber has a lower population variance? Use $\alpha = .05$.

11.3 As part of a detailed driver training program, school officials require teenagers to take a depth-perception test. In one phase of this test, the student is asked to judge the distance between a parked vehicle and a pedestrian stationed a given distance from the student. The recorded distances in feet are listed below for 15 driver education students.

5	8	7	7	10	6	4	11
6	8	4	9	9	6	5	

Use these data to construct a 99% confidence interval for σ^2, the variance of the population of depth-perception distances.

11.3 ESTIMATION AND TESTS FOR COMPARING σ_1^2 AND σ_2^2

checking equal variance assumption

One of the major applications of a test for the equality of two population variances is for **checking the validity of the equal variance assumption** (i.e., $\sigma_1^2 = \sigma_2^2$) for a two-sample t-test. First we hypothesize two populations of measurements that are normally distributed. We label these populations 1 and 2, respectively. We are interested in comparing the variance of population 1, σ_1^2, to the variance of population 2, σ_2^2.

When independent random samples have been drawn from the respective populations, the ratio

$$\frac{s_1^2}{s_2^2}\bigg/\frac{\sigma_2^2}{\sigma_2^2} = \frac{s_1^2}{\sigma_1^2}\cdot\frac{\sigma_2^2}{s_2^2}$$

F-distribution

possesses a probability distribution in repeated sampling referred to as an **F-distribution.** The formula for the probability distribution is omitted here, but we will specify its properties.

Properties of the F-Distribution

1. Unlike t or z, F can assume only positive values.
2. The F-distribution, unlike the normal distribution or the t-distribution, is nonsymmetrical. (See Figure 11.2.)
3. There are many F-distributions and each one has a different shape. We specify a particular one by designating the degrees of freedom associated with s_1^2 and s_2^2. We denote these quantities by df_1 and df_2, respectively.
4. Tail values for the F-distribution are tabulated and appear in Tables 5–9 of Appendix 3.

FIGURE 11.2   Distribution of s_1^2/s_2^2, the *F*-Distribution

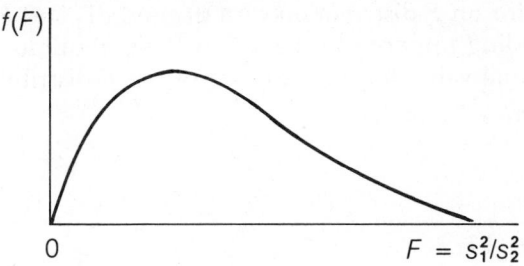

Table 5 of Appendix 3 records the upper-tail value of *F* that has an area equal to .10 to its right (see Figure 11.3). The degrees of freedom for s_1^2, designated by df_1, are indicated across the top of the table, while df_2, the degrees of freedom for s_2^2, appear in the first column to the left. Thus for $df_1 = 8$ and $df_2 = 10$, the tabulated value is 2.38. Only 10% of the measurements from an *F*-distribution with $df_1 = 8$ and $df_2 = 10$ would exceed 2.38 in repeated sampling.

FIGURE 11.3   Upper-Tail .10 Value for the *F*-Distribution; $df_1 = 8$ and $df_2 = 10$

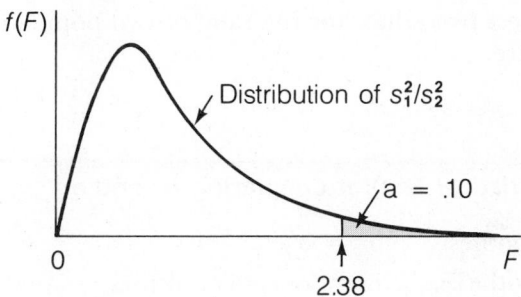

Table 6 of Appendix 3 gives the upper-tail values for the *F*-distribution with an area of .05 to its right. Thus the .05 value of the *F*-distribution with $df_1 = 6$ and $df_2 = 19$ is 2.63.

A statistical test of the null hypothesis $\sigma_1^2 = \sigma_2^2$ utilizes the test statistic s_1^2/s_2^2. When H_0 is true, s_1^2/s_2^2 follows an *F*-distribution with $df_1 = n_1 - 1$ and $df_2 = n_2 - 1$. If upper-tail and lower-tail values of *F* were given in Tables 5–9 of Appendix 3, we would have no difficulty in performing the test. Unfortunately, only upper-tail values of *F* are given. To alleviate this situation, we make use of a relation between upper and lower critical F-values. If F_{a,df_1,df_2} represents the upper-tail value corresponding to an area = a for an F-distribution with df_1 and df_2 and F_{a,df_2,df_1} represents the corresponding upper-tail value of an F-distribution with df_2 and df_1, then the lower-tail value for an area = a of the F-distribution with df_1 and df_2 is $1/F_{a,df_2,df_1}$.

Example 11.3

Determine the upper and lower .05 values for an F-distribution with $df_1 = 6$ and $df_2 = 3$.

Solution Using Table 6 of Appendix 3 and the relationship just discussed we have the following results:

upper .05 value $= F_{.05,6,3} = 8.94$

and

lower .05 value $= \dfrac{1}{F_{.05,3,6}} = \frac{1}{4.76} = 0.21$

The test procedure for the ratio of two population variances is summarized here.

A Statistical Test for Comparing σ_1^2 and σ_2^2

Null hypothesis: $\sigma_1^2 = \sigma_2^2$

Alt. hypothesis: For a one-tailed test:

　　1. $\sigma 21 > \sigma 22$

　　2. $\sigma 22 < \sigma 22$

　For a two-tailed test:

　　3. $\sigma_1^2 \neq \sigma_2^2$

Test statistic: $F = \dfrac{s_1}{s_2^2}$

(continued)

Rejection region: For a specified value of α, $df_1 = n_1 - 1$, and $df_2 = n_2 - 1$

    1. Reject H_0 if $F > F_{\alpha, df_1, df_2}$

    2. Reject H_0 if $F < F_{\alpha, df_2, df_1}$

    3. Reject H_0 if $F > F_{\alpha/2, df_1, df_2}$

       or if $F < 1/F_{\alpha/2, df_2, df_1}$

Note: Appendix Tables 5–9 give upper-tail values for F corresponding to tail areas of .10, .05, .025, .01, and .005.

Example 11.4

In Example 9.1 (p. 305), we discussed an experiment in which company officials were concerned about the length of time a particular drug product retained its potency. A random sample of 10 bottles was obtained from the production line and each bottle was analyzed to determine its potency. A second sample of 10 bottles was obtained and stored in a regulated environment for one year. Potency readings were obtained on these bottles at the end of the year. The sample data were then used to place a confidence interval on $\mu_1 - \mu_2$, the difference in mean potencies for the two time periods.

    Although we did not stress this at the time, in order to use t in the confidence interval or in a statistical test, we do require that the samples be drawn from normal populations with possibly different means *but* with a common variance. Use the sample data summarized below to test the equality of the population variances. Use $\alpha = .01$. Sample 1 data are the readings taken immediately after production and sample 2 data are the readings taken one year after production. Draw conclusions.

        Sample 1: $\bar{x}_1 = 10.37$, $s_1^2 = 0.105$

        Sample 2: $\bar{x}_2 = 9.83$, $s_2^2 = 0.058$

Solution The four parts of the statistical test of $H_0: \sigma_1^2 = \sigma_2^2$ are shown here:

    Null hypothesis: $H_0: \sigma_1^2 = \sigma_2^2$

    Alt. hypothesis: $H_a: \sigma_1^2 \neq \sigma_2^2$

Test statistic: $F = \dfrac{s_1^2}{s_2^2} = \dfrac{0.105}{0.058} = 1.81$

Rejection region: For a two-tailed test with $\alpha = .01$, we will reject H_0 if $F > F_{.005,9,9} = 6.54$ or if $F < 1/F_{.005,9,9} = .15$. Since 1.81 does not fall in the rejection region, we cannot reject H_0: $\sigma_1^2 = \sigma_2^2$.

Also, since 1.81 is not greater than $F_{.005,9,9} = 3.18$, we should feel confident that the assumption of equality of variances holds for the t-methods used with these data.

We can formulate a confidence interval for the ratio σ_1^2/σ_2^2 shown here:

General Confidence Interval for σ_1^2/σ_2^2 (or σ_1/σ_2)

$$\frac{s_1^2}{s_2^2} F_{\mathrm{L}} < \frac{\sigma_1^2}{\sigma_2^2} < \frac{s_1^2}{s_2^2} F_{\mathrm{U}}$$

where $F_{\mathrm{L}} = 1/F_{\alpha/2,\mathrm{df}_1,\mathrm{df}_2}$, $F_{\mathrm{U}} = F_{\alpha/2,\mathrm{df}_2,\mathrm{df}_1}$

$\mathrm{df}_1 = n_1 - 1$ and $\mathrm{df}_2 = n_2 - 1$

Note: The confidence interval for σ_1/σ_2 is found by taking square roots throughout.

Note that our estimation procedure for σ_1^2/σ_2^2 is appropriate for any degree of confidence. Using Tables 5–9 of Appendix 3, we can construct 80%, 90%, 95%, 98%, and 99% confidence intervals for $\sigma_1 \neq \sigma_2$.

Example 11.5

The life length of an electrical component was studied under two operating voltages, V_1 and V_2. Ten different components were randomly assigned to each of the two operating voltages. Use the data following to find a 90% confidence interval for σ_1^2/σ_2^2, the ratio of the variances in life lengths for the two populations, populations 1 and 2, corresponding to the components studied under V_1 and V_2, respectively.

Voltage V_1: $n_1 = 10$ $s_1^2 = .51$
Voltage V_2: $n_2 = 10$ $s_2^2 = .20$

Solution   Before constructing our 90% confidence interval, we must obtain $F_{.05,df_1,df_2}$ and $F_{.05,df_2,df_1}$ from Table 6 of Appendix 3. For $n_1 = n_2 = 10$, $df_1 = df_2 = 9$, and hence $F_{.05,df_1,df_2}$ and $F_{.05,df_2,df_1}$ are the same. This value is 3.18. The quantities F_L and F_U are

$$F_L = \frac{1}{3.18} \quad \text{and} \quad F_U = 3.18$$

Substituting into the confidence interval formula, we have

$$\frac{.51}{.20}\left(\frac{1}{3.18}\right) < \frac{\sigma_1^2}{\sigma_2^2} < \frac{.51}{.20}(3.18)$$

$$.80 < \frac{\sigma_1^2}{\sigma_2^2} < 8.11$$

We are 90% confident that the ratio of population variances corresponding to voltages V_1 and V_2 lies in the interval .80 to 8.11. Since this confidence interval for σ_1^2/σ_2^2 includes values greater and less than 1.0, σ_1^2 may be greater or less than σ_2^2; the sample data do not allow us to be more precise.

Example 11.6

Refer to Example 11.5. Suppose one of the components of V_1 was damaged by the experimenter midway through the test period and had to be removed from the study. Then, with $n_1 = 9$ and $n_2 = 10$, $df_1 = 8$ and $df_2 = 9$. Assuming s_1^2 and s_2^2 are as given in Example 11.5, set up a 90% confidence interval for σ_1^2/σ_2^2.

Solution   The appropriate .05 F-values can be obtained from Table 6 of Appendix 3.

$$F_{.05,8,9} = 3.23 \quad \text{and} \quad F_L = \frac{1}{3.23}$$

$$F_{.05,9,8} = 3.39 \quad \text{and} \quad F_U = 3.39$$

We then have the confidence interval

$$\frac{.51}{.20}\left(\frac{1}{3.23}\right) < \frac{\sigma_1^2}{\sigma_2^2} < \frac{.51}{.20}(3.39)$$

or $.79 < \sigma_1^2/\sigma_2^2 < 8.64$. Again, σ_1^2 may be greater or less than σ_2^2.

sensitivity of assumptions

The inferences about σ_1^2/σ_2^2 based on the F-distribution are very **sensitive** to departures from normality of the underlying distributions. The first precaution that you should take is to plot the data for each sample separately. If there is a hint that one or both of the populations may not be normal, be very careful about the inferences you make on using the F-distribution; the p-value or confidence coefficient may be substantially different from what you found using the test or confidence interval based on F.

Several alternative procedures to the F-methods are available and are discussed in detail in other textbooks (Hollander and Wolfe, 1973).

SUMMARY

In this chapter we discussed procedure for making inferences concerning a population variance and the ratio of two population variances. Estimation and statistical tests concerning σ^2 make use of the chi-square probability distribution with df $= n - 1$. Inferences concerning the ratio of two population variances utilize an F-distribution, with $df_1 = n_1 - 1$ and $df_2 = n_2 - 1$.

The need for inferences concerning one or more population variances can be traced to our discussion of numerical descriptive measures of a population (Chapter 5). To describe or make inferences about a population of measurements, we cannot always rely on the mean, a measure of central tendency. Many times in evaluating or comparing the performance of individuals on a psychological test, the consistency of manufactured products emerging from a production line, or the yields of a particular variety of corn, we gain important information by studying the population variance.

KEY TERMS

variability
s^2
point estimate for σ^2
chi-square distribution
normality assumption
checking equal variance assumption
F-distribution
sensitivity of assumptions

KEY FORMULAS

1. General confidence interval for σ^2 (or σ)

$$\frac{(n-1)s^2}{\chi_U^2} < \sigma^2 < \frac{(n-1)s^2}{\chi_L^2}$$

or

$$\sqrt{\frac{(n-1)s^2}{\chi_U^2}} < \sigma < \sqrt{\frac{(n-1)s^2}{\chi_L^2}}$$

2. Statistical test about σ^2

   Null hypothesis: $\sigma^2 = \sigma_0^2$ (σ_0^2 is specified)

   Test statistic: $\chi^2 = \dfrac{(n-1)s^2}{\sigma_0^2}$

3. General confidence interval for σ_1^2/σ_2^2 (or σ_1/σ_2)

$$\frac{s_1^2}{s_2^2} F_L < \frac{\sigma_1^2}{\sigma_2^2} < \frac{s_1^2}{s_2^2} F_U \quad \text{where} \quad F_L = \frac{1}{F_{\alpha/2,df_1,df_2}} \quad \text{and} \quad F_U = F_{\alpha/2,df_2,df_1}$$

or

$$\sqrt{\frac{s_1^2}{s_2^2} F_L} < \frac{\sigma_1}{\sigma_2} < \sqrt{\frac{s_1^2}{s_2^2} F_U}$$

4. Statistical test about σ_1^2/σ_2^2

   Null hypothesis: $\sigma_1^2 = \sigma_2^2$

   Test statistic: $F = \dfrac{s_1^2}{s_2^2}$

SUPPLEMENTARY EXERCISES

11.4   Find the value of F that locates an area a in the upper tail of the F-distribution for these conditions:
   (a) a = .05, df_1 = 7, df_2 = 12
   (b) a = .05, df_1 = 3, df_2 = 10
   (c) a = .05, df_1 = 10, df_2 = 20
   (d) a = .01, df_1 = 8, df_2 = 15
   (e) a = .01, df_1 = 13, df_2 = 25

11.5   Find approximate values for F_α for these conditions:
   (a) a = .05, df_1 = 11, df_2 = 24
   (b) a = .05, df_1 = 14, df_2 = 14
   (c) a = .05, df_1 = 35, df_2 = 22
   (d) a = .01, df_1 = 22, df_2 = 24
   (e) a = .01, df_1 = 17, df_2 = 25

   (Note: Your answers may not agree with those in the back of the book. As long as your answer is close to the recorded answer, it is satisfactory.)

11.6   Random samples of n_1 = 8 and n_2 = 10 observations were selected from populations 1 and 2, respectively. The corresponding sample variances were s_1^2 = 7.4 and s_2^2 = 12.7. Do the data provide sufficient evidence to indicate a difference between σ_1^2 and σ_2^2? Test by using α = .10. What assumptions have you made?

11.7   An experiment was conducted to determine whether there was sufficient evidence to indicate that data variation within one population, say population A, exceeded the variation within a second population, population B. Random samples of n_A = n_B = 8 measurements were selected from the two populations and the sample variances were calculated to be

$$s_A^2 = 2.87 \qquad s_B^2 = .91$$

   Do the data provide sufficient evidence to indicate that σ_A^2 is larger than σ_B^2? Test by using α = .05.

11.8   A soft drink firm is debating whether it should invest in a new type of canning machine or whether it should continue operating with the machines presently in use. The company has already determined that it will be able to fill more cans per day for the same cost if the new machines are installed. However, an important factor as yet unsolved is the variability of fills. (The company would, of course, prefer the model with the smaller variance in fills.) Let σ_1^2 and σ_2^2 denote the variances for fills from the old model and the new model, respectively. Obtaining samples of fills from the two models and utilizing the test statistics s_1^2/s_2^2, we can set up either a one-tailed or a two-tailed rejection region, using the F-distribution.

(a) What type of rejection region would be most favored by the manager of the soft drink company? Why?

(b) What type of rejection region would be most favored by the salesman for the company manufacturing the model presently in use? Why?

11.9 Refer to Exercise 11.8. Suppose random samples of $n_1 = n_2 = 11$ cans from the two machines are examined to determine the amount of fill (in ounces). The means and variances are

$$\bar{x}_1 = 11.70 \qquad \bar{x}_2 = 11.60$$
$$s_1^2 = .06 \qquad s_2^2 = .022$$

Do these data present sufficient evidence to indicate less variability of fills for the new model? Use $\alpha = .10$.

11.10 In a gasoline economy study ten 1-gallon samples of a particular brand of gasoline were used for each of two cars (A and B). Both cars averaged approximately 17 mpg, but the sample standard deviations were .95 and 1.56 for cars A and B, respectively. Use these data to test the hypothesis that the variances in miles per gallon for the two cars are identical. Use $\alpha = .05$.

11.11 Two consumer research groups are vying for a large government contract. Since subjective evaluations of consumer products will be made by judges during the study, government officials prefer to award the contract to a company that utilizes judges with consistent ratings. Of course other qualifications are also evaluated before awarding the contract. One measure of consistency is the variability of judges' scores on the same item.

Before the contract is issued, a test is conducted in which 25 judges from each company are asked to rate a single item. The sample variances are given here:

$$\text{company A: } s_1^2 = .50 \qquad \text{company B: } s_2^2 = .15$$

Use these data to test the hypothesis that the variances of the judges' ratings are the same for the two populations. The alternative hypothesis is that the variances are different. Use $\alpha = .01$.

11.12 The variability in the potency of 5-grain aspirin tablets differs from one brand to another. An interested research group would like to compare brand C, a new product recently released, to brand B, the current bestseller. Random samples of 41 tablets are obtained from bottles of each of the brands; the potency results are given here. Use these data to test the hypothesis that the population variances of brands B and C, σ_1^2 and σ_2^2, respectively, are equal. The alternative hypothesis is that the bestseller has more variability in potency; that is, σ_1^2 is larger than σ_2^2. Use $\alpha = .01$. What assumptions have you made?

	Brand B	Brand C
Sample size	41	41
Sample mean	60.2	60.5
Sample variance	2.20	.98

11.13 A consumer protection magazine was interested in comparing tires purchased from two different companies, each claiming their tires would last the same number of miles. A sample of 5 tires of each brand was obtained and tested under simulated road conditions. The number of miles before significant deterioration in tread was recorded for all tires. The data are given here (in 1000 miles):

Brand 1	40.6	35.9	48.5	36.4	38.3
Brand 2	40.9	40.2	42.5	39.1	42.6

(a) Construct a 95% confidence interval for the ratio of the two population variances.

(b) How does the confidence interval change if we're interested in σ_1/σ_2 rather than σ_1^2/σ_2^2?

11.14 A random sample of 20 patients, each of whom has suffered from depression, was selected from a mental hospital, and each patient was administered the Brief Psychiatric Rating Scale. The scale consists of a series of adjectives that the patient scores according to his or her mood. Extensive testing in the past has shown that ratings in certain mood adjectives tend to be similar and hence are grouped together as jointly measuring one or more components of one's mood. For example, a group consisting of certain adjectives seems to be measuring depression. Let us suppose that the mean and standard deviation of the scores for the 20 patients in the group are 13.2 and 4.6, respectively.

(a) Place a 99% confidence interval on σ^2, the variance of the population of patients' scores from which this sample was drawn.

(b) What's the critical assumption underlying the inference? Do you have any inclination as to whether this assumption is valid for these data?

11.15 Refer to Exercise 11.14. Suppose that extensive testing in a large number of depressed patients throughout the country has indicated that the population standard deviation of scores for the depression adjectives is 5.9.

Use the sample data of Exercise 11.14 to test the research hypothesis that the standard deviation for all patients who might be treated for depression in this hospital is less than 5.9. Give an approximate p-value for these data and draw conclusions.

11.16 A pharmaceutical company manufactures a particular brand of antihistamine tablets. In the quality control division, certain tests are routinely performed to determine whether the product being manufactured meets specific performance criteria prior to its release onto the market. In particular, the company requires that the potencies of the tablets lie in the range of 90% to 110% of the labeled drug amount.

(a) If the company is manufacturing 25-mg tablets, within what limits must tablet potencies lie?

(b) A random sample of 30 tablets is obtained from a recent batch of antihistamine tablets. The data for the potencies of the tablets are given below. Construct a histogram to see if the assumption of normality is unwarranted for inferences about the population variance.

(c) Translate the company's 90% to 110% specifications on the range of the product potency into a statistical test concerning the population variance for potencies. Draw conclusions based on $\alpha = .05$.

24.1	27.2	26.7	23.6	26.4	25.2
25.8	27.3	23.2	26.9	27.1	26.7
22.7	26.9	24.8	24.0	23.4	25.0
24.5	26.1	25.9	25.4	22.9	24.9
26.4	25.4	23.3	23.0	24.3	23.8

11.17 A study was conducted to compare the variabilities in strengths of 1-in.^{2} sections of a synthetic fiber produced under two different procedures. A random sample of 9 squares from each process was obtained and tested.

(a) Plot the data for each sample separately.

(b) Is the assumption of normality unwarranted?

(c) If permissible from part (b), use the data (psi) below to test the research hypothesis that the population variances corresponding to the two procedures are different. Use $\alpha = .10$.

Procedure 1	74	90	103	86	75	102	97	85	69
Procedure 2	59	66	73	68	70	71	82	69	74

11.18 Forty-two undergraduate students were randomly assigned to one of two sections (each of size 21) for the same course. The same instructor was assigned to teach both sections. Unannounced quizzes were used throughout the quarter in section 2 but were not used in section 1. The final section averages were the same; but the variations in final grades for the two sections, as measured by the sample variances, were different:

$$s_1^2 = 9.3 \qquad s_2^2 = 4.1$$

Use these data to test the hypothesis of equality of the population variances for the two methods of teaching (with and without unannounced quizzes) against the alternative hypothesis that unannounced quizzes constitute a teaching method that causes less variability in students' achievement. Use $\alpha = .05$.

11.19 Would your conclusion to Exercise 11.18 have changed if you had employed a two-tailed test with $\alpha = .10$? With $\alpha = .01$?

11.20 An important consideration in examining the potency of a pharmaceutical product is the amount of drop in potency for a specific shelf life (time on a pharmacist's shelf). In particular, the variability of these drops in potency is very important. Researchers studied the drops in potency for two different drug products over a 6-month period. These data are summarized in the accompanying table. Suppose that drug 1 is an experimental drug product and drug 2 a marketed product. Use a one-tailed test with $\alpha = .01$ to determine whether the data suggest that drug 1 has more variability in potency drop than drug 2.

	Drug 1	Drug 2
Sample size	10	10
Sample mean	58	56
Sample variance	82	23

11.21 Refer to Exercise 11.20. Would your result have changed if you had used a two-tailed test with $\alpha = .10$? Why might a two-tailed test be important?

11.22 Blood cholesterol levels for randomly selected patients with similar histories were compared for two diets, one a low-fat-content diet and the other a normal diet. The summary data appear in the accompanying table.

	Low-fat content	Normal
Sample size	19	24
Sample mean	170	196
Sample variance	198	435

(a) Do these data present sufficient evidence to indicate a difference in cholesterol level variabilities for the two diets? Use $\alpha = .10$.

(b) What other test might be of interest in comparing the two diets?

11.23 Sales from weight-reducing agents marketed in the United States represent sizable chunks of income for many of the companies that manufacture these products. Psychological as well as physical effects often contribute to how well a person responds to the recommended therapy. Consider a comparison of two weight-reducing agents, A and B. In particular, consider the variabilities in the lengths of times people remain on the therapy. A total of 26 overweight males, matched as closely as possible physically, were randomly divided into two groups. Those in group 1 received preparation A while those assigned to group 2 received preparation B. Use the summary data to compare the variabilities associated with the lengths of time on therapy. Use a two-tailed test with $\alpha = .10$.

	Preparation A	**Preparation B**
Sample size	13	13
Sample mean	25 days	35 days
Sample variance	50	16

11.24 Refer to Exercise 11.23. What might the null and alternative hypotheses have been if preparation A had been a placebo (no active medication) and preparation B a marketed product known to be an effective weight-reducing agent?

11.25 A chemist at an iron ore mine suspects that the variance in the amount (weight, in ounces) of iron oxide per pound of ore tends to increase as the mean amount of iron oxide per pound increases. To test this theory, ten 1-pound specimens of iron ore are selected at each of two locations, one, location 1, containing a much higher mean content of iron oxide than the other, location 2. The amounts of iron oxide contained in the ore specimens are shown in the accompanying table.

Location 1	8.1	7.4	9.3	7.5	7.1	8.7	9.1	7.9	8.4	8.8
Location 2	3.9	4.4	4.7	3.6	4.1	3.9	4.6	3.5	4.0	4.2

(a) Do the data provide sufficient evidence to indicate that the amount of iron oxide per pound of ore is more variable at location 1 than at location 2? Use $\alpha = .05$.

11.26 Refer to Example 11.2 on p. 378. Construct a 95% confidence interval for σ^2, and use this interval to help interpret the findings of the consumer group. Does it appear that the test of Example 11.2 had much chance of detecting an increase in σ^2 of 25% over the claimed value? Explain.

11.27 The risk of an investment is measured in terms of the variance in the return that could be observed. Random samples of ten yearly returns were obtained from two different portfolios:

| | Portfolio | |
Return (000)	1	2
Sample mean	132	146
Sample variance	10.9	25.6
Sample size	10	10

Does Portfolio 2 have a higher risk? Give an approximate p-value for your test.

$ 11.28 Refer to Exercise 11.27. Are there any differences in the average returns for the two portfolios? Comment on the method you used in arriving at a conclusion and why you used it.

$ 11.29 Two different modeling techniques for assessing the value of resale houses were considered. A random sample of twelve existing listings was taken and each house was valued using the two techniques. These data are shown here.

| | Technique | |
Listing	1	2
1	155	138
2	137	128
3	248	230
4	136	146
5	102	95
6	87	82
7	63	67
8	129	134
9	144	149
10	270	292
11	157	150
12	51	48

(a) Plot the data. Does it appear that the two modeling techniques give similar results?

(b) Give an estimate of the mean and standard error of the difference between estimates for the two methods.

$ 11.30 Refer to Exercise 11.29. Place a 90% confidence interval on the variance for the difference between estimates. Give the corresponding interval for σ.

$ 11.31 A personnel officer was planning to use a t-test to compare the mean number of monthly unexcused absences for two divisions of a multinational company, but then noticed a possible difficulty. The variation in the number of unexcused absences per month seemed to differ for the two groups. As a check, a random sample of five months was selected at each division, and for each month, the number of unexcused absences was obtained.

Category A	20	14	19	22	25
Category B	37	29	51	40	26

(a) What assumption seemed to bother the personnel officer?

(b) Do the data provide sufficient evidence to indicate that the variances differ for the populations of absences for the two employee categories? Use $\alpha = .05$.

11.32 A researcher was interested in weather patterns in Phoenix and Seattle. As part of the investigation, the researcher took a random sample of 20 days in July and observed the daily average temperatures. The data were collected over several years to assure independence of daily temperatures. The data collected produced the following information:

	Phoenix daily average temperature	Seattle daily average temperature
Sample size	20	20
Sample mean	95.3	63.3
Sample standard deviation	5.1	7.6

Do the data suggest that there is a difference in the variability of average daily temperatures during July for the two cities? Is there a difference in mean temperatures for the two cities during July? Use $\alpha = .05$ for both tests.

EXERCISES FROM THE DATA BASE

11.33 Refer to the clinical trial data base in Appendix 1 to calculate the sample variances for the anxiety scores within each treatment group. Use these data to run separate tests comparing each of the treatments A, B, and C to the placebo group D. Use two-sided tests with $\alpha = .05$.

11.34 Do any of these tests in Exercise 11.33 negate the possibility of comparing treatment means for groups A, B, and C to the treatment mean for the placebo group using t-tests? Explain.

11.35 Use the sleep disturbance scores from the clinical trial data base to give a 98% confidence interval for σ_B^2/σ_C^2. Do the same for σ_B^2/σ_A^2.

REGRESSION AND CORRELATION

12.1 Introduction ▪ **12.2** Scatter plots and the freehand regression line ▪ **12.3** Method of least squares ▪ **12.4** Correlation ▪ **12.5** Multiple regression (optional) ▪ **12.6** Using computers ▪ Summary ▪ Key terms ▪ Key formulas ▪ Supplementary exercises ▪ Exercises from the data base

12.1 INTRODUCTION

We have already touched on some ways to summarize data from more than one variable (Chapter 5). In particular we used the scatter plot as a way to examine the relationship between two quantitative variables based on sample data. Now, we will extend the concept of a scatter plot to better understand the relationship between two quantitative variables. In Chapter 13 we deal with the next step in making sense of data—analyzing sample data on two quantitative variables in order to make inferences about the population from which the sample data were drawn. We begin with several examples of problems dealing with two (or more) quantitative variables.

A problem of considerable interest to high school seniors, freshmen entering college, their parents, and university administrations is the expected academic achievement of a particular student after he or she has enrolled in a university. For example, we might wish to estimate what a student's grade point average (GPA) will be at the end of the freshman year before the student has been accepted or enrolled in the university. At first glance this task seems difficult.

The statistical approach to this problem is, in many respects, a formalization of the procedure we might follow intuitively. Suppose data were available giving the high school academic grades, psychological and sociological information, as well as the grades attained at the end of the college freshman year for a large number of students. Then we might categorize the students into groups possessing similar characteristics. For example, highly motivated students who earned a high rank in their high school class, graduated from a high school with superior academic standards, and so forth, should achieve, on the average, a higher GPA at the end of their college freshman year than students who lack motivation and who achieved only moderate success in high school. Carrying this line of thought a little further, we would expect the GPA of a student to be related to many variables that define the individual's psychological and physical characteristics as well as to those that define the academic and social environment to which he or she will be exposed. Ideally we would like to obtain a mathematical equation that relates a student's GPA to all these variables, so it could be used for prediction.

You will observe that the problem we have defined is of a very general nature. We are interested in some quantitative random variable y that is related to a number of quantitative variables $x_1, x_2, x_3, \ldots$. Generally, the random variable y is called the *dependent variable* and the x variables are designated as *independent variables*. We are interested in obtaining an equation that relates y to the independent variables. The variable y for our example is the student's grade point average at the end of the freshman year. The independent variables might be

$x_1 =$ rank in high school class
$x_2 =$ score on a mathematics achievement test
$x_3 =$ score on a verbal achievement test

and so on. The ultimate objective is to measure $x_1, x_2, x_3, \ldots$ for a particular student, substitute these values into the mathematical equation, and thereby predict the student's grade point average. In order to accomplish this, we must first determine the related variables $x_1, x_2, x_3, \ldots$, and measure the strength of their relationship to y. Then we must construct a **prediction equation** good **prediction equation** that will express y in terms of the selected independent variables.

Other practical examples of our prediction problem are numerous throughout business, industry, and the sciences. To illustrate this point, let's look at several more examples in the following paragraphs.

Executives of a large oil company would like to relate the performance of a gasoline blend to a number of key ingredients in the gasoline mixture. By varying the proportions of the ingredients in the overall blend and using the performance information from these many different blends, it would be possible to obtain a prediction equation relating performance to the proportions of the blend ingredients. By substituting different values of the independent variables (proportions of the key ingredients) into the prediction equation, the company could determine the proportions that would provide the best blend in terms of gasoline performance.

A biologist would like to relate physical characteristics such as height, weight, blood pressure, pulse, age, and so forth to the amount of secretion from a gland in humans. By observing these variables (characteristics) and the glandular secretion for many different people, he could obtain a prediction equation relating the amount of secretion to these physical characteristics.

A political analyst may wish to predict the success of a candidate in a political primary based on a number of important variables. Success would probably be measured by the number of votes cast for the candidate. Variables affecting the outcome of the primary might be the amount of money spent on television advertising, the size of the campaign organization, and the amount of money spent advertising in papers and magazines. Ideally the political analyst would like to study the effects of these variables on the outcomes of previous elections to aid in predicting the best strategy for a future campaign.

First we consider the problem of predicting y based on a single independent quantitative variable x. Then we observe that the solution for a *multivariable problem*, where y is related to more than one quantitative independent variable, is based on a generalization of our technique. Since the methodology for the multivariable predictor is fairly complex, we use computer programs for solutions to these problems.

12.2   SCATTER PLOTS AND THE FREEHAND REGRESSION LINE

Consider the prediction of a student's GPA at the end of the freshman year based on his or her high school GPA. The objective is to obtain an equation that will predict the achievement of college freshmen and hence will be of assistance to admissions officers in identifying potentially successful students. The GPA data for a sample of 11 students are shown in Table 12.1.

Following the data collection stage, the next step in making sense of the sample data in Table 12.1 is to summarize them using graphical or numerical methods. In Chapter 5 we introduced the scatter plot as a means for summarizing the relationship between two quantitative variables. Recall that to construct a scatter plot, we make vertical and horizontal axes of approximately equal length. Generally the independent variable x is labeled along the horizontal axis, and the dependent variable y is labeled along the vertical axis. For our example the independent variable is the high school GPA.

TABLE 12.1   Data for High School and College GPAs
(Based on a 4.0 System)

Student	High school GPA, x	Freshman GPA, y
1	2.00	1.60
2	2.25	2.00
3	2.60	1.80
4	2.65	2.80
5	2.80	2.10
6	3.10	2.00
7	2.90	2.65
8	3.25	2.25
9	3.30	2.60
10	3.60	3.00
11	3.25	3.10

Having labeled the axes, we then draw scales along the axes in such a way that all measurements can easily be plotted along the appropriate axis. For our example both the independent and dependent variables range between 0 and 4.0. After the axes are drawn, labeled, and scaled, we plot the data of Table 12.1. Each dot on the figure represents the information concerning one student and can be obtained by plotting the freshman GPA, y, against the corresponding high school value, x. The dot circled in Figure 12.1 corresponds to student 1. Note that the dot is placed at a point on the graph corresponding to $x = 2.00$ and $y = 1.60$.

Based on the scatter plot of Figure 12.1, it appears that the freshman GPA, y, increases as the high school GPA, x, increases. In fact many of the dots seem to be on a straight line. We call a line running through the dots of a scatter plot a *trend line,* or a *regression line.* When the regression line is a straight line, we say there is a linear relationship between x and y. See Figures 12.2(a) and (b). However, not all regression lines are linear. When the trend line is curved we say there is a curvilinear relationship between x and y. See Figures 12.2(c) and (d). Trend lines as shown here provide further means for summarizing the relationship between two quantitative variables.

freehand regression line

There are many methods for obtaining a trend line relating y to x. The first is called an "eyeball fit," or a **freehand regression line,** which can be obtained by placing a ruler on the graph (Figure 12.1) and moving it about until we have minimized the distances from the points to the fitted line. This has been done in Figure 12.3. The resulting trend line

FIGURE 12.1 High School and College GPA Data

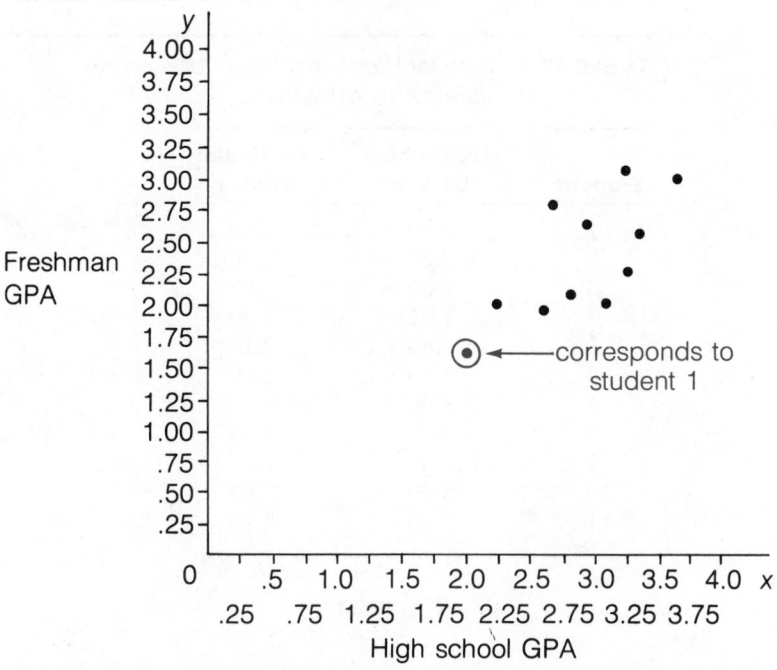

FIGURE 12.2 Different Types of Regression Lines: (a) Linear
Relationship Between x and y; (b) Linear Re-
lationship Between x and y; (c) Curvilinear
Relationship Between x and y; (d) Curvilinear
Relationship Between x and y

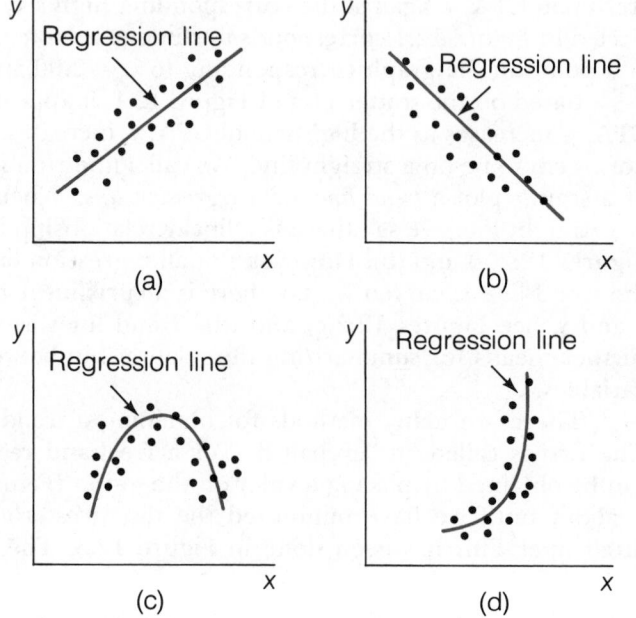

can be used to make a prediction about a freshman's GPA based on high school achievement. To predict y when $x = 2.5$, refer to the graph and note that the y coordinate for the point corresponding to $x = 2.5$ is $y = 2.05$ (see the arrows on Figure 12.3).

FIGURE 12.3 Freehand Regression Line for High School and College GPA Data

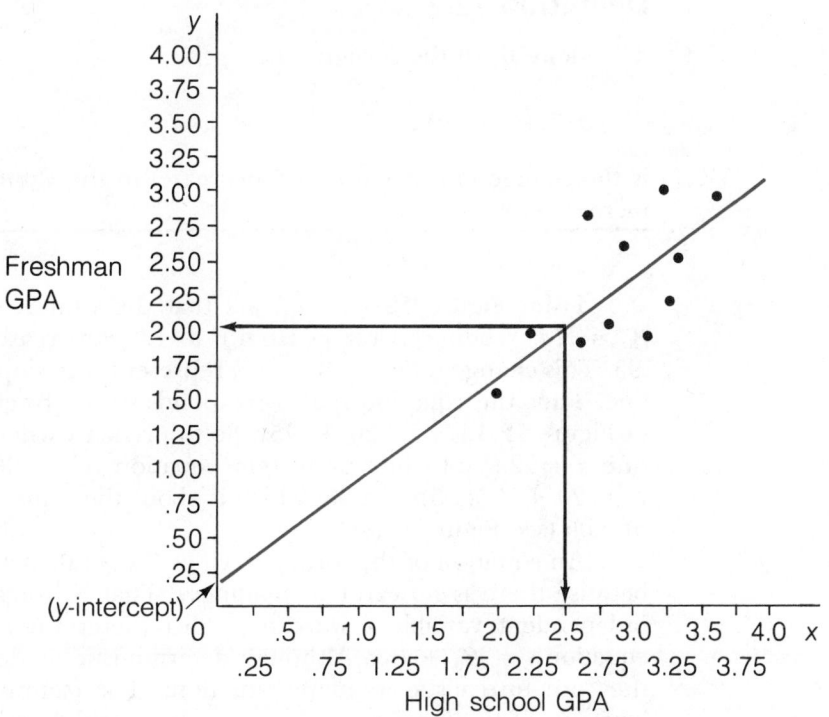

The freehand regression line shown in Figure 12.3 can be represented by a **linear equation** of the form

linear equation

$$y = \beta_0 + \beta_1 x$$

The two constants, β_0 and β_1, in the equation determine the location and the slope of the line, respectively. The constant β_0 is the y-intercept—that is, the value of y when the line crosses the vertical axis (y-axis). The constant β_1 is the slope of the line—that is, the change in y that corresponds to a one-unit increase in x.

y-intercept β_0

Definition 12.1

The **y-intercept β_0** for the straight line

$$y = \beta_0 + \beta_1 x$$

is the value of y at the point where the line crosses the y-axis.

slope β_1

Definition 12.2

The **slope β_1** of the straight line

$$y = \beta_0 + \beta_1 x$$

is the change (an increase or a decrease) in the value of y for a one-unit increase in x.

From Figure 12.4 it appears that the y-intercept is approximately $\beta_0 = .20$. When x increases from 0 to 1.0, y increases from about .20 to .95. This change $(.95 - .20 = .75)$ represents the slope (β_1) of the straight line. Thus the equation that corresponds to the freehand regression line of Figure 12.4 is $y = .20 + .75x$. So to predict y when $x = 2.5$, we substitute $x = 2.5$ into our prediction equation, $y = .20 + .75x$, to obtain $y = .20 + .75(2.5) = 2.08$, which is about the same as the guessed value of 2.05 (see Figure 12.3).

An equation of the form $y = \beta_0 + \beta_1 x$ is called a deterministic model because there is no error in reading y. That is, for a given value of the independent variable x we can predict (determine) y exactly using the equation $y = \beta_0 + \beta_1 x$. Although deterministic models are simple to use, they are unrealistic in many situations. For example, even though an equation of the form $y = \beta_0 + \beta_1 x$ adequately describes the trend in the GPA data, it cannot be used to predict a student's freshman GPA *exactly* based on his or her high school GPA.

A model that allows for the possibility that all observations do not fall on a straight line is the model

$$y = \beta_0 + \beta_1 x + \epsilon$$

where ϵ is a random error. In this model, ϵ represents the difference between a measurement y and a point on the line $\beta_0 + \beta_1 x$. The random error takes into account all unpredictable and unknown factors that are not included in the model. For example, a freshman's GPA could be affected by such factors as the student's rank in high school class, score on the Scholastic Aptitude Test, motivation, and many other factors. The

FIGURE 12.4   Slope β_1 and Intercept β_0 for the Freehand
Regression Line

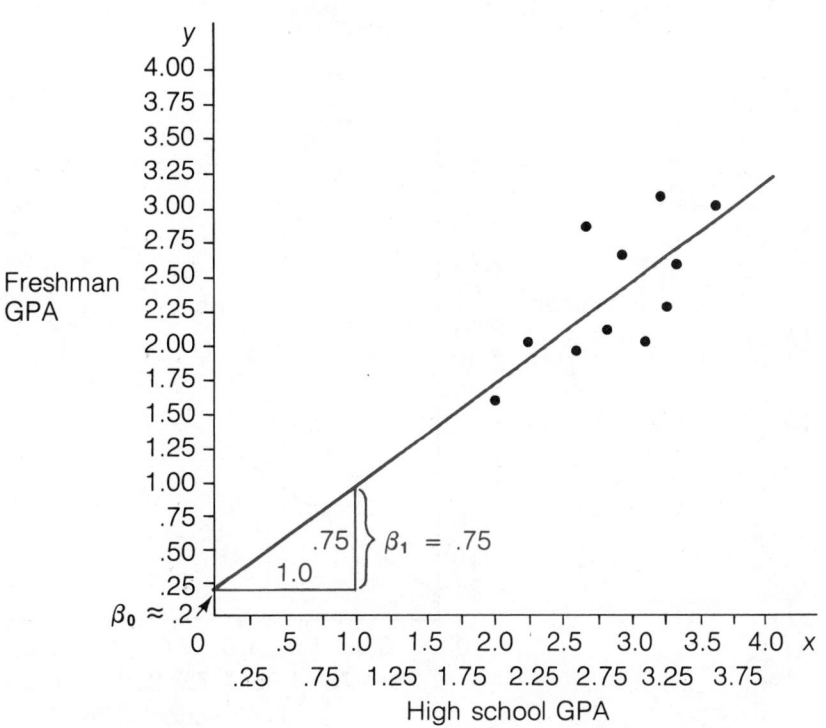

combined effects of all these and other factors not included in the model contribute to ϵ.

One assumption made concerning the random error is that the average value of ϵ for a given value of x is 0. Thus, since β_0 and β_1 are constants, the average value of y (often called the **expected value of y**) for a fixed value of x is $\beta_0 + \beta_1 x$. This line, denoted by

**expected
value of y**

$$E(y) = \beta_0 + \beta_1 x$$

is shown in Figure 12.5. The difference between a sample data point and the expected value of y (a point on the line $\beta_0 + \beta_1 x$) is ϵ. The random errors associated with the 11 data points listed in Table 12.1 are pictured in Figure 12.5.

The freehand regression line, which we discussed previously, provided us with an equation by which we can predict y from x. But, since each one of us might determine a different equation for the same data, we need a more precise way to obtain estimates of the constants β_0 and β_1 in the model $y = \beta_0 + \beta_1 x + \epsilon$. The method we will use is called the *method of least squares* and will be discussed in the next section.

FIGURE 12.5   The Mean Value of y Plotted for Various Values of x

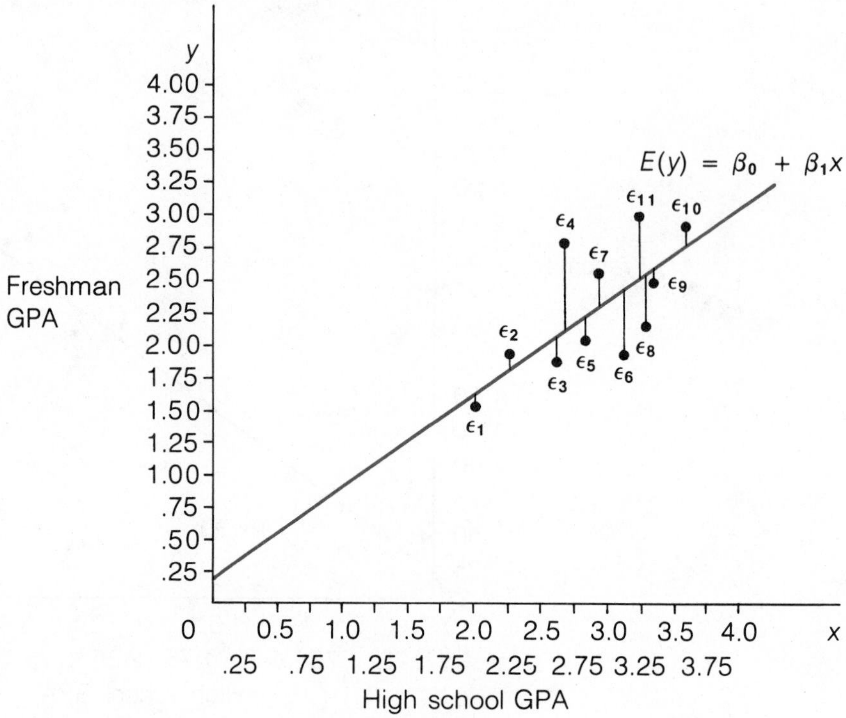

EXERCISES

12.1   Plot the data shown here in a scatter diagram.

x	5	10	12	15	18	24
y	10	19	21	28	34	40

12.2   Refer to the data of Exercise 12.1.

(a) Use an eyeball fit to construct a freehand regression line.

(b) Identify the intercept and slope for your regression line.

(c) Predict y when $x = 20$.

12.3   Use the equation $y = 1.8 + 2.0x$.

(a) Predict y when $x = 3$.

(b) Plot the equation on a graph with the horizontal axis scaled from 0 to 5 and the vertical axis scaled from 0 to 12.

$ 12.4 Results of a pharmaceutical pricing survey were reported recently relating the influence of wholesale discounts and other factors on prescription ingredient costs. One problem considered in the study was the relationship between the annual prescription volume y and the percentage of the drug ingredient purchases that were made directly from the drug manufacturers. Suppose that a sample of 10 independent pharmacies yielded the results shown in the accompanying table. Plot these data on a scatter diagram.

Annual volume (× $1,000), y	Percentage of purchases, x	Annual volume (× $1,000), y	Percentage of purchases, x
25	10	138	63
55	18	90	42
50	25	60	30
75	40	10	5
110	50	100	55

$ 12.5 Refer to the data of Exercise 12.4.

(a) Use an eyeball fit to determine a freehand regression line for predicting the annual prescription volume of an independent pharmacy based on the percentage of ingredients purchased directly from drug manufacturers.

   (i) What is the y-intercept for your regression line?

   (ii) What is the slope of the line?

(b) Predict the annual prescription volume, using the freehand regression line from part (a), when a pharmacy buys 45% of its ingredients directly from drug manufacturers.

👥 12.6 Using data from the 1980 U.S. Census, a sample of 10 states showed the per capita income, x, and public education expenditure per student (in dollars), y, given in the table here.

State	Per capita income, x	Public education expenditure per student, y
Arkansas	6785	1219
Idaho	7446	1345
Michigan	9269	1990
Mississippi	6167	1189
Missouri	8132	1673
Nebraska	8341	1746
New Jersey	9702	2576
North Dakota	7774	1607
Ohio	8775	1777
Washington	9435	2079

(a) Plot the data on a scatter diagram.

(b) Using an eyeball fit, draw a freehand regression line for the data.

(c) Identify the slope and intercept for your regression line. Write the equation corresponding to your regression line.

12.3 METHOD OF LEAST SQUARES

method of least squares

The statistical procedure for finding the prediction equation—the **method of least squares**—is, in many respects, an objective way to obtain an eyeball fit to the points. For example, when we fit a line by eye to a set of points, we move the ruler until we think that we have minimized the distances from the points to the fitted line. This same minimizing technique is used in the statistical procedure.

predicted value of y

We denote the **predicted value of y** for a given value of x (obtained from the fitted line) as $\hat{y}$. The prediction equation obtained from the method of least squares is denoted by

$$\hat{y} = \hat{\beta}_0 + \hat{\beta}_1 x$$

residual

where $\hat{\beta}_0$ and $\hat{\beta}_1$ are *estimates* of the unknown intercept β_0 and slope β_1 in the linear regression model $y = \beta_0 + \beta_1 x + \epsilon$. The error of prediction (sometimes called the **residual**) is $y - \hat{y}$, the difference between the actual value of y and what we predict it to be. See Figure 12.6. The method of

FIGURE 12.6   Least Squares Fit to the Data in Table 12.1

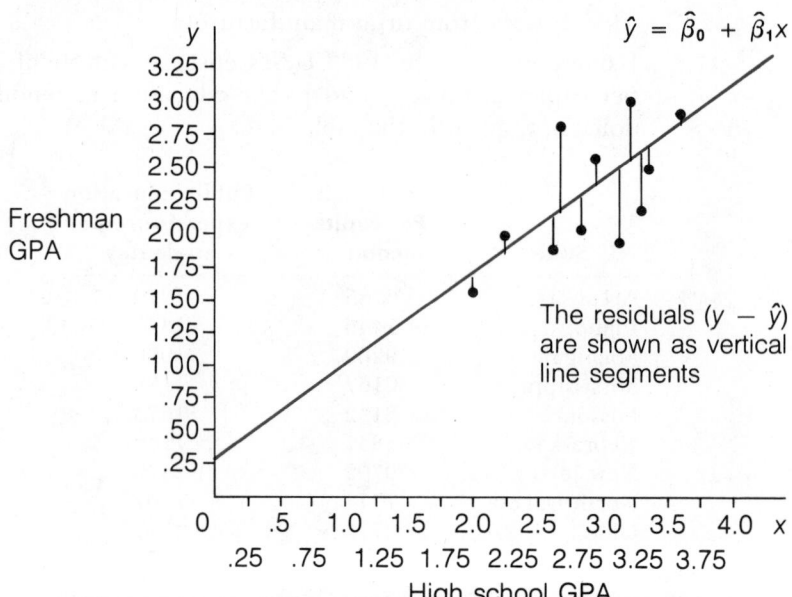

The residuals $(y - \hat{y})$ are shown as vertical line segments

**sum of squares
for error**

least squares chooses the prediction equation that minimizes the sum of the squared errors of prediction for all sample measurements. The sum of squared errors (also referred to as the **sum of squares for error**) is denoted by SSE and can be written as

$$\text{SSE} = \Sigma \, (y - \hat{y})^2$$

where y is an observed response, and $\hat{y}$ is a point on the prediction equation

$$\hat{y} = \hat{\beta}_0 + \hat{\beta}_1 x$$

Substituting for $\hat{y}$ in SSE, we have

$$\text{SSE} = \Sigma \, [y - (\hat{\beta}_0 + \hat{\beta}_1 x)]^2$$

The method of least squares chooses those values for the estimates $\hat{\beta}_0$ and $\hat{\beta}_1$ that make SSE a minimum. Derivation of these values is beyond the scope of this text, but they can be found by using the formulas that follow.

Least Squares Estimates for the Regression Line
$$\hat{y} = \hat{\beta}_0 + \hat{\beta}_1 x$$

$$\hat{\beta}_1 = \frac{S_{xy}}{S_{xx}}$$

and

$$\hat{\beta}_0 = \bar{y} - \hat{\beta}_1 \bar{x}$$

where

$$S_{xx} = \Sigma \, x^2 - \frac{(\Sigma x)^2}{n}$$

$$S_{xy} = \Sigma \, xy - \frac{(\Sigma x)(\Sigma y)}{n}$$

and $\bar{y}$ and $\bar{x}$ denote the sample means for the y-values and x-values, respectively; n is the number of y-values.

The use of these formulas for finding $\hat{\beta}_0$, $\hat{\beta}_1$, and the least squares line is illustrated in the next example.

Example 12.1

Obtain the regression line $\hat{y} = \hat{\beta}_0 + \hat{\beta}_1 x$ for the GPA data in Table 12.1 by using the method of least squares.

Solution The calculation of the least squares estimates $\hat{\beta}_0$ and $\hat{\beta}_1$ is greatly simplified by using Table 12.2.

TABLE 12.2 Calculations for the GPA Data

	x	y	x^2	xy	y^2
	2.00	1.60	4.0000	3.2000	2.5600
	2.25	2.00	5.0625	4.5000	4.0000
	2.60	1.80	6.7600	4.6800	3.2400
	2.65	2.80	7.0225	7.4200	7.8400
	2.80	2.10	7.8400	5.8800	4.4100
	3.10	2.00	9.6100	6.2000	4.0000
	2.90	2.65	8.4100	7.6850	7.0225
	3.25	2.25	10.5625	7.3125	5.0625
	3.30	2.60	10.8900	8.5800	6.7600
	3.60	3.00	12.9600	10.8000	9.0000
	3.25	3.10	10.5625	10.0750	9.6100
Total	31.70	25.90	93.6800	76.3325	63.5050

Substituting the values in Table 12.2 into the formulas, we obtain

$$S_{xx} = \Sigma x^2 - \frac{(\Sigma x)^2}{n} = 93.68 - \frac{(31.7)^2}{11} = 2.326$$

$$S_{xy} = \Sigma xy - \frac{(\Sigma x)(\Sigma y)}{n} = 76.3325 - \frac{(31.7)(25.9)}{11} = 1.693$$

$$\bar{y} = \frac{\Sigma y}{n} = \frac{25.9}{11} = 2.355$$

$$\bar{x} = \frac{\Sigma x}{n} = \frac{31.7}{11} = 2.882$$

Hence

$$\hat{\beta}_1 = \frac{S_{xy}}{S_{xx}} = \frac{1.693}{2.326} = .728$$

$$\hat{\beta}_0 = \bar{y} - \hat{\beta}_1 \bar{x} = 2.355 - .728(2.882) = .257$$

Thus the least squares prediction equation relating the freshman GPA, y, to the corresponding high school value, x, is

$$\hat{y} = .257 + .728x$$

Notice that the least squares prediction equation is similar to the equation for the freehand line for the same data set, obtained in Figure 12.4.

Algebraically, it can be shown from a least squares fit of the model $y = \beta_0 + \beta_1 x + \epsilon$ that the difference between an individual y-value (say y) and the sample mean $\bar{y}$ is

$$y - \bar{y} = (y - \hat{y}) + (\hat{y} - \bar{y})$$

where $\hat{y}$ is the predicted value of y from the least square prediction equation. It can also be shown that

$$\Sigma (y - \bar{y})^2 = \Sigma (y - \hat{y})^2 + \Sigma (\hat{y} - \bar{y})^2$$

While the proof of this equality is beyond the scope of this text, we can obtain an intuitive understanding of this relationship by considering the following situation.

Suppose that we use the model $y = \beta_0 + \epsilon$. In this model β_0 represents the population mean for the variable y, and intuitively we would estimate its value using the sample mean $\bar{y}$. (You can confirm this result by using the formula for the estimated intercept $\hat{\beta}_0$ in a linear model.) Since $\hat{y} = \bar{y}$ for this model, the sum of the squared errors of prediction is $\Sigma (y - \bar{y})^2$.

Now suppose the variable y is related to an independent variable x. From our previous work, we could fit the model $y = \beta_0 + \beta_1 x + \epsilon$ to obtain

$$\hat{y} = \hat{\beta}_0 + \hat{\beta}_1 x$$

For this model the sum of the squared prediction errors is

$$\Sigma (y - \hat{y})^2$$

In Figure 12.7 we present two prediction equations: $\hat{y} = \bar{y}$ for the model $y = \beta_0 + \epsilon$, and $\hat{y} = \hat{\beta}_0 + \hat{\beta}_1 x$ for the model $y = \beta_0 + \beta_1 x + \epsilon$. Note that we can express the distance between an observation y and the sample mean $\bar{y}$ as the sum of two components, $(\hat{y} - \bar{y})$ and $(y - \hat{y})$. The quantity $(\hat{y} - \bar{y})$ represents that portion of the overall distance that can be attributed to the independent variable x (through the prediction equation $\hat{y} = \hat{\beta}_0 + \hat{\beta}_1 x$). The quantity $(y - \hat{y})$ represents that portion

FIGURE 12.7 Relationship Between $\Sigma (y - \bar{y})^2$ and $\Sigma (y - \hat{y})^2$

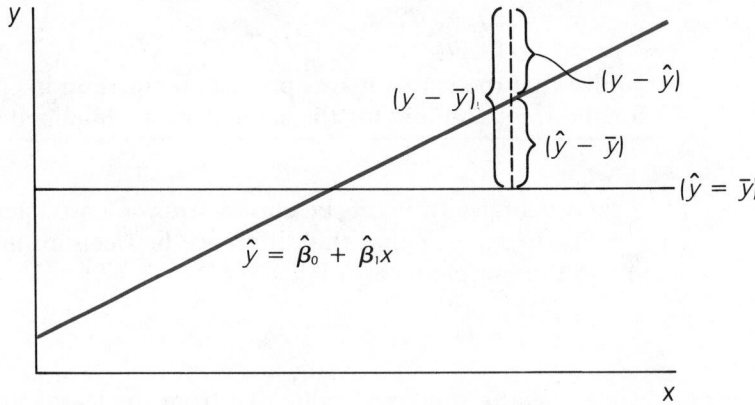

$$\hat{y} = \hat{\beta}_0 + \hat{\beta}_1 x$$

of the distance between y and $\bar{y}$ that cannot be accounted for by the independent variable x (and that we attribute to error). Combining this information for all sample observations, we can express the total variability in the sample measurements about the sample mean, $\Sigma (y - \bar{y})^2$, called the **sum of squares about the mean,** as the sum of the squared deviations of the predicted values from $\bar{y}$, called the **sum of squares due to regression,** $\Sigma (\hat{y} - \bar{y})^2$, and the sum of the squared errors of prediction, $\Sigma (y - \hat{y})^2$, called the **sum of squares for error.** Thus we have

sum of squares about the mean

sum of squares due to regression

sum of squares for error

Sum of squares about the mean = Sum of squares due to regression
+ Sum of squares for error

There is another way to view this equation. It can be shown that the sample mean $\bar{y}$ is also the average of the fitted values. Therefore the sum of squares due to regression $\Sigma (\hat{y} - \bar{y})^2$ depicts variability in the fitted values. Similarly $\Sigma (y - \hat{y})^2$ represents variability in the y-values about the fitted values. As a result, the total variability in the y-values can be written as

$$\underset{\substack{\text{total variability} \\ \text{in } y\text{-values}}}{\Sigma (y - \bar{y})^2} = \underset{\substack{\text{variability} \\ \text{explained by model}}}{\Sigma (\hat{y} - \bar{y})^2} + \underset{\substack{\text{unexplained} \\ \text{variability}}}{\Sigma (y - \hat{y})^2}$$

Figure 12.7 shows how the total variability is partitioned into the two components.

Obviously, if we're interested in predicting y based on the independent variable x, the larger the explained variability is relative to the unexplained variability, the better the model "fits" the data; and this should lead to more precise prediction of y based on x.

Example 12.2

Consider the five data points listed in columns 1 and 2 of Table 12.3.

(a) Fit the model

$$y = \beta_0 + \beta_1 x + \epsilon$$

(b) Verify that

$$\Sigma (y - \bar{y})^2 = \Sigma (y - \hat{y})^2 + \Sigma (\hat{y} - \bar{y})^2$$

Solution

(a) For these data, it can be shown that

$$S_{xx} = \Sigma x^2 - \frac{(\Sigma x)^2}{n} = 66 - \frac{(16)^2}{5} = 14.8$$

$$S_{xy} = \Sigma xy - \frac{(\Sigma x)(\Sigma y)}{n} = \frac{145 - (16)(38)}{5} = 23.4$$

$$\hat{\beta}_1 = 1.58$$

and

$$\hat{\beta}_0 = \bar{y} - \hat{\beta}_1 \bar{x} = 7.6 - 1.58(3.2) = 2.54$$

(b) For each x-value we then compute $\hat{y}$ from the least squares prediction equation. We also compute the quantities $(y - \bar{y})$, $(y - \hat{y})$, and $(\hat{y} - \bar{y})$. These quantities are displayed in Table 12.3.

TABLE 12.3 Data and Computations for Example 12.2

x	y	$\hat{y}$	$y - \bar{y}$	$y - \hat{y}$	$\hat{y} - \bar{y}$
1	4	4.1216	−3.6000	−.1216	−3.4784
2	6	5.7027	−1.6000	.2973	−1.8973
3	7	7.2838	−.6000	−.2838	−.3162
4	9	8.8649	1.4000	.1351	1.2649
6	12	12.0271	4.4000	−.0271	4.4271

From columns 4, 5, and 6 in Table 12.3, we have

$$\Sigma (y - \bar{y})^2 = 37.2000$$
$$\Sigma (y - \hat{y})^2 = .2027$$
$$\Sigma (\hat{y} - \bar{y})^2 = 36.9982$$

Note that, except for rounding errors,

$$\Sigma\,(y - \bar{y})^2 = \Sigma\,(y - \hat{y})^2 + \Sigma\,(\hat{y} - \bar{y})^2$$

Intuitively, since the explained variability $\Sigma\,(\hat{y} - \bar{y})^2 = 36.9982$ accounts for almost all of the total variability $\Sigma\,(y - \bar{y})^2 = 37.2000$, the model appears to fit the data very well. A scatter plot of y versus x with the prediction equation superimposed shows how tightly the data group about the prediction equation. See Figure 12.8.

FIGURE 12.8   *Scatter Plot with Superimposed Regression Line for the Data of Table 12.3*

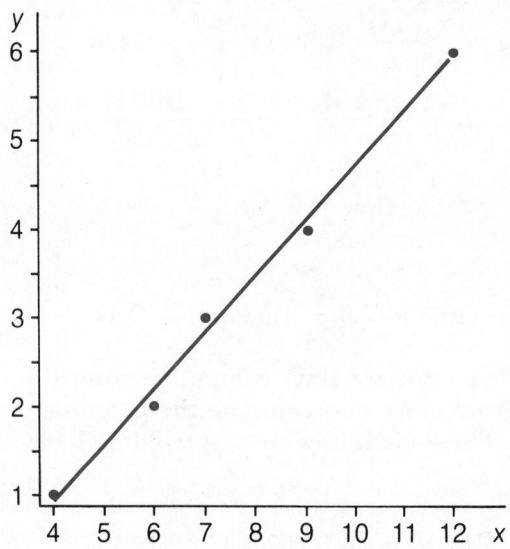

EXERCISES

12.7   Use the accompanying data to determine the least squares prediction equation.

x	1	2	3	4	5
y	2	4	6	7	9

12.8   Use the accompanying data to answer (a) and (b).

x	1	3	5	7	9
y	1	4	8	9	12

(a) Determine the least squares prediction equation.

(b) Use the least squares prediction equation to predict y when $x = 6$.

12.9   Refer to the data of Exercise 12.1. Find the least squares prediction equation and compare it to the freehand regression line you found in Exercise 12.2.

12.10   A computer solution using SAS for the least squares prediction equation to the data of Exercise 12.4 is shown here.

```
OPTIONS NODATE NONUMBER PS=60 LS=78;
DATA ONE;
   INPUT X Y;
   CARDS;
   10      25
   18      55
   25      50
   40      75
   50     110
   63     138
   42      90
   30      60
    5      10
   55     100
;
PROC GLM;
   MODEL Y=X/P;
TITLE1 '    ';
RUN;
```

```
Model: MODEL1
Dependent Variable: Y

                        Analysis of Variance

                         Sum of        Mean
     Source      DF      Squares       Square      F Value    Prob>F

     Model        1   13230.96994   13230.96994    162.560    0.0001
     Error        8     651.13006      81.39126
     C Total      9   13882.10000

          Root MSE        9.02171     R-square     0.9531
          Dep Mean       71.30000     Adj R-sq     0.9472
          C.V.           12.65317
```

(continued)

```
                      Parameter Estimates

                 Parameter      Standard    T for HO:
     Variable  DF   Estimate        Error  Parameter=0    Prob > |T|

     INTERCEP   1   4.697852    5.95202071       0.789        0.4527
     X          1   1.970478    0.15454842      12.750        0.0001

                     Dep Var    Predict
               Obs      Y        Value   Residual

                 1   25.0000    24.4026    0.5974
                 2   55.0000    40.1665   14.8335
                 3   50.0000    53.9598   -3.9598
                 4   75.0000    83.5170   -8.5170
                 5   110.0      103.2      6.7783
                 6   138.0      128.8      9.1620
                 7   90.0000    87.4579    2.5421
                 8   60.0000    63.8122   -3.8122
                 9   10.0000    14.5502   -4.5502
                10   100.0      113.1    -13.0741

Sum of Residuals              1.207923E-13
Sum of Squared Residuals         651.1301
Predicted Resid SS (Press)      1065.6517
```

(a) Plot the data.

(b) Determine the least squares prediction equation from the output here and draw the regression line in the data plot in (a).

(c) Does the prediction equation seem to adequately represent the data?

(d) Predict y (annual prescription volume) for $x = 35$.

12.11 Refer to Exercise 12.6. Obtain the least squares prediction equation relating a state's public education expenditure per student to the per capita income of the state. Compare the least squares prediction equation to your prediction equation of Exercise 12.6. Use the least squares equation to predict the expenditure of a state having a per capita income of $7,500.

12.12 Compare the least squares prediction equation for Exercise 12.11 to the SAS computer output shown here for the same problem.

```
OPTIONS NODATE NONUMBER PS=60 LS=78;
DATA ONE;
  INPUT X Y;
  CARDS;
       6785    1219
       7446    1345
       9269    1990
       6167    1189
       8132    1673
```

```
      8341   1746
      9702   2576
      7774   1607
      8775   1777
      9435   2079
PROC REG;
  MODEL Y=X/P;
TITLE1 '   ';
RUN;
```

Model: MODEL1
Dependent Variable: Y

Analysis of Variance

Source	DF	Sum of Squares	Mean Square	F Value	Prob>F
Model	1	1434433.5301	1434433.5301	59.590	0.0001
Error	8	192573.36994	24071.67124		
C Total	9	1627006.9000			

Root MSE	155.15048	R-square	0.8816
Dep Mean	1720.10000	Adj R-sq	0.8668
C.V.	9.01985		

Parameter Estimates

Variable	DF	Parameter Estimate	Standard Error	T for H0: Parameter=0	Prob > |T|
INTERCEP	1	-1090.116490	367.33428016	-2.968	0.0179
X	1	0.343438	0.04448989	7.719	0.0001

Obs	Dep Var Y	Predict Value	Residual
1	1219.0	1240.1	-21.1109
2	1345.0	1467.1	-122.1
3	1990.0	2093.2	-103.2
4	1189.0	1027.9	161.1
5	1673.0	1702.7	-29.7220
6	1746.0	1774.5	-28.5006
7	2576.0	2241.9	334.1
8	1607.0	1579.8	27.2288
9	1777.0	1923.6	-146.6
10	2079.0	2150.2	-71.2219

Sum of Residuals   2.50111E-12
Sum of Squared Residuals   192573.3699
Predicted Resid SS (Press)   379959.2084

12.13 Family income and annual savings data are displayed here for a sample of nine families.

Annual savings ($000)	Annual income ($000)
1	36
2	39
2	42
5	45
5	48
6	51
7	54
8	56
7	59

(a) Graph the data using a scatter plot.

(b) Determine an eyeball fit to the data. Predict y (annual savings, $000) based on an annual income of $x = \$45,000$.

12.14 Refer to Exercise 12.13.

(a) Determine the least squares prediction equation.

(b) Compute $\Sigma (y - \hat{y})^2$ for the least squares fit and for your eyeball fit. Note that SSE for the least squares fit will be less than or equal to that for the eyeball fit.

12.15 The following data were obtained in a study of sales volume (per district) as a fraction of the number of client contacts per month.

Sales volume ($1,000) y	Average number of client contacts per month x
15	10
26	15
28	17
30	20
32	23
86	46
109	53
95	48
130	59
160	65

(a) Plot the data.

(b) Eyeball a linear fit to the data and guess the value of the intercepts and slope.

(c) Predict sales for $x = 50$.

12.16 Refer to Exercise 12.15.

(a) Obtain the linear regression equation $\hat{y} = \hat{\beta}_0 + \hat{\beta}_1 x$ using the method of least squares. Compare this line to the one obtained in Exercise 12.15.

(b) Predict sales for $x = 50$ and compare to Exercise 12.15(c).

⑂ 12.17 As one part of a study of commercial bank branches, data are obtained on the number of independent businesses, x, located in sample ZIP code areas and the number of bank branches, y, located in these areas. The commercial centers of cities are excluded.

x:	92	116	124	210	216	267	306	378	415	502	615	703
y:	3	2	3	5	4	5	5	6	7	7	9	9

$$\Sigma x = 3944 \qquad \Sigma y = 65 \qquad \Sigma xy = 26{,}208$$
$$\Sigma x^2 = 1{,}732{,}524 \qquad \Sigma y^2 = 409 \qquad n = 12$$

(a) Plot the data. Does a linear equation relating y to x appear plausible?

(b) Calculate the regression equation (with y as the dependent variable).

12.4 CORRELATION

In this section we will extend our study of the relationships between two or more variables. Not only might we like to predict the value of one variable (the dependent variable) based on information on one or more independent variables, as we have done in previous sections, but we might also wish to provide a measure of the strength of the relationship between these variables.

One measure of the strength of the relationship between two variables x and y is called the *coefficient of linear correlation*, or, simply, the **correlation coefficient**. The stronger the correlation, the better x predicts y. Given n pairs of observations (x_i, y_i), we can compute the **sample correlation coefficient** $\hat{\rho}$ (greek lower-case rho) as

correlation coefficient

sample correlation coefficient $\hat{\rho}$

$$\hat{\rho} = \frac{S_{xy}}{\sqrt{S_{xx}S_{yy}}}$$

where

$$S_{xx} = \Sigma x^2 - \frac{(\Sigma x)^2}{n}$$

$$S_{xy} = \Sigma xy - \frac{(\Sigma x)(\Sigma y)}{n}$$

and

$$S_{yy} = \Sigma y^2 - \frac{(\Sigma y)^2}{n}$$

We will illustrate the computation of $\hat{\rho}$ with an example and then explain its practical significance.

Example 12.3

An engineer is interested in calibrating a flow meter to be used on a liquid-soap production line. For the test the engineer observes the following 10 meter readings for 10 known flow rates.

Observed meter reading y	Actual flow rate x
1.4	1
2.3	2
3.1	3
4.2	4
5.1	5
5.8	6
6.8	7
7.6	8
8.7	9
9.5	10

(a) Plot the sample data. Do the data look linear?

(b) Compute the sample correlation coefficient $\hat{\rho}$.

Solution

(a) The data plot shown:

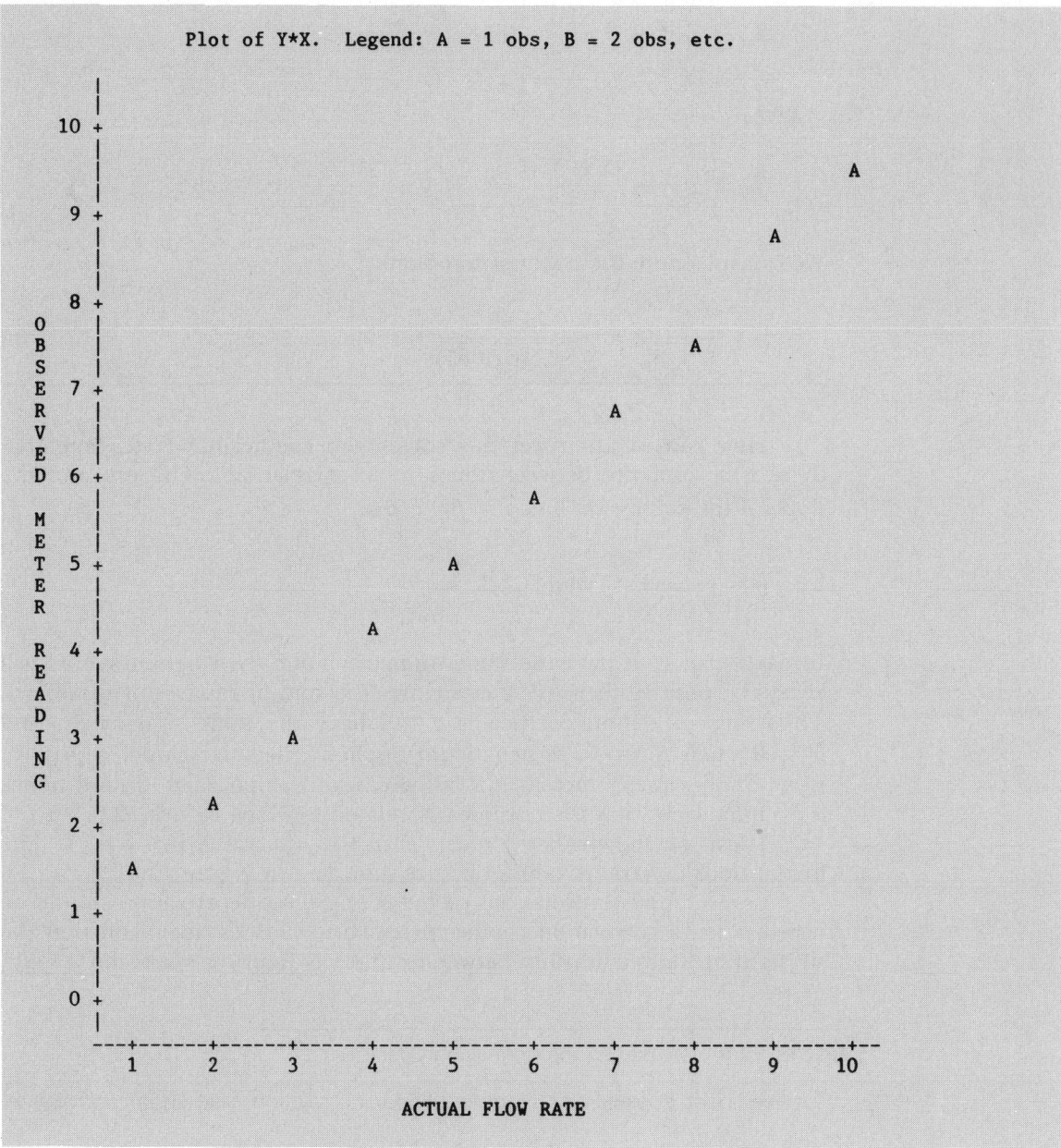

Plot of Y*X. Legend: A = 1 obs, B = 2 obs, etc.

(b) For these data, it can be shown that $\Sigma\,x = 55$, $\Sigma\,x^2 = 385$, $\Sigma\,y = 54.5$, $\Sigma\,y^2 = 364.09$, and $\Sigma\,xy = 374.1$. It follows that

$$S_{yy} = \Sigma\,y^2 - \frac{(\Sigma\,y)^2}{n} = 364.09 - \frac{(54.5)^2}{10} = 67.065$$

$$S_{xx} = \Sigma\, x^2 - \frac{(\Sigma\, x)^2}{n} = 385 - \frac{(55)^2}{10} = 82.5$$

and

$$S_{xy} = \Sigma\, xy - \frac{(\Sigma\, x)(\Sigma\, y)}{n} = 374.1 - \frac{55(54.5)}{10} = 74.35$$

We can substitute these values to obtain

$$\hat{\rho} = \frac{S_{xy}}{\sqrt{S_{xx}S_{yy}}} = \frac{74.35}{\sqrt{82.5(67.065)}} = .9996$$

How can we interpret this correlation coefficient? First, note that there is a similarity between the sample correlation coefficient and the slope of the regression line $\hat{y} = \hat{\beta}_0 + \hat{\beta}_1 x$:

$$\hat{\rho} = \frac{S_{xy}}{\sqrt{S_{xx}S_{yy}}} \qquad \text{and} \qquad \hat{\beta}_1 = \frac{S_{xy}}{S_{xx}}$$

In particular, both have the same numerator, and their denominators will always be positive (because they involve the sum of squares of numbers). Thus the correlation coefficient $\hat{\rho}$ will have the same sign as $\hat{\beta}_1$, and $\hat{\rho} = 0$ when $\hat{\beta}_1 = 0$. A negative $\hat{\rho}$ implies a negative slope; a positive value implies that y increases as x increases. A slope $\hat{\beta}_1 = 0$ (and hence $\hat{\rho} = 0$) indicates that x is not linearly related to y. See Figure 12.9.

It can be shown that $\hat{\rho}$ must lie in the interval $-1 \le \hat{\rho} \le 1$. The implications of various values of $\hat{\rho}$ are indicated in Figure 12.10.

Several misinterpretations of the coefficient of correlation should be noted. First, a correlation coefficient equal to .5 does not mean that the strength of the relationship between y and x is "halfway" between no cor-

FIGURE 12.9 Interpreting the Correlation Coefficient

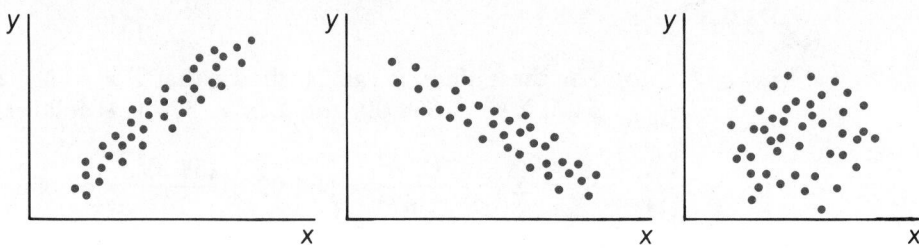

Positive linear correlation   Negative linear correlation   No linear correlation

FIGURE 12.10 *Possible Values for $\hat{\rho}$, the Sample Correlation Coefficient*

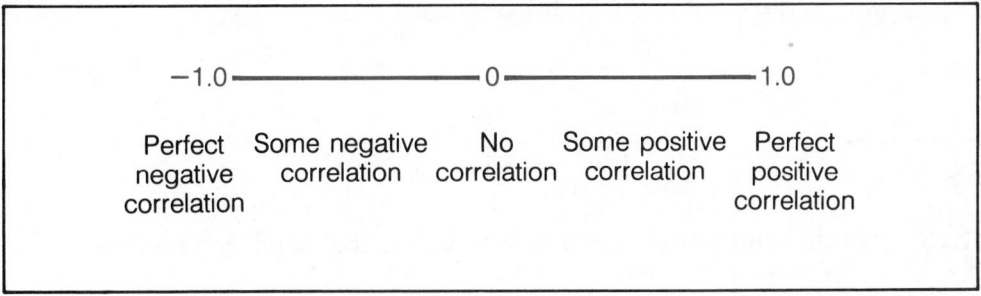

relation and perfect correlation. If we designate $S_{yy} = \Sigma\, (y - \bar{y})^2$ as the total variability of the y-values about their sample mean, it can be shown that an amount equal to $\hat{\rho}^2$ of this total variability can be explained by the variable x. The more closely x and y are linearly related, the more the variability in the y-values can be explained by variability in the x-values and the closer $\hat{\rho}^2$ will be to 1. If $\hat{\rho} = .5$, the independent variable x is accounting for $\hat{\rho}^2 = .25$ of the total variation in the y-values about $\bar{y}$. The quantity $\hat{\rho}^2$ is called a **coefficient of determination.**

coefficient of determination

Second, y and x could be perfectly related in some way other than in a linear manner when $\hat{\rho} = 0$ or some very small value. This is seen in the perfect curvilinear fit of Figure 12.11.

FIGURE 12.11 *Perfect Curvilinear Fit; $\hat{\rho} = 0$*

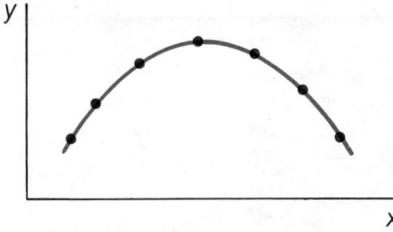

And finally, note that we cannot add correlations. If the simple linear correlations between y and x_1, y and x_2, and y and x_3 are .1, .3, and .2, respectively, it *does not* follow that x_1, x_2, and x_3 account for $\hat{\rho}_1^2 + \hat{\rho}_1^2 + \hat{\rho}_3^2 = (.1)^2 + (.3)^2 + (.2)^2$ of the variability of the y-values about their sample mean. Indeed, x_1, x_2, and x_3 may be highly correlated and contribute the same information for the prediction of y. The relationship between y and several independent variables should not be studied by computing simple correlation coefficients for each of the independent variables. Rather, we should relate y to x_1, x_2, and x_3 by using a single multivariable model. This topic is discussed briefly in the next section.

EXERCISES

12.18 Plot the sample data shown here, compute the correlation coefficient, and interpret your findings.

x	1	2	3	4	6	9	10
y	2	4	5	7	8	12	13

12.19 Refer to Exercise 12.18. Suppose the first three y-values are 16, 12, and 10.

(a) Plot the data and guess a value for $\hat{\rho}$.

(b) Compute the sample correlation coefficient, and compare the computed value to the guessed value.

(c) Why do the correlation coefficients differ for Exercises 12.18 and 12.19?

12.20 A corporation examined the relationship between profits ($000) and the percentage of operating capacity being used for each of 12 plants.

Profits ($000) y	% of operating capacity x
2.5	50
6.2	57
3.1	61
4.6	68
2.3	77
4.5	80
6.1	82
11.6	85
10.0	89
14.2	91
16.1	95
19.5	99

(a) Plot the data. Are the data nearly linear?

(b) Compute the correlation coefficient.

(c) Compute the coefficient of determination and use it to help interpret the value of $\hat{\rho}$.

 **12.21** An experiment was conducted to investigate the amplitude of the shock wave recorded on sensors placed at different distances from an explosive charge. The charge was to be detonated underground, with three sensors placed at each of the three different distances from the charge, as illustrated in Figure 12.12. The shock-wave amplitudes were recorded and summarized according to the distance from the explosion. These data are given below.

FIGURE 12.12 Location of Sensors from the Charge for the Shock-Wave Experiment of Exercise 12.21

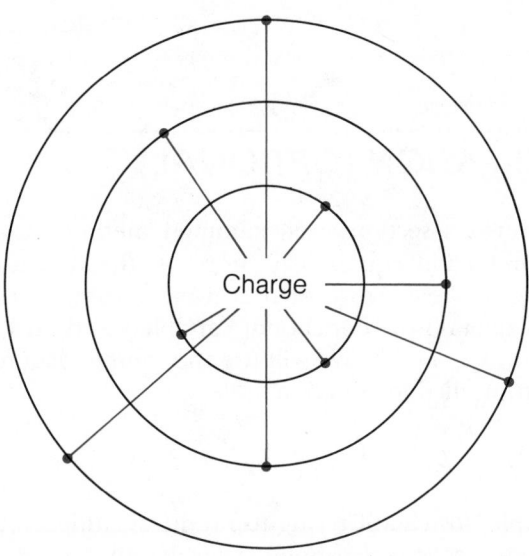

Distance, x	5	5	5	10	10	10	15	15	15
Amplitude, y	8.6	8.2	8.1	5.8	6.2	6.1	5.2	4.8	4.7

(a) Plot the sample data.

(b) Compute the correlation coefficient and interpret its value.

 **12.22** A forester was interested in training an assistant to estimate the timber volume of a standing tree. Having trained her, the forester calibrated the assistant's estimates against known timber volumes. Perhaps a better way to quantify the assistant's estimate would be to base it on an objective reading, such as the basal area of the tree. If, indeed, volume is related to

basal area, the assistant would have an objective way to estimate the timber volume of a tree. A random sample of 12 trees was obtained. For each tree included in the sample, the basal area x was recorded along with the cubic-foot volume after the tree was felled. These data appear below.

Tree	1	2	3	4	5	6	7	8	9	10	11	12
Basal area, x	.3	.5	.4	.9	.7	.2	.6	.5	.8	.4	.8	.6
Volume, y	6	9	7	19	15	5	12	9	20	9	18	13

(a) Plot the data.

(b) Compute and interpret the correlation coefficient between basal area and timber volume.

12.5   MULTIPLE REGRESSION (OPTIONAL)

In the previous sections we examined methods for obtaining the least squares prediction equation, $\hat{y} = \hat{\beta}_0 + \hat{\beta}_1 x$, based on values from two variables y and x. Now we turn to a more complicated situation in which we collect data on the dependent variable y and on k ($k > 1$) independent variables $x_1, x_2, \ldots, x_k$. We will use the sample data to obtain the multiple regression prediction equation

$$\hat{y} = \hat{\beta}_0 + \hat{\beta}_1 x + \hat{\beta}_2 x_2 + \cdots + \hat{\beta}_k x_k$$

For example, instead of trying to predict a student's GPA (y) at the end of the student's college freshman year based on the corresponding high school GPA, x_1, we may wish to use information on several additional variables, such as rank in high school class, x_2, college board verbal score, x_3, and college board math achievement score, x_4. To do this, we would have to examine the records of a sample of n freshmen to obtain information on the variables y, x_1, x_2, x_3, and x_4. Then the method of least squares can be used to obtain the least squares multiple regression equation

$$\hat{y} = \hat{\beta}_0 + \hat{\beta}_1 x_1 + \hat{\beta}_2 x_2 + \hat{\beta}_3 x_3 + \hat{\beta}_4 x_4$$

Although there are computational formulas for the least squares estimates of the parameters $\beta_0, \beta_1, \beta_2, \ldots, \beta_k$, the formulas are difficult to work with algebraically beyond the simplest case, $\hat{y} = \hat{\beta}_0 + \hat{\beta}_1 x$ (discussed in Section 12.3). Rather than spending time developing expressions for the least squares estimates of the parameters in a multiple regression equation, we will make use of available computer software packages to do the work for us for a particular problem.

Example 12.4

The sales profit, y, of a product in a sales territory is thought to be related to the total population, x_1, of the territory as well as to the advertising expenditure per person, x_2, in the territory. The data from eight different sales territories are shown in Table 12.4. Use these data to find least squares estimates for the coefficients in the multiple regression equation

$$\hat{y} = \hat{\beta}_0 + \hat{\beta}_1 x_1 + \hat{\beta}_2 x_2$$

Predict sales profit for a territory with a population of $x_1 = 2.8$ and an advertising expenditure of $x_2 = .25$.

TABLE 12.4   Data for Example 12.4

Sales territory	Sales profit per person (dollars), y	Total population (millions), x_1	Advertising per person (dollars), x_2
1	3.6	2.4	.16
2	2.5	1.3	.21
3	4.2	5.1	.12
4	4.1	4.9	.14
5	4.0	3.2	.26
6	5.1	6.7	.10
7	4.3	3.2	.41
8	11.5	.7	.11

Solution   Since it is important for you to be able to locate and identify the least squares estimates from a computer printout, a portion of some sample output follows. The least squares estimates in the printout are shown in color. Note that the least squares prediction equation, $\hat{y} = \hat{\beta}_0 + \hat{\beta}_1 x_1 + \hat{\beta}_2 x_2$, is

$$\hat{y} = 9.04 - .58x_1 - 11.27x_2$$

Substituting into this equation, we find that the predicted sales profit per person for a territory with a population of $x_1 = 2.8$ and an advertising expenditure of $x_2 = .25$ per person is

$$\hat{y} = 9.04 - .58(2.8) - 11.27(.25) = \$4.60$$

```
OPTIONS NODATE NONUMBER PS=60 LS=78;
DATA ONE;
  INPUT Y X1 X2;
  CARDS;
    3.6   2.4   0.16
    2.5   1.3   0.21
    4.2   5.1   0.12
    4.1   4.9   0.14
    4.0   3.2   0.26
    5.1   6.7   0.10
    4.3   3.2   0.41
   11.5   0.7   0.11
PROC GLM;
  MODEL Y=X1 X2/P;
TITLE1 '    ';
RUN;
```

Model: MODEL1
Dependent Variable: Y

Analysis of Variance

Source	DF	Sum of Squares	Mean Square	F Value	Prob>F
Model	2	14.97894	7.48947	0.976	0.4387
Error	5	38.36981	7.67396		
C Total	7	53.34875			

Root MSE	2.77019	R-square	0.2808	
Dep Mean	4.91250	Adj R-sq	-0.0069	
C.V.	56.39067			

Parameter Estimates

Variable	DF	Parameter Estimate	Standard Error	T for H0: Parameter=0	Prob > |T|
INTERCEP	1	9.044327	3.11645121	2.902	0.0337
X1	1	-0.583257	0.53071152	-1.099	0.3218
X2	1	-11.268245	10.29793401	-1.094	0.3237

Obs	Dep Var Y	Predict Value	Residual
1	3.6000	5.8416	-2.2416
2	2.5000	5.9198	-3.4198
3	4.2000	4.7175	-0.5175
4	4.1000	4.6088	-0.5088
5	4.0000	4.2482	-0.2482
6	5.1000	4.0097	1.0903
7	4.3000	2.5579	1.7421
8	11.5000	7.3965	4.1035

Sum of Residuals	-3.55271E-15
Sum of Squared Residuals	38.3698
Predicted Resid SS (Press)	188.9178

The purpose of Example 12.4 is to illustrate that we can rely on available software packages to obtain least squares prediction equations for multiple regression problems. We should realize that the multiple regression equation is often the first step in examining the relationship between a dependent variable y and several independent variables. However, inferences related to multiple regression prediction equations are beyond the scope of this text.

12.6 USING COMPUTERS

Computer output for data plots and linear regression problems has been used already in this chapter. In this section we'll provide some Minitab and SAS programs for your use.

The data of Table 12.1 were used to illustrate how SAS and Minitab can solve linear regression and correlation problems. The Minitab PLOT procedure was used to give a scatter plot for the data. The Minitab REGRESS procedure computes a number of quantities, including the least squares estimates of β_0 and β_1, $\hat{\rho}^2$ (actually $\hat{\rho}^2 \times 100$), and a table of original values, predicted values, and residuals. Finally, the Minitab CORRELATION procedure computes the sample correlation coefficient $\hat{\rho}$.

SAS output for the same data is shown here also. Note how PROC PLOT, PROC REG, and PROC CORR are used to analyze the linear regression data and to obtain the correlation coefficient. You should compare SAS and Minitab outputs to identify where the important results of a regression or correlation analysis are located and how they are labeled. These same procedures can then be used to solve other linear regression and correlation problems.

Additional details shown in the regression and correlation output of Minitab and SAS will become more familiar to you after you study the inferential methods for linear regression and correlation in Chapter 13. Several other examples of SAS and Minitab output are shown in the remaining exercises of this chapter.

Minitab Output

```
MTB > READ INTO C1 C2
DATA> 2.00    1.60
DATA> 2.25    2.00
DATA> 2.60    1.80
DATA> 2.65    2.80
DATA> 2.80    2.10
DATA> 3.10    2.00
DATA> 2.90    2.65
DATA> 3.25    2.25
```

(continued)

```
DATA> 3.30    2.60
DATA> 3.60    3.00
DATA> 3.25    3.10
DATA> END

MTB > NAME C1='HS_GPA'
MTB > NAME C2='FR_GPA'
MTB > PRINT C1 C2

 ROW   HS_GPA   FR_GPA

   1     2.00     1.60
   2     2.25     2.00
   3     2.60     1.80
   4     2.65     2.80
   5     2.80     2.10
   6     3.10     2.00
   7     2.90     2.65
   8     3.25     2.25
   9     3.30     2.60
  10     3.60     3.00
  11     3.25     3.10

MTB > PLOT C2 VS C2 1


        -                                        *
  3.00+                                                   *
        -
FR_GPA  -                         *
        -                               *
        -                                        *
  2.50+
        -
        -                                    *
        -
        -                    *
  2.00+          *                          *
        -
        -               *
        -
        -     *
  1.50+
      ------+---------+---------+---------+---------+---------+HS_GPA
          2.10      2.40      2.70      3.00      3.30      3.60


MTB > BRIEF 3
MTB > REGRESS C2 ON 1, C1;
SUBC> PREDICT C1.

The regression equation is
FR_GPA = 0.257 + 0.728 HS_GPA

Predictor      Coef       Stdev      t-ratio       p
Constant      0.2568     0.7243       0.35       0.731
HS_GPA        0.7279     0.2482       2.93       0.017

s = 0.3785     R-sq = 48.9%     R-sq(adj) = 43.2%
```

(continued)

```
Analysis of Variance

SOURCE        DF         SS          MS          F        p
Regression    1       1.2327      1.2327      8.60     0.017
Error         9       1.2896      0.1433
Total        10       2.5223

Obs.   HS_GPA    FR_GPA       Fit  Stdev.Fit   Residual   St.Resid
  1     2.00     1.600      1.713    0.247      -0.113      -0.39
  2     2.25     2.000      1.895    0.194       0.105       0.32
  3     2.60     1.800      2.149    0.134      -0.349      -0.99
  4     2.65     2.800      2.186    0.128       0.614       1.72
  5     2.80     2.100      2.295    0.116      -0.195      -0.54
  6     3.10     2.000      2.513    0.126      -0.513      -1.44
  7     2.90     2.650      2.368    0.114       0.282       0.78
  8     3.25     2.250      2.623    0.146      -0.373      -1.07
  9     3.30     2.600      2.659    0.154      -0.059      -0.17
 10     3.60     3.000      2.877    0.212       0.123       0.39
 11     3.25     3.100      2.623    0.146       0.477       1.37


    Fit   Stdev.Fit        95% C.I.            95% P.I.
  1.713    0.247    ( 1.154,  2.271)  ( 0.690,  2.735)
  1.895    0.194    ( 1.456,  2.333)  ( 0.932,  2.857)
  2.149    0.134    ( 1.847,  2.452)  ( 1.241,  3.058)
  2.186    0.128    ( 1.897,  2.475)  ( 1.282,  3.090)
  2.295    0.116    ( 2.033,  2.557)  ( 1.399,  3.191)
  2.513    0.126    ( 2.228,  2.799)  ( 1.610,  3.416)
  2.368    0.114    ( 2.109,  2.626)  ( 1.473,  3.262)
  2.623    0.146    ( 2.292,  2.953)  ( 1.704,  3.541)
  2.659    0.154    ( 2.310,  3.008)  ( 1.734,  3.584)
  2.877    0.212    ( 2.398,  3.356)  ( 1.896,  3.859)
  2.623    0.146    ( 2.292,  2.953)  ( 1.704,  3.541)


MTB > CORRELATION COEFFICIENT BETWEEN C1 AND C2

Correlation of HS_GPA and FR_GPA = 0.699

MTB > STOP
```

SAS Output

```
OPTIONS NODATE NONUMBER PS=60 LS=78;
DATA A1;
  INPUT STUDENT HS_GPA FR_GPA;
  LABEL STUDENT='STUDENT'
        HS_GPA ='HIGH SCHOOL GPA'
        FR_GPA ='FRESHMAN GPA';
  CARDS;
  1    2.00    1.60
  2    2.25    2.00
  3    2.60    1.80
  4    2.65    2.80
  5    2.80    2.10
```

(continued)

```
     6    3.10   2.00
     7    2.90   2.65
     8    3.25   2.25
     9    3.30   2.60
    10    3.60   3.00
    11    3.25   3.10
;
PROC PRINT N;
TITLE1 "DATA FOR HIGH SCHOOL AND COLLEGE GPA'S";
TITLE2 'LISTING OF THE DATA';

PROC PLOT DATA=A1;
  PLOT FR_GPA*HS_GPA;
TITLE2 'EXAMPLE OF PROC PLOT';

PROC REG DATA=A1;
  MODEL FR_GPA=HS_GPA;
TITLE2 'EXAMPLE OF PROC REG';

PROC CORR DATA=A1;
  VAR  FR_GPA HS_GPA;
TITLE2 'EXAMPLE OF PROC CORR';
RUN;




DATA FOR HIGH SCHOOL AND COLLEGE GPA'S
        LISTING OF THE DATA

   OBS    STUDENT    HS_GPA    FR_GPA

    1        1        2.00      1.60
    2        2        2.25      2.00
    3        3        2.60      1.80
    4        4        2.65      2.80
    5        5        2.80      2.10
    6        6        3.10      2.00
    7        7        2.90      2.65
    8        8        3.25      2.25
    9        9        3.30      2.60
   10       10        3.60      3.00
   11       11        3.25      3.10

           N = 11
```

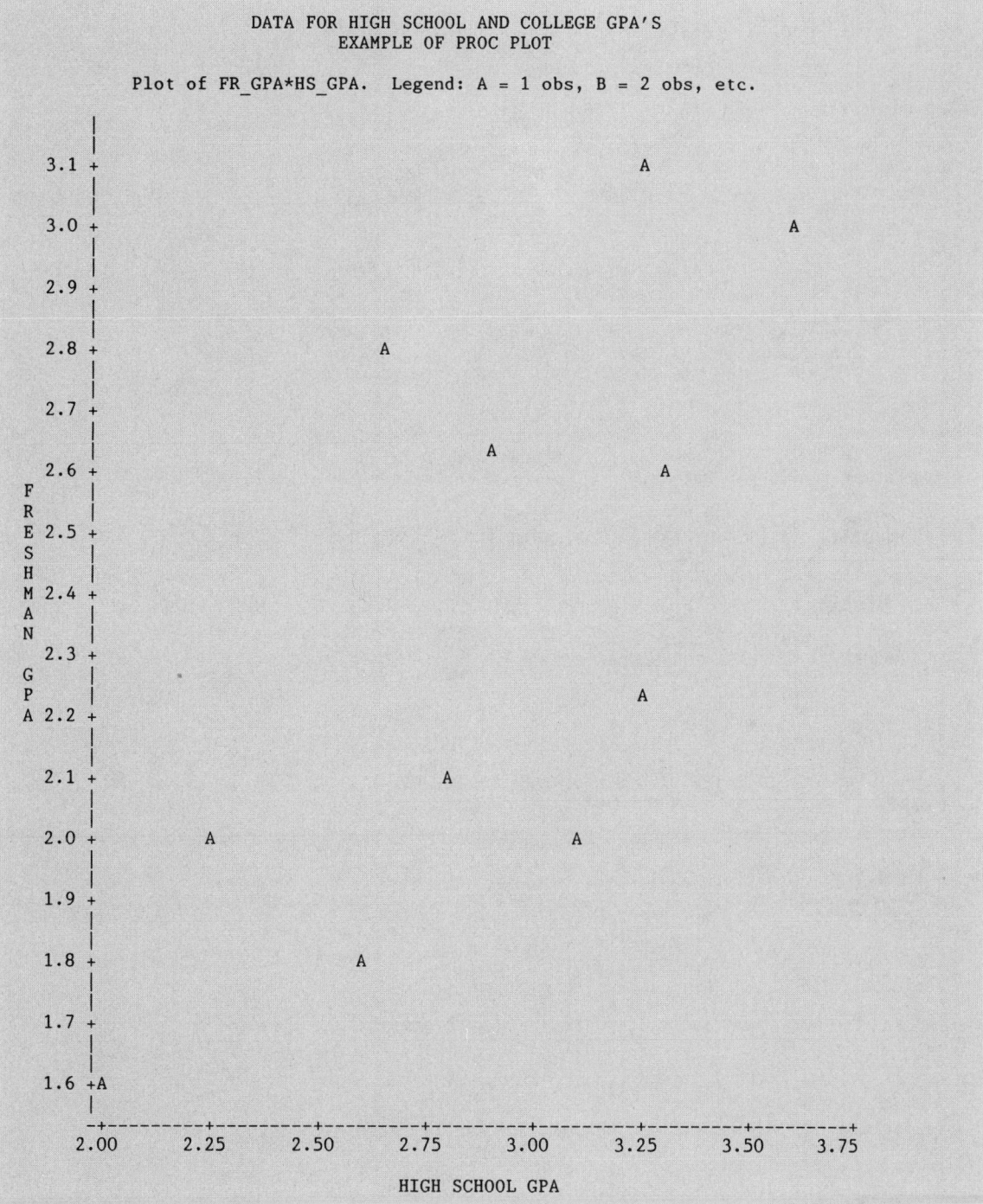

DATA FOR HIGH SCHOOL AND COLLEGE GPA'S
EXAMPLE OF PROC PLOT

Plot of FR_GPA*HS_GPA.   Legend: A = 1 obs, B = 2 obs, etc.

```
                      DATA FOR HIGH SCHOOL AND COLLEGE GPA'S
                              EXAMPLE OF PROC REG

Model: MODEL1
Dependent Variable: FR_GPA    FRESHMAN GPA

                           Analysis of Variance

                              Sum of         Mean
           Source      DF     Squares       Square    F Value    Prob>F

           Model       1      1.23267       1.23267    8.603     0.0167
           Error       9      1.28960       0.14329
           C Total     10     2.52227


                Root MSE      0.37854     R-square     0.4887
                Dep Mean      2.35455     Adj R-sq     0.4319
                C.V.         16.07681

                           Parameter Estimates

                      Parameter      Standard     T for H0:
           Variable  DF   Estimate       Error    Parameter=0    Prob > |T|

           INTERCEP   1   0.256809    0.72426138     0.355        0.7311
           HS_GPA     1   0.727921    0.24818083     2.933        0.0167

                              Variable
           Variable  DF      Label

           INTERCEP   1   Intercept
           HS_GPA     1   HIGH SCHOOL GPA




                      DATA FOR HIGH SCHOOL AND COLLEGE GPA'S
                              EXAMPLE OF PROC CORR

                           CORRELATION ANALYSIS

                     2 'VAR' Variables:   FR_GPA   HS_GPA


                          Simple Statistics

  Variable             N         Mean        Std Dev          Sum

  FR_GPA              11       2.35455       0.50222       25.90000
  HS_GPA              11       2.88182       0.48232       31.70000

                          Simple Statistics

  Variable       Minimum       Maximum     Label

  FR_GPA         1.60000       3.10000     FRESHMAN GPA
  HS_GPA         2.00000       3.60000     HIGH SCHOOL GPA
```

(continued)

```
Pearson Correlation Coefficients / Prob > |R| under Ho: Rho=0 / N = 11

                           FR_GPA              HS_GPA

        FR_GPA            1.00000             0.69908
        FRESHMAN GPA      0.0                 0.0167

        HS_GPA            0.69908             1.00000
        HIGH SCHOOL GPA   0.0167              0.0
```

SUMMARY

In this chapter we presented an introduction to regression and correlation. First we examined the relationship between a dependent variable y and a single independent variable x. A scatter plot was used to provide a graphical display of the data, and the method of least squares was used to obtain the regression equation $\hat{y} = \hat{\beta}_0 + \hat{\beta}_1 x$.

The strength of the linear relationship between y and x can be measured by the sample correlation coefficient $\hat{\rho}$ or by the coefficient of determination $\hat{\rho}^2$.

The fitting of multivariable predictors to experimental data is a very powerful and valuable method of inference. The computer prediction of election-eve outcomes uses this technique. Multivariable predictors are also employed in business forecasting, industrial production, medicine, and many areas of science. In this chapter we discussed, and gave an example to illustrate, how to obtain the least squares regression line $\hat{y} = \hat{\beta}_0 + \hat{\beta}_1 x_1 + \hat{\beta}_2 x_2$ when y is related to only two independent variables. Solutions to multiple regression problems for this situation, as well as for the situation in which we are concerned with more than two independent variables, are conveniently obtained by using some of the standard statistical software packages. Students are encouraged to pursue with their professors the computer opportunities available at their institution. Additional details on how to obtain solutions to multiple regression problems can be found in some of the references at the end of this book, particularly Mendenhall (1987) and Ott (1988).

KEY TERMS

prediction equation
freehand regression line
linear equation
y-intercept β_0 (*def*)
slope β_1 (*def*)

expected value of y
method of least squares
predicted value of y
residual
sum of squares for error

sum of squares about the mean   correlation coefficient
sum of squares due to regression   sample correlation coefficient $\hat{\rho}$
sum of squares for error   coefficient of determination

KEY FORMULAS

1. Least squares prediction equation

$$\hat{y} = \hat{\beta}_0 + \hat{\beta}x$$

where

$$\hat{\beta}_1 = \frac{S_{xy}}{S_{xx}}, \qquad \hat{\beta}_0 = \bar{y} - \hat{\beta}_1\bar{x}, \qquad S_{xx} = \Sigma\, x^2 - \frac{(\Sigma\, x)^2}{n}$$

and

$$S_{xy} = \Sigma\, xy - \frac{(\Sigma\, x)(\Sigma\, y)}{n}$$

2. Sample correlation coefficient

$$\hat{\rho} = \frac{S_{xy}}{\sqrt{S_{xx}S_{yy}}}, \qquad -1 \le \hat{\rho} \le 1$$

where $\quad S_{yy} = \Sigma\, y^2 - \dfrac{(\Sigma\, y)^2}{n}$

SUPPLEMENTARY EXERCISES

12.23 An investigator was interested in examining the effect of different doses of a new drug on the pulse rates of human subjects. Four doses of the drug were used in the experiment (1.5, 2.0, 2.5, and 3.0 mL/kg of body weight). Three persons were randomly assigned to each of the four drug doses. After a prestudy pulse rate was recorded for each individual, subjects were injected with the appropriate drug dose. One hour later, pulse rates were again recorded. The changes in pulse rates are listed in the accompanying table.

Change in pulse rate, y	20, 21, 19	16, 17, 17	15, 13, 14	8, 10, 8
Drug dose, x	1.5, 1.5, 1.5	2.0, 2.0, 2.0	2.5, 2.5, 2.5	3.0, 3.0, 3.0

(a) Plot the sample data.

(b) Find the least squares line for these data using the accompanying SAS output.

(c) Predict the change in pulse rate that would accompany a drug dose of 2.3 mg/kg of body weight. (Note: a dose of $x = 2.3$ with no y-value was included as another observation so the software would compute the predicted value for $x = 2.3$.)

```
OPTIONS NODATE NONUMBER LS=78 PS=60;
DATA RAW;
  INPUT DOSE CHANGE;
  LABEL DOSE  ='DRUG DOSE'
        CHANGE='CHANGE IN PULSE RATE';
  CARDS;
  1.5   20
  1.5   21
  1.5   19
  2.0   16
  2.0   17
  2.0   17
  2.3    .
  2.5   15
  2.5   13
  2.5   14
  3.0    8
  3.0   10
  3.0    8
;
PROC PRINT N;
TITLE1 'EFFECT OF DIFFERENT DOSES ON PULSE RATES';
TITLE2 'LISTING OF THE DATA';

PROC PLOT DATA=RAW;
  PLOT CHANGE*DOSE;
TITLE2 'PLOT OF THE DATA';

PROC REG DATA=RAW;
  MODEL CHANGE=DOSE / P;
TITLE2 'EXAMPLE OF PROC REG WITH THE P OPTION (PREDICTED VALUES)';
RUN;




EFFECT OF DIFFERENT DOSES ON PULSE RATES
        LISTING OF THE DATA

        OBS    DOSE    CHANGE

          1    1.5      20
          2    1.5      21
          3    1.5      19
          4    2.0      16
          5    2.0      17
          6    2.0      17
          7    2.3       .
          8    2.5      15
          9    2.5      13
```

(continued)

```
          10      2.5      14
          11      3.0       8
          12      3.0      10
          13      3.0       8

                N = 13
```

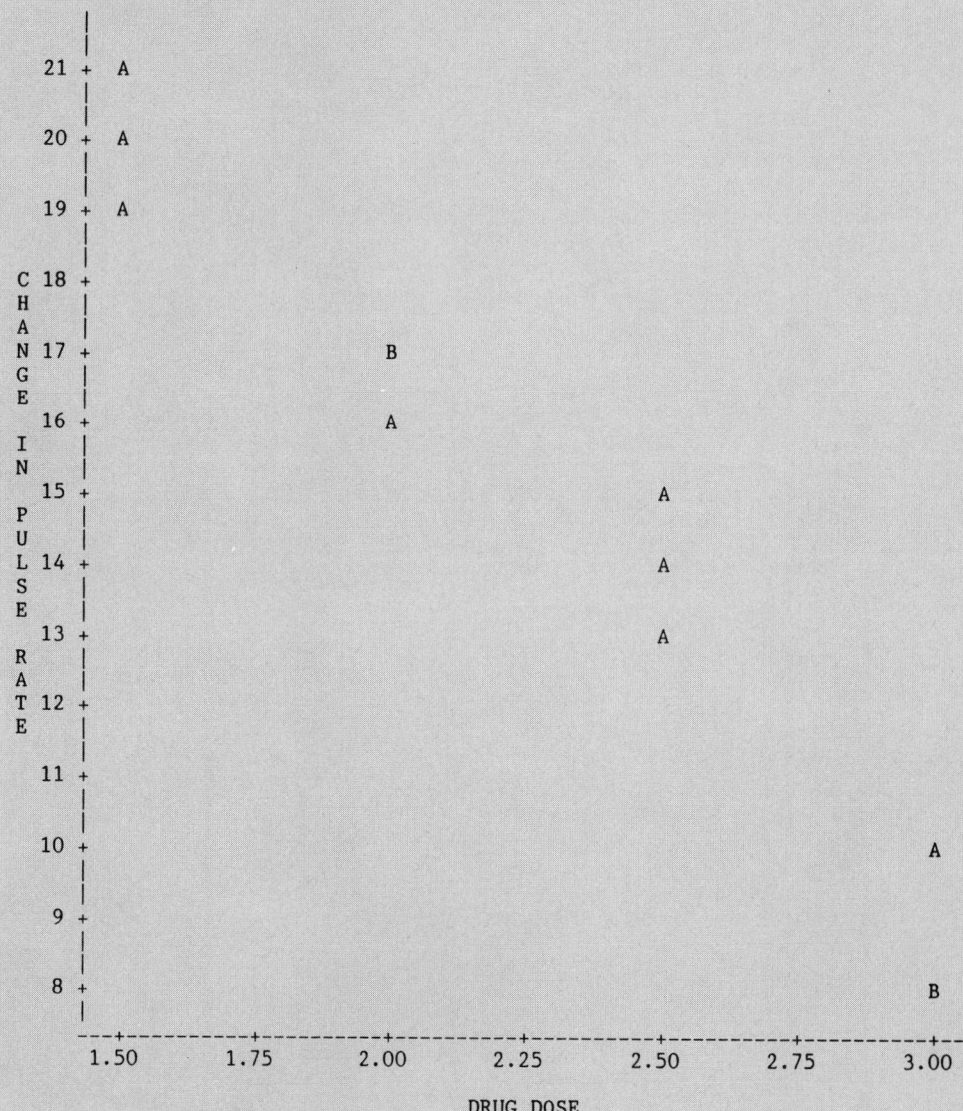

EFFECT OF DIFFERENT DOSES ON PULSE RATES
PLOT OF THE DATA

Plot of CHANGE*DOSE.   Legend: A = 1 obs, B = 2 obs, etc.

```
          |
          |
       21 +  A
          |
          |
       20 +  A
          |
          |
       19 +  A
          |
C      18 +
H         |
A         |
N      17 +              B
G         |
E         |
       16 +              A
I         |
N         |
       15 +                      A
P         |
U         |
L      14 +                      A
S         |
E         |
       13 +                      A
R         |
A         |
T      12 +
E         |
          |
       11 +
          |
          |
       10 +                                      A
          |
          |
        9 +
          |
          |
        8 +                                      B
          |
        ---+----------+----------+----------+----------+----------+----------+--
          1.50       1.75       2.00       2.25       2.50       2.75       3.00
                                  DRUG DOSE
```

NOTE: 1 obs had missing values.

```
                    EFFECT OF DIFFERENT DOSES ON PULSE RATES
              EXAMPLE OF PROC REG WITH THE P OPTION (PREDICTED VALUES)

Model: MODEL1
Dependent Variable: CHANGE      CHANGE IN PULSE RATE

                         Analysis of Variance

                             Sum of         Mean
         Source        DF    Squares        Square      F Value    Prob>F

         Model          1   201.66667    201.66667      168.056    0.0001
         Error         10    12.00000      1.20000
         C Total       11   213.66667


            Root MSE      1.09545     R-square       0.9438
            Dep Mean     14.83333     Adj R-sq       0.9382
            C.V.          7.38502

                         Parameter Estimates

                     Parameter      Standard     T for H0:
      Variable  DF    Estimate        Error     Parameter=0    Prob > |T|

      INTERCEP   1    31.333333     1.31148770      23.891        0.0001
      DOSE       1    -7.333333     0.56568542     -12.964        0.0001

                      Variable
      Variable  DF    Label

      INTERCEP   1   Intercept
      DOSE       1   DRUG DOSE
```

```
                    EFFECT OF DIFFERENT DOSES ON PULSE RATES
              EXAMPLE OF PROC REG WITH THE P OPTION (PREDICTED VALUES)

                        Dep Var    Predict
              Obs       CHANGE      Value    Residual

                1      20.0000    20.3333    -0.3333
                2      21.0000    20.3333     0.6667
                3      19.0000    20.3333    -1.3333
                4      16.0000    16.6667    -0.6667
                5      17.0000    16.6667     0.3333
                6      17.0000    16.6667     0.3333
                7          .       14.4667      .
                8      15.0000    13.0000     2.0000
                9      13.0000    13.0000    -178E-17
               10      14.0000    13.0000     1.0000
               11       8.0000     9.3333    -1.3333
               12      10.0000     9.3333     0.6667
               13       8.0000     9.3333    -1.3333

Sum of Residuals            1.953993E-14
Sum of Squared Residuals      12.0000
Predicted Resid SS (Press)    17.7709
```

12.24   Refer to Exercise 12.23. Calculate the correlation coefficient and the coefficient of determination for the data. Interpret your results.

12.25   A production foreman was concerned about the quality of the outgoing product from his department. He felt very strongly that the percentage of defective items passing through his assembly line during a 30-minute period increased throughout the day. At nine 30-minute periods throughout the day, the assembly line was closely examined to determine the number of defectives being produced. For each of these 30-minute periods, the number of hours that workers had been working (from 8:00 AM) was also recorded. The data are given in the accompanying table.

Number of defectives, y	Number of hours, x, that workers are on the job
13	1.0
14	1.5
16	2.5
14	2.0
15	3.5
20	4.5
18	4.0
18	5.5
20	6.0

Total $\Sigma y = 148$; $\Sigma y^2 = 2490$ $\Sigma x = 30.5$; $\Sigma x^2 = 128.25$; $\Sigma xy = 535.5$

(a) Plot the data.

(b) Write a linear model relating the number of defectives, y, to the number of hours on the job, x, for these data.

(c) Use the method of least squares to fit the model.

(d) Predict the number of items that would be defective in a 30-minute period if the workers had just completed 5 hours of work.

12.26   Refer to Exercise 12.25. Compute the sample correlation coefficient to measure the strength of the linear relationship between x and y. Interpret your answer.

12.27   A chain of grocery stores conducted a study to determine the relationship between the amount of money, x, spent on advertising and the weekly volume, y, of sales. Six different levels of advertising expenditure were tried in a random order over a six-week period. The accompanying data were observed (in units of $100).

Weekly sales volume, y	10.2	11.5	16.1	20.3	25.6	28.0
Amount spent on advertising, x	1.0	1.25	1.5	2.0	2.5	3.0

(a) Plot these data on a scatter diagram.

(b) Use the method of least squares to find the regression equation $\hat{y} = \hat{\beta}_0 + \hat{\beta}_1 x$.

(c) Use the prediction equation of part (b) to estimate sales volume for an expenditure of $220 in advertising.

12.28 Refer to Exercise 12.27. Compute the correlation coefficient between the sales volume and advertising volume. Does there appear to be a strong linear relationship between x and y?

12.29 Suppose that the following data were collected on emphysema patients; the number of years, x, the patient smoked and inhaled and a physician's evaluation, y, of the patient's lung capacity (measured on a scale of 0 to 100). The results for a sample of ten patients appear in the accompanying table. (Note: $S_{xx} = 876.9$, $S_{yy} = 2510$, and $S_{xy} = 1148$.)

Patient	Years smoking, x	Lung capacity, y
1	25	55
2	36	60
3	22	50
4	15	30
5	48	75
6	39	70
7	42	70
8	31	55
9	28	30
10	33	35

(a) Plot the data on a scatter diagram.

(b) Use the method of least squares to find the regression line $\hat{y} = \hat{\beta}_0 + \hat{\beta}_1 x$.

(c) Does there appear to be a positive linear relationship between x and y?

(d) Calculate the correlation coefficient between the variables "lung capacity" and "number of years smoking."

(e) Predict a person's lung capacity after 30 years of smoking.

12.30 An experiment was conducted to measure the strength of the linear relationship between two variables: a student's emotional stability (as measured by a guidance counselor's subjective judgment after an encounter session) and the student's score on an achievement test administered to children entering the first grade. The variable of emotional stability was measured on a scale of 0 to 40 (from low to high), and the achievement test was also measured from 0 to 40. Use the accompanying data from a random sample of 15 children to calculate the correlation coefficient. (Note: $S_{xx} = 485.33$, $S_{yy} = 522.93$, and $S_{xy} = 316.67$.)

Student	Emotional stability, x	Achievement, y	Student	Emotional stability, x	Achievement, y
1	23	31	9	32	33
2	21	23	10	29	35
3	31	34	11	16	21
4	34	29	12	29	22
5	26	29	13	23	24
6	22	27	14	27	28
7	14	21	15	25	15
8	18	17			

12.31 **Class Exercise** Conduct a study to determine whether there is a correlation between a social science major's performance in a math or a statistics course and his or her performance in a course in the social sciences. For example, you may wish to visit the department of sociology to obtain a random sample of 30 senior sociology majors. Contact these students to determine their (numerical) grades in a specific sociology course (such as "Introductory Sociology") and a mathematics (or statistics) course required for graduation. Let x denote a student's sociology grade and y his or her mathematics (statistics) grade.

(a) Identify the population from which the sample was drawn.

(b) Find the least squares prediction equation, $\hat{y} = \hat{\beta}_0 + \hat{\beta}_1 x$.

(c) Calculate the coefficient of linear correlation.

(d) Describe the strength of the relationship between the two sets of scores.

12.32 Use the computer output for the data in Table 12.5 to answer the following.

(a) Determine the least squares regression line

$$\hat{y} = \hat{\beta}_0 + \hat{\beta}_1 x_1 + \hat{\beta}_2 x_2 + \hat{\beta}_3 x_3 + \hat{\beta}_4 x_4$$

(b) Predict y (population increase) for birth rate $x_1 = 30$, death rate $x_2 = 15$, life expectancy $x_3 = 65$, and per capita GNP $x_4 = 1,100$.

(c) Does the regression seem to fit the data? Explain.

TABLE 12.5   Demographic Characteristics of Ten Nations in 1980

Nation	Projected popultion increase for the year 2000, by % (y)	Birth rate (x_1)	Death rate (x_2)	Life expectancy (x_3)	Per capita GNP (x_4)
Bolivia	67.9	44	19	47	$ 510
Cuba	27.0	18	6	72	810
Cyprus	16.7	19	8	73	2,110
Egypt	54.2	38	10	55	400
Ghana	81.2	48	17	48	390
Jamaica	27.3	29	7	70	1,110
Nigeria	93.1	50	18	42	560
South Africa	63.0	38	10	60	1,460
South Korea	33.8	23	7	62	1,160
Turkey	53.0	35	10	58	1,210

```
MTB > BRIEF 3
MTB > REGRESS C1 ON 4, C2 C3 C4 C5

The regression equation is
PROJ_POP = 81.9 + 1.26 BRTHRATE - 0.42 DEATHRTE - 1.19 LIFE_EXP + 0.00177 GNP

Predictor      Coef      Stdev     t-ratio        p
Constant      81.89      70.47        1.16     0.298
BRTHRATE     1.2553     0.6508        1.93     0.112
DEATHRTE     -0.425      1.335       -0.32     0.763
LIFE_EXP    -1.1937     0.8479       -1.41     0.218
GNP        0.001773   0.006621        0.27     0.800

s = 7.729      R-sq = 94.8%    R-sq(adj) = 90.6%

Analysis of Variance

SOURCE        DF          SS         MS        F        p
Regression     4      5434.4     1358.6    22.74    0.002
Error          5       298.7       59.7
Total          9      5733.1


SOURCE        DF      SEQ SS
BRTHRATE       1      5295.4
DEATHRTE       1        15.2
LIFE_EXP       1       119.6
GNP            1         4.3
```

(continued)

```
Obs.BRTHRATE    PROJ POP        Fit Stdev.Fit   Residual    St.Resid
  1    44.0      67.90        73.86      5.81       -5.96      -1.17
  2    18.0      27.00        17.43      6.15        9.57       2.04R
  3    19.0      16.70        18.95      6.65       -2.25      -0.57
  4    38.0      54.20        60.40      5.00       -6.20      -1.05
  5    48.0      81.20        78.32      4.58        2.88       0.46
  6    29.0      27.30        33.73      5.60       -6.43      -1.21
  7    50.0      93.10        87.87      4.83        5.23       0.87
  8    38.0      63.00        56.31      5.35        6.69       1.20
  9    23.0      33.80        35.84      6.29       -2.04      -0.45
 10    35.0      53.00        54.49      3.77       -1.49      -0.22
```

R denotes an obs. with a large st. resid.

```
MTB > PRINT C1-C5

ROW   PROJ_POP   BRTHRATE   DEATHRTE   LIFE_EXP    GNP

  1     67.9        44         19         47       510
  2     27.0        18          6         72       810
  3     16.7        19          8         73      2110
  4     54.2        38         10         55       400
  5     81.2        48         17         48       390
  6     27.3        29          7         70      1110
  7     93.1        50         18         42       560
  8     63.0        38         10         60      1460
  9     33.8        23          7         62      1160
 10     53.0        35         10         58      1210

MTB > BRIEF 3
MTB > REGRESS C1 ON 1, C2
```

The regression equation is
PROJ_POP = - 19.8 + 2.09 BRTHRATE

```
Predictor      Coef       Stdev     t-ratio        p
Constant    -19.778       7.635      -2.59      0.032
BRTHRATE     2.0906      0.2125       9.84      0.000

s = 7.397      R-sq = 92.4%     R-sq(adj) = 91.4%
```

Analysis of Variance

```
SOURCE      DF        SS          MS          F         p
Regression   1     5295.4      5295.4      96.77     0.000
Error        8      437.8        54.7
Total        9     5733.1

Obs.BRTHRATE    PROJ POP        Fit Stdev.Fit   Residual    St.Resid
  1    44.0      67.90        72.21      3.13       -4.31      -0.64
  2    18.0      27.00        17.85      4.16        9.15       1.50
  3    19.0      16.70        19.94      3.99       -3.24      -0.52
  4    38.0      54.20        59.66      2.47       -5.46      -0.78
  5    48.0      81.20        80.57      3.75        0.63       0.10
  6    29.0      27.30        40.85      2.59      -13.55      -1.96
  7    50.0      93.10        84.75      4.09        8.35       1.35
  8    38.0      63.00        59.66      2.47        3.34       0.48
  9    23.0      33.80        28.31      3.34        5.49       0.83
 10    35.0      53.00        53.39      2.35       -0.39      -0.06
```

12.33 Refer to the data in Table 12.5. Use the Minitab output shown here to answer (a) and (b).

```
MTB > REGRESS C1 ON 2, C2 C3

The regression equation is
PROJ_POP = - 18.8 + 1.88 BRTHRATE + 0.57 DEATHRTE

Predictor        Coef       Stdev      t-ratio         p
Constant      -18.828       8.241        -2.28     0.056
BRTHRATE       1.8754      0.4840         3.87     0.006
DEATHRTE        0.572       1.142         0.50     0.632

s = 7.770      R-sq = 92.6%    R-sq(adj) = 90.5%

Analysis of Variance

SOURCE         DF          SS           MS        F         p
Regression      2      5310.5       2655.3    43.98     0.000
Error           7       422.6         60.4
Total           9      5733.1

SOURCE         DF      SEQ SS
BRTHRATE        1      5295.4
DEATHRTE        1        15.2

Obs.BRTHRATE   PROJ_POP        Fit Stdev.Fit  Residual  St.Resid
  1      44.0     67.90       74.56      5.74     -6.66     -1.27
  2      18.0     27.00       18.36      4.49      8.64      1.36
  3      19.0     16.70       21.38      5.08     -4.68     -0.80
  4      38.0     54.20       58.16      3.97     -3.96     -0.59
  5      48.0     81.20       80.92      4.00      0.28      0.04
  6      29.0     27.30       39.56      3.74    -12.26     -1.80
  7      50.0     93.10       85.24      4.41      7.86      1.23
  8      38.0     63.00       58.16      3.97      4.84      0.72
  9      23.0     33.80       28.31      3.51      5.49      0.79
 10      35.0     53.00       52.53      3.00      0.47      0.07


MTB > REGRESS C1 ON 3, C2 C3 C4

The regression equation is
PROJ_POP = 77.7 + 1.26 BRTHRATE - 0.36 DEATHRTE - 1.11 LIFE_EXP

Predictor        Coef       Stdev      t-ratio         p
Constant        77.71      63.18         1.23     0.265
BRTHRATE       1.2560      0.5983         2.10     0.081
DEATHRTE       -0.363       1.208        -0.30     0.774
LIFE_EXP      -1.1053      0.7182        -1.54     0.175

s = 7.106      R-sq = 94.7%    R-sq(adj) = 92.1%

Analysis of Variance

SOURCE         DF          SS           MS        F         p
Regression      3      5430.1       1810.0    35.84     0.000
Error           6       303.0         50.5
Total           9      5733.1
```

(continued)

```
SOURCE          DF      SEQ SS
BRTHRATE        1       5295.4
DEATHRTE        1         15.2
LIFE_EXP        1        119.6

Obs.BRTHRATE   PROJ_POP      Fit  Stdev.Fit  Residual   St.Resid
  1     44.0     67.90     74.13     5.25     -6.23      -1.30
  2     18.0     27.00     18.56     4.11      8.44       1.46
  3     19.0     16.70     17.98     5.14     -1.28      -0.26
  4     38.0     54.20     61.02     4.08     -6.82      -1.17
  5     48.0     81.20     78.78     3.92      2.42       0.41
  6     29.0     27.30     34.22     4.87     -6.92      -1.34
  7     50.0     93.10     87.56     4.30      5.54       0.98
  8     38.0     63.00     55.49     4.02      7.51       1.28
  9     23.0     33.80     35.53     5.68     -1.73      -0.41
 10     35.0     53.00     53.93     2.89     -0.93      -0.14
```

(a) Find the following least squares regression equations

$$\hat{y} = \hat{\beta}_0 + \hat{\beta}_1 x_1$$

$$\hat{y} = \hat{\beta}_0 + \hat{\beta}_1 x_1 + \hat{\beta}_2 x_2$$

$$\hat{y} = \hat{\beta}_0 + \hat{\beta}_1 x_1 + \hat{\beta}_2 x_2 + \hat{\beta}_3 x_3$$

(b) Refer to Exercise 12.32 and the output. Does it appear that all 4 variables are needed to predict y? Do we need the variables x_1, x_2, and x_3? The variables x_1 and x_2? Or might knowledge of x_1 (birth rate) be sufficient? Explain.

12.34 Social adjustment and perceived self-image tests were administered to $n = 6$ teenagers who had recently completed a STRAIGHT program (for drug rehabilitation).

(a) Plot the data.

(b) Use these data to compute the correlation coefficient.

(c) Use the coefficient of determination and the data plot to interpret the value of $\hat{\rho}$.

Social adjustment score	Perceived self-image	Social adjustment score	Perceived self-image
55	35	28	18
37	23	52	31
61	42	70	45

12.35 Quantity discount is a usual practice in business; the larger the quantity purchased, the smaller is the unit price. The following data illustrate this fact.

	No. of units (in 000) purchased x	Unit price ($00) y
	1	100
	2	80
	3	70
	4	60
	5	40
Totals	15	350

(a) Compute the y intercept $\hat{\beta}_0$ and the slope $\hat{\beta}_1$.

(b) Predict y when $x = 6$. Would you be uncomfortable predicting y when $x = 10$? Why or why not?

12.36 A study was conducted to examine the efficiencies of various manufacturing sites of a large corporation. At each site, the average number of acceptable cartons of manufactured goods per month was recorded, as was the average number of hours of assembly-line operation per month. These data are shown here.

Location	Average number of acceptable cartons (000) y	Average number of hours of line operation x
1	12	20
2	11	38
3	15	40
4	16	45
5	20	57
6	18	68
7	22	74
8	26	79
9	20	81
10	21	86
11	27	93
12	32	104
13	33	110
14	34	120
15	31	138

(a) Plot the sample data.

(b) Obtain a least squares fit to these data, using a linear regression model.

(c) Plot the least squares prediction equation on the graph of part (a). Does this model appear to fit the data adequately?

12.37   Refer to the data of Exercise 12.36. Suppose the last four data points corresponding to locations 12, 13, 14, and 15 were as shown here, rather than as indicated in the previous exercise.

Location	y	x
12	25	104
13	23	110
14	20	120
15	15	138

(a) Plot the entire new data set for the 15 locations.

(b) Would a linear regression model still provide a good fit to the data?

12.38   A sociologist working for the government of a large city collected data on the number of nonviolent crimes (in 1000s) reported and the increase of all crimes over the previous reporting period. Quarterly data are shown here:

Quarter	Nonviolent crimes	Increase in all crimes
1	7.2	14.1
2	6.4	14.5
3	6.6	13.3
4	7.3	13.6
5	7.5	15.2
6	6.9	15.7
7	7.1	15.3
8	7.4	14.8
9	7.6	16.1
10	7.3	16.6
11	7.1	16.2
12	7.0	15.9

(a) Plot the nonviolent crime data versus quarter. Also plot the increase versus quarter on the same graph.

(b) Does there appear to be a relationship between the two crime variables?

(c) Compute the correlation coefficient between nonviolent crimes and the increase in all crimes.

12.39 The fuel of a new four-cylinder diesel engine was studied under various external, controlled operating temperatures. For each setting, two different engines were studied and the fuel consumption recorded.

Observation	External temperature (°F) x	Fuel consumption (gallons) y
1	20	25
2	20	26
3	30	28
4	30	27
5	40	32
6	40	35
7	50	42
8	50	46
9	60	55
10	60	53
11	70	55
12	70	57
13	80	60
14	80	58
15	90	61
16	90	58

(a) Plot the sample data. Do you think a linear regression line will be an adequate model?

(b) Fit the least squares regression model $y = \beta_0 + \beta_1 x + \epsilon$, and draw the prediction equation on the graph of part (a).

(c) What is the sample correlation coefficient for these data?

12.40 We are given the following scatter diagrams.

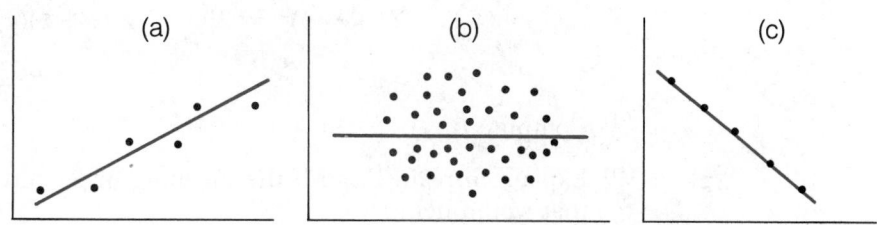

(a) Which of the following relationships *best* describes diagram (a)?

Strong positive relationship   Little or no relationship
Strong negative relationship   Perfect positive relationship
Rather weak positive relationship   Perfect negative relationship
Rather weak negative relationship

(b) Which relationship best describes diagram (b)?

(c) Which best describes (c)?

12.41 A study was conducted in a poverty region to determine the effect that the level of education had on the level of family income. For ten families in the region, information was collected on the number of grades of school completed by the head of each family and the annual income of the family. Family income is thought to be linearly related to the amount of schooling. The data are given next.

Family	Grades of school completed x	Family income ($000) y
1	6	21
2	5	19
3	10	31
4	7	25
5	8	28
6	12	33
7	5	20
8	9	29
9	7	22
10	11	32

Summary of data:

$$\Sigma x = 80 \qquad\qquad \Sigma y = 260$$
$$\Sigma x^2 = 694 \quad \Sigma (y - \bar{y})^2 = 250 \quad \Sigma (x - \bar{x})(y - \bar{y}) = 113$$

(a) Compute $\hat{\beta}_1$.

(b) Explain in plain English the meaning of $\hat{\beta}_0$ and $\hat{\beta}_1$. (Use a sketch if that would help.)

12.42 For the following graphs, write the regression equation for each of the lines, (a), (b), (c), and (d).

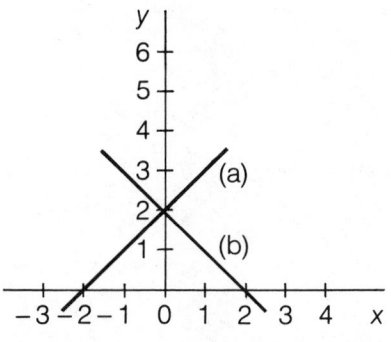

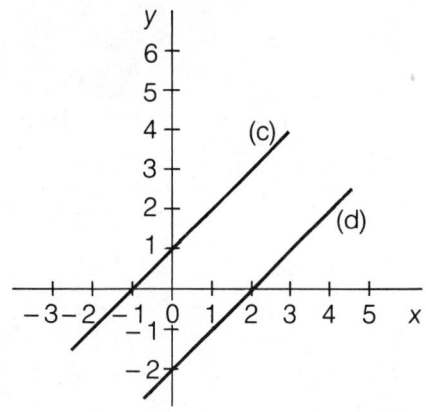

12.43 Fill in the blanks. The correlation coefficient $\hat{\rho}$ has an upper limit of
_____. If all points are exactly on the
straight line, $\hat{\rho}$ will be _____ or _____ depending on whether the relationship is _____ or _____.
If the points on the scatter diagram are randomly scattered, $\hat{\rho}$ will be near
_____. The better the fit, the _____ (larger, smaller)
the magnitude of $\hat{\rho}$.

12.44 English, mathematics, and social studies achievement test scores (labeled
ENGLACT, MATHACT, and SOCSACT, respectively) are displayed for 20
grade school children in the Minitab output shown here. Determine pairwise correlation coefficients for ENGLACT, MATHACT, and SOCSACT.

```
MTB > READ INTO C1 C2 C3
DATA> 16      15       8
DATA> 15      27      24
DATA> 17      11      19
DATA> 18      14      25
DATA> 18       8      12
DATA> 11      10       7
DATA> 13      14      11
DATA> 17       8      23
DATA> 21      23      26
DATA> 17      12      13
DATA> 12      18      11
DATA>  9      11       5
DATA> 26      28      26
DATA> 18      19      15
DATA> 19      17      14
DATA> 22      27      27
DATA> 15      19      11
DATA> 23      28      21
DATA> 19      14      13
DATA> 24      22      15
DATA> END
```

(continued)

```
MTB > NAME C1='ENGLACT'
MTB > NAME C2='MATHACT'
MTB > NAME C3='SOCSACT'
MTB > PRINT C1-C3

 ROW  ENGLACT  MATHACT  SOCSACT

   1       16       15        8
   2       15       27       24
   3       17       11       19
   4       18       14       25
   5       18        8       12
   6       11       10        7
   7       13       14       11
   8       17        8       23
   9       21       23       26
  10       17       12       13
  11       12       18       11
  12        9       11        5
  13       26       28       26
  14       18       19       15
  15       19       17       14
  16       22       27       27
  17       15       19       11
  18       23       28       21
  19       19       14       13
  20       24       22       15


MTB > DESCRIBE C1-C3

                N      MEAN    MEDIAN    TRMEAN     STDEV    SEMEAN
ENGLACT        20    17.500    17.500    17.500     4.371     0.977
MATHACT        20    17.25     16.00     17.17      6.69      1.50
SOCSACT        20    16.30     14.50     16.33      6.99      1.56

              MIN       MAX        Q1        Q3
ENGLACT     9.000    26.000    15.000    20.500
MATHACT     8.00     28.00     11.25     22.75
SOCSACT     5.00     27.00     11.00     23.75

MTB > CORRELATION COEFFICIENT BETWEEN C1-C3

         ENGLACT   MATHACT
MATHACT    0.595
SOCSACT    0.668     0.567

MTB > REGRESS C1 ON 1, C2

The regression equation is
ENGLACT = 10.8 + 0.389 MATHACT

Predictor       Coef      Stdev    t-ratio         p
Constant      10.791      2.283       4.73     0.000
MATHACT       0.3889     0.1238       3.14     0.006
```

(continued)

```
s = 3.609        R-sq = 35.4%      R-sq(adj) = 31.8%

Analysis of Variance

SOURCE          DF            SS          MS        F        p
Regression       1         128.54      128.54     9.87    0.006
Error           18         234.46       13.03
Total           19         363.00

Obs.  MATHACT    ENGLACT        Fit  Stdev.Fit   Residual   St.Resid
  1      15.0     16.000     16.625     0.854     -0.625      -0.18
  2      27.0     15.000     21.292     1.452     -6.292      -1.90
  3      11.0     17.000     15.069     1.118      1.931       0.56
  4      14.0     18.000     16.236     0.902      1.764       0.50
  5       8.0     18.000     13.902     1.401      4.098       1.23
  6      10.0     11.000     14.680     1.207     -3.680      -1.08
  7      14.0     13.000     16.236     0.902     -3.236      -0.93
  8       8.0     17.000     13.902     1.401      3.098       0.93
  9      23.0     21.000     19.736     1.076      1.264       0.37
 10      12.0     17.000     15.458     1.036      1.542       0.45
 11      18.0     12.000     17.792     0.812     -5.792      -1.65
 12      11.0      9.000     15.069     1.118     -6.069      -1.77
 13      28.0     26.000     21.681     1.556      4.319       1.33
 14      19.0     18.000     18.181     0.836     -0.181      -0.05
 15      17.0     19.000     17.403     0.808      1.597       0.45
 16      27.0     22.000     21.292     1.452      0.708       0.21
 17      19.0     15.000     18.181     0.836     -3.181      -0.91
 18      28.0     23.000     21.681     1.556      1.319       0.41
 19      14.0     19.000     16.236     0.902      2.764       0.79
 20      22.0     24.000     19.347     0.999      4.653       1.34
```

12.45   Earnings from a particular stock are listed here for the past 8 years.

Year	1988	1987	1986	1985	1984	1983	1982	1981
Earnings per share	2.30	1.80	1.50	1.20	1.05	1.00	.99	.97

(a) Plot the data.

(b) Use the method of least squares to fit the model $y = \hat{\beta}_0 + \hat{\beta}_1 x + \epsilon$. Hint: Use $x = $ (year $-$ 1980) to simplify your calculations.

(c) What earnings per share would you predict for 1989? Do you have reservations about using the prediction equation?

12.46   The accompanying data are mean weights and mean waist sizes for state representatives participating in the Miss America Pageant during the years of 1972–1983. Compute the correlation coefficient between weight and waist size. Weights are measured in pounds and waist sizes in inches.

Year	Mean weight	Mean waist size	Year	Mean weight	Mean waist size
1972	119.4	24.1	1978	116.2	23.9
1973	118.1	24.0	1979	114.7	23.5
1974	118.6	24.0	1980	115.0	23.6
1975	118.0	23.9	1981	116.7	23.7
1976	116.0	23.6	1982	115.1	23.5
1977	115.9	23.6	1983	114.0	23.2

12.47 In the following table, average expenditure per student and average teacher salary for public education in selected cities of Texas are presented. Considering teachers' salaries to be a major cause of expenditures in a school system, determine the linear relationship, using the method of least squares.

Place	Average expenditure per student	Average teacher salary
Abilene	$1,192	$12,468
Amarillo	1,342	12,108
Brownsville	988	10,944
Dallas-Fort Worth	1,236	12,420
Galveston	1,548	13,248
Houston	1,332	13,272
Kileen	1,118	10,704
Lubbock	1,173	11,292
San Antonio	1,185	12,252

12.48 (a) Plot the data given in the table below.

(b) Estimate the association between units sold in April and units sold in December for the salespeople listed.

Salesperson	April sales	December sales
Mr Hinshaw	23	34
Mrs. Fabares	18	22
Ms. Mills	30	36
Mr. Haggar	22	31
Mr. Redburn	19	26
Ms. Blitzer	28	46

12.49 Suppose that the following data were collected on emphysema patients: the number of years the patient smoked and inhaled and a physician's subjective evaluation of the extent of lung damage. The latter variable is measured on a scale from 0 to 100. Measurements taken on 10 patients are as follows:

Patient	Years smoking	Lung damage
1	25	55
2	36	60
3	22	50
4	15	30
5	48	75
6	39	70
7	42	70
8	31	58
9	28	57
10	33	59

(a) Calculate the sample correlation coefficient and interpret.

(b) Find the equation of the regression line.

(c) Will this equation give fairly accurate predictions? Briefly explain your answer.

(d) Predict the lung damage reading for a 30-year smoker.

12.50 A retailer of satellite dishes would like to know the impact that advertising has on his sales. For six months he records the number of ads run in the newspaper and the number of sales. The results are as follows:

	March	April	May	June	July	August
No. of ads	0	3	5	8	10	9
No. of sales	5	7	12	15	17	15

(a) Calculate the correlation coefficient for the sample.

(b) Find the regression line equation.

(c) Predict the number of sales if the retailer ran six ads.

12.51 Make a two-dimensional scatter diagram on graph paper of these data, which list the distance five trucks have traveled and their maintenance costs.

Distance traveled (kilometers)	Maintenance costs ($)
50,000	9,000
35,000	5,000
70,000	14,000
43.000	4,500
61,000	10,500

⚓ 12.52 A manufacturing company has developed an aptitude test to be used when hiring new employees. The test is used to predict how well the candidate would perform on the job, if hired.

Let x = score on the aptitude test (range 0 to 50).

Let y = score on job evaluation at the end of one year's employment (range 0 to 10).

To test how well x predicts y, a study was done involving five randomly selected employees after their first year of service. The following sums were obtained.

$$\Sigma x = 150, \quad \Sigma x^2 = 5500, \quad \Sigma y = 25, \quad \Sigma y^2 = 161, \quad \Sigma xy = 930$$

(a) Find the equation of the regression line.

(b) If the company wishes to hire people who will have a predicted job evaluation score of at least $y = 8$, what minimum score, x, should be used on the aptitude test to give this predicted score?

(c) Find the correlation coefficient for x and y.

(d) What percentage of the variation in the job evaluation score, y, is explained by the aptitude score, x?

12.53 Address the following general statements:

(a) Even if there is a direct cause-and-effect relationship between two variables x and y, their correlation coefficient need not be anywhere near 1 or -1. Give two examples of how this could happen.

(b) Even if the correlation between two variables x and y is very strong (coefficient near 1 or -1), there need not be any cause-and-effect relationship at all between x and y. Give an example of how this could happen.

12.54 Modified True/False Questions. Circle T or F. If your answer is false, change the statement so that it becomes true. Only change the words that are underlined.

(a) T   F   The correlation coefficient for a set of paired data points will always be a number between <u>zero</u> and <u>one</u> (inclusive).

(b) T   F   If two different scatter diagrams have exactly the same regression lines, then they will <u>also have the same</u> correlation coefficient.

(c) T   F   Truncating the range of the x-values in a scatter diagram may cause a significant <u>decrease</u> in the correlation coefficient.

(d) T   F   If the points of a scatter diagram all lie exactly along a straight line, then their correlation coefficient will always be <u>exactly equal to one</u>.

(e) T   F   If the correlation coefficient for the paired variables x and y is positive, then y will tend to <u>decrease</u> as x <u>decreases</u>.

(f) T   F   If the correlation coefficient for a set of paired variables x and y is zero, then there is <u>no relationship</u> between x and y.

EXERCISES FROM THE DATA BASE

12.55   Refer to the clinical trials data base in Appendix 1 to determine the correlation coefficient between the HAM-D total score and the HAM-D anxiety score.

12.56   Determine the correlation coefficient between the HAM-D total score and the Hopkins OBRIST cluster total.

12.57   For the clinical trials data, determine the correlation coefficient between age and daily consumptions of coffee and tea.

CHAPTER 13

INFERENCES RELATED TO LINEAR REGRESSION AND CORRELATION

13.1   INTRODUCTION

In Chapter 12 we gave formulas for finding the least squares estimates for β_0 and β_1 in the linear regression model

$$y = \beta_0 + \beta_1 x + \epsilon$$

Now we would like to use these estimates to make inferences about the relationship between y and x. For example, suppose that x represents the amount of force applied to a one-foot section of steel and y denotes the corresponding increase in width of the steel sample. Applying the results of Chapter 12 we would obtain a random sample of n observations and compute the least squares estimates for β_0 and β_1 as

$$\hat{\beta}_1 = \frac{S_{xy}}{S_{xx}} \quad \text{and} \quad \hat{\beta}_0 = \bar{y} - \hat{\beta}_1 \bar{x}$$

It might be of interest in this problem to test to see if there is a positive linear relationship between x and y. To do this we could conduct a statistical test of the null hypothesis $H_0: \beta_1 = 0$ against the alternative hypothesis $H_a: \beta_1 > 0$. The estimate $\hat{\beta}_1$ will be used in this test.

13.2 INFERENCES ABOUT β_0 AND β_1

Before we can make any inferences about parameters in the linear regression model we need to expand on the assumptions we have for the model. Previously we have assumed that the random error term ϵ associated with observation y has an expected value of zero. In addition, we will assume the following:

1. The ϵs are independent of each other.
2. For a given setting of the independent variable x, ϵ is normally distributed with mean 0 and variance σ_ϵ^2. The variance σ_ϵ^2 is constant for all settings of x.

These two assumptions imply that y_i is normally distributed with mean $\beta_0 + \beta_1 x_i$ and constant variance σ_ϵ^2 and that the ys are independent. See Figure 13.1. For example, y_1 is normally distributed with mean $\beta_0 + \beta_1 x_1$ and variance σ_ϵ^2. Similarly y_2 is normally distributed with mean $\beta_0 + \beta_1 x_2$ and variance σ_ϵ^2; also, y_1 and y_2 are independent.

FIGURE 13.1 Normality Assumption for ϵ in Linear Regression

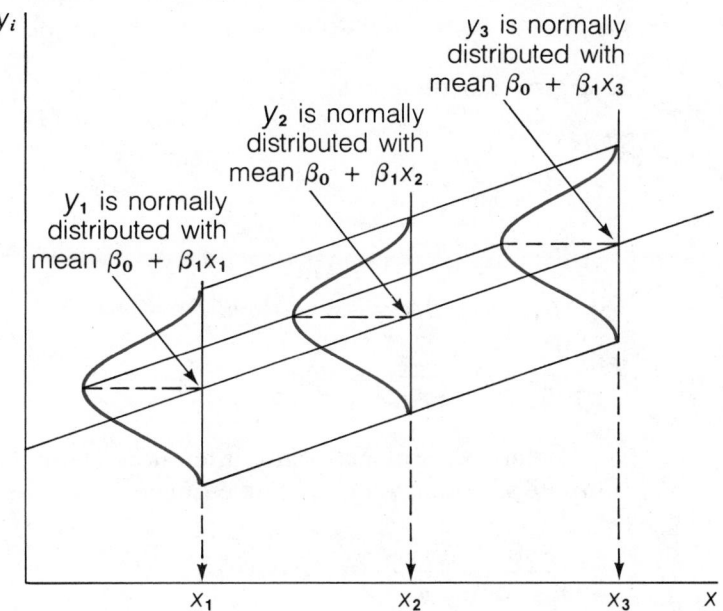

sampling distributions for $\hat{\beta}_0$ and $\hat{\beta}_1$

expected values for $\hat{\beta}_0$ and $\hat{\beta}_1$

Under these assumptions, $\hat{\beta}_0$ and $\hat{\beta}_1$ have **sampling distributions** that are normal with means (called **expected values**) and standard errors as shown here.

Expected Values and Standard Errors for β_0 and $\hat{\beta}_1$ in Linear Regression

$$\mu_{\hat{\beta}_0} = \beta_0 \qquad\qquad \mu_{\hat{\beta}_1} = \beta_1$$

$$\sigma_{\hat{\beta}_0} = \sigma_\epsilon \sqrt{\frac{\sum x^2}{n S_{xx}}} \qquad \sigma_{\hat{\beta}_1} = \frac{\sigma_\epsilon}{\sqrt{S_{xx}}}$$

Based on knowledge of the sampling distributions for $\hat{\beta}_0$ and $\hat{\beta}_1$, we can make inferences (e.g., run a test or construct a confidence interval) about the unknown intercept β_0 and slope β_1.

Example 13.1

In examining the weight loss of a compound, a chemist hypothesized that weight loss y (in pounds) is linearly related to the relative humidity x of the room in which the process operates. From a sample of 12 observations we find $\bar{y} = 4.8$, $\bar{x} = 5.9$, $S_{xy} = 138.2$, $S_{xx} = 126.0$, $\sum x^2 = 544.6$. Compute $\hat{\beta}_0$, $\hat{\beta}_1$, and their standard error.

Solution   For these data

$$\hat{\beta}_1 = \frac{S_{xy}}{S_{xx}} = 1.09$$

$$\sigma_{\hat{\beta}_1} = \frac{\sigma_\epsilon}{\sqrt{S_{xx}}} = \frac{\sigma_\epsilon}{11.26}$$

$$\hat{\beta}_0 = \bar{y} - \hat{\beta}_1 \bar{x} = 4.8 - 1.09(5.9) = -1.63$$

$$\sigma_{\hat{\beta}_0} = \sigma_\epsilon \sqrt{\frac{\sum x^2}{n S_{xx}}} = .60 \sigma_\epsilon$$

Before we can make any inferences about β_0 and β_1, we need to compute an estimate of σ_ϵ^2. The estimate we will use is

$$\frac{\text{SSE}}{n-2} = \frac{\sum (y - \hat{y})^2}{n-2}$$

This quantity, which we designate as s_ϵ^2, is based on $n - 2$ degrees of freedom for linear regression problems.

Estimate of σ_ϵ^2, and σ_ϵ for Linear Regression

$$s_\epsilon^2 = \frac{\Sigma (y - \hat{y})^2}{n - 2} = \frac{\text{SSE}}{n - 2}$$

$$s_\epsilon = \sqrt{\frac{\text{SSE}}{n - 2}}$$

Note: If computer software is not available and calculations are done by hand or using a calculator, use the shortcut formula $\text{SSE} = S_{yy} - \hat{\beta}_1 S_{xy}$.

Example 13.2

The yield per plot in bushels of corn was observed on $n = 10$ plots that had been fertilized in varying degrees. Let the independent variable x denote the amount of fertilizer applied. The data and the coded fertilizer values are recorded in Table 13.1. Use these sample data to obtain the least squares prediction equation $\hat{y}$ for the linear regression model $y = \beta_0 + \beta_1 x + \epsilon$. Also, calculate estimates of σ_ϵ^2 and σ_ϵ.

TABLE 13.1   Corn Yield Data for Example 13.2

Yield (in bushels)	Fertilizer (in pounds per plot)
12	2
13	2
13	3
14	3
15	4
15	4
14	5
16	5
17	6
18	6

Solution   For these data we can find that

$$n = 10 \qquad \Sigma x = 40$$
$$\Sigma y = 147 \qquad \bar{x} = 4.0$$

$$\bar{y} = 14.7 \qquad \Sigma\, x^2 = 180$$
$$\Sigma\, y^2 = 2193 \qquad \Sigma\, xy = 611$$

Substituting into appropriate linear regression formulas we find the least squares estimates for β_1 and β_0 to be, respectively,

$$\hat{\beta}_1 = \frac{S_{xy}}{S_{xx}} = \frac{611 - \dfrac{40(147)}{10}}{180 - \dfrac{(40)^2}{10}} = \frac{23}{20} = 1.15$$

and $\hat{\beta}_0 = \bar{y} - \hat{\beta}_1 \bar{x} = 14.7 - 1.15(4.0) = 10.10$

So the least squares prediction equation is $\hat{y} = 10.10 + 1.15x$.
The estimate for σ_ϵ^2 (and σ_ϵ) requires that we find

$$S_{yy} = \Sigma y^2 - \frac{(\Sigma y)^2}{n} = 2193 - \frac{(147)^2}{10} = 32.10$$

and

$$SSE = S_{yy} - \hat{\beta}_1 S_{xy} = 32.10 - 1.15(23) = 5.65$$

The estimate for σ_ϵ^2, from the data, is

$$s_\epsilon^2 = \frac{SSE}{n-2} = \frac{5.65}{8} = 0.71$$

Hence $s_\epsilon = 0.84$.

Example 13.3

Refer to Example 13.2 to compute the estimated standard error for $\hat{\beta}_0$ and $\hat{\beta}_1$.

Solution The formulas for the standard errors are, respectively,

$$\sigma_{\hat{\beta}_0} = \sigma_\epsilon \sqrt{\frac{\Sigma x^2}{nS_{xx}}} \qquad\qquad \sigma_{\hat{\beta}_1} = \frac{\sigma_\epsilon}{\sqrt{S_{xx}}}$$

Using $s_\epsilon = 0.84$ as the estimate of σ_ϵ and substituting into the formulas we obtain the estimated standard error for $\hat{\beta}_0$ and $\hat{\beta}_1$.

$$s_{\hat{\beta}_0} = s_\epsilon \sqrt{\frac{\sum x^2}{n S_{xx}}} = 0.84 \sqrt{\frac{180}{10(20)}} = 0.80$$

$$\text{and} \quad s_{\hat{\beta}_1} = \frac{s_\epsilon}{\sqrt{S_{xx}}} = \frac{0.84}{4.47} = 0.19$$

You will note that by substituting s_ϵ for σ_ϵ in the formulas for $\sigma_{\hat{\beta}_1}$ and $\sigma_{\hat{\beta}_0}$ we obtain the estimated standard errors for $\hat{\beta}_1$ and $\hat{\beta}_0$. In practice, since we will never know σ_ϵ and hence will always substitute an estimate of its value in the standard error formula, we will drop the word "estimated" and simply call $s_{\hat{\beta}_1}$ and $s_{\hat{\beta}_0}$ the standard errors for $\hat{\beta}_1$ and $\hat{\beta}_0$, respectively.

Using the normality assumption for ϵ, we can construct confidence intervals and statistical tests for β_0 and β_1. If we assume that the ϵ_i from the linear regression model

$$y_i = \beta_0 + \beta_1 x_i + \epsilon_i$$

are normally distributed, we can specify a confidence intervals for β_0 and β_1 using the formula: estimate $\pm t$ standard error.

General Confidence Intervals for β_0 and β_1 in Linear Regression

$$\hat{\beta}_0 \pm t s_\epsilon \sqrt{\frac{\sum x^2}{n S_{xx}}} \quad \text{and} \quad \hat{\beta}_1 \pm t \frac{s_\epsilon}{\sqrt{S_{xx}}}$$

$$\text{where} \quad s_\epsilon = \sqrt{\frac{SSE}{n-2}}$$

and t is based on df $= n - 2$

Example 13.4

Use the data from Example 13.2 to develop a 95% confidence interval for β_0 and β_1.

Solution   The calculations from Example 13.2 yielded the linear regression equation $\hat{y} = 10.10 + 1.15x$. The $t_{.025}$ value for df $= 8$ is 2.306, $s_\epsilon = 0.84$, $S_{xx} = 20$, and $\sum x^2 = 180$. Substituting these values into the appropriate formulas, we obtain the 95% confidence intervals shown here.

$$\beta_0: \quad 10.10 \pm 2.306 \, (0.84) \sqrt{\frac{180}{10(20)}} \quad \text{or} \quad 10.10 \pm 1.84$$

$$\beta_1: \quad 1.15 \pm 2.306 \, \frac{(0.84)}{\sqrt{20}} \quad \text{or} \quad 1.15 \pm 0.43$$

In other words we are 95% confident that the true value of the intercept β_0 lies somewhere in the interval $8.26 \le \beta_0 \le 11.94$. Similarly, we are 95% confident that the true value of the slope β_1 lies somewhere in the interval $0.72 \le \beta_1 \le 1.58$.

Example 13.5

A restaurant operating on a "reservations only" basis would like to use the number of advance reservations x to predict the number of dinners y to be prepared. Data on reservations and number of dinners served for one day chosen at random from each week in a 100-week period gave the following results:

$$\bar{x} = 150 \qquad\qquad \bar{y} = 120$$
$$\Sigma \, (x - \bar{x})^2 = 90{,}000 \qquad \Sigma \, (y - \bar{y})^2 = 70{,}000$$
$$\Sigma \, (x - \bar{x})(y - \bar{y}) = 60{,}000$$

(a) Find the least squares estimates $\hat{\beta}_0$ and $\hat{\beta}_1$ for the linear regression line $\hat{y} = \hat{\beta}_0 + \hat{\beta}_1 x$.

(b) Predict the number of means to be prepared if the number of reservations is 135.

(c) Construct a 90% confidence interval for the slope. Does information on x (number of advance reservations) help in predicting y (number of dinners prepared)?

Solution

(a) The least squares estimates are given by

$$\hat{\beta}_1 = \frac{S_{xy}}{S_{xx}} = \frac{60{,}000}{90{,}000} = 0.67$$

and

$$\hat{\beta}_0 = \bar{y} - \hat{\beta}_1 \bar{x} = 120 - 0.67(150) = 19.50$$

(b) The predicted number of means required for the number of advance reservations equal to 135 is

$$\hat{y} = 19.50 + 0.67(135) = 109.95 \quad \text{or} \quad 110$$

(c) The 90% confidence interval for β_1 uses the formula

$$\hat{\beta}_1 \pm t \text{ standard error}$$

where the standard error is s_ϵ/S_{xx}.

Although Table 3 in Appendix 3 does not list a t-value for a = .05 and df = 98, we'll use the t-value for the next higher df (df = 120); this value is 1.658.

The standard deviation s_ϵ can be computed using the summary sample data

$$s_\epsilon^2 = \frac{\text{SSE}}{n - 2}$$

where

$$\begin{aligned}
\text{SSE} &= S_{yy} - \hat{\beta}_1 S_{xy} \\
&= 70,000 - 0.67(60,000) \\
&= 29,800
\end{aligned}$$

Thus

$$s_\epsilon = \sqrt{\frac{29.800}{98}} = \sqrt{304.08} = 17.44$$

and the 90% confidence interval for β_1 is

$$0.67 \pm 1.658 \frac{(17.44)}{\sqrt{90,000}}$$

or

$$0.67 \pm .10$$

Since we are 90% confident that the true value of β_1 lies somewhere in the interval $.57 \le \beta_1 \le .77$ and since $\beta_1 = 0$ does not lie in this interval, it appears that the number of advance reservations is a useful predictor of the number of meals to be prepared in the context of a linear regression model, $y = \beta_0 + \beta_1 x + \epsilon$.

Consider the problem of conducting a statistical test about the slope of a linear regression model. If β_1 is different from zero and the population regression line slopes upward (or downward), as shown in Figure 13.2(a), knowledge of x will help us to predict values of y. When x is large, we predict large values of y; when x is small, we predict small values of y.

FIGURE 13.2 Can x Be Used to Predict y?

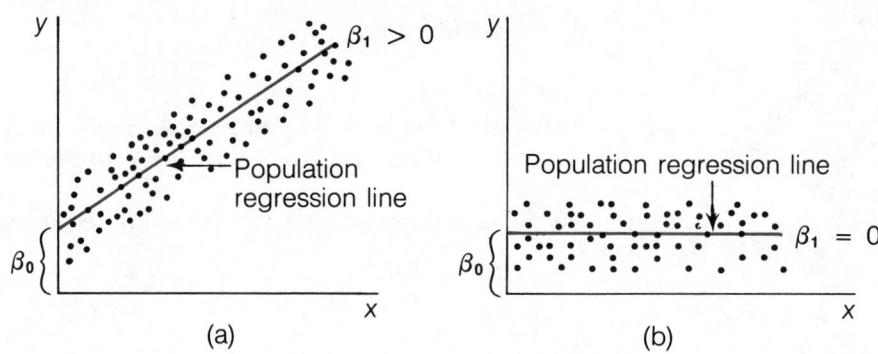

However, suppose that the slope of the population regression line is equal to 0 and the line, therefore, appears as shown in Figure 13.2(b). When $\beta_1 = 0$, knowledge of x will be of no help in predicting y.

In our statistical test for β_1 we ask the question, "Does x contribute information for the prediction of y?" Or, restated, "Does the slope β_1 differ from 0?" Or is it possible the apparent linear arrangement of the data points has occurred solely by chance?

A statistical test of H_0: $\beta_1 = 0$ is summarized next. In order to use this test, we assume that for each value of x there is a population of y-values that can be represented by a normal distribution and that regardless of the value of x, the population of y-values will have the same variance.

A Statistical Test About the Slope β_1

Null Hypothesis: $\beta_1 = 0$

Alt. hypothesis: For a one-tailed test:

  1. $\beta_1 > 0$

  2. $\beta_1 < 0$

  For a two-tailed test:

  3. $\beta_1 \neq 0$

Test statistic: $t = \dfrac{\hat{\beta}_1}{\sqrt{s_\epsilon^2/S_{xx}}}$ where $S_\epsilon^2 = \dfrac{S_{yy} - \hat{\beta}_1 S_{xy}}{n - 2}$

and $S_{yy} = \Sigma\, y^2 - (\Sigma\, y)^2/n$

(continued)

Rejection region: For a given value of α, for df $= (n - 2)$, and for a one-tailed test:

      1. Reject H_0 if $t > t_a$ with a $= \alpha$.

      2. Reject H_0 if $t < -t_a$ with a $= \alpha$.

      For a given value of α, for df $= (n - 2)$, and for a two-tailed test:

      3. Reject H_0 if $|t| > t_a$ with a $= \alpha/2$.

We illustrate the test procedure with an example.

Example 13.6

Use the data in Table 12.1 on p. 399 to determine whether there is evidence to indicate that β_1, the population slope between high school and college freshman GPAs, is positive. Use $\alpha = .05$.

Solution   The four parts of the statistical test are as follows:

Null hypothesis: $\beta_1 = 0$

Alt. hypothesis: $\beta_1 > 0$

Test statistic:

$$t = \frac{\hat{\beta}_1}{\sqrt{s_\epsilon^2 / S_{xx}}}$$

Rejection region: For $\alpha = .05$ and df $= (n - 2)$, reject H_0 if $t > t_a$, where a $= .05$.

In Example 12.1 we computed

$$\hat{\beta}_1 = .728 \qquad S_{xx} = 2.326 \qquad S_{xy} = 1.693$$

In order to calculate s_ϵ^2 we must first compute S_{yy}. For the data we find

$$S_{yy} = \Sigma y^2 - \frac{(\Sigma y)^2}{n} = 63.505 - \frac{(25.9)^2}{11} = 2.522$$

Substituting, we have

$$s_\epsilon^2 = \frac{S_{yy} - \hat{\beta}_1 S_{xy}}{n - 2} = \frac{2.522 - (.728)(1.693)}{9} = \frac{1.289}{9} = .1432$$

The test statistic is then

$$t = \frac{.728}{\sqrt{.1432/2.326}} = 2.93$$

The t-value in Table 3 of Appendix 3 for a $= .05$ and df $= 9$ is 1.833. Since the computed value of t exceeds 1.833, we reject H_0 and conclude that the slope β_1 for the linear regression model $y = \beta_0 + \beta_1 x + \epsilon$ is positive, and hence that the variable x (high school GPA) is useful in predicting y (a college freshman's GPA).

Example 13.7

The laboratory of a hospital participating in the clinical trial of an antibiotic drug had to be validated to see that the laboratory personnel could accurately assay blood samples "spiked" with fixed amounts of the antibiotic. The validation consisted of the following experiment. Thirteen spiked samples (with amounts known only to the study investigator) were sent to the laboratory to be assayed for the amount of the antibiotic present. The results of the validation experiment are shown here. (Note: The spiked samples with known amounts added were supplied in a blinded fashion to the laboratory.) The amounts found are the assay results found by the hospital laboratory.

Amount added (μg/mL) x	Amount found (μg/mL) y
0	0
5	4.5
5	5.0
5	4.8
10	8.9
10	8.9
10	8.9
20	17.0
20	18.2
20	15.4
40	32.6
40	36.1
40	31.5

(a) Plot the sample data.

(b) Fit these data to a linear regression model.

(c) Test the null hypothesis H_0: $\beta_1 = 1$ versus H_a: $\beta_1 \neq 1$. Give the p-value for your test.

Solution

(a) A plot of the sample data is shown in Figure 13.3.

FIGURE 13.3

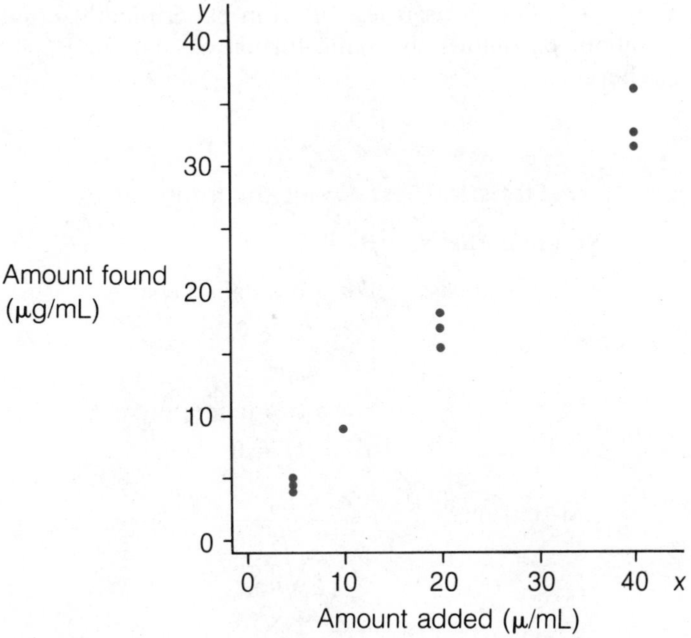

(b) For the data, it can be shown that $\hat{\beta}_1 = .822$ with a standard error of $s_\epsilon/\sqrt{S_{xx}} = .024$; the estimate of β_0 is $\hat{\beta}_0 = .529$ with a standard error of .537. The linear regression equation is then

$$\hat{y} = .529 + .822x$$

(c) The statistical test for $H_0: \beta_1 = 1$ is a slight variation of the test for $H_0: \beta_1 = 0$. The test statistic for this variation is

$$t = \frac{\hat{\beta}_1 - \beta_1}{s_\epsilon/\sqrt{S_{xx}}}$$

where β_1 is the hypothesized value of β_1 under H_0. The test of $H_0: \beta_1 = 1$ is shown here.

Null hypothesis: $\beta_1 = 1$
Alt. hypothesis: $\beta_1 \neq 1$

Test statistic: $t = \dfrac{.822 - 1}{.024} = -7.42$

Rejection region: Based on df $= 11$ and Table 3 of Appendix 3, the p-value for the test result is $p < .001$.

Although used less often in experimental situations, a statistical test about β_0 follows the same format as that for β_1. The details are shown here.

A Statistical Test About the Intercept β_0

Null hypothesis: $\beta_0 = 0$

Alt. hypothesis: For a one-tailed test:

    1. $\beta_0 > 0$

    2. $\beta_0 < 0$

For a two-tailed test:

    3. $\beta_0 \neq 0$

Test statistic: $t = \dfrac{\hat{\beta}_0}{s_\epsilon \sqrt{\dfrac{\Sigma x^2}{n S_{xx}}}}$ where $s_\epsilon^2 = \dfrac{S_{yy} - \hat{\beta}_1 S_{xy}}{n - 2}$

and $S_{yy} = \Sigma y^2 - (\Sigma y)^2/n$

Rejection region: For a given value of α, for df $= (n - 2)$, and for a one-tailed test:

    1. Reject H_0: if $t > t_a$ with a $= \alpha$.

    2. Reject H_0: if $t < -t_a$ with a $= \alpha$.

For a given value of α, for df $= (n - 2)$, and for a two-tailed test:

    3. Reject H_0 if $|t| > t_a$ with a $= \alpha/2$.

EXERCISES

13.1   A biologist is interested in studying the growth rate of a bacteria culture over a period of time. In a laboratory experiment five different bacterial cultures were chosen. One culture was randomly selected and assigned to an incubation time of 1 hour, one to an incubation time of 3 hours, and one each to the incubation times 5, 7, and 9 hours. The growth rate y was measured on each culture after the required incubation period. Let x denote the incubation time.

(a) Use the sample data of the table below and the method of least squares to obtain the regression line

$$\hat{y} = \hat{\beta}_0 + \hat{\beta}_1 x$$

(b) Conduct a test of significance to determine if there is a linear relationship between the mean growth rate and time. Use $\alpha = .05$.

Incubation time	Growth rate
x	y
1	10.0
3	10.3
5	12.2
7	12.6
9	13.9

13.2 An experiment was conducted to examine the relationship between the weight gain of chickens whose diets had been supplemented by different amounts of the amino acid lysine and the amount of lysine ingested. Since the percentage of lysine is known and we can monitor the amount of feed consumed, we can determine the amount of lysine eaten. A random sample of twelve 2-week-old chickens was selected for the study. Each was caged separately and allowed to eat at will from feed composed of a base supplemented with lysine. The sample data summarizing weight gains and amounts of lysine eaten over the test period are given here. (In the data, y represents weight gain in grams, and x represents the amount of lysine ingested in grams.)

(a) Plot the data in a scatter diagram. Does a linear model seem appropriate?

(b) Fit the linear regression model $y = \beta_0 + \beta_1 x + \epsilon$

Chick	y	x	Chick	y	x
1	14.7	.09	7	17.2	.11
2	17.8	.14	8	18.7	.19
3	19.6	.18	9	20.2	.23
4	18.4	.15	10	16.0	.13
5	20.5	.16	11	17.8	.17
6	21.1	.23	12	19.4	.21

13.3 Refer to Exercise 13.2.

(a) Compute an estimate of σ_ϵ^2.

(b) Identify the standard error of $\hat{\beta}_1$.

(c) Conduct a statistical test of the research hypothesis that for this diet preparation and length of study, there is a direct (positive) linear relationship between weight gain and the amount of lysine eaten.

13.4 Refer to Exercise 13.2. Use the sample data to construct a 95% confidence interval for the intercept β_0. Does the interval include zero as a possible value? Does the validation experiment conform to theory in regard to the intercept.

13.5 Refer to Exercises 12.6 and 12.11 (pp. 405 and 414). Conduct a statistical test to show that the slope of the population regression line relating public education expenditure per student for a state to the per capita income of the state is positive. Use $\alpha = .05$.

13.6 Research in dentistry over the past 20 years has indicated that plaque from different locations in the mouth can differ in chemical composition. Since the quantity of plaque at a given site might be quite small, it is necessary to have a sensitive procedure in order to study the chemical composition of plaque. One such procedure relates the DNA content (one important chemical component) of plaque to the weight of plaque.

In order to study the relationship between weight of plaque and DNA content, ten male volunteers (ages 18–20) were selected at random from a group of volunteers. Over a four-day period each person consumed his normal diet supplemented by 30 grams of sucrose per day. No tooth brushing was allowed. The four-day accumulation of plaque for each person was weighed and analyzed for DNA content. These sample data are summarized in the accompanying table.

Person	Plaque weight (mg) x	DNA (μg) y
1	42.7	260
2	52.3	303
3	24.6	175
4	33.4	214
5	41.8	226
6	36.7	246
7	27.0	181
8	47.3	251
9	31.4	154
10	33.9	247

(a) Graph these data in a scatter plot.

(b) Use the computer output shown here to determine the least squares regression line.

```
The regression equation is
Y = 62.8 + 4.39 X

Predictor        Coef      Stdev     t-ratio         p
Constant        62.81      35.69        1.76     0.116
X              4.3895     0.9381        4.68     0.000

s = 24.88       R-sq = 73.2%     R-sq(adj) = 69.9%

Analysis of Variance

SOURCE          DF          SS          MS         F         p
Regression       1       13552       13552     21.89     0.000
Error            8        4952         619
Total            9       18504

Obs.      X         Y      Fit  Stdev.Fit  Residual   St.Resid
  1    42.7    260.00   250.24       9.46      9.76       0.42
  2    52.3    303.00   292.38      16.28     10.62       0.56
  3    24.6    175.00   170.79      14.13      4.21       0.21
  4    33.4    214.00   209.42       8.60      4.58       0.20
  5    41.8    226.00   246.29       9.01    -20.29      -0.87
  6    36.7    246.00   223.90       7.88     22.10       0.94
  7    27.0    181.00   181.32      12.32     -0.32      -0.01
  8    47.3    251.00   270.43      12.38    -19.43      -0.90
  9    31.4    154.00   200.64       9.52    -46.64      -2.03R
 10    33.9    247.00   211.61       8.42     35.39       1.51

R denotes an obs. with a large st. resid.
```

13.7 Refer to Exercise 13.6. If the least squares prediction equation is $\hat{y} = 62.81 + 4.39x$, test to see if there is a significant linear relationship between the DNA content and the plaque weight. That is, use the sample data to test whether the population slope β_1 differs from zero. Use $\alpha = .05$. (Note: $S_{yy} = 18,504.1$, $S_{xy} = 3,087.43$, and $S_{xx} = 703.37$.)

13.8 Use the output of Exercise 13.6 to compare the results of your test in Exercise 13.7.

13.3 INFERENCES ABOUT $E(y)$

The methods of previous sections can be expanded to include inferences concerning the average value of y for a given setting of the independent variable. For example, in evaluating the effects of different levels of advertising expenditure x on sales y, it may be of interest to estimate the average sales per month for a given level of expenditure x. The estimate of $E(y)$ for a specific setting of x can be obtained by evaluating the prediction equation.

$$\hat{y} = \hat{\beta}_0 + \hat{\beta}_1 x$$

sampling distribution for $\hat{y}$

at that setting. It can be shown that in repeated sampling at a particular setting of x, the **sampling distribution for $\hat{y}$** has a mean

$$E(y) = \beta_0 + \beta_1 x$$

and standard error given by

$$s_\epsilon \sqrt{\frac{1}{n} + \frac{(x - \bar{x})^2}{S_{xx}}}$$

Again assuming the the ϵ_is are normally distributed, a general confidence interval for $E(y)$ is given by the formula that follows.

Confidence Interval for $E(y)$

$$\hat{y} \pm ts_\epsilon \sqrt{\frac{1}{n} + \frac{(x - \bar{x})^2}{S_{xx}}}$$

where

$$s_\epsilon^2 = \frac{SSE}{n - 2}$$

and the t-value is based on df $= n - 2$.

Example 13.8

Use the data of Example 13.2 to give a 90% confidence interval for the mean corn yield when 5 pounds of fertilizer are applied to a plot.

Solution   The prediction equation in Example 13.2 was

$$\hat{y} = 10.10 + 1.15x$$

where x is the amount of fertilizer applied. For our example we have $x = 5$ so that

$$\hat{y} = 10.10 + 1.15(5) = 15.85$$

The standard error of $\hat{y}$ can be computed by using $S_{xx} = 20$, $s_\epsilon = 0.84$, $\bar{x} = 4$, and $n = 10$. The t-value in Table 3 of Appendix 3 for a $= .05$ and df $= n - 2 = 8$ is 1.86. Hence the appropriate confidence interval for the average corn yield per plot when 5 pounds of fertilizer are applied is

$$15.85 \pm 1.86(.84) \sqrt{\frac{1}{10} + \frac{(5-4)^2}{20}} \qquad \text{or} \qquad 15.85 \pm .61$$

that is, 15.24 to 16.46.

Example 13.9

In Example 13.8 we've constructed a 90% confidence interval for the mean corn yield when 5 pounds of fertilizer are applied. Use the same sample data to construct a 90% confidence interval on $E(y)$ for any specific value of fertilizer in the range from 2 to 6. Graph your results.

Solution   Using the results from Example 13.8, $\hat{y} = 10.10 + 1.15x$, $s_\epsilon = .84$, and

$$\sqrt{\frac{1}{n} + \frac{(x - \bar{x})^2}{S_{xx}}} = \sqrt{.1 + \frac{(x-4)^2}{20}}$$

Our 90% confidence interval for $E(y)$, then, is of the form

$$\hat{y} \pm 1.86(.85) \sqrt{.1 + \frac{(x-4)^2}{20}}$$

All we need to do is substitute a specific value of x in this form to determine a confidence interval. For fertilizer settings of 2, 3, 4, 5, and 6 the 90% confidence limits are given below.

x	90% confidence interval
2	11.54 to 13.26
3	12.94 to 14.16
4	14.21 to 15.19
5	15.24 to 16.46
6	16.14 to 17.86

   Plotting the endpoints of the confidence intervals and connecting the points, we get the general 90% confidence interval on $E(y)$ for any value of x between 2 and 6. The graph in Figure 13.4 displays 90% confidence bands for $E(y)$. Notice how the width of the confidence interval (the vertical distance between the two curves of the graph) varies for different values of x. Since the confidence width is narrower for the "central" values of x, it follows that $E(y)$ is estimated more precisely for values of x in the

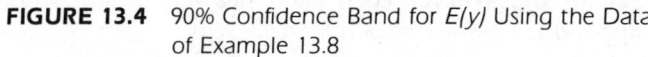

FIGURE 13.4 90% Confidence Band for E(y) Using the Data of Example 13.8

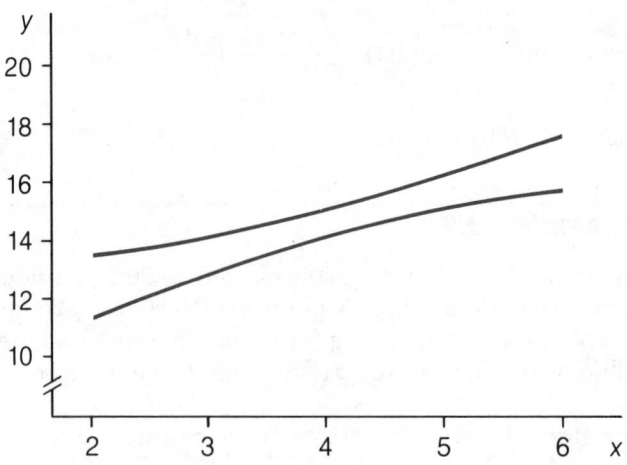

center of the experimental region. The widening of the gap between the bands at the extremities of the experimental region indicates that it would be unwise to extrapolate [try to estimate $E(y)$] beyond the region of experimentation.

A statistical test about $E(y)$ for a given setting of the independent variable x in linear regression can also be formulated using the test procedure shown here.

Statistical Test About E(y)

Null hypothesis: $E(y) = \mu_0$ (where μ_0 is specified)

Alt. hypothesis: For a one-tailed test:

  1. $E(y) > \mu_0$

  2. $E(y) < \mu_0$

  For a two-tailed test:

  3. $E(y) \neq \mu_0$

Test statistic: $t = \dfrac{\hat{y} - \mu_0}{s_\epsilon \sqrt{\dfrac{1}{n} + \dfrac{(x - \bar{x})^2}{S_{xx}}}}$

(continued)

where $s_\epsilon^2 = \dfrac{S_{yy} - \hat{\beta}_1 S_{xy}}{n - 2}$

Rejection region: For a given value of α, df $= n - 2$, and a one-tailed test:

1. Reject H_0 if $t > t_a$ with a $= \alpha$.
2. Reject H_0 if $t < -t_a$ with a $= \alpha$.

For a given value of α, df $= n - 2$, and a two-tailed test:

3. Reject H_0 if $|t| > t_a$ where a $= \alpha/2$.

Example 13.10

An experiment was run to examine the rate of growth of a particular type of bacteria. The growth rate y was determined for two different cultures at five equally spaced time intervals (1, 2, 3, 4, and 5 hours past culture seeding).

	Time (hr), x				
	1	**2**	**3**	**4**	**5**
Growth rate, y	8.0	9.0	9.1	10.2	10.4
	8.5	9.2	9.3	9.8	10.1

(a) Use the computer output that follows to determine the least squares fit to the linear regression model $y = \beta_0 + \beta_1 x + \epsilon$.

(b) Conduct a test of H_0: $E(y) = 9.5$ when $x = 3.5$. Use $\alpha = .05$ for a two-tailed test.

```
                       LINEAR REGRESSION

                 General Linear Models Procedure

Dependent Variable: RATE   GROWTH RATE
                              Sum of           Mean
Source                 DF    Squares          Square    F Value    Pr > F

Model                   1    4.80200000    4.80200000     70.88    0.0001
Error                   8    0.54200000    0.06775000
Corrected Total         9    5.34400000
```

(continued)

	R-Square	C.V.	Root MSE	RATE Mean
	0.898578	2.780858	0.260288	9.36000000

Source	DF	Type I SS	Mean Square	F Value	Pr > F
TIME	1	4.80200000	4.80200000	70.88	0.0001

Source	DF	Type III SS	Mean Square	F Value	Pr > F
TIME	1	4.80200000	4.80200000	70.88	0.0001

Parameter	Estimate	T for H0: Parameter=0	Pr > |T|	Std Error of Estimate
INTERCEPT	7.890000000	40.87	0.0001	0.19303497
TIME	0.490000000	8.42	0.0001	0.05820223

Observation	Observed	Predicted Residual	Lower 95% CLM Upper 95% CLM
1	8.00000000	8.38000000	8.05123956
		-0.38000000	8.70876044
2	8.50000000	8.38000000	8.05123956
		0.12000000	8.70876044
3	9.00000000	8.87000000	8.63753127
		0.13000000	9.10246873
4	9.20000000	8.87000000	8.63753127
		0.33000000	9.10246873
5	9.10000000	9.36000000	9.17019007
		-0.26000000	9.54980993
6	9.30000000	9.36000000	9.17019007
		-0.06000000	9.54980993
7 *	.	9.60500000	9.40367617
		.	9.80632383
8	10.20000000	9.85000000	9.61753127
		0.35000000	10.08246873
9	9.80000000	9.85000000	9.61753127
		-0.05000000	10.08246873
10	10.40000000	10.34000000	10.01123956
		0.06000000	10.66876044
11	10.10000000	10.34000000	10.01123956
		-0.24000000	10.66876044

* Observation was not used in this analysis

Sum of Residuals	-0.00000000
Sum of Squared Residuals	0.54200000
Sum of Squared Residuals - Error SS	0.00000000
Press Statistic	0.88400885
First Order Autocorrelation	-0.20885609
Durbin-Watson D	2.04501845

Solution

(a) From the computer output (under ESTIMATE), the least squares fit is

$$\hat{y} = 7.89 + 0.49x$$

where y is growth rate and x is time.

(b) The parts of the statistical test are shown here:

H_0: $E(y) = 9.5$
H_a: $E(y) \neq 9.5$

Test statistic:　　$t = \dfrac{\hat{y} - \mu_0}{s_\epsilon \sqrt{\dfrac{1}{n} + \dfrac{(x - \bar{x})^2}{S_{xx}}}}$

Rejection region:　For $\alpha = .05$ and df $= n - 2 = 8$, the desired *t*-value from Table 3 of Appendix 3 is 2.306. So, we will reject H_0 if $|t| > 2.306$.

Using the least squares equation, the predicted value of *y* when $x = 3.5$ is

$$\hat{y} = 7.89 + 0.49(3.5) = 9.605$$

Also it can be shown that $S_{xx} = 20$ and $s_\epsilon = 0.26$ (see ROOT MSE in the output). When we substitute these values into the test statistic, we find that

$$t = \frac{9.605 - 9.5}{0.26 \sqrt{\dfrac{1}{10} + \dfrac{(3.5 - 3)^2}{20}}}$$

which simplifies to $t = 1.20$. Since $t = 1.20$ does not exceed $t = 2.306$, we do not have sufficient evidence to suggest that $E(y)$ differs from 9.5 when $x = 3.5$.

13.4　PREDICTING *y* FOR A GIVEN VALUE OF *x*

In Section 13.3 we were concerned with estimating the expected value of *y* for a given value of *x*. Suppose, however, that after obtaining a least squares prediction equation for the general linear model, an investigator would like to predict the actual value of *y* (say the next measurement) for a given value of the independent variable *x*. Note that this problem differs from the problem discussed in the previous section in that we do not want to estimate the average value of *y* for a given value of *x*; rather, we wish to predict what a particular observation will be for that same setting of *x*.

We still use the least squares equation $\hat{y}$ as our predictor, but the corresponding interval about the observation *y* is called a prediction interval. (Prediction intervals are constructed about variables, whereas confidence intervals are constructed about parameters.)

General $100(1 - \alpha)\%$ Prediction Interval

$$\hat{y} \pm t_{\alpha/2}s_\epsilon \sqrt{1 + \frac{1}{n} + \frac{(x - \bar{x})^2}{S_{xx}}}$$

where

$$s_\epsilon^2 = \frac{\text{SSE}}{n - 2}$$

and $t_{\alpha/2}$ is based on df $= n - 2$.

Note the similarity between the confidence interval for $E(y)$ and the prediction interval for the variable y. The only difference is that the above prediction interval has a 1 added to the quantity under the square root sign. This makes the interval wider to account for the fact that we're predicting a variable (future value of y) rather than a constant $E(y)$.

Example 13.11

Use the data of Example 13.2 to predict the actual crop yield for a plot fertilized with 5 pounds of fertilizer. Place a 90% prediction interval about the actual value of y.

Solution   Using our previous work from Example 13.8 the predicted value of y (using $\hat{y}$) at $x = 5$ is $\hat{y} = 15.85$. Also, for $x = 5, \bar{x} = 4, n = 10$, and $S_{xx} = 20$,

$$\frac{1}{n} + \frac{(x - \bar{x})^2}{S_{xx}} = .15$$

The corresponding t-value for a $= .05$ and df $= n - 2 = 8$ is 1.86. Hence the 90% prediction interval is

$$15.85 \pm 1.86(.84) \sqrt{1 + .15} \qquad \text{or} \qquad 15.85 \pm 1.68$$

that is, 14.17 to 17.53.

Note that the above interval is almost three times wider than the corresponding interval for $E(y)$ of Example 13.8. This is to be expected since here we are placing an interval about a quantity that may vary, while in Example 13.8 we were placing an interval about $E(y)$, which cannot vary. Since both intervals are called 90% intervals, the prediction interval must be wider to have the same fraction of intervals (.90) covering y in repeated sampling.

Example 13.12

(a) Refer to the data of Example 13.2. Construct a general 90% prediction interval for y when x takes the values 2, 3, 4, 5, and 6.

(b) Graph your results to show a 90% prediction band for y when $2 \le x \le 6$.

(c) On the graph of part (b) superimpose the graph of the 90% confidence band for $E(y)$.

Solution

(a) From previous calculations, $\hat{y} = 10.10 + 1.15x$, $s_{\epsilon} = .84$, and

$$\sqrt{1 + \frac{1}{n} + \frac{(x - \bar{x})^2}{S_{xx}}} = \sqrt{1 + \frac{1}{10} + \frac{(x - 4)^2}{20}}$$

Hence a general 90% prediction interval is of the form

$$\hat{y} \pm 1.86(.84) \sqrt{1.1 + \frac{(x - 4)^2}{20}}$$

Substituting the values $x = 2, 3, 4, 5$, and 6 into this form we have the intervals given here.

x	90% Prediction Interval
2	10.618 to 14.182
3	11.874 to 15.226
4	13.061 to 16.339
5	14.174 to 17.526
6	15.218 to 18.782

(b, c) Plotting the endpoints of the prediction intervals and connecting the dots, we obtain the 90% prediction bands, shown by the solid lines in Figure 13.5. The dotted lines indicate the corresponding 90% confidence bands for $E(y)$. Notice that the prediction bands are wider than the corresponding confidence bands to allow for the fact that we are predicting the value of a random variable rather than estimating a parameter.

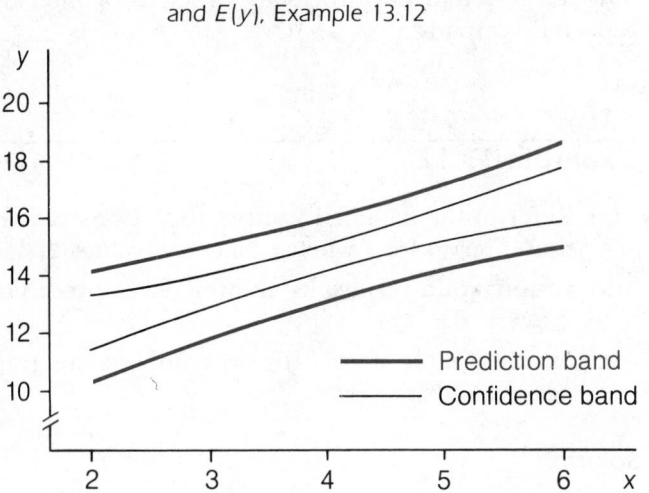

FIGURE 13.5   90% Prediction and Confidence Bands for y and $E(y)$, Example 13.12

A study of Figure 13.5 and the formulas for these confidence and prediction intervals should suggest factors that may influence the precision of our confidence intervals for $E(y)$ and prediction intervals for y. First, the plus or minus terms in the confidence interval and prediction interval involve $t_{\alpha/2}$, s_ϵ, n, $(x - \bar{x})^2$, and S_{xx}. Forgetting about $t_{\alpha/2}$, and s_ϵ for the moment, the width of these intervals will decrease as n increases, $(x - \bar{x})^2$ decreases, and S_{xx} increases. Obviously, if we take more observations, n increases. The quantity $(x - \bar{x})^2$ can be made smaller by making predictions at values of x closer to the mean of the x-values ($\bar{x}$). This is seen in Figure 13.5, where the confidence (and prediction) interval is wider for values of x farther away from the center of the region $\bar{x} = 4$. Finally, we can increase $S_{xx} = \Sigma\ (x_i - \bar{x})^2$ and hence improve the confidence and prediction intervals based on our model by increasing the spread of the x-values in our sample. This is certainly an important point when the experimenter has control of the x-values. However, there is a point of diminishing returns. The width of the confidence interval for $E(y)$ and the prediction interval for y are adequate measures of precision assuming the model adequately fits the data. If the x-values are spread too far making S_{xx} quite large (say, the values $x = 1$, 5, 6, 7, and 11 in Example 13.12), a linear regression model may no longer adequately describe the relation between x and y, thus rendering invalid the confidence interval for $E(y)$ and prediction interval for y discussed in this section.

The point is that one should understand the factors affecting the precision of the confidence and prediction intervals based on a linear regression model. However, these methods should not be applied blindly.

Plot the data; see how well the model fits the data; plot the confidence and prediction intervals to see the penalty associated with predictions away from $\bar{x}$, and so on.

EXERCISES

13.9 The extent of disease transmission can be affected greatly by the viability of infectious organisms suspended in the air. Because of the infectious nature of the disease under study, the viability of these organisms must be studied in an airtight chamber. One way to do this is to disperse an aerosol cloud, prepared from a solution containing the organisms, into the chamber. The biological recovery at any particular time is the percentage of the total number of organisms suspended in the aerosol that are viable. The data below are the biological recovery percentages computed from thirteen different aerosol clouds. For each of the clouds, recovery percentages were determined at different times.

Cloud	Time (in minutes) x	Biological recovery %
1	0	70.6
2	5	52.0
3	10	33.4
4	15	22.0
5	20	18.3
6	25	15.1
7	30	13.0
8	35	10.0
9	40	9.1
10	45	8.3
11	50	7.9
12	55	7.7
13	60	7.7

For the least squares equation $\hat{y} = \hat{\beta}_0 + \hat{\beta}_1 x$ estimate the mean log biological recovery percentage at 30 minutes, using a 95% confidence interval.

13.10 Refer to Exercise 13.2. Estimate the mean weight gain for chickens fed on a diet supplemented with lysine if .19 grams of lysine were ingested over a study period of the same duration. Use a 95% confidence interval.

13.11 Refer to Exercise 13.10. Construct a 95% prediction interval for the weight gain of a chick chosen at random and observed to ingest .19 grams of lysine. Compare your results to the confidence interval of Exercise 13.10.

13.12   Using the data of Exercise 13.9, construct a 95% prediction interval for the log biological recovery percentage at 30 minutes. Compare your result to the confidence interval on $E(y)$ of Exercise 13.9.

13.5   A STATISTICAL TEST ABOUT ρ

Before concluding our discussion of linear regression, we present a test of significance for ρ. We may be interested in a test of the null hypothesis H_0: $\rho = 0$ (i.e., there is no linear correlation between x and y). Although the results and conclusions drawn from this test are identical to those for H_0: $\beta_1 = 0$, we present the test of ρ for completeness. The assumptions required for this test are identical to those for a test of H_0: $\beta_1 = 0$.

> **A Statistical Test About ρ**
>
> Null hypothesis: $\rho = 0$
>
> Alt. hypothesis: For a one-tailed test:
>   **1.** $\rho > 0$
>   **2.** $\rho < 0$
>   For a two-tailed test:
>   **3.** $\rho \neq 0$
>
> Test statistic: $t = \hat{\rho} \sqrt{\dfrac{n-2}{1-\hat{\rho}^2}}$
>
> Rejection region: For a specified value of α, for df $= (n-2)$, and for a one-tailed test:
>   **1.** Reject H_0 if $t > t_a$ where a $= \alpha$.
>   **2.** Reject H_0 if $t < -t_a$ where a $= \alpha$.
>   For a specified value of α, for df $= (n-2)$, and for a two-tailed test:
>   **3.** Reject H_0 if $|t| > t_a$ where a $= \alpha/2$.

13.6   USING COMPUTERS

The Minitab and SAS outputs shown in Chapter 12 (pp. 427–433), those shown here for the corn yield data of Table 13.1, and output for Exercises

13.29 and 13.32 provide ample illustrations of how easy it is to obtain the necessary calculations for the descriptive and inferential problems of linear regression and correlation presented in Chapters 12 and 13. In particular, note the t-test for β_1 shown in the following Minitab and SAS output. We could also compute a confidence interval for β_1 using the standard error from the output and the formula

$$\hat{\beta}_1 \pm t \text{ standard error}$$

The only inferences for linear regression that have not been solved yet by way of Minitab and SAS are those discussed in Section 13.3 related to $E(y)$. Later in this chapter we will show how to use SAS to develop 95% confidence limits for $E(y)$.

Refer to the User's Manuals for Minitab and SAS if you want to become more involved with the details of these two software systems. Both systems can do much, much more than we have illustrated in the text. The Minitab and SAS outputs that follow show the calculations necessary for drawing inferences about β_0 and β_1 based on the corn yield data of Table 13.1. Check these results versus the hand-computed calculations we have done on these data throughout the chapter and become familiar with where to locate computations in Minitab and SAS needed for inferences in linear regression.

Minitab Output

```
MTB > READ INTO C1 C2
DATA> 12   2
DATA> 13   2
DATA> 13   3
DATA> 14   3
DATA> 15   4
DATA> 15   4
DATA> 14   5
DATA> 16   5
DATA> 17   6
DATA> 18   6
DATA> END
     10 ROWS READ
MTB > NAME=  C1='YIELD'
MTB > NAME C2='FERTLIZR'
MTB > PLOT C1 VS C2
```

(continued)

```
 18.0+                                                          *
     -
YIELD -                                                         *
     -
     -
 16.0+                                                  *
     -
     -                              2
     -
 14.0+                  *                            *
     -
     -        *         *
     -
     -
 12.0+        *
     -
         --------+---------+---------+---------+---------+---------FERTLIZR
               2.40      3.20      4.00      4.80      5.60
```

```
MTB > REGRESS C1 ON 1, C2;
SUBC> DW.

The regression equation is
YIELD = 10.1 + 1.15 FERTLIZR

Predictor       Coef        Stdev      t-ratio        p
Constant      10.1000      0.7973       12.67      0.000
FERTLIZR       1.1500      0.1879        6.12      0.000

s = 0.8404       R-sq = 82.4%      R-sq(adj) = 80.2%

Analysis of Variance

SOURCE        DF          SS           MS          F         p
Regression     1       26.450       26.450      37.45     0.000
Error          8        5.650        0.706
Total          9       32.100

Unusual Observations
Obs.FERTLIZR      YIELD       Fit Stdev.Fit  Residual   St.Resid
   7    5.00      14.000    15.850     0.325    -1.850     -2.39R

R denotes an obs. with a large st. resid.

Durbin-Watson statistic = 2.30
MTB > STOP
```

SAS Output

```
OPTIONS NODATE NONUMBER PS=60 LS=78;
DATA RAW;
  INPUT YIELD FERTLIZR @@;
  LABEL YIELD   ='YIELD (BUSHELS)'
        FERTLIZR='FERTILIZER (POUNDS/PLOT)';
  CARDS;
  12 2 13 2 13 3 14 3 15 4 15 4 14 5 15 5 17 6 18 6
```

```
;
PROC PRINT N;
TITLE1 'RELATIONSHIP BETWEEN YIELD AND THE AMOUTN OF FERTILIZER';
TITLE2 'LISTING OF THE DATA';

PROC REG;
  MODEL YIELD=FERTLIZR;
TITLE2 'EXAMPLE OF PROC REG';
RUN;
```

                RELATIONSHIP BETWEEN YIELD AND THE AMOUTN OF FERTILIZER
                              LISTING OF THE DATA

                        OBS   YIELD   FERTLIZR

                         1   12   2
                         2   13   2
                         3   13   3
                         4   14   3
                         5   15   4
                         6   15   4
                         7   14   5
                         8   15   5
                         9   17   6
                        10   18   6

                              N = 10

                RELATIONSHIP BETWEEN YIELD AND THE AMOUTN OF FERTILIZER
                              EXAMPLE OF PROC REG

Model: MODEL1
Dependent Variable: YIELD   YIELD (BUSHELS)

                              Analysis of Variance

Source	DF	Sum of Squares	Mean Square	F Value	Prob>F
Model	1	24.20000	24.20000	31.226	0.0005
Error	8	6.20000	0.77500		
C Total	9	30.40000			

        Root MSE   0.88034   R-square   0.7961
        Dep Mean   14.60000   Adj R-sq   0.7706
        C.V.   6.02973

                              Parameter Estimates

Variable	DF	Parameter Estimate	Standard Error	T for HO: Parameter=0	Prob > |T|
INTERCEP	1	10.200000	0.83516465	12.213	0.0001
FERTLIZR	1	1.100000	0.19685020	5.588	0.0005

Variable	DF	Variable Label
INTERCEP	1	Intercept
FERTLIZR	1	FERTILIZER (POUNDS/PLOT)

SUMMARY

The material in this chapter follows closely that presented in Chapter 12, where we dealt with linear regression and correlation in a descriptive sense; we discussed the notion of relating a response y to a variable using a regression education, we presented an objective way to estimate the intercept and slope of the linear regression model $y = \beta_0 + \beta_1 x + \epsilon$, and we discussed the correlation coefficient and interpreted it as a measure of the strength of the *linear* relationship between two variables y and x. In this chapter we dealt with statistical inferences related to linear regression and correlation. First we presented the formula for computing s_ϵ^2, the estimate of σ_ϵ^2, the error variance in linear regression. Using this estimate we presented a statistical test and confidence interval for the slope β_1 and for $E(y)$, the expected value of y. All of these procedures followed the general format for a confidence interval or statistical test based on t.

Inferences based on multiple regression are beyond the scope of this text. If interested in additional material on multiple regression see Ott (1984) and Hildebrand and Ott (1987).

KEY TERMS

sampling distributions for $\hat{\beta}_0$ and $\hat{\beta}_1$
expected values for $\hat{\beta}_0$ and $\hat{\beta}_1$
sampling distribution for $\hat{y}$

KEY FORMULAS

1. Expected value and standard error of the sampling distribution for $\hat{\beta}_1$ in linear regression

   Expected value: β_1

   Standard error: $\dfrac{\sigma_\epsilon}{\sqrt{S_{xx}}}$

2. Estimate of σ_ϵ^2 (or σ_ϵ)

   $$s_\epsilon^2 = \frac{\text{SSE}}{n-2} \quad \text{where} \quad \text{SSE} = \Sigma\,(y - \hat{y})^2 = S_{yy} - \hat{\beta}_1 S_{xy}$$

   $$s_\epsilon = \sqrt{\frac{\text{SSE}}{n-2}}$$

3. General confidence interval for β_1

$$\hat{\beta}_1 \pm t\frac{s_\epsilon}{\sqrt{S_{xx}}}$$

4. Statistical test for β_1

$$H_0\colon \beta_1 = 0$$

$$\text{Test statistic: } t = \frac{\hat{\beta}_1}{\sqrt{s_\epsilon^2 / S_{xx}}}$$

5. Expected value and standard error of the sampling distribution for $\hat{\beta}_0$ in linear regression

Expected value: β_0

Standard error: $\sigma_\epsilon \sqrt{\dfrac{\Sigma x^2}{nS_{xx}}}$

6. General confidence interval for β_0

$$\hat{\beta}_0 \pm ts_\epsilon \sqrt{\frac{\Sigma x^2}{nS_{xx}}}$$

7. Statistical test for β_0

$$H_0\colon \beta_0 = 0$$

$$\text{Test statistic: } t = \frac{\hat{\beta}_0}{s_\epsilon \sqrt{\dfrac{\Sigma x^2}{nS_{xx}}}}$$

8. General confidence interval for $E(y)$

$$\hat{y} \pm ts_\epsilon \sqrt{\frac{1}{n} + \frac{(x - \bar{x})^2}{S_{xx}}}$$

9. Statistical test for $E(y)$

$$H_0\colon E(y) = \mu_0$$

$$\text{Test statistic: } t = \frac{\hat{y} - \mu_0}{s_\epsilon \sqrt{\dfrac{1}{n} + \dfrac{(x - \bar{x})^2}{S_{xx}}}}$$

10. General prediction interval for y

$$\hat{y} \pm ts_\epsilon \sqrt{1 + \frac{1}{n} + \frac{(x - \bar{x})^2}{S_{xx}}}$$

11. Statistical test for ρ

$$H_0: \rho = 0$$

Test statistic $t = \hat{\rho} \sqrt{\dfrac{n - 2}{1 - \hat{\rho}^2}}$

SUPPLEMENTARY EXERCISES

13.13 Refer to the pulse rate data of Exercise 12.23. Use the data reproduced here to test the research hypothesis that a change in pulse rate y is linearly related to the drug dose x. Use $\alpha = .05$.

Change in pulse rate, y	20, 21, 19	16, 17, 17	15, 13, 14	8, 10, 8
Drug dose, x	1.5, 1.5, 1.5	2.0, 2.0, 2.0	2.5, 2.5, 2.5	3.0, 3.0, 3.0

13.14 Refer to Exercise 13.13. Suppose you had been asked to test the research hypothesis $H_a: \rho \neq 0$. What would your conclusion have been?

13.15 Crude birth rates and infant mortality rates for each of nine nations are shown here.

Nation	Crude birth rate	Infant mortality rate
Benin	49	149
Canada	15	12
Chile	22	38
Dominican Republic	37	96
Guyana	28	46
Haiti	42	130
Hong Kong	17	13
Japan	14	8
Mexico	33	70

(a) Plot the sample data, using crude birth rate as x and infant mortality rate as y.

(b) Compute the sample correlation coefficient.

13.16 Refer to Exercise 13.15. Assume that these data represent a random sample from all the nations of the world.

(a) Use these data to test $H_0: \rho = 0$ versus $H_a: \rho \neq 0$.

(b) Draw conclusions.

13.17 Given the following data relating x (gross margin) to y (total sales), answer (a) through (c).

x (gross margin) in $000	y (sales in $000)
15	20
19	38
20	25
22	28
24	31
27	40
23	35
21	24
20	25
5	7

(a) Plot the data.

(b) Compute the least square estimates of β_0 and β_1 for the model $y = \beta_0 + \beta_1 x + \epsilon$.

(c) Predict y when $x = 25$.

13.18 Refer to Exercise 13.17. Give a 95% confidence interval for β_1.

13.19 Refer to Exercise 13.17. Use these data to construct a 95% confidence interval for $E(y)$ when $x = 25$.

13.20 Refer to Exercise 13.17. Can you compute a confidence interval for $E(y)$ when $x = 40$? Any problems?

13.21 The amount of heat loss was examined for a new brand of thermal panes for different controlled outdoor temperatures. Nine panes were tested, with the results shown here.

Outdoor temperature (F)	Heat loss
20	86
20	80
20	77
40	68
40	74
40	65
60	43
60	33
60	38

(a) Plot the data, with outdoor temperature on the x axis and heat loss on the y axis.

(b) Use these data to obtain the prediction equation $\hat{y} = \hat{\beta}_0 + \hat{\beta}_1 x$.

13.22 Refer to Exercise 13.21. Use the sample data to test $H_0: \beta_1 = 0$ versus $H_a: \beta_1 < 0$. Set $\alpha = .05$.

13.23 Refer to Exercise 13.21. Conduct a statistical test of $H_0: E(y) = 55$ when $x = 50$ using a two-tailed test with $\alpha = .05$.

13.24 Earlier we examined the relationship between annual income and annual savings. The data are shown here.

Annual savings ($000)	Annual income ($000)
1	36
2	39
2	42
5	45
5	48
6	51
7	54
8	56
7	59

(a) Compute the sample correlation coefficient and interpret its value.

(b) Conduct a statistical test of $H_0: \rho = 0$ versus $H_a: \rho > 0$. Use $\alpha = .05$. Draw conclusions.

13.25 Refer to Example 12.2. Use the sample data shown here to conduct a two-tailed statistical test of $H_0: \rho = 0$ with $\alpha = .05$. Draw conclusions.

Observed meter reading y	Actual flow rate x
1.4	1
2.3	2
3.1	3
4.2	4
5.1	5
5.8	6
6.8	7
7.6	8
8.7	9
9.5	10

13.26 Refer to Exercise 13.25. Give a 95% confidence interval for $E(y)$ when $x = 5$. How would this interval change if we used a 90% confidence coefficient? Would a 95% confidence interval for $E(y)$ at $x = 9$ be wider or narrower than the corresponding interval at $x = 5$?

13.27 In an earlier exercise we computed the correlation coefficient between profit (y) and percent of operating capacity for a sample of twelve plants. If $\hat{\rho} = .81$, conduct a statistical test of $H_0: \rho = 0$ versus $H_a: \rho > 0$ using $\alpha = .05$.

13.28 The advertising expense data from an earlier exercise are shown here. Use these data to give a 90% confidence interval for β_1, the slope of the linear regression model, $y = \beta_0 + \beta_1 x + \epsilon$.

Weekly sales volume y	Amount spent on advertising x
10.2	1.0
11.5	1.25
16.1	1.5
20.3	2.0
25.6	2.5
28.0	3.0

13.29 Refer to Exercise 13.28 and the SAS output shown here to find a 95% confidence interval for $E(y)$ when $x = 1.75$. Note how confidence limits for $E(y)$ are also given for all x-values in the sample data as well as for the specified x-value, 1.75. Locate the confidence limits for $E(y)$ when $x = 1.25$.

```
OPTIONS NODATE NONUMBER LS=78 PS=60;
DATA RAW;
  INPUT SPENT VOLUME;
  LABEL SPENT ='AMOUNT SPENT ON ADVERTISING'
        VOLUME='WEEKLY SALES VOLUME';
  CARDS;
  1.00    10.2
  1.25    11.5
  1.50    16.1
  1.75     .
  2.00    20.3
  2.50    25.6
  3.00    28.0
;
PROC PRINT N;
TITLE1 'RELATIONSHIP BETWEEN ADVERTISING EXPENSE';
TITLE2 'AND WEEKLY SALES VOLUME';
TITLE3 'LISTING OF THE DATA';

PROC REG;
  MODEL VOLUME = SPENT / P CLM;
TITLE3 'REGRESSION ANALYSIS';
RUN;
```

```
        RELATIONSHIP BETWEEN ADVERTISING EXPENSE
              AND WEEKLY SALES VOLUME
               LISTING OF THE DATA

          OBS     SPENT     VOLUME

           1      1.00      10.2
           2      1.25      11.5
           3      1.50      16.1
           4      1.75       .
           5      2.00      20.3
           6      2.50      25.6
           7      3.00      28.0


              N = 7
```

```
        RELATIONSHIP BETWEEN ADVERTISING EXPENSE
              AND WEEKLY SALES VOLUME
                REGRESSION ANALYSIS
```

Model: MODEL1
Dependent Variable: VOLUME   WEEKLY SALES VOLUME

Analysis of Variance

Source	DF	Sum of Squares	Mean Square	F Value	Prob>F
Model	1	261.96637	261.96637	190.453	0.0002
Error	4	5.50196	1.37549		
C Total	5	267.46833			

Root MSE	1.17281	R-square	0.9794
Dep Mean	18.61667	Adj R-sq	0.9743
C.V.	6.29980		

Parameter Estimates

Variable	DF	Parameter Estimate	Standard Error	T for H0: Parameter=0	Prob > |T|
INTERCEP	1	1.003509	1.36312865	0.736	0.5025
SPENT	1	9.393684	0.68067860	13.800	0.0002

Variable	DF	Variable Label
INTERCEP	1	Intercept
SPENT	1	AMOUNT SPENT ON ADVERTISING

(continued)

Obs	Dep Var VOLUME	Predict Value	Std Err Predict	Lower95% Mean	Upper95% Mean	Residual
1	10.2000	10.3972	0.764	8.2755	12.5189	-0.1972
2	11.5000	12.7456	0.640	10.9673	14.5239	-1.2456
3	16.1000	15.0940	0.543	13.5876	16.6005	1.0060
4	.	17.4425	0.486	16.0923	18.7926	.
5	20.3000	19.7909	0.486	18.4407	21.1410	0.5091
6	25.6000	24.4877	0.640	22.7094	26.2660	1.1123
7	28.0000	29.1846	0.903	26.6771	31.6920	-1.1846

```
Sum of Residuals            3.197442E-14
Sum of Squared Residuals        5.5020
Predicted Resid SS (Press)     16.2666
```

💲 13.30 The earnings per share data from Exercise 12.45 are shown here.

Year	1988	1987	1986	1985	1984	1983	1982	1981
Earnings per share	2.30	1.80	1.50	1.20	1.05	1.00	.99	.97

Conduct a statistical test of $H_0: \beta_1 = 0$ versus $H_a: \beta_1 > 0$ using $\alpha = .05$. Recall that we let $x = $ (year $- 1980$)

💲 13.31 Refer to Exercise 13.30. Discuss the results of a statistical test of $H_0: \rho = 0$ versus $H_a: \rho > 0$.

🚜 13.32 A study was conducted to investigate the effect of different levels of nitrogen on the yield of lettuce plants. Sample data are shown here.

Nitrogen (pounds per plot) x	Yield (emergent stalks per plot) y
2.0	21
2.8	18
3.0	24
3.6	26
4.1	32
4.3	29

Use the SAS output shown here to

(a) Observe the sample data.

(b) Locate the least squares estimates of β_0 and β_1.

(c) Predict yield when 4 pounds of nitrogen are applied to a plot.

```
OPTIONS NODATE NONUMBER LS=78 PS=60;
DATA RAW;
  INPUT NITROGEN YIELD;
  LABEL NITROGEN='NITROGEN (POUNDS PER PLOT)'
        YIELD   ='EMERGENT STALKS PER PLOT';
  CARDS;
  2.0      21
  2.8      18
  3.0      24
  3.6      26
  4.0      .
  4.1      32
  4.3      29
;
PROC PRINT N;
TITLE1 'RELATIONSHIP BETWEEN DIFFERENT LEVELS';
TITLE2 'OF NITROGEN AND YIELD OF LETTUCE PLANTS';
TITLE3 'LISTING OF THE DATA';

PROC PLOT;
  PLOT YIELD*NITROGEN;
TITLE3 'PLOT OF THE DATA';

PROC REG;
  MODEL YIELD = NITROGEN / P;
TITLE3 'REGRESSION ANALYSIS';
RUN;
```

```
            RELATIONSHIP BETWEEN DIFFERENT LEVELS
          OF NITROGEN AND YIELD OF LETTUCE PLANTS
                  LISTING OF THE DATA

             OBS     NITROGEN     YIELD

              1        2.0         21
              2        2.8         18
              3        3.0         24
              4        3.6         26
              5        4.0         .
              6        4.1         32
              7        4.3         29

                      N = 7
```

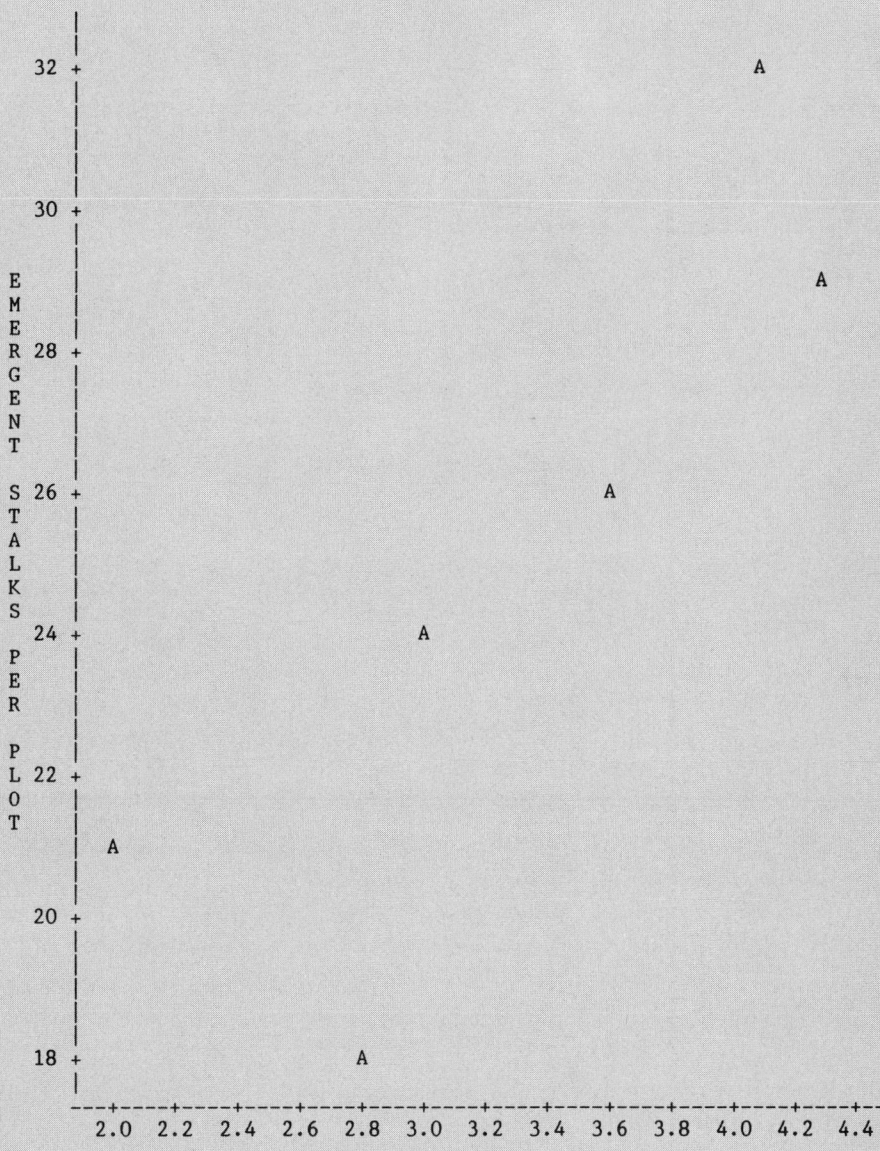

RELATIONSHIP BETWEEN DIFFERENT LEVELS
OF NITROGEN AND YIELD OF LETTUCE PLANTS
PLOT OF THE DATA

Plot of YIELD*NITROGEN.   Legend: A = 1 obs, B = 2 obs, etc.

NITROGEN (POUNDS PER PLOT)

NOTE: 1 obs had missing values.

RELATIONSHIP BETWEEN DIFFERENT LEVELS
OF NITROGEN AND YIELD OF LETTUCE PLANTS
REGRESSION ANALYSIS

Model: MODEL1
Dependent Variable: YIELD   EMERGENT STALKS PER PLOT

Analysis of Variance

Source	DF	Sum of Squares	Mean Square	F Value	Prob>F
Model	1	95.00266	95.00266	10.271	0.0327
Error	4	36.99734	9.24934		
C Total	5	132.00000			

Root MSE	3.04127	R-square	0.7197	
Dep Mean	25.00000	Adj R-sq	0.6496	
C.V.	12.16509			

Parameter Estimates

Variable	DF	Parameter Estimate	Standard Error	T for HO: Parameter=0	Prob > |T|
INTERCEP	1	8.412234	5.32261107	1.580	0.1892
NITROGEN	1	5.026596	1.56841626	3.205	0.0327

Variable	DF	Variable Label
INTERCEP	1	Intercept
NITROGEN	1	NITROGEN (POUNDS PER PLOT)

Obs	Dep Var YIELD	Predict Value	Residual
1	21.0000	18.4654	2.5346
2	18.0000	22.4867	-4.4867
3	24.0000	23.4920	0.5080
4	26.0000	26.5080	-0.5080
5	.	28.5186	.
6	32.0000	29.0213	2.9787
7	29.0000	30.0266	-1.0266

Sum of Residuals   3.552714E-15
Sum of Squared Residuals   36.9973
Predicted Resid SS (Press)   102.0687

13.33 Refer to Exercise 13.32. Use the sample data to construct a 90% confidence interval for β_1.

13.34 The efficiency of a set of production workers was observed, after which they attended a one-week efficiency improvement course. The efficiency of the workers after the course was observed. Data presented below show the before-the-course rating and the after-the-course rating for the workers. Is there a linear relationship between the before-the-course ratings and the after-the-course ratings (that is, can one predict what the after-the-course efficiency rating will be based on the before-the-course efficiency rating)?

Rating before the course	Rating after the course
32	44
30	52
37	53
44	63
43	44
33	50
36	45
46	53
30	48

Test your linear relationship using a two-tailed test with $\alpha = .05$. If a before-the-course rating was 35, what do you predict the after-the-course rating will be?

13.35 In a preliminary study to appraise the behavior of professional football teams, an analyst wanted to determine whether there was an association between points scored in the first half by winning teams and points scored in the second half. A random sample of 12 games was taken with the following results:

Points scored in first half	Points scored in second half	Points scored in first half	Points scored in second half
12	3	28	5
17	7	13	7
10	10	14	3
17	10	6	16
17	7	19	10
14	3	10	17

Based on $\alpha = .05$, is there an association (positive or negative) between points scored in the first half by winning teams and points scored in the second half by winning teams?

13.36 The number of residential customers (in thousands) and kilowatt hours (in millions) for Public Service Company of New Mexico for recent years are listed below. Considering kilowatt hours of electricity to be a function of the number of residential customers, determine the linear relationship and predict kilowatt hours, given that the number of residential customers is 100,000. Using $\alpha = .01$, determine if the slope is greater than 0.

Year	Kilowatt hr	Customers	Year	Kilowatt hr	Customers
1988	1135	91	1983	957	40
1987	1105	81	1982	917	32
1986	1090	73	1981	875	29
1985	1068	66	1980	828	23
1984	1001	51	1979	786	21

13.37 The thermal pollution of automobiles was studied for 1987 and 1988 models. The data relating automobile weight to BTU (in 1000s) per vehicle mile are shown in the table below.

(a) Compute s_ϵ^2.

(b) Give a 95% confidence interval for β_1 and discuss the meaning of your finding.

Weight (in 1000 lb) x	BTU per vehicle mile (in 1000s) y
1.8	4
2.6	5.2
4.2	8.5
5.0	11.6
4.8	10.1
3.4	6.3

13.38 Labor data (in terms of manhours) are presented here for the number of orders processed per month by a large manufacturing center.

Month	Orders processed x	Manhours required to process orders y
1	3000	8000
2	3400	9200
3	4000	10000
4	2800	7500
5	2000	5800
6	1700	5000
7	1400	4400
8	1300	3700
9	1000	3100
10	600	2220
11	1500	4100
12	2200	5500
13	3300	8100
14	3600	9400
15	4100	10600
16	3200	7900

(a) Examine the plot of orders versus months. Are there cyclical patterns?

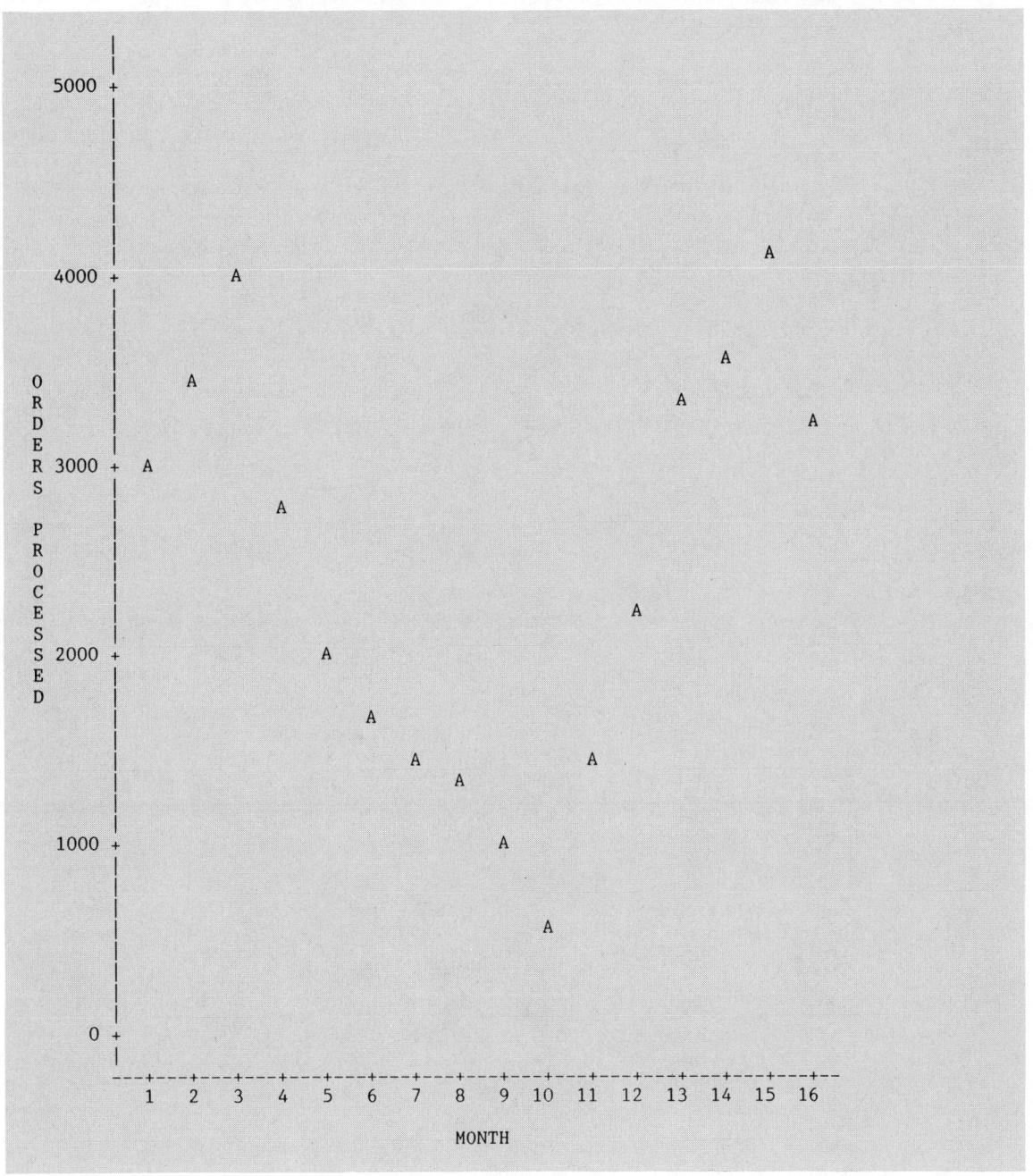

(b) From the plot of orders (x) versus manhours (y), ignoring the apparent cyclical effect of part (a), what regression model might adequately describe the data?

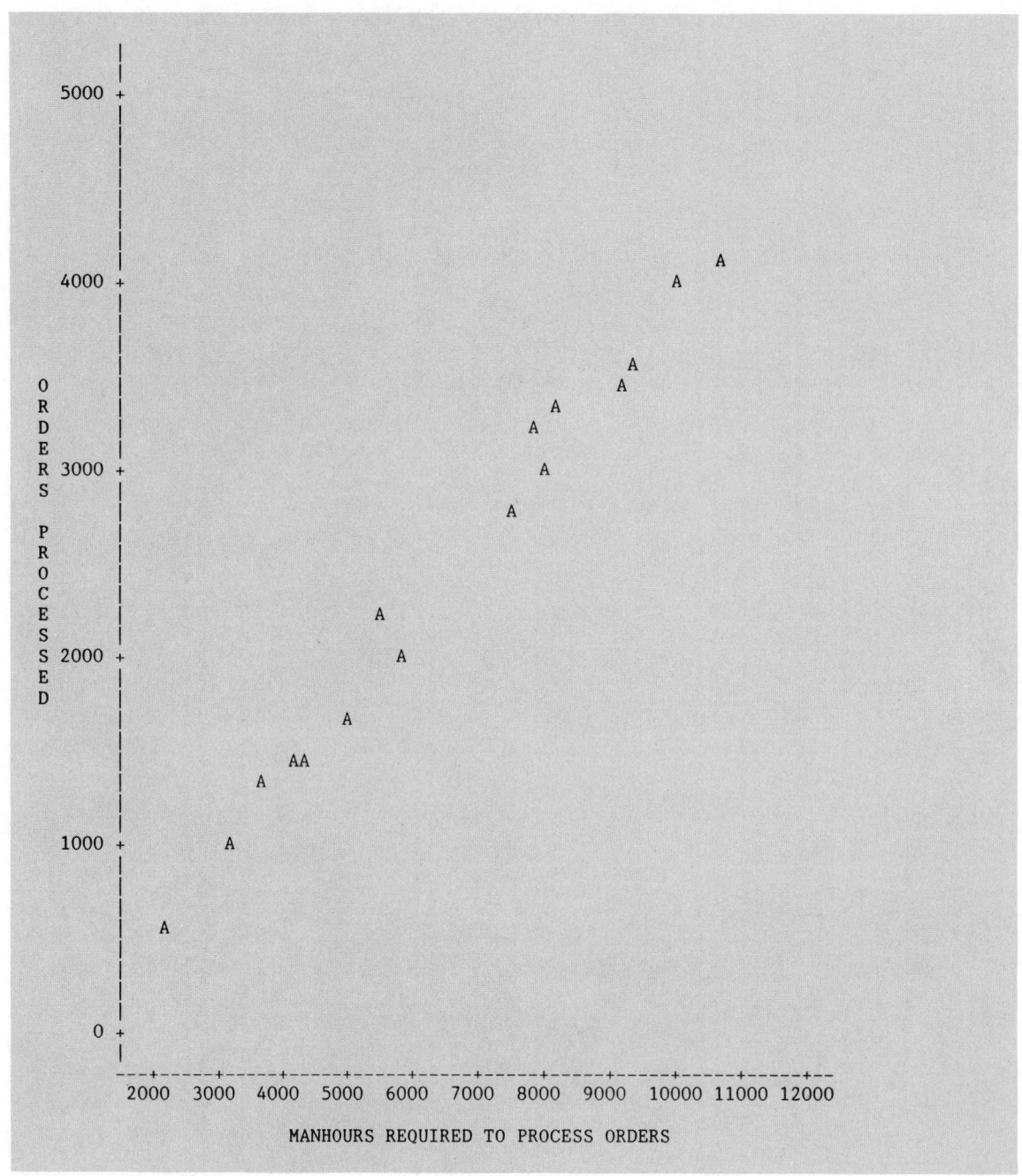

13.39 Refer to Exercise 13.38. Fit the linear regression model $y = \beta_0 + \beta_1 x + \epsilon$, and draw conclusions about the slope and intercept.

§ 13.40   A random sample of fourteen pharmacies was used to examine the relation between sales volume and the profit before tax (PBT). These data are
shown here:

Pharmacy	Sales Volume ($1000) x	PBT ($1000) y
1	38	1.3
2	20	2.1
3	48	2.2
4	44	2.6
5	56	3.3
6	39	4.0
7	65	4.1
8	84	4.2
9	82	5.5
10	105	5.7
11	126	7.0
12	52	7.5
13	80	7.7
14	101	7.9

(a) Plot the sample data.

(b) Calculate the sample correlation coefficient.

(c) Is there a significant linear trend between sales (x) and PBT (y)?

EXERCISES FROM THE DATA BASE

13.41   Refer to Exercise 12.55 (p. 455) to conduct a two-sided test of H_0: $\rho = 0$
for the HAM-D total score and HAM-D anxiety score data. Give the ρ-
value for these data.

13.42   Refer to Exercise 12.56 (p. 455). Is there a significant positive correlation
between the HAM-D total score and the Hopkins OBRIST cluster total?
Use $\alpha = .05$ for your test.

13.43   Refer to Exercise 12.57 (p. 455). Can age be used to predict daily consumption of coffee and tea? Explain.

CHAPTER 14

THE DESIGN OF AN EXPERIMENT: GETTING MORE INFORMATION FOR YOUR MONEY

14.1 Introduction ■ **14.2** Choosing the sample size to estimate μ, $\mu_1 - \mu_2$, or π ■ **14.3** The paired-difference experiment: An example of a designed experiment ■ **14.4** Using computers ■ Summary ■ Key terms ■ Key formulas ■ Supplementary exercises

14.1 INTRODUCTION

design of an experiment

The **design of an experiment** is essentially a plan for purchasing a specified quantity of information. And naturally one would hope to make the purchase at the lowest possible cost. The amount of information available in a sample for making an inference about a population parameter can be measured by the width (or half-width) of the confidence interval, which can be calculated from the sample data. The smaller the interval width, the more is known about the population parameter of interest.

The width of the confidence interval for most parameters depends on the population standard deviation σ and the sample size n. In Section 8.2 we presented the confidence interval for μ when σ is known. A 95% confidence interval for μ has the form $\bar{x} \pm 1.96\sigma/\sqrt{n}$. The width of this interval is $2(1.96\sigma/\sqrt{n})$. Thus we obtain more information about μ (a smaller confidence interval width) either by increasing the sample size n or by decreasing σ, the parameter that measures the variability of the sampled data. Experimenters can control the quantity of information contained in the sample by determining the number of observations they will include in their sample and by selecting the sample in such a way as to reduce the variation in the data. Although this is an oversimplification of the problems encountered in designing an experiment, it does summarize the essential points.

A strong similarity exists between audio communication and statistics. Both are concerned with the transmission of a message or signal from one point to another. For example, a stockbroker is responsible for transmitting a verbal message (perhaps the result of a stock sale) from her office in New York City to her client's residence in New Orleans. Or a speaker may wish to communicate with a large audience. If static or background noise is sizable, either in the case of the speaker or the stockbroker, the message may arrive in garbled form and hence represent only part of the complete signal. The receiver must use this partial information to infer the nature of the complete message.

Similarly, scientific experimentation is conducted to confirm certain theories about natural phenomena. Sometimes we want simply to explore some aspect of nature and to deduce, exactly or with a good approximation, the relationships between certain variables. We can think of experimentation as communication between nature and the scientist. The message consisting of less than complete information about the natural phenomenon is contained in the experimenter's sample data. Imperfections in the measuring instruments, nonhomogeneity of experimental material, and numerous other factors contribute background noise or static, which tend to obscure nature's signal and cause the observed response to vary in a random manner.

For both the communications engineer and the statistician, two factors affect the quantity of information in an experiment: the magnitude of the background noise (the variation) and the volume of the signal (the size of the sample). The greater the noise (the variation), the less information will be contained in the sample. On the other hand, the louder the signal (the larger the sample size), the greater the amplification will be and the greater the chance that the message will penetrate the noise and be received.

The design of experiments is a very broad area of statistics that is primarily concerned with acquiring a fixed amount of information at the lowest possible cost. To this end one can study methods of sampling that will reduce background variation in an experiment and/or amplify nature's signal. Despite the complexity of the subject, some of the important considerations in the design of good experiments can be easily understood and should be presented in an introductory course. We will illustrate how to determine the sample size (volume of the signal) necessary for estimating a population mean μ, a binomial proportion π, or the difference between two means ($\mu_1 - \mu_2$) using a confidence interval of a desired width. Then we will show how one can control the background variation by using a particular experimental design, the paired-difference experiment.

14.2   CHOOSING THE SAMPLE SIZE TO ESTIMATE μ, $\mu_1 - \mu_2$, or π

How can we determine the number of observations to include in a sample? The implications of such a question are clear. Observations cost money. If the sample is too large, time and talent are wasted. Conversely, it is also wasteful if the sample is too small because inadequate information has been purchased for the time and effort expended, and it may be impossible to increase the sample size at a later time. Hence the number of observations the experimenters should buy will depend on the amount of information they want.

Suppose we want to estimate the mean value of accident claims filed against an insurance company. To determine how many claims to examine (or sample), we would have to determine how accurate the company wants us to be. For example, the company might indicate that the tolerable error is to be ± 5 units or less; then the confidence interval would be of the form $\bar{x} \pm 5$.

Recall that the standard error of the sampling distribution for the sample mean $\bar{x}$ is $\sigma_{\bar{x}} = \sigma/\sqrt{n}$. If a 95% confidence interval is to be of the form $\bar{x} \pm E$, we must solve the expression

$$1.96\sigma_{\bar{x}} = E$$

for n.

In general, if we want to estimate μ using a confidence interval of the form $\bar{x} \pm E$, where E is specified, we solve the equation

$$z\sigma_{\bar{x}} = E$$

for n. The solution is shown here.

Sample Size for Estimating μ Using a Confidence Interval of the Form $\bar{x} \pm E$

$$n = \frac{z^2\sigma^2}{E^2}$$

where the value of z used depends on the degree of confidence required for the confidence interval; $z = 1.645$ for a 90% confidence interval, $z = 1.96$ for a 95% confidence interval, and $z = 2.58$ for a 99% confidence interval. (Note: We usually round the computed value of n upward to the next interger.)

The procedure for determining a sample size to estimate μ requires knowledge of the population variance σ^2 (or standard deviation σ). We can obtain an approximate sample size by estimating σ^2, using one of two methods:

1. Employ information from a prior experiment to calculate a sample variance s^2. Use this value to approximate σ^2.

2. Use information on the range of the observations (and the Empirical Rule) to obtain an estimate of σ. (This method was employed in Section 5.5 as a check on the calculation of s.)

Then we would substitute the estimated value of σ^2 in the equation

$$n = \frac{z^2\sigma^2}{E^2}$$

to determine an approximate sample size n.

We illustrate the procedure for choosing a sample size in the following examples.

Example 14.1

Union officials are concerned about reports of inferior wages paid to employees in an industry under their jurisdiction. It is decided to take a random sample of wage sheets from n employees within the industry to estimate the industry mean hourly wage. It is known that wages within the industry have a range of $10 per hour. Determine the sample size required to estimate μ, the mean hourly wage, using 95% confidence interval of the form $\bar{x} \pm .50$.

Solution   The tolerable error for the confidence interval is $E = \$.50$. Before using the equation for n, we must estimate the population variance σ^2. We do so using a *range estimate* of σ. Since the range of hourly wages is $10, an estimate of σ (using the method of Section 5.5) is

$$s \approx \frac{\text{range}}{4} = \frac{10}{4} = 2.5$$

Then an approximate sample size can be found by substituting $(2.5)^2$ for σ^2 in the equation n. Substituting $z = 1.96$ and $\sigma = 2.5$, we have

$$n = \frac{(1.96)^2(2.5)^2}{(.5)^2} = 96.04$$

Thus the union officials would have to sample 97 wage sheets to estimate the mean hourly wage using a 95% confidence interval of the form $\bar{x} \pm .50$.

Example 14.2

A federal agency has decided to investigate the advertised weight displayed on cartons of a certain brand of cereal. The company in question periodically samples cartons of cereal coming off the production line to check their weight. A summary of 1500 of the weights made available to the agency indicates a mean weight of 11.80 ounces per carton and a standard deviation of .75 ounce. Use the company's information to determine the number of cereal cartons the federal agency must examine in order to estimate the average weight of cartons currently produced to within .25 ounce using a 99% confidence interval.

Solution   The federal agency has specified that the confidence interval must be of the form $\bar{x} \pm .25$; so $E = .25$ ounce. To determine the sample size required to achieve this bound, we need an estimate of the population variance σ^2. Assuming the weights made available to the federal agency by the company were randomly selected from its production, we can use the given standard deviation of these weights to form an estimate of σ^2. Thus

$$\sigma^2 \approx (.75)^2 = .5625$$

The appropriate z-value for a 99% confidence interval is 2.58. Now an approximate sample size can be found by using the equation

$$n = \frac{(2.58)^2(.75)^2}{(.25)^2} = 59.91$$

Hence the federal agency must obtain a random sample of $n = 60$ cereal cartons to estimate the mean weight to within .25 ounce.

Selecting the sample size to estimate a binomial parameter π is accomplished in a manner similar to the method employed when estimating μ. You will recall that $\hat{\pi} = x/n$ is our point estimate of π and that $\sigma_{\hat{\pi}} = \sqrt{\dfrac{\pi(1 - \pi)}{n}}$ is the standard error of the sampling distribution of $\hat{\pi}$. Then, for a confidence interval of the form $\hat{\pi} \pm E$, we solve the expression

$$z\sigma_{\hat{\pi}} = E$$

for n. The solution is shown next.

> **Sample Size for Estimating π Using a Confidence Interval of the Form $\pi \pm E$**
>
> $$n = \frac{z^2 \pi (1 - \pi)}{E^2}$$
>
> where E is the specified tolerable error for the confidence interval and z is determined by the degree of confidence desired for the confidence interval.

Note that we must know π to solve for n and that this requirement creates a circular problem, since our final objective is to estimate π. Actually, it is not as complicated as it appears. We often know before the experiment begins that π will lie in a fairly narrow range. For example, the fraction of popular vote for a presidential candidate in a national election is often close to .5. Thus we can substitute for π the value dictated by experience. A second method for finding π is to estimate π by using data collected from a prior experiment. Finally, if there is no prior information, we can use $\pi = .5$. Substituting .5 for π will yield the largest possible sample size for the tolerable error E that you have specified and thus will give a conservative answer to the required sample size. The sample size you calculate this way will likely be larger than required, but you will be on the safe side.

We illustrate the selection of the sample size for estimating a binomial parameter π with an example.

Example 14.3

In a national election poll we want to estimate the proportion π of voters in favor of the Republican candidate. How many people should be polled so that we can estimate π to within .02 (i.e., to within 2 percentage points) using a 95% confidence interval?

Solution   The pollster has specified that the interval is to be of the form $\hat{\pi} \pm .02$, and the z-value for a 95% confidence interval is 1.96. Then the sample size n necessary to achieve the desired interval is found by substituting into the equation

$$n = \frac{z^2 \pi (1 - \pi)}{E^2}$$

To solve for n, first we must obtain an approximation for π. If a similar survey has been run recently, we can use the sample proportion from that survey to estimate π. Otherwise, we substitute $\pi = .5$ to obtain a conservative* sample size (one that is likely to be larger than required). For the case in which no prior survey was run, the sample size is

$$n = \frac{(1.96)^2(.5)(.5)}{(.02)^2} = 2401$$

That is, 2401 potential voters must be polled in order to estimate the proportion favoring the Republican candidate using a 95% confidence interval of the form $\hat{\pi} \pm .02$.

We have discussed choosing the sample size required to estimate either μ or π, using a confidence interval with tolerable error E. For either case the sample size can be determined by setting z standard errors of the appropriate sample statistic equal to the desired error E and solving for n. When estimating μ, we must supply some estimate of the population variance σ^2. This estimate can usually be obtained by using a value of s^2 from a prior experiment or by using a range estimate to approximate σ. When estimating π, we must supply some value for the binomial parameter appearing in the standard error of $\hat{\pi}$. If data are available from a previous study, we could use this information to estimate π in the formula; otherwise, we substitute $\pi = .5$ into the formula for n to obtain a conservative sample size (a sample size at least as large as necessary to achieve the specified tolerable error E). The following example shows you how to apply this same procedure to find the sample sizes for a comparison of two population means.

Example 14.4

Suppose we want to estimate the difference between two population means $(\mu_1 - \mu_2)$, based on independent random samples from the two populations, using a 95% confidence interval of the form $(\bar{x}_1 - \bar{x}_2) \pm .5$. If we know that $\sigma_1^2 \approx \sigma_2^2 \approx 3$, and we plan to select equal sample sizes, $n_1 = n_2 = n$, find n.

Solution   The sample-size formula has the familiar form z (standard error) $= E$. The sampling distribution of $(\bar{x}_1 - \bar{x}_2)$ has a standard error of

$$\sigma_{\bar{x}_1 - \bar{x}_2} = \sqrt{\frac{\sigma_1^2}{n_1} + \frac{\sigma_2^2}{n_2}}$$

---

*$\pi = .5$ gives the largest possible value of $\pi(1 - \pi)$, namely .25. Since $\pi(1 - \pi)$ appears in the numerator of the formula for n, $\pi = .5$ will yield the largest value for n.

Then, letting $n_1 = n_2 = n$ and $\sigma_1^2 = \sigma_2^2 = 3$, we would solve the expression

$$z\sigma_{\bar{x}_1 - \bar{x}_2} = E$$

or

$$1.96 \sqrt{\frac{3}{n} + \frac{3}{n}} = .5$$

for n.

Solving for n, we obtain

$$1.96 \sqrt{\frac{6}{n}} = .5$$

$$\sqrt{n} = \frac{1.96 \sqrt{6}}{.5}$$

$$n = 92.20$$

Thus we should select approximately 93 observations in each sample. The resulting 95% confidence interval would be of the form $(\bar{x}_1 - \bar{x}_2) \pm .5$.

EXERCISES

14.1 A forester wants to estimate the mean diameter of pine trees in a 100-acre tree plantation. A preliminary survey of the area suggests that the diameters can be as small as 9 inches and as large as 27 inches. The forester wishes the estimate to be within .5 inch of the trees' mean diameter, using a 95% confidence interval. How many trees should be included in a random sample?

14.2 Refer to Exercise 14.1. How would n change for a 90% confidence interval with $E = .5$? What about a 99% confidence interval?

14.3 Suppose you want to estimate the mean fair market price of homes, all of similar size and age, in a particular development and you want your estimate to be within $10,000 of the actual mean market price, using a 95% confidence interval of the form $\bar{x} \pm$ $10,000. Further, suppose that you are fairly certain that no house will sell for less than $80,000 or more than $200,000. If you plan to sample recent sales prices at your local courthouse, approximately how many should be included in your sample? What assumption must you make concerning the sales in order that your estimate be valid?

$ 14.4 A certified public accountant plans to audit the financial records of a company to estimate the proportion of account ledger pages that are in error. If the accountant believes that the ledger page error rate is in the neighborhood of 1% and wants to estimate the true rate correct to within 1%, approximately how many ledger pages should be included in the sample if a 90% confidence interval is used? What sample size would be needed for a 95% confidence interval?

14.5 Suppose that you plan to sample the opinions of a particular hospital's nurses to estimate the proportion who are satisfied with working conditions. If you wish to estimate the proportion with a 95% confidence interval of the form $\hat{\pi} \pm .05$, how many nurses should be included in your random sample? (Hint: Since you have no prior information on the approximate value of π, find the approximate sample size by using $\pi = .5$.)

$ 14.6 The giant size of a new "tough cleaning" laundry detergent has a listed net weight of 42 oz. If the variability in weight has a standard deviation of 2 oz, how many boxes must be sampled to estimate the average fill weight to within ±.25 oz using a 95% confidence interval?

14.7 Refer to Exercise 14.6. Determine the effect of a 90% and a 99% confidence level on the required sample size.

14.8 Investigators would like to estimate the average annual taxable income of apartment dwellers in a city to within $500 using a 95% confidence interval. If we assume the annual incomes range from $0 to $40,000, determine the number of observations that should be included in the sample.

14.9 A study is conducted to compare the average number of years of service at the age of retirement for military personnel in 1970 versus 1980. A random sample of career records is to be obtained for each year. Previous information suggests that the common population standard deviation is approximately four years. Determine the number of observations to be collected from records for each year if we wish to estimate the mean difference in age using a 95% confidence interval with a half width of .5 years.

14.10 Refer to Exercise 14.8. Determine the required sample size if the desired error in a 95% confidence interval is $E = 250$. Do the same for $E = 1000$. Compare your results to those of Exercise 14.8.

$ 14.11 As part of a much larger study of trends in long-distance telephone usage, a study is to be conducted of residential homes with married couples between 25 and 40 years of age. How large a sample should be taken if the mean number of long-distance calls for one month is to be estimated to within one call using a 90% confidence interval? Assume $\sigma \approx 4.0$.

14.12 National public opinion polls are based on interviews of as few as 1500 persons in a random sampling of public sentiment toward one or more issues. These interviews are commonly done in person, because mail returns have poor response and telephone interviews tend to reach older people, thus biasing the results. Suppose that a random sample of 1500 persons is surveyed to determine the proportion of the adult public in agreement with recent energy conservation proposals.

(a) If 560 indicate they favor the policies set forth by the current administration, estimate π, the proportion of adults holding a "favor" opinion. How many persons must be surveyed to have a 95% confidence interval with a half-width of .01?

14.3 THE PAIRED-DIFFERENCE EXPERIMENT: AN EXAMPLE OF A DESIGNED EXPERIMENT

noise reducers

volume increasers

paired-difference experiment

We observed that two factors affect the quantity of information in an experiment that has been designed for the purpose of making inferences about a population parameter: namely, the variation in the population (as measured by σ) and the size of the sample. Some experimental designs are classified as either **noise reducers** or **volume increasers,** depending on whether the primary effect on the quantity of information in an experiment is to reduce the variation (σ) in the data or to increase the volume of information.

The **paired-difference experiment** is an example of a noise-reducing (variation-reducing) design. Suppose that we want to test the difference between the average wear of two types of automobile tires, using a 20,000-mile road test. If only the rear wheels of each automobile are to be used and five automobiles are in the experiment, two methods of design might be suggested. First, we could randomly assign five tires of type A and five tires of type B to the 10 rear wheels. Then we might have a random assignment of tires to rear wheels as demonstrated in Figure 14.1. The automobiles would then be driven over the 20,000-mile test course and the amount of wear would be recorded for each tire.

FIGURE 14.1   *Random Assignment of Five Tires of Type A and Type B*

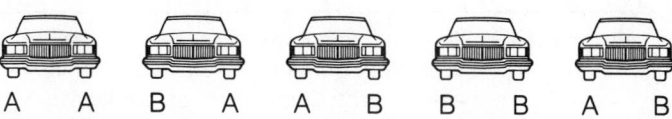

A   A   B   A   A   B   B   B   A   B

But there is a disadvantage to this design. Since each automobile has a different driver, we would expect the wear measurements to vary greatly from automobile to automobile, depending upon the test driver, his method of accelerating and braking, the balance of the wheels, and the road surface to which the tires were exposed. Thus some wear measurements on tires A or B would be extremely high and others would be extremely low, making the respective sample variances, s_A^2 and s_B^2, large.

Recall that the t-statistic for testing the equality of two population means μ_A and μ_B is

$$t = \frac{\bar{x}_A - \bar{x}_B}{s_p \sqrt{\dfrac{1}{n_A} + \dfrac{1}{n_B}}}$$

where

$$s_p^2 = \frac{(n_A - 1)s_A^2 + (n_B - 1)s_B^2}{n_A + n_B - 2}$$

We reject the null hypothesis for large (positive or negative) values of t. If the sample variances s_A^2 and s_B^2 are large, we require large differences in the sample means $(\bar{x}_A - \bar{x}_B)$ to declare a difference in the mean wear for the two types of tires. Hence the automobile-to-automobile variability, which inflates the sample variances s_A^2 and s_B^2, makes it difficult to detect a difference in the population mean wears $(\mu_A - \mu_B)$ and thereby reduces the quantity of information in the experiment.

We can improve on the design in Figure 14.1 be reducing the variability in the sample data. To do this, we make a comparison of the wear for the two tire types under controlled driver-road conditions. In this way we filter out the variability due to automobiles (drivers) by making comparisons of tire types A and B on each automobile. We do this by randomly assigning a type A and a type B tire to the rear wheels of each automobile. Note that this randomization scheme is restricted, since we require both tire types to appear on a car. A typical assignment is shown in Figure 14.2. Note that one tire of type A and one of type B are randomly assigned and mounted on the rear wheels of each automobile. The five automobiles could then be driven over the 20,000-mile test course, after which the amount of wear would be recorded for each tire.

FIGURE 14.2   Restricted Random Assignment of Tires to Rear Wheels that Eliminates Automobile Variability in the Comparison of Tire Types A and B

A   B   B   A   B   A   A   B   B   A

difference

The results of a wear test using the restricted random assignment of Figure 14.2 are given in Table 14.1. The wears for both an A and a B tire, x_A and x_B, are shown in the second and third columns for each automobile. The **difference** between each pair, denoted by the symbol d, is recorded in column four.

TABLE 14.1   Results of a Tire Wear Test

| | Tire wear | | |
Automobile	x_A	x_B	$d = (x_A - x_B)$
1	10.6	10.2	.4
2	9.8	9.4	.4
3	12.3	11.8	.5
4	9.7	9.1	.6
5	8.8	8.3	.5
	$\bar{x}_A = 10.24$	$\bar{x}_B = 9.76$	$\bar{d} = .48$

The averages for x_A, x_B, and d are shown at the bottom of the table. Note that $\bar{d} = (\bar{x}_A - \bar{x}_B)$. We would like to use these data to determine if there is sufficient evidence to indicate a difference in the mean wear for the two types of tires, $(\mu_A - \mu_B)$.

At first glance we maybe tempted to employ the methods of comparing two means (Section 9.4). However, this would not be the proper analysis because one of the assumptions of the t-test has been violated—the two samples are not independent because the pairs of observations are linked; they read high or low, depending on the driver-automobile combination to which the pair of tires was assigned. Hence the data analysis for an experiment is dictated by the design used. If the design employed was the one obtained by random assignment of the five tires of types A and B to the 10 rear wheels (Figure 14.1), we would analyze the data by using the methods for comparing two population means presented in Section 9.4. But the experimenter realized that the wear measurements would vary greatly from car to car; thus he could filter or block out this variability if he assigned one tire of type A and one tire of type B to the rear wheels of each car. Hence the restricted random assignment (Figure 14.2) dictates that we must perform another analysis to determine whether the mean wear differs for the two types of tires. The appropriate method is to compare the wear between the tire types A and B by using the five difference measurements given in Table 14.1.

The proper analysis utilizes the five difference measurements to test the hypothesis that the mean difference between tires A and B is equal to zero. (This hypothesis is equivalent to the hypothesis that the difference in the means for the two types of tires $(\mu_A - \mu_B)$ is equal to zero.) The testing procedure is identical to the procedure for a statistical test of a hypothesis concerning a population mean μ when σ is unknown (Section 8.6). Thus we will use the sample of differences and the one-sample Student's t-test to test the null hypothesis $\mu_d = 0$, where μ_d, the mean of the population of differences, is equal to $(\mu_A - \mu_B)$. Elements of the paired-difference test are as follows:

Paired-Difference Test

Null hypothesis: $\mu_d = (\mu_A - \mu_B) = 0$

Alt. hypothesis: For a one-tailed test:

1. $\mu_d > 0$

2. $\mu_d < 0$

For a two-tailed test:

3. $\mu_d \neq 0$

Test statistic: $t = \dfrac{\bar{d}}{s_d/\sqrt{n}}$ where $\bar{d} = \dfrac{\Sigma d}{n}$

$$s_d = \sqrt{\frac{\Sigma d^2 - (\Sigma d)^2/n}{n-1}}$$

and n is the number of differences.

Rejection region: For a given value of α, for df $= (n - 1)$, and for a one-tailed test:

1. Reject H_O if $t > t_a$, with $a = \alpha$.

2. Reject H_O if $t < -t_a$, with $a = \alpha$.

For a given value of α, for df $= (n - 1)$, and for a two-tailed test:

3. Reject H_O if $|t| > t_a$, with $a = \alpha/2$.

The notation in the t-statistic is slightly different from that used for a statistical test about μ in Section 8.6. Here we use the symbol d to denote the difference between the observations (x_A and x_B) made on the same pair. Then the n differences are used to compute

$$\bar{d} = \frac{\Sigma d}{n}$$

the mean of the sample differences.

Keep in mind that for the paired-difference experiment, n refers to the number of sample differences (or number of pairs) rather than to the total number of measurements, and s_d refers to the standard deviation of the sample differences:

$$s_d = \sqrt{\frac{\Sigma\, d^2 - (\Sigma\, d)^2/n}{n-1}}$$

Now let's take the data in Table 14.1 and test the null hypothesis that $\mu_d = 0$ against the alternative hypothesis that μ_d is not equal to zero. First we must compute $\bar{d}$ and s_d. For these data,

$$\Sigma\, d = .4 + .4 + .5 + .6 + .5 = 2.4$$
$$\Sigma\, d^2 = (.4)^2 + (.4)^2 + (.5)^2 + (.6)^2 + (.5)^2 = 1.18$$

Hence for $n = 5$ differences,

$$\bar{d} = \frac{\Sigma\, d}{n} = \frac{2.4}{5} = .48$$
$$s_d^2 = \frac{\Sigma\, d^2 - (\Sigma\, d)^2/n}{n-1} = \frac{(1.18) - (2.4)^2/5}{4}$$
$$= \frac{1.18 - 1.152}{4} = .007$$
$$s_d = \sqrt{.007} = .0837$$

Substituting into the test statistic t, we have

$$t = \frac{\bar{d}}{s_d/\sqrt{n}} = \frac{.48}{.0837/\sqrt{5}} = \frac{(.48)(2.236)}{.0837} = 12.8$$

The rejection region for a two-tailed test can be located by using Table 3 in Appendix 3. Let's suppose we want $\alpha = .05$. The value of t for $a = .025$ and $(n-1) = 4$ df is 2.776. Hence we reject the null hypothesis if the observed value of t is either greater than 2.776 or less than -2.776. Since $t = 12.8$ is greater than 2.776, we reject the null hypothesis that the average difference μ_d is equal to zero. That is, we reject the hypothesis that the difference in mean wear $(\mu_A - \mu_B)$ is equal to zero. We conclude that a difference exists in mean wear for the two types of tires. Indeed, because $\bar{x}_A$ exceeds $\bar{x}_B$, we would venture to state that type A tires show more wear over a comparable test course than type B tires.

A general confidence interval for the mean difference can be obtained by using the general confidence interval for a single population mean (Section 8.6).

> **Confidence Interval for $\mu_d = (\mu_1 - \mu_2)$, Based on a Paired-Difference Experiment**
>
> $$\bar{d} \pm t \frac{s_d}{\sqrt{n}}$$
>
> The value of t corresponding to a 90%, a 95%, or a 99% confidence interval is found in Table 3 of Appendix 3 for df = $(n - 1)$ and a = .05, .025, or .005, respectively. For this situation n represents the number of differences.

Example 14.5

Find a 95% confidence interval for the difference in mean wear for tire types A and B, Table 14.1.

Solution   For this example $n = 5$ and a = .025. The value of t corresponding to a = .025 and df = 4 is 2.776. The 95% confidence interval for $\mu_d = (\mu_A - \mu_B)$ is

$$\bar{d} \pm t \frac{s_d}{\sqrt{n}}$$

$$.48 \pm (2.776) \left(\frac{.0837}{\sqrt{5}} \right)$$

$$.48 \pm .10$$

Thus we estimate the difference in mean tire wear for tire types A and B to be in the interval .38 to .58. The confidence coefficient for this interval estimate is .95. This means that when we use this estimation procedure, 95% of the time the constructed intervals will enclose μ_d.

---

randomized block design

The statistical design employed in the tire experiment represents a simple example of a **randomized block design.** Several points about this design should be emphasized. First, pairing of the measurements occurred when the experiment was planned (actually it occurred in the restricted random assignment of tires to wheels) and not after the data were obtained. Comparisons of wear were then made within each car to eliminate the variability between automobiles (drivers). Second, pairing will not always provide more information for testing the difference between two population means. Recall that the objective of the paired-difference experiment is to reduce the "background noise" (car-to-car variability). If there were absolutely no differences between the cars, we would lose information by pairing because, instead of having ten sample measurements (one measurement for each of five tires of types A and B), we would have only five differences. This reduces the number of degrees of freedom for the *t*-test statistic and so makes it more difficult to detect a difference in means when one exists. Thus we recommend pairing only if we can filter out undesirable background noise. The paired-difference experiment is only one of a class of noise-reducing designs referred to as block designs.

Volume-increasing experimental designs are concerned with shifting information in an experiment to display more clearly the parameter(s) of interest. They involve a study of the effect of a set of variables on a response measurement *y*, and the choice of the combination of variables to be used in the experiment. This study can become quite complex. Our only exposure to controlling the volume of information in an experiment in the test is through the sample size calculations presented in Section 14.2.

EXERCISES

14.13 Consider the accompanying data for a paired-difference experiment.

Pair	x_A	x_B	$d = (x_A - x_B)$
1	21	29	-8
2	28	30	-2
3	17	21	-4
4	24	25	-1
5	27	33	-6
6	18	22	-4
7	20	19	1
8	23	29	-6
9	28	26	2

(a) How many degrees of freedom are associated with the *t*-statistic?

(b) Do the data provide sufficient evidence to indicate a difference between μ_A and μ_B? Test by using $\alpha = .05$.

(c) Construct a 95% confidence interval for $(\mu_A - \mu_B)$. Interpret the interval.

14.14 A group of psychologists want to examine the amount of learning exhibited by schizophrenics after taking a specified dose of a tranquilizer. Ten schizophrenics are randomly selected from a patient ward, and before the drug is administered, each is given a standard examination. The amount of time required to complete the exam is recorded for each individual. One hour after the specified drug dose is administered, each patient is given the same standard exam as previously. Again completion times are recorded. If these patients exhibit any learning, we would expect the mean completion time prior to receiving the drug to be more than that needed one hour after receiving the drug. This is our alternative hypothesis.

Use the accompanying data to test the null hypothesis that the mean difference between the before-dose and the after-dose test scores is zero (that is, H_0: $\mu_d = 0$) against the alternative hypothesis H_a: $\mu_d > 0$. Use $\alpha = .05$.

Patient	Before x_1	After x_2	Difference $x_1 - x_2$
1	10	8	2
2	15	13	2
3	30	29	1
4	29	25	4
5	26	21	5
6	28	28	0
7	19	15	4
8	13	10	3
9	14	12	2
10	21	17	4

14.15 Refer to Exercise 14.7. Construct a 95% confidence interval for $(\mu_1 - \mu_2)$. Interpret the interval.

14.16 Suppose that two independent random samples of $n_1 = n_2 = 10$ observations have been selected from two populations to test the null hypothesis $(\mu_1 - \mu_2) = 0$. After the measurements have been recorded in two adjacent columns, side-by-side observations are paired for the $n = 10$ pairs. Will a paired-difference analysis provide more information to test H_0: $\mu_d = 0$ than the unpaired t-test of Section 9.4? Explain.

14.17 Two analysts, A and B, supposedly of identical abilities, are used to measure the parts per million of a certain type of chemical impurity in drinking water. It is claimed that analyst A tends to give higher readings than analyst B. To test this theory, each of six water samples is divided and then analyzed by A and B. The data are shown here.

Water sample	Analyst A	Analyst B
1	31.4	28.1
2	37.0	37.1
3	44.0	40.6
4	28.8	27.3
5	59.9	58.4
6	37.6	38.9

Do the data provide sufficient evidence to indicate that the mean reading for analyst A exceeds the mean reading for analyst B? Test by using $\alpha = .05$.

14.18 Refer to Exercise 14.17. Construct a 90% confidence interval for $(\mu_A - \mu_B)$, the difference in the mean readings for analysts A and B. Interpret the interval.

14.19 An agronomist compared the mean yield of two new varieties of wheat using side-by-side, 10-acre plots at each of eight different farm locations. One 10-acre plot at a location was planted with variety A; the other was planted with variety B. The mean yield per acre for each of the plots is recorded in the accompanying table.

Location	Variety A	Variety B
1	76	77
2	69	83
3	73	76
4	79	78
5	68	80
6	81	75
7	70	73
8	75	85

(a) Do the data provide sufficient evidence to indicate a difference in the mean yield per acre for the two varieties? Test by using $\alpha = .10$.

(b) Construct a 90% confidence interval for the difference in mean yields for the two varieties of wheat. Interpret the interval.

14.4 USING COMPUTERS

The data from Table 14.1 have been used to illustrate how Minitab and SAS can be used to generate a t-test or confidence interval for $\mu_1 - \mu_2$ based on paired data. Minitab gives the confidence interval and t-test directly using the TTEST and TINTERVAL procedures on the difference data; this is shown in the accompanying output. SAS, on the other hand, gives the t-test directly using PROC MEANS (with the indicated options), but we would have to form the corresponding confidence interval using the sample mean $(\bar{d})$ and the standard error for $\bar{d}$ shown in the SAS output given here.

Additional examples of computer solutions for problems posed in this chapter are shown in Supplementary Exercises 14.25, 14.28, and 14.38. The programs written for these examples and exercises show how to use SAS and Minitab for similar problems.

Minitab Output

```
MTB > READ INTO C1 C2
DATA> 10.6     10.2
DATA>  9.8      9.4
DATA> 12.3     11.8
DATA>  9.7      9.1
DATA>  8.8      8.3
DATA> END
      5 ROWS READ
MTB > NAME C1='TIRE A'
MTB > NAME C2='TIRE B'
MTB > LET C3 = C1 - C2
MTB > NAME C3='DIFF'
MTB > PRINT C1-C3

 ROW  TIRE A  TIRE B   DIFF

   1   10.6    10.2    0.4
   2    9.8     9.4    0.4
   3   12.3    11.8    0.5
   4    9.7     9.1    0.6
   5    8.8     8.3    0.5

MTB > TTEST ON DATA IN C3

TEST OF MU = 0.0000 VS MU N.E. 0.0000

             N      MEAN    STDEV   SE MEAN        T    P VALUE
DIFF         5    0.4800   0.0837   0.0374    12.83     0.0002

MTB > TINTERVAL 95 FOR DATA IN C3

             N      MEAN    STDEV   SE MEAN    95.0 PERCENT C.I.
DIFF         5    0.4800   0.0837   0.0374   ( 0.3761,  0.5839)

MTB > STOP
```

SAS Output

```
OPTIONS NODATE NONUMBER LS=78 PS=60;
DATA RAW;
  INPUT AUTO TIRE_A TIRE_B;
  DIFF = TIRE_A - TIRE_B;
  LABEL  AUTO   ='AUTOMOBILE'
         TIRE_A ='AMOUNT OF WEAR FOR TIRE A'
         TIRE_B ='AMOUNT OF WEAR FOR TIRE B'
         DIFF   ='DIFFERENCE IN WEAR BETWEEN TIRES';
  CARDS;
   1   10.6    10.2
   2    9.8     9.4
   3   12.3    11.8
   4    9.7     9.1
   5    8.8     8.3
;
PROC PRINT N;
TITLE1 'RESULTS OF A TIRE WEAR TEST ON 5 AUTOMOBILES';
TITLE2 'LISTING OF THE DATA';                              (continued)
```

```
PROC MEANS N MEAN STD STDERR T PRT;
  VAR DIFF;
  TITLE2 'EXAMPLE OF PROC MEANS WITH THE T AND PRT OPTIONS';
RUN;
```

```
              RESULTS OF A TIRE WEAR TEST ON 5 AUTOMOBILES
                          LISTING OF THE DATA

              OBS     AUTO    TIRE_A    TIRE_B    DIFF

               1       1       10.6      10.2     0.4
               2       2        9.8       9.4     0.4
               3       3       12.3      11.8     0.5
               4       4        9.7       9.1     0.6
               5       5        8.8       8.3     0.5

                             N = 5
```

Analysis Variable : DIFF DIFFERENCE IN WEAR BETWEEN TIRES

N Obs	N	Mean	Std Dev	Std Error	T	Prob>|T|
5	5	0.4800000	0.0836660	0.0374166	12.8285396	0.0002

SUMMARY

The design of an experiment is actually a plan by which an experimenter can purchase a specified quantity of information about one or more population parameters. The information pertinent to a given parameter can be measured by the width of the confidence interval. The factors affecting the quantity of information in an experiment are data variation and sample size or, making an analogy to audio communication, noise and amplification. Good experimental designs are those that filter out unwanted data variation and, at the same time, increase the volume of the signal by increasing the sample size. Proper experimental design can increase the quantity of information per fixed cost or, alternatively, reduce the cost of a specified quantity of information.

A signal is amplified by shifting information in an experiment to (in a sense) increase n and thereby increase the quantity of information about a given parameter. Selecting the sample size to achieve a specified tolerable error for a confidence interval illustrates the effect of the sample size on the quantity of information in an experiment. The larger n is, the smaller the width of the confidence interval will be.

The paired-difference experiment, a simple case of a randomized block design, was used to illustrate the concept of noise reduction. One source of unwanted variation (the differences in automobiles and test drivers) was eliminated from the data, thereby unveiling a clear difference in the wearing quantities of tire types A and B. Many more observations would have been required to reveal the difference in mean wear for the two tire types if independent random samples had been employed. Thus filtering out the effect of differences in test drivers and automobiles reduced the cost of a specified amount of information concerning the difference in the wearing qualities of the two types of tires.

The study of experimental design is a separate course in itself, requiring more space than we can allot here. But it is important to note that good designs are those that simultaneously achieve both noise reduction and signal amplification, thus helping the experimenter to acquire a specified amount of information at minimum cost. You will find additional information on this subject in the References at the end of the text.

KEY TERMS

design of an experiment   paired-difference experiment
noise reducers   difference
volume increasers   randomized block design

KEY FORMULAS

1. Sample size for estimating μ using a confidence interval of the form $\bar{x} \pm E$

$$n = \frac{z^2 \sigma^2}{E^2}$$

(Note: $z = 1.645$, 1.96, or 2.58 for a 90%, 95%, or 99% confidence interval, respectively.)

2. Sample size for estimating π using a confidence interval of the form $\hat{\pi} \pm E$

$$n = \frac{z^2 \pi(1 - \pi)}{E^2}$$

3. Statistical test about $\mu_d = \mu_1 - \mu_2$ from a paired-difference experiment

Null hypothesis: $\mu_d = 0$

Test statistic: $t = \dfrac{\bar{d}}{s_d/\sqrt{n}}$

4. Confidence interval for $\mu_d = \mu_1 - \mu_2$ from a paired-difference experiment

$$\bar{d} \pm \frac{ts_d}{\sqrt{n}}$$

SUPPLEMENTARY EXERCISES

14.20 List two factors that affect the quantity of information in an experiment.

14.21 Explain what is meant by noise-reducing experimental designs and by volume-increasing experimental designs.

14.22 Give an example of noise-reducing experiment.

14.23 Will the paired-difference test always provide more information for testing the difference between two populations than the methods of Section 9.4? Explain.

14.24 Consider the accompanying data for a paired-difference experiment.

Pair	x_A	x_B	d
1	4.1	3.9	.2
2	4.0	4.0	0
3	3.8	3.6	.2
4	4.2	4.1	.1
5	3.9	3.8	.1
6	3.9	4.0	−.1

Do the data provide sufficient evidence to indicate a difference between μ_A and μ_B? Test by using $\alpha = .05$.

14.25 Refer to Exercise 14.24 and the computer output here to answer parts (a) and (b).

(a) Construct a 95% confidence interval for $(\mu_A - \mu_B)$. Interpret the interval.

(b) Suppose that the data had resulted from independent random sampling of the two populations—that is, no pairing. Use the method of

Section 9.4 to construct a 95% confidence interval for $(\mu_A - \mu_B)$. Notice the difference in the interval widths for parts (a) and (b).

```
MTB > READ INTO C1 C2
DATA>   4.1     3.9
DATA>   4.0     4.0
DATA>   3.8     3.6
DATA>   4.2     4.1
DATA>   3.9     3.8
DATA>   3.9     4.0
DATA> END

      6 ROWS READ
MTB > LET C3 = C1 - C2

MTB > PRINT C1-C3

  ROW     C1     C2        C3

    1     4.1    3.9    0.200000
    2     4.0    4.0    0.000000
    3     3.8    3.6    0.200000
    4     4.2    4.1    0.100000
    5     3.9    3.8    0.100000
    6     3.9    4.0   -0.100000

MTB > TINTERVAL 95 FOR DATA IN C3

              N     MEAN    STDEV   SE MEAN    95.0 PERCENT C.I.
C3            6    0.0833   0.1169   0.0477    ( -0.0394,  0.2061)

MTB > TWOSAMPLE-T FOR C1 VS C2

TWOSAMPLE T FOR C1 VS C2
       N     MEAN    STDEV   SE MEAN
C1  6     3.983    0.147    0.0601
C2  6     3.900    0.179    0.0730

95 PCT CI FOR MU C1 - MU C2: (-0.1307, 0.2973)

TTEST MU C1 = MU C2 (VS NE): T= 0.88  P=0.40  DF=  9

MTB > STOP
```

14.26 (a) If you have conducted a paired-difference experiment, is it valid to analyze the data in an unpaired manner (i.e., by using the method of Section 9.4)?

(b) Suppose that you have conducted the experiment by selecting independent samples of equal size from the two populations. In a similar situation some experimenters might pair the observations, one observation from each sample in each pair, and then analyze the data by using the paired-difference test. What is wrong with this procedure?

14.27 An agricultural experiment station was interested in comparing the yields for two new varieties of corn. Because it was felt that there might be a great deal of variability in yield from one farm to another, each variety was randomly assigned to a different 1-acre plot on each of seven farms. The 1-acre plots were planted and the corn was harvested at maturity. The results of the experiment (in bushels of corn) are shown in the accompanying table.

Farm	Variety A	Variety B
1	28.2	21.5
2	24.6	20.1
3	29.7	24.0
4	20.5	21.2
5	34.6	29.8
6	27.1	21.7
7	31.4	26.8

(a) Show that the sample standard deviation of the differences $(x_A - x_B)$ is $s_d = 2.39$.

(b) Use the data to test the null hypothesis that there is no difference in mean yields for the two varieties of corn. Use $\alpha = .05$.

14.28 A pharmaceutical firm decides to test the effectiveness of a new drug on reducing cigarette smoking. Seven subjects are selected at random, and before the drug is administered each subject records the number of cigarettes consumed over the last twenty-four hours. One week after the specified drug dosage has been administered, each subject again records the number of cigarettes consumed for the same time period. Use the accompanying computer output to determine whether the drug is effective in reducing the number of cigarettes smoked.

Subject	Before drug	After drug
1	15	13
2	22	16
3	35	30
4	9	10
5	16	15
6	42	37
7	21	25

```
MTB > READ INTO C1 C2
DATA>    15    13
DATA>    22    16
DATA>    35    30
DATA>     9    10
DATA>    16    15
```

(continued)

```
DATA>    42    37
DATA>    21    25
DATA>  END

      7 ROWS READ
MTB > LET C3 = C2 - C1
MTB > NAME C1='BEFORE'
MTB > NAME C2='AFTER'
MTB > NAME C3='DIFF'
MTB > PRINT C1-C3

 ROW   BEFORE   AFTER    DIFF

   1       15      13      -2
   2       22      16      -6
   3       35      30      -5
   4        9      10       1
   5       16      15      -1
   6       42      37      -5
   7       21      25       4

MTB > TTEST ON DATA IN C3;
SUBC> ALTERNATIVE = -1.

TEST OF MU = 0.00 VS MU L.T. 0.00

              N      MEAN    STDEV    SE MEAN        T    P VALUE
DIFF          7     -2.00     3.65       1.38    -1.45      0.099

MTB > STOP
```

14.29 Data have been collected by a student for a required course in her curriculum. Collection of the data presented no real problem but she is at a loss regarding the analysis. She knows you are about to complete a course in statistics, and she has asked you to help her. She explains to you that she randomly chose ten subjects from the population that was of interest to her and observed their pulse rates before and after administering a certain medication. Results were as follows.

	Pulse rate	
Subject	Before	After
1	47	49
2	70	71
3	65	69
4	80	83
5	102	107
6	97	107
7	85	83
8	92	91
9	75	78
10	69	74

Construct a 95% confidence interval for the difference in mean pulse rates.

14.30 A study of sex differences in learning ability was made on college freshmen. When the mean learning ability (as measured by a standardized test) of a random sample of 25 men was compared with the mean learning ability of a random sample of 25 women, no significant difference in learning ability was found by statistician A. Statistician B performed essentially the same experiment, except that 25 pairs of freshmen were studied, each pair consisting of a man and a woman of approximately the same IQ. Statistician B found a significant difference in learning ability between men and women.

(a) Which of the two experiments would you be more likely to accept as the "better" of the two?

(b) How many degrees of freedom did statistician A have in the test of hypothesis? How many did statistician B have?

14.31 Insurance adjusters were concerned about the high repair estimates that they were getting from garage A. To verify their suspicions, each of ten damaged cars was taken to garage A to obtain a repair estimate and then to another, more reliable garage, B. The results are given in the accompanying table (in hundreds of dollars).

Damaged car	Garage A	Garage B	Damaged car	Garage A	Garage B
1	2.1	2.0	6	5.4	5.0
2	4.5	3.8	7	7.3	6.5
3	6.3	5.9	8	9.5	8.6
4	3.0	2.8	9	10.1	9.0
5	1.2	1.3	10	7.6	7.2

(a) Show that the sample standard deviation of the differences $(x_A - x_B)$ is $s_d = .38$.

(b) Test the null hypothesis that there is no difference in mean repair estimates for the two garages against the alternative that, on the average, estimates from garage A are higher than those from garage B. Use $\alpha = .01$.

14.32 Improperly filled orders are a costly problem for mail-order houses. To estimate the mean loss per incorrectly filled order, a large mail-order house plans to randomly sample n incorrectly filled orders and to analyze the added cost associated with each. It is estimated that the added cost of the return and replacement of the incorrect order can never be less than $40 or more than $400. If the company wants to estimate the mean additional cost per incorrect order correct to within $20, using a 95% confidence interval, how many incorrect orders must it analyze?

14.33 A manufacturer of dishwashers claims that its machines will operate repair-free for 4 years. A quick examination of repair slips on models of this dishwasher indicates that the dates of first repairs have ranged anywhere from 0 to 9 years. Use this information to approximate the number of observations (of repair records) required to estimate the mean time to first repair of the dishwashers, using a 90% confidence interval of the form $\bar{x} \pm .25$ years (3 months).

14.34 How many repair records (Exercise 14.33) must be examined to estimate the mean time to first repair to within 6 months using a 90% confidence interval?

14.35 A study is to be conducted to estimate the mean gain in weight for chicks on a specified ration during their first four weeks of growth. Experience has shown that the standard deviation in chick weight gain during the first four weeks of growth is in the neighborhood of 6 grams. If the experimenter wishes the estimate of mean weight gain to be correct within 1 gram using a 95% confidence interval, how many chicks should be included in the sample?

14.36 Two brands of latex outdoor paint were to be compared for durability at each of six different geographic locations. At each location one-half of a strip of exterior wood was painted with brand A, the other half with brand B. Durability readings were then obtained after a one-year trial period.

Location	Brand A	Brand B
1	3.1	3.2
2	2.9	2.8
3	3.6	3.5
4	4.2	4.0
5	3.8	3.6
6	2.7	2.4

(a) Show that $s_d = .14$.

(b) Use the experimental data to determine if a difference exists between the mean durability for the two brands. Let $\alpha = .05$.

14.37 What advantage(s) does the design proposed in Exercise 14.36 have over the corresponding unpaired experiment?

14.38 Random samples of 100 oranges for each of two varieties were collected at each of nine locations and the total sugar content (in pounds) contained in each batch of 100 oranges was recorded. The data are shown here.

Location	Variety A	Variety B	Location	Variety A	Variety B
1	2.1	2.0	6	2.3	2.3
2	1.9	1.7	7	2.0	1.7
3	2.1	1.9	8	2.1	2.2
4	1.9	2.0	9	2.3	2.1
5	2.2	2.1			

Use the following computer output to determine whether the data provide sufficient evidence to indicate a difference in the mean sugar content per 100 oranges between the two varieties. Test by using $\alpha = .05$.

```
MTB > READ INTO C1 C2
DATA>   2.1   2.0
DATA>   1.9   1.7
DATA>   2.1   1.9
DATA>   1.9   2.0
DATA>   2.2   2.1
DATA>   2.3   2.3
DATA>   2.0   1.7
DATA>   2.1   2.2
DATA>   2.3   2.1
DATA> END

      9 ROWS READ
MTB > LET C3 = C1 - C2
MTB > NAME C1='A'
MTB > NAME C2='B'
MTB > NAME C3='DIFF'
MTB > PRINT C1-C3

  ROW     A      B        DIFF

    1    2.1    2.0    0.100000
    2    1.9    1.7    0.200000
    3    2.1    1.9    0.200000
    4    1.9    2.0   -0.100000
    5    2.2    2.1    0.100000
    6    2.3    2.3    0.000000
    7    2.0    1.7    0.300000
    8    2.1    2.2   -0.100000
    9    2.3    2.1    0.200000

MTB > TTEST ON DATA IN C3

TEST OF MU = 0.0000 VS MU N.E. 0.0000

               N     MEAN    STDEV   SE MEAN      T    P VALUE
DIFF           9   0.1000   0.1414   0.0471    2.12    0.067

MTB > STOP
```

$ **14.39**   A service station manager would like to decrease the price charged for regular gasoline. Before doing this, the manager must estimate the average number of gallons sold per week to determine if the decrease would be profitable. Previous records suggest that sales run anywhere from 1000 to 6000 gallons per week. Use this information to determine the sample size required to estimate μ, the average number of gallons sold per week, with a 95% confidence interval of the form $\bar{x} \pm 200$ gallons.

14.40 Student government would like to estimate the proportion of students in favor of converting from the semester to the quarter system. Determine the sample size required to estimate π, the proportion of students favoring the change, with a 99% confidence interval of the form $\hat{\pi} \pm .03$. (Hint: since π is unknown, use $\pi = .5$ to find the approximate sample size.)

14.41 A large corporation conducted a study and found that approximately 25% of its potential customers were aware of the corporation's product. To increase this percentage, an extensive advertising campaign was carried out. At the end of the campaign, a sample of potential customers was to be contacted to determine the proportion aware of the product. How many potential customers must be contacted to estimate π, the fraction aware of the corporation's product, to within .02 using a 95% confidence interval? (Hint: Since π is unknown, use $\pi = .5$ to find the approximate sample size.)

14.42 The city council has enough money to erect one traffic light and must choose one of two locations. The number of vehicles was counted simultaneously for specified time periods at each of the locations. They have asked you as a "statistician" to analyze the results for them and to make a recommendation.

Time period	Location 1	Location 2
7:00–7:15	92	90
7:30–7:45	135	111
8:00–8:15	150	141
8:30–8:45	125	112
9:00–9:15	114	82
4:00–4:15	90	70
4:30–4:45	121	105
5:00–5:15	138	120
5:30–5:45	135	114
6:00–6:15	116	95

(a) What are the null and alternative hypotheses?

(b) What is the test statistic?

(c) How many degrees of freedom do you use in looking up the table value for specifying the rejection region?

(d) What would you recommend to the city council, stated in words they could understand as "nonstatisticians"?

14.43 To obtain data for a thesis, a psychology major plans to interview people to determine if their reaction to a certain situation is positive or negative.

(a) The student wants to estimate the true proportion favoring the situation to within .04 using a 95% confidence interval. If a pilot study showed 60 of 180 people with positive reactions, how many should be interviewed?

(b) Give the psychology major some help in getting the needed random sample from a student body of roughly 14,000 students. Outline how you would draw the random sample.

14.44 Suppose you were asked to estimate the proportion of students on your campus who fully support themselves while attending school. Let us say that the percentage is expected to be somewhere near 10% based on a prior study. Assume your client has asked you to estimate the true population proportion with a tolerable error of no more than 2% with a desired confidence level of 95%. What sample size would you recommend?

14.45 University employees are disturbed about the extremely long time it takes to be reimbursed for out-of-pocket travel expenses. To back up their complaints, they decide to take a sample of travel expense requests filed during the next six months and determine the average amount of time it takes before an individual is reimbursed. Previous experience suggests that these requests take from 7 to 60 days. Determine the sample size needed to estimate μ, the average time for reimbursement, to within 2 days using a 90% confidence interval.

14.46 A psychiatrist would like to estimate the average time it takes schizophrenics to react to a specified stimulus. Previous work in this field suggests that the standard deviation for reaction times of schizophrenics is approximately .5 second. Determine the sample size required to estimate μ using a 95% confidence interval of the form $\bar{x} \pm .2$ second.

14.47 A small company was concerned about the percentage of its accounts that were overdue on the last day of the previous month. It would be too time-consuming to examine all accounts, since records were not stored in a computer. Hence it was decided to draw a random sample of accounts to estimate the fraction that were overdue on the last day of the previous month. Previous samples have shown that approximately 10% of all accounts were overdue at the end of a month. How large a sample of accounts should be taken in order to estimate π to within .05 using a 99% confidence interval?

14.48 Refer to Exercise 14.24. Suppose you want to conduct a new paired-difference experiment to estimate the difference between the population means. If you wish to estimate $(\mu_A - \mu_B)$ with a 95% confidence interval of the form $\bar{d} \pm .03$, how many pairs must you sample?

14.49 An experimenter plans to compare the mean number of particles of effluent in water collected at two different points in a water treatment system. The means are expected to be in the neighborhood of 30 and it is known that the variances for the particle counts in samples taken at the two locations are approximately equal to 30. If the experimenter wishes to estimate the difference in the mean counts at the two locations correct to within 1 particle, with a 95% confidence interval, how many water samples should be analyzed at each location? Assume that $n_1 = n_2$.

14.50 An instructor in an introductory statistics class was interested in determining if exam grades were higher on the second exam in the course than on the first exam. The instructor was interested because of having had two "help" sessions for students prior to the second exam, but no "help" sessions for the first exam. The instructor's sample consisted of the 10 students who attended both "help" sessions. The scores on the two exams were as follows:

Student	Exam #1	Exam #2	Student	Exam #1	Exam #2
McNamara	39	44	Sgarlata	40	43
Sturm	30	27	Fairchild	33	52
Coonce	44	60	Adams	32	46
Shermulis	17	28	Winiarz	53	54
Turban	36	49	Hansen	8	31

Make the appropriate test, using $\alpha = .05$; your null hypothesis should be that there was no improvement.

14.51 A list of twelve items were purchased from each of two supermarkets. Only items stocked by both stores were considered, and the sample of twelve items was randomly selected. The following results were achieved via the sampling process.

Item	Price at store #1	Price at store #2	Item	Price at store #1	Price at store #2
Lettuce	0.69	0.89	Milk	1.59	1.59
Hamburger	1.65	1.79	TV dinner	1.48	1.69
Pizza	3.15	3.19	Napkins	0.79	0.89
Toothpaste	1.09	1.35	Canned beans	0.55	0.62
Bread	0.98	1.09	Six-pack of cola	1.69	1.89
Ice cream	2.98	2.79	Cereal	1.39	1.59

Estimate the difference in mean prices charged at the two stores using a 95% confidence interval.

14.52 Given that the standard deviation of annual electricity bills for residential users in New Mexico is approximately $68 and that you want to estimate the mean annual bill to within $20 using a 90% confidence interval, what size random sample should you take?

14.53 **Class Exercise** Scan available newspapers, magazines, and professional journals for the results of a survey in which the experimenter estimates the proportion (or percentage) of items in a population possessing a specified attribute. For example, surveys are frequently conducted to estimate the proportion of people in the United States who think that the president is doing an acceptable job. For the survey you choose, perform the following:

(a) Identify the sample size.

(b) Use the methods of Chapter 10 to give a point estimate and construct a 95% confidence interval for π.

(c) Indicate how you might have improved the design of the survey.

(d) If, prior to running the survey, you wish to estimate π correct to within .02 using a 95% confidence interval, how large a sample size would be necessary?

CHAPTER 15

ANALYSIS OF VARIANCE

15.1 Introduction ▪ **15.2** The logic behind an analysis of variance ▪ **15.3** A statistical test about more than two population means: An example of an analysis of variance ▪ **15.4** Checking on the equal variance assumption ▪ **15.5** Using computers ▪ Summary ▪ Key terms ▪ Key formulas ▪ Supplementary exercises ▪ Exercises from the data base

15.1   INTRODUCTION

We discussed methods for comparing two population means in Chapter 9. Very often the two-sample problem is a simplification of what is encountered in real life. That is, frequently we want to compare more than two population means.

For example, suppose that we want to compare the mean incomes of steelworkers for three different ethnic groups, say black, white, and Spanish Americans, in a certain city. Independent random samples of steelworkers would be selected from each of the three ethnic groups (the three populations). We would have to consult the personnel files of the steel companies in the city, list steelworkers in each ethnic group, and select a random sample from each. On the basis of the three sample means, we want to decide whether the population mean incomes differ and, if so, by how much. Note that the sample means most likely will differ, but this does not necessarily imply a difference in mean income for the three ethnic groups. Even if the population mean incomes were identical, the sample means most probably would differ. How, then, do you decide whether the differences among the sample means are large enough to imply a difference among the corresponding population means? We will answer this question by using a technique known as an analysis of variance.

15.2 THE LOGIC BEHIND AN ANALYSIS OF VARIANCE

The reason we call the method an analysis of variance can be seen more easily with an example. Assume that we want to compare three population means based on three samples of five observations each. The data for the three samples are shown in Table 15.1. Do the data present sufficient evidence to indicate a difference among the three population means? A brief visual analysis of the data in Table 15.1 leads us to a rapid, intuitive "yes." A glance at each of the three samples indicates that there is very little variation within each sample; that is, the variation in the measurements within each sample is very small. In contrast, the spread or variation among the sample means is so large in comparison to the within-sample variation that we intuitively conclude that a real difference does exist among the population means.

TABLE 15.1   A Comparison of Three Population Means (Small Amount of Within-Sample Variation)

Sample 1	Sample 2	Sample 3
29.0	25.1	20.1
29.2	25.0	20.0
29.1	25.0	19.9
28.9	24.9	19.8
28.8	25.0	20.2
$\bar{x}_1 = 29.0$	$\bar{x}_2 = 25.0$	$\bar{x}_3 = 20.0$
$s_1 = .16$	$s_2 = .07$	$s_3 = .16$

How does our intuition work when a larger within-sample variation is present? (See Table 15.2.) In this case the variation among the sample means is not large relative to the variation within samples. Hence it would be difficult to conclude that the samples were drawn from populations with different means.

The variations in the observations for the two sets of data (Tables 15.1 and 15.2) are shown graphically in Figure 15.1. The strong evidence to indicate a difference in population means for the data of Table 15.1 is apparent in Figure 15.1(a). The lack of evidence to indicate a difference in population means for the data of Table 15.2 is indicated by the overlapping of data points for the samples in Figure 15.1(b).

TABLE 15.2   A Comparison of Three Population Means
(Large Amount of Within-Sample Variation)

Sample 1	Sample 2	Sample 3
29.0	33.1	15.2
14.2	7.4	39.3
45.1	17.6	14.8
48.9	44.2	25.5
7.8	22.7	5.2
$\bar{x}_1 = 29.0$	$\bar{x}_2 = 25.0$	$\bar{x}_3 = 20.0$
$s_1 = 18.19$	$s_2 = 14.18$	$s_3 = 12.96$

FIGURE 15.1   Dot Diagrams for the Data of Tables 15.1 and
15.2

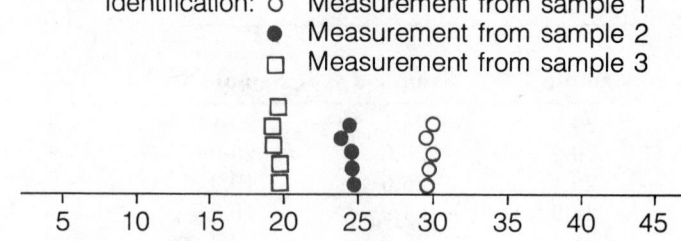

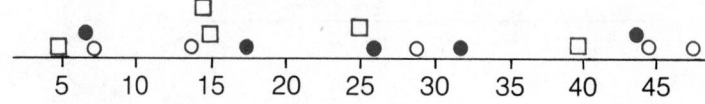

analysis of
variance

Figure 15.1 and the data of Tables 15.1 and 15.2 indicate very clearly, then, what we mean by an **analysis of variance.** All differences in sample means are judged statistically significant or not significant by comparing them with a measure of the random variation within the population data. Recall that we will measure the variability of the population data by the population standard deviation σ (or, equivalently, the population variance σ^2).

15.3 A STATISTICAL TEST ABOUT MORE THAN TWO POPULATION MEANS: AN EXAMPLE OF AN ANALYSIS OF VARIANCE

Earlier (Chapter 9) we presented a method for testing the equality of two population means based on independent random samples from the two populations. We hypothesized two normal populations (1 and 2), with means denoted by μ_1 and μ_2, respectively. To test the null hypothesis that $\mu_1 = \mu_2$, independent random samples of sizes n_1 and n_2 were drawn from the two populations. The sample data were then used to compute the value of the test statistic

$$t = \frac{\bar{x}_1 - \bar{x}_2}{s_p \sqrt{\dfrac{1}{n_1} + \dfrac{1}{n_2}}}$$

where

$$s_p^2 = \frac{(n_1 - 1)s_1^2 + (n_2 - 1)s_2^2}{(n_1 - 1) + (n_2 - 1)} = \frac{(n_1 - 1)s_1^2 + (n_2 - 1)s_2^2}{n_1 + n_2 - 2}$$

is the pooled estimate of the common population variance σ^2. The rejection region for a specified value of α, the probability of a type I error, was then found by using Table 3 in Appendix 3.

Now suppose that we wish to extend this method to test the equality of more than two population means. The test procedure described above applies only to two means and therefore is inappropriate. Hence we will employ a more general method of data analysis known as the analysis of variance. We illustrate its use with the following example.

Students from 5 different campuses throughout the country were surveyed to determine their attitudes toward industrial pollution. Each student sample was asked a specific set of questions and then given a total score for the interview (Table 15.3). Suppose that 9 students were surveyed at each of 5 campuses and that we want to compare the average student scores for the 5 campuses.

If we are interested in testing the equality of the population means (i.e., $\mu_1 = \mu_2 = \mu_3 = \mu_4 = \mu_5$), we might be tempted to perform all possible pairwise comparisons of the population means. Hence if we assume that the five distributions are approximately normal, with a common variance σ^2, we could conduct the ten t-tests comparing the five means, two at a time, as listed in Table 15.4 (see Section 9.4).

TABLE 15.3   Summary of the Sample Results for Five
Populations

	Population (campus)				
	1	**2**	**3**	**4**	**5**
Sample mean	$\bar{x}_1$	$\bar{x}_2$	$\bar{x}_3$	$\bar{x}_4$	$\bar{x}_5$
Sample variance	s_1^2	s_2^2	s_3^2	s_4^2	s_5^2

TABLE 15.4   All Possible Null Hypotheses for Comparing
Two Means from Five Populations

$\mu_1 = \mu_2$	$\mu_1 = \mu_4$	$\mu_2 = \mu_3$	$\mu_2 = \mu_5$	$\mu_3 = \mu_5$
$\mu_1 = \mu_3$	$\mu_1 = \mu_5$	$\mu_2 = \mu_4$	$\mu_3 = \mu_4$	$\mu_4 = \mu_5$

One obvious disadvantage to this test procedure is that it is tedious and time consuming. But the more important and less apparent disadvantage of running multiple t-tests to compare means is that the probability of incorrectly rejecting at least one of the hypotheses increases as the number of t-tests increases. Thus although we may have the probability of a type I error fixed at $\alpha = .05$ for each individual test, the probability of incorrectly rejecting H_0 on at least one of these tests is larger than .05. In other words, the combined probability of a type I error for the set of ten hypotheses would be larger than the value .05 set for each individual test. Indeed, it could be as large as .40.

What we need, then, is a single test of the hypothesis "all five population means are equal" that will be less tedious than the individual t-tests and that can be performed with a specified probability of a type I error (say $\alpha = .05$). First we assume that the five sets of measurements are normally distributed with means given by μ_1, μ_2, μ_3, μ_4, and μ_5 and with a common variance σ^2. Consider the quantity

$$s_W^2 = \frac{(n_1 - 1)s_1^2 + (n_2 - 1)s_2^2 + (n_3 - 1)s_3^2 + (n_4 - 1)s_4^2 + (n_5 - 1)s_5^2}{(n_1 - 1) + (n_2 - 1) + (n_3 - 1) + (n_4 - 1) + (n_5 - 1)}$$

$$= \frac{(n_1 - 1)s_1^2 + (n_2 - 1)s_2^2 + (n_3 - 1)s_3^2 + (n_4 - 1)s_4^2 + (n_5 - 1)s_5^2}{n_1 + n_2 + n_3 + n_4 + n_5 - 5}$$

Note that this quantity is merely an extension of

$$s_p^2 = \frac{(n_1 - 1)s_1^2 + (n_2 - 1)s_2^2}{n_1 + n_2 - 2}$$

variability within populations

which is used as the estimator of the common variance for two populations in a test of the hypothesis $\mu_1 = \mu_2$ (Section 9.4). Thus s_W^2 represents a pooled estimate of the common variance σ^2 and measures the **variability** of the observations **within** the five **populations.** (The subscript W refers to the within-population variability.)

Next we consider a quantity that measures the variability between or among the population means. If the null hypothesis $\mu_1 = \mu_2 = \mu_3 = \mu_4 = \mu_5$ is true, the populations are identical, with mean μ and variance σ^2. Drawing single samples from the five populations is then equivalent to drawing five different samples from the same population. What kind of variation might be expected for these sample means? If the variation is too great, we would reject the hypothesis that $\mu_1 = \mu_2 = \mu_3 = \mu_4 = \mu_5$.

To assess the variation from sample mean to sample mean, we need to know the distribution of the mean of a sample of nine observations in repeated sampling. From the Central Limit Theorem (Chapter 7) we know that the sampling distribution of $\bar{x}$ based on nine observations will have a mean of $\mu_{\bar{x}} = \mu$ and a variance of $\sigma_{\bar{x}}^2 = \sigma^2/9$. Since we have five different samples of nine observations each, we can estimate the variance $\sigma^2/9$ of the sampling distribution by computing the sample variance for the five sample means:

$$\text{sample variance (of the means)} = \frac{\Sigma \bar{x}^2 - (\Sigma \bar{x})^2/5}{5 - 1}$$

Note that we merely consider the $\bar{x}$s as a sample of five observations and calculate the sample variance. This quantity is an estimate of $\sigma^2/9$; so 9 times the sample variance of the means is an estimate of σ^2. We designate this estimate of σ^2 by s_B^2, where the subscript B denotes a measure of the **variability among** (between) the **sample means.** For this problem

variability among sample means

$$s_B^2 = 9 \times \text{(sample variance of the means)}$$

Under the null hypothesis that all five population means are identical, we have two estimates of σ^2, namely s_W^2 and s_B^2. Suppose that the ratio s_B^2/s_W^2 is used as a test statistic to test the hypothesis that $\mu_1 = \mu_2 = \mu_3 = \mu_4 = \mu_5$. What would the distribution of this quantity be if we were to repeat the experiment over and over again, each time calculating s_B^2 and s_W^2? As you might surmise from Chapter 11, s_B^2/s_W^2 follows an F-distribution, with degrees of freedom that can be shown to be $df_1 = 4$

for s_B^2 and $df_2 = 40$ for s_W^2. The proof of these remarks is beyond the scope of this text. However, we make use of the result for testing the null hypothesis $\mu_1 = \mu_2 = \mu_3 = \mu_4 = \mu_5$.

The test statistic used to test equality of population means is

$$F = \frac{s_B^2}{s_W^2}$$

When the null hypothesis is true, both s_B^2 and s_W^2 are estimates of σ^2, and F would be expected to assume a value near $F = 1$. When the hypothesis of equality is false, s_B^2 will tend to be larger than s_W^2 due to the differences among the population means. Hence we will reject the null hypothesis in the upper tail of the distribution of $F = s_B^2/s_W^2$. For α, the probability of a type I error, equal to .10, .05, .025, .01, or .005, we can locate the rejection region for the one-tailed test by using Tables 5 through 9 in Appendix 3, with $df_1 = 4$ and $df_2 = 40$. Thus for $\alpha = .05$ the tabulated value of F is 2.61 (see Figure 15.2). If the calculated value of $F = s_B^2/s_W^2$ falls in the rejection region, we conclude that at least one of the population means is different from the others.

FIGURE 15.2   Rejection Region of F for $a = .05$, $df_1 = 4$, $df_2 = 40$

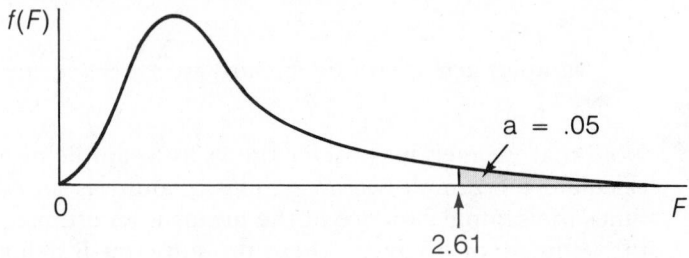

This procedure can be generalized, with only slight modification in the formulas, to test the equality of k (where k is an integer equal to or greater than two) population means from normal populations, with a common variance σ^2. Independent random samples of size n_1, $n_2, \ldots, n_k$ are drawn from the respective populations. Then we compute the sample means and variances. The null hypothesis $\mu_1 = \mu_2 = \cdots = \mu_k$ is tested against the alternative hypothesis that at least one of the population means is different from the others. The test procedure is given next.

An Analysis of Variance for Testing the Equality of *k* Population Means

Null hypothesis: $\mu_1 = \mu_2 = \cdots = \mu_k$

Alt. hypothesis: At least one of the population means is different from the others.

Test statistic: $F = \dfrac{s_B^2}{s_W^2}$

where

$$s_W^2 = \frac{(n_1 - 1)s_1^2 + (n_2 - 1)s_2^2 + \cdots + (n_k - 1)s_k^2}{n_1 + n_2 + \cdots + n_k - k}$$

$$s_B^2 = \frac{\sum n_i \bar{x}_i^2 - (\sum n_i \bar{x}_i)^2/n}{k - 1}$$

and $n = n_1 + n_2 + \cdots + n_k$.

Rejection region: Reject H_0 if $F > F_{\alpha, df_1, df_2}$ where $df_1 = (k - 1)$, and $df_2 = (n_1 + n_2 + \cdots + n_k - k)$. See Tables 5–9 of Appendix 3 corresponding to $\alpha = .10$, .05, .025, .01, and .005.

Note: When $n_1 = n_2 = \cdots = n_k$, the formula for s_B^2 simplifies to

$$s_B^2 = n' \left[\frac{\sum \bar{x}^2 - (\sum \bar{x})^2/k}{k - 1} \right] \quad \text{where } n' \text{ is the common sample size}$$

We illustrate these ideas in an example.

Example 15.1

A group of psychologists was interested in studying the effect of anxiety on learning as measured by student performance on a series of tests. On the basis of a prestudy test, 27 students were classified into one of three anxiety groups. Group 1 students were those who scored extremely low on a scale measuring anxiety. Those placed in group 3 were students who scored extremely high on the anxiety scale. The remaining students were placed in group 2. The results of the prestudy anxiety test indicated that 6 students were in group 1, 12 were in group 2, and 9 were in group 3.

Following the prestudy assignment of students to groups, the same battery of tests was given to each of the 27 students. The sample mean and variance of these test scores (based on a total of 100 points) are summarized in Table 15.5 for each group. Use the sample data to test the hypothesis that the average test score for low-, middle-, and high-anxiety students is identical (that is, that anxiety has no effect on a student's performance on this battery of tests). Use $\alpha = .01$.

TABLE 15.5 Summary of Test Scores

Group 1 (low)	Group 2 (medium)	Group 3 (high)
$n_1 = 6$	$n_2 = 12$	$n_3 = 9$
$\bar{x}_1 = 88$	$\bar{x}_2 = 82$	$\bar{x}_3 = 78$
$s_1^2 = 10.1$	$s_2^2 = 14.8$	$s_3^2 = 13.9$

Solution   First we hypothesize a population for each anxiety group that corresponds to all possible test scores for students who could have been included in the study. We will assume the measurements in each population are approximately normally distributed, with the mean of population 1 equal to μ_1, the mean of 2 equal to μ_2, and the mean of 3 equal to μ_3. In addition, we will assume that the populations have a common variance σ^2. From these populations, independent random samples of size $n_1 = 6$, $n_2 = 12$, and $n_3 = 9$ students were obtained and assigned to the respective groups.

To test the null hypothesis of equality of the population means, $\mu_1 = \mu_2 = \mu_3$, we first compute s_W^2 and s_B^2. We calculate s_W^2 directly from the sample data.

$$s_W^2 = \frac{(n_1 - 1)s_1^2 + (n_2 - 1)s_2^2 + (n_3 - 1)s_3^2}{n_1 + n_2 + n_3 - 3}$$

$$= \frac{5(10.1) + 11(14.8) + 8(13.9)}{6 + 12 + 9 - 3} = = \frac{324.5}{24} = 13.52$$

However, before obtaining s_B^2, we must first compute $\Sigma\, n_i \bar{x}_i$ and $\Sigma\, n_i \bar{x}_i^2$. From Table 15.5 we determine that

$$\Sigma\, n_i \bar{x}_i = 6(88) + 12(82) + 9(78) = 2{,}214$$

$$\Sigma\, n_i \bar{x}_i^2 = 6(88)^2 + 12(82)^2 + 9(78)^2 = 181{,}908$$

Hence for $k = 3$ and $n = n_1 + n_2 + n_3 = 27$, we have

$$s_B^2 = \frac{\Sigma\, n_i \bar{x}_i^2 - (\Sigma\, n_i \bar{x}_i)^2 / n}{k - 1}$$

$$= \frac{181{,}908 - (2{,}214)^2 / 27}{2} = \frac{181{,}908 - 181{,}548}{2} = 180$$

The test statistic for the null hypothesis $\mu_1 = \mu_2 = \mu_3$ is

$$F = \frac{s_B^2}{s_W^2} = \frac{180.0}{13.52} = 13.31$$

Using the probability of a type I error, $\alpha = .01$, we can locate the upper-tail rejection region for this one-tailed test by using Table 8 of Appendix 3, with

$$df_1 = k - 1 = 2$$
$$df_2 = n_1 + n_2 + n_3 - 3 = 24$$

The tabulated value of F is 5.61. Since the observed value of F is greater than 5.61, we reject the hypothesis of equality of the population means (i.e., we conclude that at least one of the means is different from the rest). Although the F-test does not tell us which of the population means are different, the sample data suggest that anxiety has a detrimental effect on a student's performance in the battery of tests.

After completing the F-test in an analysis of variance, the results of a study are usually summarized in an **analysis of variance table.** The format of an analysis of variance (ANOVA) table is shown in Table 15.6.

analysis of variance table

TABLE 15.6 An Example of an ANOVA Table

Source	Sum of squares	Degrees of freedom	Mean square	F-test
Between samples	SSB	$k - 1$	s_B^2	s_B^2/s_W^2
Within samples	SSW	$n - k$	s_W^2	
Total	SSB + SSW	$n - 1$		

The analysis of variance table lists the sources of variability in the first column. The second column lists quantities called sums of squares. For our purposes, the sums of squares between samples (SSB) and within samples (SSW) are the numerators of the expressions for s_B^2 and s_W^2, respectively. So,

$$SSB = \Sigma n_i \bar{x}_i^2 - (\Sigma n_i \bar{x}_i)^2/n$$
$$SSW = (n_i - 1)s_1^2 + (n_2 - 1)s_2^2 + \cdots + (n_k - 1)s_k^2$$

mean square

The third column of the ANOVA table gives the degrees of freedom associated with each source of variability and the fourth column displays the **mean squares.** Historically, s_B^2 and s_W^2 have become known as mean squares; s_B^2 is called the mean square between samples and s_W^2 is called the mean square within samples. Note that a mean square is merely a sum of squares divided by its degrees of freedom. Finally, the last column of the ANOVA table gives the value of the F-statistic for the analysis of variance.

Example 15.2

Construct an analysis of variance table for the data of Example 15.1.

Solution Using the previous results, we obtain the following ANOVA table.

Source	Sum of squares	Degress of freedom	Mean square	F-test
Between groups	360	2	180	13.31
Within groups	324.5	24	13.52	
Total	684.5	26		

15.4 CHECKING ON THE EQUAL VARIANCE ASSUMPTION

The assumption of equal population variances, like the assumption of normality of the populations, has been made in several places in the text, such as for the t-test when comparing two population means and for the analysis of variance F-test in a completely randomized design.

Let us now consider an experiment where we wish to compare t population means based on independent random samples from each of the populations. Recall that we assume we are dealing with normal populations with a common variance σ_ϵ^2 and possibly different means. If there were just two populations of interest, we could verify the assumption of equality of the two population variances using the F-test of Chapter 11. However, with $t > 2$, rather than making all pairwise F-tests, we seek a single test that can be used to verify the assumption of equality of the population variances.

Hartley's test

The one test, **Hartley's test**, we will use in this text for the null hypothesis

$$H_0\colon \sigma_1^2 = \sigma_2^2 = \cdots = \sigma_t^2$$

represents a logical extension to the F-test for $t = 2$. If s_i^2 denotes the sample variance computed from the ith sample, the test statistic is

F_{max}

$$F_{\mathbf{max}} = \frac{s_{\max}^2}{s_{\min}^2}$$

where $s_{\max}^2$ and $s_{\min}^2$ are the largest and smallest of the s_i^2s, respectively. Critical values of the $F_{\max}$ test statistic were tabulated by H. O. Hartley (see Table 12 of Appendix 3). The test procedure is summarized here.

Hartley's Test for Homogeneity of Population Variances

Null hypothesis: $\sigma_1^2 = \sigma_2^2 = \cdots = \sigma_t^2$ (i.e., homogeneity of variances)

Alt. hypothesis: Not all population variances are the same.

Test statistic: $F_{max} = \dfrac{s_{\max}^2}{s_{\min}^2}$

Rejection region: For a specified value of α, reject H_0 if F_{max} exceeds the tabulated F-value (Table 12 of Appendix 3) for a $= \alpha$, t, and $df_2 = n - 1$, where n is the number of observations in each sample.

Note that theoretically we required the sample sizes to all be the same. In practice, if the sample sizes are nearly equal, the largest n_i can be used for running the test of homogeneity. This procedure will result in the probability of a type I error being slightly more than the nominal value α.

Hartley's test has some drawbacks and therefore most practitioners do not routinely run it. One reason is that the test is extremely sensitive to departures from normality. So, in checking one assumption (constant variance), one must be very careful about departures from another analysis of variance assumption (normality of the populations). Fortunately, as we mentioned in Chapter 9, the assumption of homogeneity (equality) of population variances is less critical when the sample sizes are substantially different. When the sample sizes are nearly equal, the variances can be markedly different, and the p-values for an analysis of variance will still be only mildly distorted. Thus we recommend that Hartley's test be used only for the more extreme cases. In these extreme situations where homogeneity of the population variances is a problem, a transformation of the data may help to stabilize the variances. Then inferences can be made from an analysis of variance.

15.5 USING COMPUTERS

Calculations for an analysis of variance can be greatly simplified by using available software. Output from SAS and from Minitab illustrate how these two software systems can be used to run an analysis of variance for the data of Table 15.1. These same programs can be used for other data sets that you may need to analyze.

Minitab Output

```
MTB > READ INTO C1-C3
DATA> 29.0    25.1    20.1
DATA> 29.2    25.0    20.0
DATA> 29.1    25.0    19.9
DATA> 28.9    24.9    19.8
DATA> 28.8    25.0    20.2
DATA> END;
        5 ROWS READ
MTB > PRINT C1-C3

  ROW     C1      C2      C3

    1    29.0    25.1    20.1
    2    29.2    25.0    20.0
    3    29.1    25.0    19.9
    4    28.9    24.9    19.8
    5    28.8    25.0    20.2


MTB > AOVONEWAY C1-C3

ANALYSIS OF VARIANCE
SOURCE      DF         SS         MS        F         p
FACTOR       2   203.3333   101.6667   5545.41    0.000
ERROR       12     0.2200     0.0183
TOTAL       14   203.5533
                                    INDIVIDUAL 95 PCT CI'S FOR MEAN
                                    BASED ON POOLED STDEV
LEVEL        N       MEAN     STDEV   ----+---------+---------+---------+--
C1           5     29.000     0.158                                   (*
C2           5     25.000     0.071                       *)
C3           5     20.000     0.158   (*
                                      ----+---------+---------+---------+--
POOLED STDEV =      0.135            21.0      24.0      27.0      30.0
MTB > STOP
```

SAS Output

```
OPTIONS NODATE NONUMBER LS=78 PS=60;
DATA RAW;
  INPUT SAMPLE RESPONSE @@;
```

```
CARDS;
1    29.0   1    29.2   1    29.1   1    28.9   1    28.8
2    25.1   2    25.0   2    25.0   2    24.9   2    25.0
3    20.1   3    20.0   3    19.9   3    19.8   3    20.2
;
PROC PRINT N;
TITLE1 'A COMPARISON OF THREE POPULATION MEANS';
TITLE2 'LISTING OF THE DATA';

PROC ANOVA;
  CLASS SAMPLE;
  MODEL RESPONSE = SAMPLE;
  MEANS SAMPLE;
TITLE2 'EXAMPLE OF PROC ANOVA';
RUN;
```

```
                A COMPARISON OF THREE POPULATION MEANS
                        LISTING OF THE DATA

                   OBS      SAMPLE      RESPONSE

                    1         1          29.0
                    2         1          29.2
                    3         1          29.1
                    4         1          28.9
                    5         1          28.8
                    6         2          25.1
                    7         2          25.0
                    8         2          25.0
                    9         2          24.9
                   10         2          25.0
                   11         3          20.1
                   12         3          20.0
                   13         3          19.9
                   14         3          19.8
                   15         3          20.2

                           N = 15


                A COMPARISON OF THREE POPULATION MEANS
                        EXAMPLE OF PROC ANOVA


                     Analysis of Variance Procedure

Dependent Variable: RESPONSE
                                 Sum of          Mean
Source               DF          Squares        Square      F Value     Pr > F

Model                 2       203.3333333    101.6666667    5545.45     0.0001
Error                12         0.2200000      0.0183333
Corrected Total      14       203.5533333

                   R-Square          C.V.        Root MSE       RESPONSE Mean
                   0.998919        0.548922      0.135401         24.6666667
```

(continued)

Source	DF	Anova SS	Mean Square	F Value	Pr > F
SAMPLE	2	203.3333333	101.6666667	5545.45	0.0001

Level of SAMPLE	N	-----------RESPONSE---------- Mean	SD
1	5	29.0000000	0.15811388
2	5	25.0000000	0.07071068
3	5	20.0000000	0.15811388

SUMMARY

The test procedure described in this chapter is called an analysis of variance because testing the null hypothesis of equality of the means relies on a test statistic composed of a measure of variability between populations, s_B^2, and a measure of variability within populations, s_W^2. The rejection region can be fixed so that the probability α of a type I error is some specified value (our tables allow α to be either .05 or .01). The conclusions that are drawn relate to all the means and not to individual ones. Thus if the test statistic falls in the rejection region, we conclude that at least one of the means is different from the others. However, we cannot specify exactly which means are different from the others based on an ANOVA. In this text, following an ANOVA in which we reject H_0, we have asked you to look at the data and determine which means appear to be different. This will have to suffice for now. Formal procedures for comparing pairs of means following rejection of H_0 in an ANOVA are discussed in Ott (1988).

The analysis of variance is analogous to the use of a floodlight. Large objects (the differences among means) may be recognizable, whereas smaller objects (differences between individual means) may not. In contrast, the t-test is more like a spotlight, which has a smaller field of vision but provides a more intensified light. Thus we are better able to detect individual differences using the Student's t-test.

We conclude from this discussion that we should perform an analysis of variance to indicate overall differences. If a significant value of the F-statistic is obtained, we can run a few t-tests for detecting individual differences. Thus, in the study of attitudes toward industrial pollution on five campuses, we might want to compare the mean total scores for campuses 1 and 5, 2 and 5, and so forth. However, we should limit ourselves to only a few comparisons. Otherwise we increase the risk of detecting a difference that does not exist.

In this chapter we have presented an explanation and an example of an analysis of variance. We needed such a procedure because we were unable to test the equality of more than two population means by using a single test statistic. The analysis of variance is a very general test procedure and can be applied to solve many different problems. However, because of the complexity of the subject, we will go no further. Several useful references are Hicks (1973), Mendenhall (1968), and Ott (1988) for those interested in extending their knowledge of analysis of variance.

KEY TERMS

analysis of variance   mean square
variability within populations   Hartley's test
variability among sample means F_{max}
analysis of variance table

KEY FORMULAS

Analysis of variance F-test

$$F = \frac{s_B^2}{s_W^2}$$

where $s_W^2 = \frac{(n_1 - 1)s_1^2 + (n_2 - 1)s_2^2 + \cdots + (n_k - 1)s_k^2}{n_1 + n_2 + \cdots + n_k - k}$

and $s_B^2 = \frac{\sum n_i \bar{x}_i^2 - (\sum n_i \bar{x}_i)^2/n}{k - 1}$

SUPPLEMENTARY EXERCISES

15.1 Examine the logic behind an analysis of variance.

15.2 Elasticity readings for random samples of size 6 drawn from three different plastic processes (A, B, and C) are given in the accompanying table.

	Process A	Process B	Process C
	4.2	5.6	3.2
	1.1	5.1	2.5
	3.7	4.4	2.9
	2.6	4.2	3.6
	2.1	4.2	3.2
	3.7	5.1	4.1
Sample mean	2.90	4.77	3.25
Sample variance	1.39	.34	.31

Perform an analysis of variance for the experiment to determine if there are differences in the mean elasticity readings among the three plastic processes. Use $\alpha = .05$.

15.3 Refer to Exercise 15.2. Use the sample data and the results of the analysis of variance to construct a 95% confidence interval for $(\mu_B - \mu_A)$. (Hint: Use s_W^2 for the pooled sample variance.)

15.4 The length of life of an electronics component was to be studied under five different operating voltages, V_1, V_2, V_3, V_4, and V_5. Ten different components were randomly assigned to each of the five operating voltages. A summary of the resulting lengths of life is recorded for each of the five groups in the accompanying table. Perform an analysis of variance to test the hypothesis that $\mu_1 = \mu_2 = \mu_3 = \mu_4 = \mu_5$. Use $\alpha = .05$.

Voltage V_1	Voltage V_2	Voltage V_3	Voltage V_4	Voltage V_5
$n_1 = 10$	$n_2 = 10$	$n_3 = 10$	$n_4 = 10$	$n_5 = 10$
$\bar{x}_1 = 3.2$	$\bar{x}_2 = 3.8$	$\bar{x}_3 = 4.1$	$\bar{x}_4 = 4.0$	$\bar{x}_5 = 3.7$
$s_1^2 = .46$	$s_2^2 = .51$	$s_3^2 = .39$	$s_4^2 = .20$	$s_5^2 = .28$

15.5 Refer to Exercise 15.4. Suppose that two components assigned to voltage V_1 and three assigned to voltage V_3 were found to be damaged prior to experimentation. Assuming that no new components can be found to replace the damaged ones, run an analysis of variance to test equality of the means, using the revised data in the accompanying table. Use $\alpha = .05$.

Voltage V_1	Voltage V_2	Voltage V_3	Voltage V_4	Voltage V_5
$n_1 = 8$	$n_2 = 10$	$n_3 = 7$	$n_4 = 10$	$n_5 = 10$
$s_1^2 = .59$	$s_2^2 = .51$	$s_3^2 = .45$	$s_4^2 = .20$	$s_5^2 = .28$

15.6 A clinical pharmacology lab conducted an experiment to compare two pain-relieving drug products. Six subjects were allotted at random to drug A and eight to drug B. The figures below give the number of hours of pain relief provided by each drug:

Drug A	Drug B
4	9
5	6
5	4
7	8
5	6
4	9
	8
	6

(continued)

	Drug A		**Drug B**
Total	30	Total	56
Mean	5	Mean	7
Variance	1.20	Variance	3.14

Perform an analysis of variance and construct an ANOVA table. Draw conclusions using $\alpha = .05$. Are the drug products different with regard to the number of hours of pain relief?

15.7 Refer to Exercise 15.6.

(a) Since there are only two drug products, compare the mean hours of relief provided by drugs A and B using a t-test with $\alpha = .05$ (two-sided). Draw a conclusion.

(b) Did your conclusion in (a) agree with the one we made in Exercise 15.6? They should agree.

15.8 A clinical psychologist wanted to compare three methods for reducing hostility levels in university students. A certain psychological test (HLT) was used to measure the degree of hostility. High scores on this test indicate great hostility. Eleven students obtaining high, nearly equal scores were used in the experiment. Five were selected at random from among the 11 problem cases and treated by method A. Three were taken at random from the remaining 6 students and treated by method B. The other 3 students were treated by method C. All treatments continued throughout a semester. Each student was given the HLT test at the end of the semester, with the score results as shown in the accompanying table. Give the level of significance for a test of the null hypothesis "there is no difference in mean HLT test score for the three methods after treatment." (Hint: Use Tables 5–9 of Appendix 3 to give an approximate level of significance, such as $p > .05$.)

	Method A	**Method B**	**Method C**
	80	70	63
	92	81	76
	87	74	70
	83		
	78		
Sample mean	84.0	75.0	69.7

15.9 Refer to Exercise 15.8. Use the sample data to construct a 99% confidence interval for $(\mu_A - \mu_B)$.

15.10 Three different methods of instruction in speed-reading were to be compared with respect to the mean level of comprehension. A total of 13 students volunteered for the study; 4 were randomly assigned to instructional procedure 1, 4 to procedure 2, and 5 to procedure 3. After a one-week training period all students were asked to read an identical passage on a film, which was delivered at the rate of 300 words per minute. Students were then asked to answer questions on the film passage. Comprehension grades on these questions are listed in the accompanying table. Perform an analysis of variance to test the hypothesis that all three instructional groups have the same average level of comprehension. Use $\alpha = .05$.

	Procedure 1	Procedure 2	Procedure 3
	82	71	91
	80	79	93
	81	78	84
	83	74	90
			88
Sample mean	81.5	75.5	89.2

15.11 An experiment was conducted to compare the effectiveness of three mouthwashes (A, B, and C) in the treatment of morning halitosis. Although ads were run in the local newspapers at the study site, only 16 people responded and qualified for entrance into the study. Four of these 16 did not complete the study; consequently, only the results for the 12 who completed the study are shown in the computer printout that follows. Scores in the data represent results for individual participants on a pleasurable-nonpleasurable scale. (Higher scores imply greater pleasure.) Use the computer printout to respond to the following statements.

(a) Identify the three sample means and standard deviations.

(b) Compute the standard error of the mean for mouthwash B. Compare this value to the value in the output.

(c) State the results of an F-test on the equality of the population mean scores.

(d) Based on the analysis of variance and the magnitudes of the sample means, which mouthwashes appear to be different?

```
ONE WAY ANOVA
MOUTHWASH A                                        59.00000
                                                   66.00000
                                                   58.00000
                                                   61.00000

NO. OBSERVATIONS          4
MEAN                      61.00000
```

(continued)

```
STANDARD DEVIATION
                               3.55903
STANDARD ERROR                 1.77951
MOUTHWASH B                                    53.00000
                                               56.00000
                                               50.00000


NO. OBSERVATIONS               3
MEAN                           53.00000
STANDARD DEVIATION             3.00000
STANDARD ERROR                 1.73205


MOUTHWASH C                                    57.00000
                                               61.00000
                                               72.00000
                                               66.00000
                                               68.00000


NO. OBSERVATIONS               5
MEAN                           64.80000
STANDARD DEVIATION
                               5.89067
STANDARD ERROR                 2.63439
```

ANOVA TABLE

SOURCE	SS	d.f.	MS	F	P
GROUP	262.11667	2	131.05833	6.05506	0.02157
ERROR	194.80000	9	21.64444		
TOTAL	456.91667	11			

15.12 A psychologist used a class of 88 students as subjects for an experiment to study the effects of preconditioning upon accuracy of recall of details of observed events. The professor divided the group at random into four equal groups of 22. Group A were the controls and were not given any preconditioning; group B were privately told to watch for something unusual during the next Wednesday morning's class, but were not told what the unusual event would be; group C were privately told that between 9:25 and 9:30 the professor's lecture would be interrupted by a messenger, who would enter the classroom, hand the professor an envelope, and depart quietly; group D were given the same information given to group C, but were also privately told that they would later be asked to describe the messenger's personal appearance in detail. Subjects were asked not to discuss the matter with anyone. A questionnaire was administered to all subjects after the event, in which they were asked to describe certain physical characteristics, items of clothing, and general appearance of the messenger. The resulting scores, on a scale ranging from 0 to 100, were analyzed by the analysis of variance. Mean scores for the four groups were as follows:

Group	Mean score	Sample variance
A	80	21
B	83	25
C	85	30
D	92	26

Complete the following analysis of variance table for the data above:

Source	Sums of squares	Degrees of freedom	Mean squares	F-test
Between groups				
Within groups				
Total	17676			

15.13 Refer to Exercise 15.12.

(a) What hypothesis is tested by the F-test above?

(b) What is your conclusion from the F-test above? Give the table F-value.

(c) The above analysis requires random samples. What populations are sampled here? Are the samples really random? How might nonrandomness affect your conclusions?

(d) One quantity in the ANOVA table is closely related to the s_p^2 one computes in the t-test. Give the numerical value of that quantity.

(e) Compute the 95% confidence interval for the mean difference in groups C and D (i.e., $\mu_C - \mu_D$). Is the difference significant?

15.14 Sample data from an experiment aimed at comparing the nicotine tar content of five different brands of cigarette are shown here.

Brand	n_i	$\bar{x}$ (mg)	s_i
1	10	9.6	1.3
2	10	10.2	1.4
3	10	10.8	1.1
4	10	11.5	1.2
5	10	13.6	1.5

Run an ANOVA and draw conclusions.

15.15 A horticulturist was investigating the phosphorus content of tree leaves from three different varieties of apple trees, A, B, and C. Random samples of five leaves from each of the three varieties were analyzed for phosphorus content. (See accompanying table.) Use these data to test the hypothesis of equality of the mean phosphorus levels for the three variables. Use $\alpha = .05$.

	Variety A	Variety B	Variety C
	.35	.65	.60
	.40	.70	.80
	.58	.90	.75
	.50	.84	.73
	.47	.79	.66
Sample mean	.46	.78	.71
Sample variance	.008	.010	.006

15.16 Refer to the data of Exercise 15.15. Construct a 95% confidence interval for the difference in mean phosphorus content for varieties A and C.

15.17 An experiment was conducted to compare the effect of three different paints on the corrosion of pipes. A long pipe was cut into 12 segments, which were randomly assigned to one of the three paints so that each paint would be used on four segments. The segments were painted and allowed to weather for a period of six months. The accompanying corrosion readings were than obtained. Is there evidence to indicate a difference in the mean levels of corrosion for the three paints? Use an analysis of variance with $\alpha = .01$. Summarize your findings in an ANOVA table.

	Paint 1	Paint 2	Paint 3
	10.1	13.4	12.7
	11.4	12.9	11.9
	12.1	13.3	12.5
	10.8	13.1	12.3
Sample mean	11.10	13.18	12.35
Sample variance	.727	.049	.117

15.18 To compare the water-repellent properties of four different chemical coatings, the following tests were performed. Twelve different fabric samples were obtained from the same bolt of material, with three samples randomly assigned to each of the four chemical groups (A, B, C, D). Each of the samples was then treated with the assigned chemical coating. Following the chemical treatment, a fixed amount of water was applied to the fabric and the amount of moisture penetration was recorded. These data are recorded in the computer printout that follows. Use the information from the printout to respond to the following statements.

(a) Give the sample mean and standard deviation for the amount of moisture penetration for each chemical coating.

(b) Interpret the results of an analysis of variance to compare the mean moisture penetrations for the four chemicals.

```
ONE WAY ANOVA
GROUP A                                        10.10000
                                               12.20000
                                               11.90000

NO. OBSERVATIONS        3
MEAN                    11.40000
STANDARD
DEVIATION               1.13578
STANDARD ERROR          0.65574
GROUP B                                        11.40000
                                               12.90000
                                               12.70000

NO. OBSERVATIONS        3
MEAN                    12.33333
STANDARD
DEVIATION               0.81445
STANDARD ERROR          0.47022
GROUP C                                         9.90000
                                               12.30000
                                               11.40000

NO. OBSERVATIONS        3
MEAN                    11.20000
STANDARD
DEVIATION               1.21244
STANDARD ERROR          0.70000
GROUP D                                        12.10000
                                               13.40000
                                               12.90000

NO. OBSERVATIONS        3
MEAN                    12.80000
STANDARD
DEVIATION               0.65574
STANDARD ERROR          0.37859
```

ANOVA TABLE

SOURCE	SS	d.f.	MS	F	P
GROUP	5.20000	3	1.73333	1.79931	0.22517
ERROR	7.70667	8	0.96333		
TOTAL	12.90667	11			

15.19   Use the data of Exercise 15.18 to conduct an analysis of variance by using a computer program available to you. Compare your results to those in the output of Exercise 15.18.

15.20   Because of the loss in efficiency caused by breakdowns in machinery, production records of each machine in a manufacturing plant must be closely monitored to try to anticipate when equipment is run-down and in need of repair. The data in the accompanying table give the production records for four different machines based on the outputs (in hundreds of pounds) from random samples of five shifts over the past week.

	Machine			
	1	**2**	**3**	**4**
	26.2	20.6	30.7	32.1
	32.0	26.4	35.2	34.7
	34.1	25.1	36.3	35.5
	33.6	24.9	31.9	36.8
	35.6	24.3	30.4	33.3
Sample mean	32.30	24.26	32.90	34.48
Sample variance	13.28	4.77	7.24	3.38

(a) Perform an analysis of variance to determine whether the mean outputs differ for the four machines. Use $\alpha = .05$.

(b) Which machine appears to be less productive?

15.21 Refer to the data of Exercise 15.20.

(a) Construct a 95% confidence interval for $(\mu_4 - \mu_2)$.

(b) Construct a 95% confidence interval for $(\mu_4 - \mu_1)$.

15.22 Sustained-release drug products are now marketed by many pharmaceutical companies. Even though single-dose, nonsustained-release capsules usually get more drug into the bloodstream quicker, sustained-release preparations supposedly achieve and maintain a more even level of release of the drug product over a longer period of time. Consider a study to compare three different drug products, A, B, and C. Equivalent doses of the three preparations were placed in mixtures of fluids similar to the gastric juices of the stomach and observed until 50% of the drug product was released from the capsule formulation. These release times (in minutes) appear in the accompanying table.

	Preparation		
	A	**B**	**C**
	15	38	19
	24	33	21
	20	39	27
	16	31	22
	18	26	24
	19	29	18
Sample mean	18.67	32.67	21.83

(a) Which preparation would you suspect is the sustained-release formulation?

(b) Perform an analysis of variance to test the equality of the population mean release times for the three preparations. Use $\alpha = .05$. Summarize your findings in an ANOVA table.

15.23 Students in a class in environmental engineering were assigned the project of comparing the mean dissolved oxygen contents from samples drawn at four different locations of a lake. The four locations were the center of the lake, the north and south edges, and a spot midway between the center and the east side of the lake. At each location five different vial samples were drawn and analyzed for dissolved oxygen content (in parts per million). Use the accompanying data to determine if there is sufficient evidence to indicate a difference in the mean dissolved oxygen contents for the four locations. Use $\alpha = .05$.

	Location			
	1	**2**	**3**	**4**
	4.6	6.7	6.4	5.8
	4.8	6.2	6.3	5.3
	4.3	6.4	6.6	5.7
	4.9	6.5	6.7	5.2
	4.7	6.3	6.5	5.0
Sample mean	4.66	6.42	6.50	5.40
Sample variance	.05	.04	.03	.12

15.24 A computer printout for the data of Exercise 15.23 is shown here. Compare the computer results to those you obtained in Exercise 15.23.

```
ONE WAY ANOVA
LOCATION 1                          4.60000
                                    4.80000
                                    4.30000
                                    4.90000
                                    4.70000


NO. OBSERVATIONS    5
MEAN                4.66000
STANDARD
DEVIATION           0.23022
STANDARD ERROR      0.10296
LOCATION 2                          6.70000
                                    6.20000
                                    6.40000
                                    6.50000
                                    6.30000
```

```
NO. OBSERVATIONS      5
MEAN                  6.42000
STANDARD
DEVIATION             0.19235
STANDARD ERROR        0.08602
LOCATION 3                              6.40000
                                       6.30000
                                       6.60000
                                       6.70000
                                       6.50000

NO. OBSERVATIONS      5
MEAN                  6.50000
STANDARD
DEVIATION             0.15811
STANDARD ERROR        0.07071
LOCATION 4                              5.80000
                                       5.30000
                                       5.70000
                                       5.20000
                                       5.00000

NO. OBSERVATIONS      5
MEAN                  5.40000
STANDARD
DEVIATION             0.33912
STANDARD ERROR        0.15166

ANOVA TABLE
SOURCE        SS        d.f.      MS          F           P
GROUP      11.60950      3      3.86983    67.30145    0.00000
ERROR       0.92000     16      0.05750
TOTAL      12.52950     19
```

15.25 A survey was conducted to examine the change in the prices (over the last month) of items included in a typical market basket. Six grocery stores in each of four geographic locations were sampled. The data shown in the accompanying table correspond to the increases in the price of lettuce over the past month for the sampled stores. Use the data to perform an analysis of variance. Draw appropriate conclusions.

	Geographic location		
1	**2**	**3**	**4**
10.1	15.3	11.8	16.8
11.3	14.8	12.6	9.2
8.2	10.4	14.2	17.5
8.7	9.3	13.9	18.2
12.1	10.7	8.9	10.9
10.4	15.6	7.5	14.5

(a) Give the sample mean and standard deviation for lettuce-price increases at each of the four locations.

(b) Identify all parts of a statistical test concerning the equality of the four population means.

(c) State conclusions, using the results of an analysis of variance.

(d) Which means appear to be different?

15.26 Construct a 99% confidence interval for $(\mu_4 - \mu_1)$ for the data of Exercise 15.25.

15.27 Use a computer program available to you to perform an analysis of variance for the data of Exercise 15.22. Compare the computer results to those you obtained by hand in Exercise 15.22.

15.28 **Class Exercise** Every consumer has been faced with the problem of assessing value when either buying or selling goods or services. Frequently, when a major purchase or sale is contemplated, one can enlist the services of a professional appraiser to help assess value. For example, in making arrangements to move furniture and belongings from an apartment or home to another location, one could enlist (purchase) the services of a moving company. A standard procedure is to solicit appraisals from two or three companies before reaching a decision. Similarly, in trying to sell an automobile to a used car dealer, each dealer visited would give an appraisal as to the worth of the car. For those who have had occasion to observe different appraisers, it is interesting to note the difference in the values assigned to the goods or the services. Some appraisers may tend to overvalue worth, others may tend to undervalue worth, and still others may tend to vary up and down with no apparent pattern.

Set up an experiment to compare the mean appraised values of three or more appraisers of a specific item. For example, to illustrate differences in assessing worth, you could choose three bicycle shops in town. With the help of other persons in your class, take the same five bicycles (at different times) to the bicycle shops for an assessment of worth. Try to enlist the services of the same appraiser from each shop for all bicycles. As another example, you may wish to work with three or more real estate agents on multiple listings to obtain appraisals on the same houses. For the experiment you choose, answer the following items.

(a) Perform an analysis of variance to compare the mean appraised value. Use $\alpha = .05$.

(b) Do any of the appraisers appear to overestimate (underestimate)?

(c) Which appraisers appear to agree?

15.29 A supermarket manager was interested in the effect of location in his market on the sales of Diet Coke. To measure the effect, displays of Diet Coke were located near the front entrance for one six-week period, near the dairy and meat sections in the rear of the market for another six-week period, and on a shelf along with other soft drinks for another six-week period. The number of six-packs sold per week in the three locations over the six-week periods are shown here.

Near the front entrance	Near the meat counter	On the soft drink shelf
38	30	25
44	41	18
58	43	26
51	48	30
43	43	27
54	40	31

Perform an analysis of variance test with $\alpha = .01$. What can you conclude?

15.30 At a large university in the midwest, beginning statistics was taught in a large lecture section with 250 students seated in rows A through L. In an effort to assess the row effect on student performance, random samples were taken of students in rows A, D, H, and L, and the scores noted as follows:

Row A	Row D	Row H	Row L
58	60	48	52
47	55	44	49
53	49	36	17
50	51	43	32
64	42	23	28
48	36	26	23

Perform an analysis of variance to determine whether the mean scores for the students are related to seating locations in mass lecture classes. Set $\alpha = .05$. Does the performance of students in any particular row seem to be inferior?

15.31 Ten firms were selected from four different sales-volume categories to see if the ratios of cash flow to debt were the same for the different categories. The sample data are represented in the following table.

	Small	Medium	Large	Very large
Number	10	10	10	10
Mean cash flow to debt ratio	.245	.259	.274	.201
Standard deviation	.0418	.0229	.0317	.0239

Perform an analysis of variance test for equality of means using $\alpha = .05$. What conclusions can be drawn?

15.32 An experiment similar to that described in Exercise 15.14 was conducted to compare the tar content of five different brands of cigarettes. Sample data are shown here:

Brand	$\bar{x}$ (mg)	s	n_i
1	10.6	2.4	10
2	11.2	2.5	10
3	10.5	1.8	10
4	11.8	1.9	10
5	13.7	2.2	10

(a) Based on your intuition, is there evidence to indicate any differences among the mean contents of the five brands?

(b) Run an analysis of variance to confirm or reject your conclusion of part (a).

15.33 The number of units of production was recorded for a random sample of ten hourly periods from the three bottling assembly lines of a plant. These data are shown here:

Assembly Line		
1	**2**	**3**
290	258	249
265	276	257
286	277	264
275	243	266
288	248	278
250	259	273
279	265	281
294	282	254
285	275	261
293	268	265

(a) Plot the data separately for each line. Are there any obvious differences?

(b) Identify the means and standard deviations for the three lines using the output shown here.

(c) Use the computer output shown here to construct an analysis of variance table. Draw conclusions based on the analysis of variance.

```
LISTING OF THE DATA

OBS    LINE    UNITS

 1      1      290
 2      1      265
 3      1      286
 4      1      275
 5      1      288
 6      1      250
 7      1      279              MEANS BY LINE
 8      1      294              Analysis Variable : Units
 9      1      285              --------------- LINE=1 --------------
10      1      293                                       Standard
11      2      258              N Obs   N        Mean    Deviation
12      2      276              -------------------------------------
13      2      277                 10  10   280.5000000   13.8984412
14      2      243
15      2      248              --------------- LINE=2 --------------
16      2      259                                       Standard
17      2      265              N Obs   N        Mean    Deviation
18      2      282              -------------------------------------
19      2      275                 10  10   265.1000000   12.9995726
20      2      268
21      3      249              --------------- LINE=3 --------------
22      3      257                                       Standard
23      3      264              N Obs   N        Mean    Deviation
24      3      266              -------------------------------------
25      3      278                 10  10   264.8000000   10.2610374
26      3      273              -------------------------------------
27      3      281
28      3      254
29      3      261
30      3      265
        N = 30
```

Analysis of Variance Procedure

Dependent Variable: UNITS

Source	DF	Sum of Squares	Mean Square	F Value	Pr > F
Model	2	1612.466667	806.233333	5.17	0.0125
Error	27	4207.000000	155.814815		
Corrected Total	29	5819.466667			

R-Square	C.V.	Root MSE	UNITS Mean
0.277082	4.620896	12.48258	270.133333

Source	DF	Anova SS	Mean Square	F Value	Pr > F
LINE	2	1612.466667	806.233333	5.17	0.0125

15.34   An experiment was conducted to compare the number of major defectives observed along each of five production lines in which changes were being instituted. Production was monitored continuously during the period of changes, and the number of major defectives was recorded per day for each line. These data are shown here.

Production Line

1	2	3	4	5
34	54	75	44	80
44	41	62	43	52
32	38	45	30	41
36	32	10	32	35
51	56	68	55	58

Compute $\bar{x}$ and s^2 for each sample. Does there appear to be a problem with nonconstant variances?

15.35 The yields of corn, in bushels per plot, were recorded for four different varieties of corn, A, B, C, and D. In a controlled greenhouse environment, each variety was randomly assigned to 8 of 32 plots available for the study. The yields are listed here.

A	2.5	3.6	2.8	2.7	3.1	3.4	2.9	3.5
B	3.6	3.9	4.1	4.3	2.9	3.5	3.8	3.7
C	4.3	4.4	4.5	4.1	3.5	3.4	3.2	4.6
D	2.8	2.9	3.1	2.4	3.2	2.5	3.6	2.7

Perform an analysis of variance on these data and draw your conclusions. Use $\alpha = .05$.

15.36 An experiment was conducted to test the effects of five different diets in turkeys. Six turkeys were randomly assigned to each of the five diet groups and were fed for a fixed period of time.

Group	Weight gained (lb)
Control diet	4.1, 3.3, 3.1, 4.2, 3.6, 4.4
Control diet + level 1 of additive A	5.2, 4.8, 4.5, 6.8, 5.5, 6.2
Control diet + level 2 of additive A	6.3, 6.5, 7.2, 7.4, 7.8, 6.7
Control diet + level 1 of additive B	6.5, 6.8, 7.3, 7.5, 6.9, 7.0
Control diet + level 2 of additive B	9.5, 9.6, 9.2, 9.1, 9.8, 9.1

(a) Plot the data separately for each sample
(b) Compute $\bar{x}_i$ and s_i for each sample.
(c) Is there any evidence of unequal variances? If not, run an analysis of variance and draw conclusions.

EXERCISES FROM THE DATA BASE

15.37   In previous chapters we have made pairwise comparisons among the means of the four treatment groups from the clinical trials data base (Appendix 1). Now an analysis of variance can be used to compare all four at once. Use the sample data on the HAM-D total score to compare the four treatment groups with an analysis of variance. Construct an ANOVA table to summarize the results of your analysis. Which means (if any) appear to differ?

15.38   Repeat the work of Exercise 15.37 using the Hopkins OBRIST cluster total. Do you get the same results?

15.39   "Baseline" comparisons are run in most clinical trials to examine the comparability of the treatment groups *prior* to receiving the assigned medication. Why might this be important? What baseline comparison could we make for the clinical trials data base?

15.40   Run an analysis of variance to compare the ages of the patients in the four treatment groups at the start of the study. Are the four groups comparable with respect to age? Summarize your results in an ANOVA table.

15.41   Compare the number of tablets or capsules taken for the four treatment groups using an analysis of variance. Did patients in the four groups appear to take approximately the same number of tablets? Explain.

CHAPTER 16

ANALYZING COUNT DATA

16.1 Introduction ▪ **16.2** The chi-square goodness-of-fit test ▪ **16.3** The chi-square test of independence ▪ **16.4** Using computers ▪ Summary ▪ Key terms ▪ Key formulas ▪ Supplementary exercises ▪ Exercises from the data base

---

16.1 INTRODUCTION

count data

Many experiments, particularly in the social sciences, yield **count data** (data that can be counted). For instance, the classification of people into five income brackets results in an enumeration or count corresponding to the number of persons classified in each of the five income brackets. Or we might be interested in studying the reaction of a mouse to a particular stimulus in a psychological experiment. If a mouse will react in one of three ways when a particular stimulus is applied and if a large number of mice are subjected to the stimulus, the experiment will yield a count for each category, indicating the number of mice that fall into it. Similarly, a traffic study might require a count and a classification of the type of motor vehicles using a particular section of highway. An industrial process manufactures items that fall into one of three quality classes: acceptables, seconds, and rejects. A student of the arts might classify paintings in one of k categories, according to style and period, in order to study trends in style over time. We might wish to classify ideas in a philosophical study or in the field of literature. An advertising campaign would yield classifications of consumer reaction. Indeed, many observations cannot be measured on a continuous scale and hence result in count data.

The preceding examples are only a few of the many types of problems that involve count, or enumerative, data. Many such problems are analyzed by means of a chi-square statistic developed in the early 1900s by Karl Pearson. We will illustrate the use of this method for goodness-of-fit tests and for an important group of problems involving the investigation of the dependence (or independence) between two methods of data classification. Other applications are discussed in Conover (1980) and Hollander and Wolfe (1973).

16.2   THE CHI-SQUARE GOODNESS-OF-FIT TEST

**multinomial
experiment**

The illustrations we just presented exhibit, to a reasonable degree of approximation, the following characteristics, which define a **multinomial experiment:**

> ### The Multinomial Experiment
> 1. The experiment consists of n identical trials.
> 2. The outcome of each trial falls into one of k classes, or *cells.*
> 3. The probability that the outcome of a single trial will fall in a particular cell, say cell i, is π_i ($i = 1, 2, \ldots, k$) and remains the same from trial to trial. Note that
>
> $$\pi_1 + \pi_2 + \pi_3 + \cdots + \pi_k = 1$$
>
> 4. The trials are independent.
> 5. The experimenter is interested in $n_1, n_2, n_3, \ldots, n_k$, where n_i ($i = 1, 2, \ldots, k$) is equal to the number of trials in which the outcome falls in cell i. Note that
>
> $$n_1 + n_2 + n_3 + \cdots + n_k = n$$

A multinomial experiment is analogous to tossing n balls at k boxes, where each ball must fall in one of the boxes. The boxes are arranged so that the probability that a ball will fall in a box varies from box to box but remains the same for a particular box in repeated tosses. Finally, the balls are tossed in such a way that the trials are independent. At the conclusion of the experiment, we observe n_1 balls in the first box, n_2 in the second, and n_k in the kth. The total number of balls is equal to

$$\Sigma \, n_i = n$$

Note that there is a similarity between the binomial and multinomial experiments and, in particular, that the binomial experiment represents the special case for the multinomial experiment when $k = 2$, where $\pi_1 = \pi$ and $\pi_2 = 1 - \pi$. Working with the multinomial experiment, we will make inferences about the k parameters $\pi_1, \pi_2, \ldots, \pi_k$. In this chapter, inferences about $\pi_1, \pi_2, \ldots, \pi_k$ will be expressed in terms of a statistical test of a hypothesis about their specific numerical values or their relationship to one another.

Suppose we knew the probability π_1 that a ball lands in cell 1 is .1. Then if we toss $n = 100$ balls at the cells, we would expect $100(.1) = 10$ balls to land in cell 1. Indeed, the expected number in cell i after n trials is $n\pi_i$.

Definition 16.1

expected number of outcomes of type *i*

In a multinomial experiment where each trial can result in one of k outcomes, the **expected number of outcomes of type i** in n trials is $n\pi_i$, where π_i is the probability that a given trial results in outcome i.

In 1900 Karl Pearson, a statistician, proposed the following test statistic, which is a function of the squares of the deviations of the observed cell counts from their expected value. If we let O denote the observed cell count and E the expected count, Pearson's statistic can be written as

$$\chi^2 = \sum \frac{(O - E)^2}{E}$$

Suppose that we hypothesize values for the cell probabilities, $\pi_1, \pi_2, \ldots, \pi_k$ and calculate the expected cell counts using Definition 16.1 to examine how well the data fit or agree with the hypothesized cell probabilities. Certainly if our hypothesized value of each π is true, the cell counts should not deviate greatly from the expected cell counts. Large values of chi-square would imply rejection of a null hypothesis. How large is large? The answer to this question can be found by examining the sampling distribution of chi-square.

The quantity chi-square possesses approximately a chi-square distribution in repeated sampling when n is large. For this approximation to be good, it is desirable that the expected number falling into each cell be five or more. The chi-square distribution was first introduced in Chapter 11 as part of our discussions of inferences about σ^2 (and σ). Recall that chi-square is not a symmetrical distribution and that there are many chi-square distributions; we obtain a particular one by specifying the degrees of freedom (df) for the distribution.

The chi-square test for k specified cell probabilities, often called a *chi-square goodness-of-fit test,* is based on $k - 1$ df. Upper-tail values of

$$\chi^2 = \sum \frac{(O - E)^2}{E}$$

are shown in Table 4 in Appendix 3. For example, a chi-square distribution with df $= 14$ has an area a $= .10$ to the right (above) the value 21.0642 (see Figure 16.1). The rejection region for the one-tailed test con-

FIGURE 16.1   Tabulated Value of the Chi-Square Distribution; a = .10 and df = 14

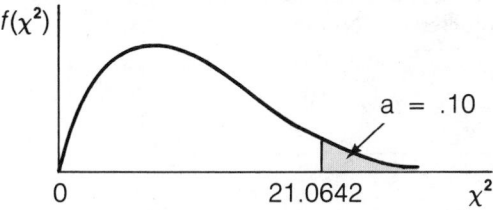

cerning k cell probabilities can be determined for a specified value of a type I error α using Table 4 in Appendix 3. If the observed value of chi-square falls in the rejection region, we reject the null hypothesis that specifies the k cell probabilities. We summarize the test procedure here.

Chi-Square Goodness-of-Fit Test

Null hypothesis: Each of the k cell probabilities is specified.

Alt. hypothesis: At least one of the cell probabilities differs from its hypothesized value.

Test statistic: $\chi^2 = \sum \dfrac{(O - E)^2}{E}$

Rejection region: Reject the null hypothesis if chi-square exceeds the tabulated value for df = $k - 1$ and a = α.

There are two assumptions that we must make in order to run this test. The first is that the observed cell counts satisfy the properties of a multinomial experiment, and the second is that the expected cell counts must be five or more.

Example 16.1

Several researchers studied a heroin epidemic in the San Francisco Bay area. A sample of 102 drug users was interviewed and each subject was asked to name the kind of drug that each first injected. These data are listed in Table 16.1. Because previous data were inadequate, it was hypothesized that the proportions of subjects reporting "type of first drug injected" would be $\frac{1}{3}$ for heroin, $\frac{1}{3}$ for speed, and $\frac{1}{3}$ for all others. Test the hypothesis that the cell probabilities for type of drug first injected do not differ from $\frac{1}{3}$. Use $\alpha = .05$.

TABLE 16.1   Type of Drug First Injected by 102 Subjects During a Heroin Epidemic

First drug injected	Number
Heroin	42
Speed	36
Other	24
Total	102

Solution   This experiment possesses the characteristics of a multinomial experiment with $n = 102$ trials and $k = 3$ outcomes:

Outcome 1: First drug injected was heroin, with probability π_1.
Outcome 2: First drug injected was speed, with probability π_2.
Outcome 3: First drug injected was other than heroin or speed, with probability π_3.

The null hypothesis is:

$$H_0: \pi_1 = \pi_2 = \pi_3 = 1/3$$

which is to be tested against the research hypothesis that at least one of these probabilities differs from ⅓. The observed and expected cell counts are given in Table 16.2. Note that the expected cell counts all exceed five and that the observed cell counts differ from the expected cell counts.

TABLE 16.2   Observed and Expected Cell Counts for Example 16.1

Drug first injected	Observed	Expected
Heroin	42	$102(⅓) = 34$
Speed	36	$102(⅓) = 34$
Other	24	$102(⅓) = 34$

The test statistic is

$$\chi^2 = \sum \frac{(O - E)^2}{E}$$
$$= \frac{(42 - 34)^2}{34} + \frac{(36 - 34)^2}{34} + \frac{(24 - 34)^2}{34} = 4.94$$

The tabulated value of chi-square for a $= .05$, with $k - 1 = 3 - 1 = 2$ df, is 5.99. Since 4.94 does not exceed 5.99, we have insufficient evidence to indicate that the cell probabilities differ from $\frac{1}{3}$.

EXERCISES

16.1 In a study of 60 premarital pregnancies, a research team reported 30 such pregnancies in the lower social class, 20 in the middle social class, and 10 in the higher social class. Under the null hypothesis that unwanted pregnancies are equally likely in all three social classes, use the sample data and a chi-square goodness-of-fit test for this hypothesis. State all parts of your test and draw a conclusion based on $\alpha = .05$.

16.2 A sample of 125 securities analysts was obtained, and each analyst was asked to select four stocks on the New York Stock Exchange that were expected to outperform the Standard and Poor's Index over a 3-month period. One theory suggests that the securities analysts would be expected to do no better than chance and that the number of correct guesses from the four selected had a multinomial distribution as shown here.

Number correct	0	1	2	3	4
Multinomial probabilities (π_i)	.0625	.2500	.3750	.2500	.0625

If the number of correct guesses from the sample of 125 analysts had a frequency distribution as shown here, use these data to conduct a chi-square goodness-of-fit test. Use $\alpha = .05$. Draw conclusions.

Number correct	0	1	2	3	4
Frequency	3	23	51	39	9

16.3 Refer to Exercise 16.2. Suppose the assumed multinomial probabilities were all .20 (i.e., all $\pi_i = .20$).

(a) How would your conclusions change?

(b) Can you suggest a problem with the chi-square goodness-of-fit test based on the results of part (a)?

16.3   THE CHI-SQUARE TEST OF INDEPENDENCE

A problem frequently encountered in the analysis of count data concerns the interdependence of two methods of classification of observed events. For example, a physician in a clinic might want to classify patients suffering from emphysema according to one of two drug products (either A or B) prescribed for their treatment and then according to the patient's response to the medication (greatly improved, improved, no change). It would be important for the physician to determine if the proportion of patients who show various degrees of improvement when treated by drug A is different from the proportion that show improvement when treated by drug B. Thus the physician must investigate a contingency (dependence) between the two methods of classification "drug type" and "patient condition."

We illustrate a test of the dependence of two methods of classification with the following example.

A total of 210 emphysema patients ($n = 210$) entering a clinic over a one-year period were treated with one of two drugs (either the standard drug A or an experimental drug B) for a period of one week. After this period of time each patient's condition was rated as either greatly improved, improved, or no change. The number of patients appearing in each category is presented in Table 16.3. This two-way classification of the data is called a **contingency table.** The physicians in the clinic want to determine whether a patient's rating is dependent on the drug product that was used in his or her treatment. If the two classifications are dependent, this would imply that the proportions of patients classified as greatly improved, improved, or no change are different for the two drugs.

contingency table

TABLE 16.3   Results of a One-Year Emphysema Study

Drug product	Patient condition			Total
	No change	Improved	Greatly improved	
A (standard)	20	35	45	100
B (experimental)	15	45	50	110
Total	35	80	95	210

The null hypothesis for our test is that the two methods of classification are independent—or, in other words, that the two drug products are equally effective (ineffective) for treating emphysema patients. A test of the null hypothesis of independence of the drug and the condition classifications makes use of the **observed cell counts** and the **expected cell counts.** The expected cell count is the number of observations we would expect to be classified in a cell if the null hypothesis of independence of the two methods of classification is true. Without proof we state that the expected cell count for a particular cell is computed by multiplying the row total by the column total and dividing by the total number of sample measurements:

$$\text{expected cell count} = \frac{(\text{row total})(\text{column total})}{n}$$

These calculations are given below for the data of Table 16.3.

Expected cell count for drug A, no change $= \dfrac{(100)\,(35)}{210} = 16.67$

Expected cell count for drug A, improved $= \dfrac{(100)\,(80)}{210} = 38.10$

Expected cell count for drug A, greatly improved

$$= \frac{(100)\,(95)}{210} = 45.23$$

Expected cell count for drug B, no change $= \dfrac{(110)\,(35)}{210} = 18.33$

Expected cell count for drug B, improved $= \dfrac{(110)\,(80)}{210} = 41.90$

Expected cell count for drug B, greatly improved

$$= \frac{(110)\,(95)}{210} = 49.77$$

The observed cell counts and the expected cell counts (the numbers in parentheses) are shown in Table 16.4 for the data of Table 16.3.

TABLE 16.4 Observed and Expected Cell Counts

	Patient condition			
Drug	**No change**	**Improved**	**Greatly improved**	**Total**
A	20(16.67)	35(38.10)	45(45.23)	100
B	15(18.33)	45(41.90)	50(49.77)	110
Total	35	80	95	210

Note in Table 16.4 that the expected cell counts in a given row may be summed to obtain the corresponding row total. For example, in the first row of Table 16.4,

$$16.67 + 38.10 + 45.23 = 100$$

additive property of expected cell counts

Similarly, for any column, the corresponding expected cell counts may be summed to obtain the appropriate column total. This **additive property of the expected cell counts** can reduce the computational labor involved in obtaining expected cell counts. Thus for Table 16.4 we would really only need to compute the two entries shown below.

	Patient condition			
Drug	**No change**	**Improved**	**Greatly improved**	**Total**
A	16.67	38.10		100
B				110
Total	**35**	**80**	**95**	**210**

The remaining expected cell counts can be obtained by subtracting from the appropriate row or column total.

We would suspect that the null hypothesis is false if the observed cell counts differ appreciably from the expected cell counts. To measure this agreement or disagreement, we compute a quantity known as the **chi-square statistic,** whose formula is

chi-square statistic

$$\chi^2 = \sum \frac{(O - E)^2}{E}$$

If the observed cell counts differ appreciably from the expected cell counts, the quantities $(O - E)^2$ will be large, and hence χ^2 will be large. Thus we will reject our hypothesis of independence of the two methods of classification for large values of χ^2. How large is large? The answer to this question can be found by obtaining the sampling distribution of the quantity χ^2.

It can be shown that when n is large, χ^2 has approximately a chi-square probability distribution with df $= (r - 1)(c - 1)$, where r is the number of categories of one variable and c is the number of categories of the second variable. Some researchers recommend that the sample size be large enough so that all expected cell counts are 5 or more. This requirement is probably too stringent. We will require that no expected cell count be less than 1 and that no more than 20% of them be less than 5.

The rejection region for the one-tailed test of independence can then be determined for a specified probability of a type I error, α, using

Table 4 of Appendix 3. If the observed value of χ^2 falls beyond the tabulated value, we reject the null hypothesis of independence of the two classifications. We summarize this test procedure next.

Chi-Square Test of Independence

Null hypothesis: The two classifications are independent.

Alt. hypothesis: The two classifications are dependent.

Test statistic: $\chi^2 = \sum \dfrac{(O - E)^2}{E}$

Rejection region: Reject the null hypothesis if χ^2 exceeds the tabulated value of χ^2 for a $= \alpha$ and df $= (r - 1)(c - 1)$.

Note: This test will always be a one-tailed, upper-tailed test.

Example 16.2

Use the sample count data in Table 16.4 to test the hypothesis of independence of the drug and the condition classifications of emphysema patients. In other words, we wish to test the null hypothesis that the two drug products are equally effective (ineffective) for treating persons suffering from emphysema. Set $\alpha = .05$.

Solution   The observed and expected cell counts are presented in Table 16.4. Note that the expected cell counts satisfy the criteria that no more than 20% are less than 5 and that none is less than 1. The value of the test statistic chi-square will be computed and compared with the tabulated value of the chi-square distribution possessing $(r - 1)(c - 1) = (1)(2) = 2$ df and for $\alpha = .05$. To find the value of the test statistic, we use the formula

$$\chi^2 = \sum \frac{(O - E)^2}{E}$$

Substituting, we obtain

$$\chi^2 = \frac{(20 - 16.67)^2}{16.67} + \frac{(35 - 38.10)^2}{38.10} + \cdots + \frac{(50 - 49.77)^2}{49.77} = 1.75$$

The rejection region for this test can be obtained from Table 4 of Appendix 3 for df $= 2$ and a $= .05$. This value is 5.99. Since the observed value of chi-square does not exceed 5.99, we conclude that there is insufficient evidence to reject the null hypothesis of independence. Thus it may not matter which drug is administered; the experimental and the standard drugs may be equally effective.

Example 16.3

A poll of 100 U.S. congressmen was taken to determine their opinions concerning a bill to raise the ceiling on our national debt. Each congressman was then classified according to political party affiliation and opinion on the policy. The survey results are listed in Table 16.5. At this point, neither the opponents nor the proponents can claim victory. Test the null hypothesis that these two classifications are independent of one another (i.e., congressmen do not hold opinions along party lines) against the alternative hypothesis that congressmen's opinions on the national debt bill are related to their political party affiliations. Use $\alpha = .10$.

TABLE 16.5 Congressional Opinion Survey Results

| | Opinion | | | |
| | | | | |
Party	Approve of bill	Do not approve of bill	No opinion yet	Total
Republican	28	14	5	47
Democrat	19	28	6	53
Total	47	42	11	100

Solution Before calculating chi-square, we must first obtain the expected cell counts, using the row and column totals of Table 16.5 and the additive property of the expected cell counts. The remaining expected cell counts shown in Table 16.6 were obtained using the additive property. Again, note that the expected cell counts satisfy the requirements that no more than 20% are less than five, and that none is less than one.

$$\text{Republican, approve} = \frac{(47)(47)}{100} = 22.09$$

$$\text{Republican, disapprove} = \frac{(47)(42)}{100} = 19.74$$

TABLE 16.6 *Expected Values for the Data of Example 16.3*

Party	Opinion			Total
	Approve of bill	**Do not approve of bill**	**No opinion yet**	
Republican	22.09	19.74	5.17	47
Democrat	24.91	22.26	5.83	53
Total	47	42	11	100

Now we compute

$$\chi^2 = \sum \frac{(O - E)^2}{E}$$

and obtain

$$\chi^2 = \frac{(28 - 22.09)^2}{22.09} + \frac{(14 - 19.74)^2}{19.74} + \cdots + \frac{(6 - 5.83)^2}{5.83} = 6.14$$

The rejection region for this test can be located by using Table 4 of Appendix 3, with $a = .10$ and df $= (r - 1)(c - 1) = (1)(2) = 2$. This tabulated value is 4.61. Since the observed value of chi-square exceeds the tabulated chi-square value, we reject the null hypothesis of independence of the classifications and conclude that congressmen do seem to hold opinions along party lines; 28 of the 47 (60%) of the Republicans approved of the bill, while only 19 of the 53 (36%) of the Democrats approved the bill.

The use of the chi-square probability distribution in analyzing enumerative data presented in a two-classification contingency table illustrates the analysis for only one of the many types of classification problems. Many other types of applications are more complicated; hence they will be omitted from this text. However, this short presentation on contingency tables has provided you with an adequate tool for evaluating and making inferences concerning count data that are summarized into two classifications. Frequently, sociological studies are summarized in two-classification contingency tables for publication in magazines or newspapers. Using what you have learned from this section, you will be in a position to determine whether two methods of classifying observed events are independent.

TABLE 16.7 Drug-Related Complaints by Sex and Marital Status

| | | | | | **Marital Status** | | | | | | | | | | |
| | **Divorced** | | | **Widowed** | | | **Married** | | | **Total** | | |
Complaint	Male	Female	Total	Male	Female	Total	Male	Female	Total	Male	Female	Total
Overdose	96	208	304	46	82	128	266	330	596	408	620	1028
Suicide	14	100	114	18	96	114	72	156	228	104	352	456
Psychiatric	38	24	62	22	10	32	116	36	152	176	70	246
Addiction	52	68	120	14	12	26	146	78	224	212	158	370
Total	200	400	600	100	200	300	600	600	1200	900	1200	2100

Sometimes we are interested in the relationship between two variables while "controlling" for one or more other, related variables. Consider the data of Table 16.7. Suppose we were interested in the relationship between the two variables "complaint" and "sex." If the third variable, "marital status," is also related to the other variables, we may get more information about the relationship between the two variables of interest by using the third variable (marital status) as a **control variable.** In this way we would examine the relationship between "complaint" and "sex" for each of the three measured levels of marital status.

control variable

partial tables

The data of Table 16.7 have been arranged in three **partial tables,** corresponding to the levels of the control variable, in Table 16.8. In examining the relationship between two variables in the presence of a control variable, we could first run separate chi-square tests of independence for each of the partial tables. These results would indicate whether the two variables of interest are independent at each of the indicated levels of the control variable. A combined test of independence of the two variables could be obtained by adding the values of chi-square computed for each of the partial tables. Here we would be testing the independence of the two variables while controlling the influence of a third variable on the two variables being studied. The degrees of freedom of the overall test would be the sum of the degrees of freedom for each of the partial tables (Table 16.8).

TABLE 16.8   Partial Tables for the Comparison of Drug-Related Complaints and Sex in the Presence of the Control Variable, Marital Status

| | Marital Status | | | | | | | | |
| | (a) Divorced | | | (b) Widowed | | | (c) Married | | |
Complaint	Male	Female	Total	Male	Female	Total	Male	Female	Total
Overdose	96	208	304	46	82	128	266	330	596
Suicide	14	100	114	18	96	114	72	156	228
Psychiatric	38	24	62	22	10	32	116	36	152
Addiction	52	68	120	14	12	26	146	78	224
Total	200	400	600	100	200	300	600	600	1200

Example 16.4

Use the data of Table 16.8 to test the independence of the two variables "complaint" and "sex," for each of the partial tables. Use $\alpha = .05$ for all tables and interpret your results.

Solution   Since we already know how to compute expected cell counts by using the formula

$$\frac{\text{(row total)(column total)}}{n}$$

we have listed the expected cell counts for each of the partial tables in Table 16.9.

TABLE 16.9   *Expected Cell Counts for the Partial Tables in Table 16.8*

| | Marital Status | | | | | | | | |
| | (a) Divorced | | | (b) Widowed | | | (c) Married | | |
Complaint	Male	Female	Total	Male	Female	Total	Male	Female	Total
Overdose	101.3	202.7	304	42.7	85.3	128	298	298	596
Suicide	38.0	76.0	114	38.0	76.0	114	114	114	228
Psychiatric	20.7	41.3	62	10.7	21.3	32	76	76	152
Addiction	40.0	80.0	120	8.6	17.4	26	112	112	224
Total	200	400	600	100	200	300	600	600	1200

You should check a few of the cell entries to convince yourself that these are correct.

Using Table 16.8 (a), we have the test procedure shown next.

Null hypothesis: For the divorced, the two variables are independent.

Alt. hypothesis: For the divorced, the two variables "complaint" and "sex" are dependent.

Test statistic:
$$\chi^2 = \sum \frac{(O - E)^2}{E}$$
$$= \frac{(96 - 101.3)^2}{101.3} + \frac{(208 - 202.7)^2}{202.7} + \cdots$$
$$+ \frac{(68 - 80)^2}{80}$$
$$= .2773 + .1386 + \cdots + 1.8000 = 50.2579$$

Rejection region: Based on the chi-square value of 7.81 from Table 4 of Appendix 3 with a = .05 and df = 3, we reject the null hypothesis and conclude that for divorced people the variables "drug-related com-

Rejection region:
(continued)

plaint" and "sex" are dependent. It appears that a higher percentage of females than males are classified in the suicide category. A clear picture of this trend is seen in Table 16.10, giving a percentage comparison of Table 16.8(a).

TABLE 16.10 *Percentage Comparison of Drug-Related Complaints by Sex, for the Divorced*

Complaint	Male	Female
Overdose	48%	52%
Suicide	7%	25%
Psychiatric	19%	6%
Addiction	26%	17%
Total	100%	100%
Sample Size	200	400

We can form the same research and null hypotheses for widowed people. The computed value of the test statistic is $\chi^2 = 39.1672$. Again comparing the computed value to 7.81 (a = .05, df = 3), we reject the null hypothesis of independence of the two variables for widowed people. As was seen for divorced people, a higher percentage of females than males have the drug-related complaint categorized as suicide. See the percentage comparison in Table 16.11. Note, however, that suicide percentages for the widowed are much higher for both males and females than for divorced people. In contrast the divorced have higher percentages for the addiction complaint.

TABLE 16.11 *Percentage Comparison of Drug-Related Complaints by Sex, for the Widowed*

Complaint	Male	Female
Overdose	46%	41%
Suicide	18%	48%
Psychiatric	22%	5%
Addiction	14%	6%
Total	100%	100%
Sample Size	100	200

Finally, for married people the computed value of chi-square is $\chi^2 = 100.5678$. Again, this result is highly significant. From the percentage comparison in Table 16.12 and the chi-square test of independence, we see that married females have higher percentages associated with the overdose and suicide complaints than males. Also note that the married percentages for males and females are very similar to the corresponding percentages for divorced, but the widowed percentages appear to be quite different from the percentages for divorced and married people.

TABLE 16.12 Percentage Comparison of Drug-Related
Complaints by Sex, for Married People

Complaint	Male	Female
Overdose	44%	55%
Suicide	12%	26%
Psychiatric	19%	6%
Addiction	24%	13%
Total	99%	100%
Sample Size	600	600

In summary, we have seen that sex and drug-related complaints at the hospital emergency room are related at each level of the control variable "marital status."

Example 16.5

Combine the separate chi-square tests for the partial tables in Table 16.9 to compute a pooled chi-square. List the parts of the statistical test and draw conclusions. Give the level of significance of the test.

Solution The four parts of the pooled test of independence are given here:

Null hypothesis: When controlling for marital status, the two variables are independent.

Alt. hypothesis: When controlling for marital status, the two variables "complaint" and "sex" are dependent.

Test statistic:	Chi-square equals the sum of the computed chi-square test statistics for the partial tables.
Rejection region:	Since we are asked to give the level of significance for our test, we do not specify a rejection region. The chi-square test statistic has df = 9, the sum of the degrees of freedom for the separate partial tables. Since $\chi^2 = 100.9928$ exceeds the a = .005 value of 23.589 in Table 4 of Appendix 3 for df = 9, the test is significant at the $p < .005$ level. Indeed, the data indicate a dependence of the two variables while controlling for marital status.

A word of caution. The pooled chi-square test just performed is *not* equivalent to—and will not necessarily lead to the same conclusion as—the chi-square test of independence between "sex" and "complaint" that *ignores* the control variable "marital status." The results for these two separate tests may lead to quite different conclusions. The pooled chi-square gives us a test of independence between two variables while controlling the influence of a third variable.

In contrast, to construct the chi-square test of independence that ignores the influence of a third variable, we form a two-way table (by summing over the cells of the third variable) and perform the chi-square test in the usual way.

16.4 USING COMPUTERS

Although the calculations required for a chi-square test of independence are not very difficult, as the number of rows and columns increases, they do become tedious. Minitab is especially easy to use, but SAS can also be used to conduct a chi-square test of independence. For Minitab, one enters the cell frequencies and obtains the test results; for SAS, if the data file has not been entered previously, the individual observations (rather than the cell frequencies) must be entered before PROC FREQ can be used to obtain the chi-square test.

Output from Minitab and SAS is shown here for the data of Table 16.3. These programs can be used as models for the analyses of other data sets. Additional output from Minitab is shown in Exercises 16.9, 16.14, 16.22, 16.26 and 16.27.

Minitab Output

```
MTB > READ INTO C1-C3
DATA> 20    35    45
DATA> 15    45    50
DATA> END

     2 ROWS READ
MTB > CHISQUARE C1-C3

Expected counts are printed below observed counts

             C1       C2       C3    Total
      1      20       35       45      100
           16.67    38.10    45.24

      2      15       45       50      110
           18.33    41.90    49.76

Total        35       80       95      210

ChiSq =  0.667 +  0.251 +  0.001 +
         0.606 +  0.229 +  0.001 = 1.755
df = 2

MTB > STOP
```

SAS Output

```
            RESULTS OF A ONE-YEAR EMPHYSEMA STUDY
                    LISTING OF THE DATA

OBS   PRODUCT      CONDITN        OBS   PRODUCT      CONDITN

  1      A      GREATLY IMPROVED   56      A          IMPROVED
  2      A      GREATLY IMPROVED   57      A          IMPROVED
  3      A      GREATLY IMPROVED   58      A          IMPROVED
  4      A      GREATLY IMPROVED   59      A          IMPROVED
  5      A      GREATLY IMPROVED   60      A          IMPROVED
  6      A      GREATLY IMPROVED   61      A          IMPROVED
  7      A      GREATLY IMPROVED   62      A          IMPROVED
  8      A      GREATLY IMPROVED   63      A          IMPROVED
  9      A      GREATLY IMPROVED   64      A          IMPROVED
 10      A      GREATLY IMPROVED   65      A          IMPROVED
 11      A      GREATLY IMPROVED   66      A          IMPROVED
 12      A      GREATLY IMPROVED   67      A          IMPROVED
 13      A      GREATLY IMPROVED   68      A          IMPROVED
 14      A      GREATLY IMPROVED   69      A          IMPROVED
 15      A      GREATLY IMPROVED   70      A          IMPROVED
 16      A      GREATLY IMPROVED   71      A          IMPROVED
 17      A      GREATLY IMPROVED   72      A          IMPROVED
 18      A      GREATLY IMPROVED   73      A          IMPROVED
 19      A      GREATLY IMPROVED   74      A          IMPROVED
 20      A      GREATLY IMPROVED   75      A          IMPROVED
 21      A      GREATLY IMPROVED   76      A          IMPROVED
```

(continued)

22	A	GREATLY IMPROVED	77	A	IMPROVED
23	A	GREATLY IMPROVED	78	A	IMPROVED
24	A	GREATLY IMPROVED	79	A	IMPROVED
25	A	GREATLY IMPROVED	80	A	IMPROVED
26	A	GREATLY IMPROVED	81	A	NO CHANGE
27	A	GREATLY IMPROVED	82	A	NO CHANGE
28	A	GREATLY IMPROVED	83	A	NO CHANGE
29	A	GREATLY IMPROVED	84	A	NO CHANGE
30	A	GREATLY IMPROVED	85	A	NO CHANGE
31	A	GREATLY IMPROVED	86	A	NO CHANGE
32	A	GREATLY IMPROVED	87	A	NO CHANGE
33	A	GREATLY IMPROVED	88	A	NO CHANGE
34	A	GREATLY IMPROVED	89	A	NO CHANGE
35	A	GREATLY IMPROVED	90	A	NO CHANGE
36	A	GREATLY IMPROVED	91	A	NO CHANGE
37	A	GREATLY IMPROVED	92	A	NO CHANGE
38	A	GREATLY IMPROVED	93	A	NO CHANGE
39	A	GREATLY IMPROVED	94	A	NO CHANGE
40	A	GREATLY IMPROVED	95	A	NO CHANGE
41	A	GREATLY IMPROVED	96	A	NO CHANGE
42	A	GREATLY IMPROVED	97	A	NO CHANGE
43	A	GREATLY IMPROVED	98	A	NO CHANGE
44	A	GREATLY IMPROVED	99	A	NO CHANGE
45	A	GREATLY IMPROVED	100	A	NO CHANGE
46	A	IMPROVED	101	B	GREATLY IMPROVED
47	A	IMPROVED	102	B	GREATLY IMPROVED
48	A	IMPROVED	103	B	GREATLY IMPROVED
49	A	IMPROVED	104	B	GREATLY IMPROVED
50	A	IMPROVED	105	B	GREATLY IMPROVED
51	A	IMPROVED	106	B	GREATLY IMPROVED
52	A	IMPROVED	107	B	GREATLY IMPROVED
53	A	IMPROVED	108	B	GREATLY IMPROVED
54	A	IMPROVED	109	B	GREATLY IMPROVED
55	A	IMPROVED	110	B	GREATLY IMPROVED

RESULTS OF A ONE-YEAR EMPHYSEMA STUDY
LISTING OF THE DATA

OBS	PRODUCT	CONDITN	OBS	PRODUCT	CONDITN
111	B	GREATLY IMPROVED	166	B	IMPROVED
112	B	GREATLY IMPROVED	167	B	IMPROVED
113	B	GREATLY IMPROVED	168	B	IMPROVED
114	B	GREATLY IMPROVED	169	B	IMPROVED
115	B	GREATLY IMPROVED	170	B	IMPROVED
116	B	GREATLY IMPROVED	171	B	IMPROVED
117	B	GREATLY IMPROVED	172	B	IMPROVED
118	B	GREATLY IMPROVED	173	B	IMPROVED
119	B	GREATLY IMPROVED	174	B	IMPROVED
120	B	GREATLY IMPROVED	175	B	IMPROVED
121	B	GREATLY IMPROVED	176	B	IMPROVED
122	B	GREATLY IMPROVED	177	B	IMPROVED
123	B	GREATLY IMPROVED	178	B	IMPROVED
124	B	GREATLY IMPROVED	179	B	IMPROVED
125	B	GREATLY IMPROVED	180	B	IMPROVED
126	B	GREATLY IMPROVED	181	B	IMPROVED

(continued)

```
127    B    GREATLY IMPROVED    182    B    IMPROVED
128    B    GREATLY IMPROVED    183    B    IMPROVED
129    B    GREATLY IMPROVED    184    B    IMPROVED
130    B    GREATLY IMPROVED    185    B    IMPROVED
131    B    GREATLY IMPROVED    186    B    IMPROVED
132    B    GREATLY IMPROVED    187    B    IMPROVED
133    B    GREATLY IMPROVED    188    B    IMPROVED
134    B    GREATLY IMPROVED    189    B    IMPROVED
135    B    GREATLY IMPROVED    190    B    IMPROVED
136    B    GREATLY IMPROVED    191    B    IMPROVED
137    B    GREATLY IMPROVED    192    B    IMPROVED
138    B    GREATLY IMPROVED    193    B    IMPROVED
139    B    GREATLY IMPROVED    194    B    IMPROVED
140    B    GREATLY IMPROVED    195    B    IMPROVED
141    B    GREATLY IMPROVED    196    B    NO CHANGE
142    B    GREATLY IMPROVED    197    B    NO CHANGE
143    B    GREATLY IMPROVED    198    B    NO CHANGE
144    B    GREATLY IMPROVED    199    B    NO CHANGE
145    B    GREATLY IMPROVED    200    B    NO CHANGE
146    B    GREATLY IMPROVED    201    B    NO CHANGE
147    B    GREATLY IMPROVED    202    B    NO CHANGE
148    B    GREATLY IMPROVED    203    B    NO CHANGE
149    B    GREATLY IMPROVED    204    B    NO CHANGE
150    B    GREATLY IMPROVED    205    B    NO CHANGE
151    B         IMPROVED       206    B    NO CHANGE
152    B         IMPROVED       207    B    NO CHANGE
153    B         IMPROVED       208    B    NO CHANGE
154    B         IMPROVED       209    B    NO CHANGE
155    B         IMPROVED       210    B    NO CHANGE
156    B         IMPROVED
157    B         IMPROVED              N = 210
158    B         IMPROVED
159    B         IMPROVED
160    B         IMPROVED
161    B         IMPROVED
162    B         IMPROVED
163    B         IMPROVED
164    B         IMPROVED
165    B         IMPROVED
```

```
              RESULTS OF A ONE-YEAR EMPHYSEMA STUDY
          EXAMPLE OF PROC FREQ WITH THE CHI-SQUARE OPTION

                   TABLE OF PRODUCT BY CONDITN

          PRODUCT      CONDITN(PATIENT CONDITION)
```

Frequency	GREATLY IMPROVED	IMPROVED	NO CHANGE	Total
A	45	35	20	100
B	50	45	15	110
Total	95	80	35	210

(continued)

```
        STATISTICS FOR TABLE OF PRODUCT BY CONDITN

    Statistic                        DF      Value      Prob
    -----------------------------------------------------------
    Chi-Square                        2      1.755      0.416
    Likelihood Ratio Chi-Square       2      1.757      0.415
    Mantel-Haenszel Chi-Square        1      0.451      0.502
    Phi Coefficient                          0.091
    Contingency Coefficient                  0.091
    Cramer's V                               0.091

    Sample Size = 210
```

SUMMARY

In this chapter we discussed the multinomial distribution and two test procedures based on the multinomial: the chi-square goodness-of-fit test and the chi-square test of independence. Both tests use the test statistic

$$\chi^2 = \frac{\Sigma(O - E)^2}{E}$$

but the expected cell counts are computed differently. For the goodness-of-fit test, the expected cell count for category i is $n\pi_{io}$ where π_{io} is the hypothesized value of the multinomial parameter π_i; in the test for independence, the expected cell count for a given cell of a two-way table is (row total) (column total)/n.

The goodness-of-fit test and the test for independence in contingency tables are just two examples of the use of a chi-square statistic in the analysis of count data, but, since more extensive coverage of such tests would take considerable time, we will limit our coverage of these methods to the two examples discussed. Further reading on the subject can be found in Siegel (1956) and Hollander and Wolfe (1973).

KEY TERMS

count data

multinomial experiment

expected number of outcomes of
   type i (*def*)

contingency table

observed and expected cell counts

additive property of
   expected cell counts

chi-square statistic

control variable

partial tables

KEY FORMULAS

1. Chi-square goodness-of-fit test

Test statistic: $\chi^2 = \sum \dfrac{(O - E)^2}{E}$

where $E = n\pi_i$ and df $= k - 1$

2. Chi-square test of independence

Test statistic: $\chi^2 = \sum \dfrac{(O - E)^2}{E}$

where $E = \dfrac{\text{(row total)(column total)}}{n}$ and   df $= (r - 1)(c - 1)$

SUPPLEMENTARY EXERCISES

16.4   State the four parts of a chi-square test of independence.

16.5   In a 2 × 3 contingency table (2 rows, 3 columns), what is the minimum number of expected values that need to be computed if the remaining ones are computed by subtraction? What is the minimum number for a 3 × 3 table?

16.6   A pharmaceutical firm was interested in testing the effectiveness of a new drug product in controlling worms in the small intestine of sheep. A pre-study test was used to select 40 sheep with approximately the same level of infection. These sheep were then randomly divided into two groups of 20. Those in the first group of sheep were given the drug product; those in the second group received no treatment. After a period of two weeks each of the 40 sheep was examined and classified as either "favorable" or "unfavorable," depending on the observed worm count. To be labeled favorable, the observed worm count had to be less than 100. The results of the study are presented in the accompanying table.

Classification	Group 1 (Drug-treated)	Group II (Control)
Favorable	15( )*	7
Unfavorable	5	13

*Open parentheses indicate the expected counts that must be computed by using the formula "(row total) times (column total) divided by n." The remaining expected counts can be computed by subtraction from the appropriate row or column total.

(a) State the four parts of the chi-square test of independence for this situation.

(b) Compute the expected cell counts required for the test of independence.

16.7 Refer to Exercise 16.6. Test the hypothesis of independence of the two methods of classification (i.e., that the drug product was not effective in controlling worms). Use $\alpha = .10$.

16.8 A preelection survey was conducted in three different districts to compare the fraction of voters favoring the incumbent governor. Random samples of 50 registered voters were polled in each of the districts. The results are presented in the accompanying table. Do these data present sufficient evidence to indicate that the fractions favoring the incumbent governor differ in the three districts? Use $\alpha = .05$.

Opinion	District 1	District 2	District 3
Favor incumbent governor	19( )	14( )	26
Do not favor incumbent governor	31	36	24

16.9 The following computer program was used to compute the chi-square test statistic for the data of Exercise 16.8.

```
MTB > READ INTO C1-C3
DATA> 19 14 26
DATA> 31 36 24
DATA> END

    2 ROWS READ
MTB > CHISQUARE C1-C3

Expected counts are printed below observed counts

           C1        C2        C3     Total
    1      19        14        26       59
         19.67     19.67     19.67

    2      31        36        24       91
         30.33     30.33     30.33

Total      50        50        50      150

ChiSq =  0.023 +  1.633 +  2.040 +
         0.015 +  1.059 +  1.322 = 6.091

df = 2

MTB > STOP
```

(a) Compare the computer output to the results you obtained in Exercise 16.8.

(b) Give the level of significance for the test. (Hint: Use Table 4 of Appendix 3 to given an approximate p-value.)

16.10 A sociological survey was conducted to compare, for different income categories, the fraction of families with more than two children. A random sample of 150 families was questioned and each one was classified into one of three income categories (under $10,000, $10,000 to $25,000, or over $25,000). The number of children was also recorded for each family. Using the accompanying data, determine if there is sufficient evidence to indicate that the fraction of families having more than two children differs among the three income categories. Use $\alpha = .05$.

| Number of children | Income category | | |
	Under $10,000	$10,000–$25,000	Over $25,000
Two or fewer	11( )	13( )	21
More than two	68	28	9

16.11 The governor of each state was polled to determine his or her opinion concerning a particular domestic policy issue. At the same time the governor's party affiliation was recorded. The data are given here. If we assume that the 50 governors represent a random sample of political leaders throughout the nation, do the data present sufficient evidence to indicate a dependence between party affiliation and the opinion expressed on the domestic policy issue? Use $\alpha = .05$.

	Republican	Democrat
Approve of policy	18	8
Do not approve	5	8
No opinion	5	6

16.12 Use the data given in the following table to determine whether a physician's concern for the need to communicate with lower-class people before spending four weeks in a general hospital affects his rating of concern after spending four weeks in the hospital. Comment.

| Concern before experience in a general hospital | Concern after experience in a general hospital | | |
	High	Low	Totals
Low	27	5	32
High	9	9	18

16.13   At a recent national party convention, pro-lifers were against having the party platform reflect a pro-abortion stance. In a poll of 150 delegates, each was classified by religious background and attitude toward abortion. Use the data shown below to examine the relationship between the two variables. Are they independent?

Believe abortion is always permissible	Religious background	
	Catholic	Protestant
Yes	10	40
No	70	30

 16.14   The marketing research group of a particular firm conducted a survey in three cities to compare the sales potential of a new soft drink. Each person contacted was asked to try the new drink and classify it as either excellent, satisfactory, or unsatisfactory. The results of the survey are summarized in the accompanying table. Use the Minitab computer output shown here to conduct a chi-square test of independence. Give the approximate level of significance for your test and draw conclusions.

Classification	City 1	City 2	City 3
Excellent	62	51	45
Satisfactory	28	30	35
Unsatisfactory	10	19	20

```
MTB > READ INTO C1-C3
DATA> 62    51    45
DATA> 28    30    35
DATA> 10    19    20
DATA> END

     3 ROWS READ
MTB > PRINT C1-C3

 ROW    C1    C2    C3

   1    62    51    45
   2    28    30    35
   3    10    19    20

MTB > CHISQUARE C1-C3

Expected counts are printed below observed counts

           C1        C2       C3    Total
     1     62        51       45      158
         52.67     52.67    52.67
```

(continued)

```
  2        28       30       35      93
         31.00    31.00    31.00

  3        10       19       20      49
         16.33    16.33    16.33

Total     100      100      100     300

ChiSq =  1.654 +  0.053 +  1.116 +
         0.290 +  0.032 +  0.516 +
         2.456 +  0.435 +  0.823 = 7.376
df = 4

MTB > STOP
```

16.15 A university conducted a self-study to satisfy the requirements for accreditation. One aspect of the self-study concerned faculty evaluations. Through the use of student evaluations of their instructors, each faculty member was classified both by rank and by ability as a teacher. Use the accompanying results to test the null hypothesis of independence of two classifications. Use $\alpha = .05$ and draw a conclusion.

Teaching evaluation	Rank			
	Instructor	Assistant professor	Associate professor	Professor
Above average	36( )	62( )	45( )	50
Average	48( )	50( )	35( )	43
Below average	30	13	20	35

16.16 A survey of student opinion concerning a proposed increase in the activities fee was taken to determine if student opinion was independent of sex. The results of 300 student interviews are recorded in the accompanying table. Test, using $\alpha = .05$.

Sex of student	Opinion		
	Favor increase	Oppose	Undecided
Male	59( )	69( )	14
Female	91	54	13

16.17 An operations manager for a manufacturing firm has collected data on the number of machine failures for 50 weeks for the year. These data are shown here.

Number of failures per week	0	1	2	3	4	≥5
Number of weeks observed	10	24	12	2	1	1

(a) Compute the percentage of weeks that each category of failure was observed.

(b) A member of the firm's research and development division told the manager to expect (through normal wear and tear) to observe 20%, 45%, 25%, 5%, 3%, and 2% of the weeks in the designated failure categories. Test this hypothesis. Use $\alpha = .05$.

☤ **16.18** Thirty patients suffering from vertigo (dizziness) were randomly assigned to one of two groups. Those in the first group were to be given an anti-vertigo product, while those in the second group were to receive an identically appearing placebo. Following a specified treatment period, each patient was asked to rate the effectiveness of his or her therapy as either effective, moderately effective, or ineffective in the treatment of vertigo. Use the sample data to run a chi-square test of independence. Use $\alpha = .05$ and draw conclusions.

	Rating		
Treatment	**Effective**	**Moderately effective**	**Ineffective**
Antivertigo	10( )	3( )	2
Placebo	2	6	7

💲 **16.19** Two comparable standard metropolitan statistical areas (SMSA) were chosen for trial advertising campaigns using two entirely different promotional plans for a new product. After a two-month trial period in each area, a random sample of 100 shoppers was obtained and each person was asked whether he or she was aware of the new product. These data are summarized in the accompanying table. Is there sufficient evidence to indicate a difference in the proportions of individuals aware of the new product for the two promotional campaigns? Use $\alpha = .05$.

	Aware of product		
Campaign	**Yes**	**No**	**Total**
1	70( )	30	100
2	62	38	100

16.20 Commuter train riders have sometimes been used to answer survey questions because they are a captive audience during the ride. In an experiment with a new schedule of express and local trains along a suburban commuter line, random samples of 50 riders were obtained from the morning rush hour crowd on two local and two express trains. Each rider was asked to classify his or her response as favorable, unfavorable, or undecided about the new schedule. The data are shown in the accompanying table. Is there evidence to indicate a difference in the proportions of responses falling into the three categories for the four train runs? Use $\alpha = .05$.

	Response			
Train run	Favorable	Unfavorable	Undecided	Total
Local 1	30( )	15( )	5	50
Local 2	32( )	16( )	2	50
Express 1	23( )	24( )	3	50
Express 2	25	21	4	50

16.21 Refer to Exercise 16.20. Suppose that the two local trains were two trains along the same run, and, similarly, the two express trains were different trains along the same run. Combine the data to compare the local and express runs for the different categories of response. Use $\alpha = .05$. Does your conclusion differ from that in Exercise 16.20?

16.22 A survey of admissions practices at a liberal arts college was conducted to determine whether there appeared to be a difference in the acceptance rates for white and minority (nonwhite) applicants. The results of this survey, which combines information from 4000 applicants, are shown in the table. Use the accompanying computer printout to conduct a chi-square test of independence. Use Table 4 of Appendix 3 to obtain an approximate level of significance for the test and draw conclusions.

Applicant accepted?	Applicant		
	Nonwhite	White	Total
Yes	38	126	164
No	362	3474	3836
Total	400	3600	4000

```
MTB > READ INTO C1-C2
DATA>    38     126
DATA>   362    3474
DATA> END
```

(continued)

```
         2 ROWS READ
MTB > CHISQUARE C1 C2

Expected counts are printed below observed counts

             C1       C2    Total
       1     38      126      164
            16.40   147.60

       2     362     3474     3836
           383.60  3452.40

Total        400     3600     4000

ChiSq = 28.449 +   3.161 +
         1.216 +   0.135 = 32.961
df = 1

MTB > STOP
```

16.23 A large number of motor vehicle accidents had been occurring along a 10-mile stretch of interstate highway that passes through a large metropolitan area. So the highway patrol safety committee conducted a study to classify these accidents by outcome (whether or not a fatality occurred) and by the single probable cause of the accident. The results of the study are shown in the accompanying table.

| | **Probable cause of accident** | | | |
Outcome	Speeding	Driving while intoxicated	Reckless driving	Other
Fatality	42	61	20	12
No fatality	88	185	100	60

(a) Compute the percentages of accidents with fatalities for each of the cause categories.

(b) Conduct a chi-square test of independence, using $\alpha = .05$.

(c) Draw conclusions.

16.24 Give the approximate level of significance for the test results in Exercise 16.23. (Hint: Use Table 4 of Appendix 3.)

16.25 A carcinogenicity study was conducted to examine the tumor potential of a drug product scheduled for initial testing in humans. A total of 300 rats (150 males and 150 females) were studied for a 6-month period. At the beginning of the study 100 rats (50 males, 50 females) were randomly assigned to the control group, 100 to the low-dose group, and the remaining 100 (50 males, 50 females) to the high-dose group. Each day of the 6-month period the rats in the control group received an injection of an inert solution while those in the drug groups received an injection of the solution plus drug. The sample data are shown in the accompanying table.

Rat group	Number of tumors	
	One or more	None
Control	10	90
Low-dose	14	86
High-dose	19	81

(a) Give the percentages of rats with one or more tumors for each of the three groups.

(b) Conduct a chi-square test of independence, using $\alpha = .05$.

(c) Does there appear to be a drug-related problem regarding tumors for this drug product? That is, as the dose is increased, does there appear to be an increase in the percentage of rats with one or more tumors?

16.26 A computer output for the data of Exercise 16.25 is shown here. Compare the Minitab output with your results in Exercise 16.25.

```
MTB > READ INTO C1 C2
DATA> 10  90
DATA> 14  86
DATA> 19  81
DATA> END

     3 ROWS READ
MTB > PRINT C1 C2

 ROW    C1    C2

   1    10    90
   2    14    86
   3    19    81

MTB > CHISQUARE C1 C2

Expected counts are printed below observed counts

             C1       C2    Total
   1         10       90      100
          14.33    85.67

   2         14       86      100
          14.33    85.67

   3         19       81      100
          14.33    85.67

 Total       43      257      300
```

(continued)

```
ChiSq =  1.310 +  0.219 +
         0.008 +  0.001 +
         1.519 +  0.254 = 3.312
df = 2

MTB > STOP
```

16.27   Refer to the data of Exercise 16.25. Since there were an equal number of male and female rats assigned to each of the treatment groups, it was also decided to examine the sample results by sex. The breakdown by sex is shown in the computer output that follows. Use the computer output to reach some overall conclusions concerning the tumor potential of the drug product.

```
        TUMORS FEMALE
MTB > READ INTO C1 C2
DATA> 4 46
DATA> 10 40
DATA> 14 36
DATA> END

    3 ROWS READ
MTB > CHISQUARE C1 C2

Expected counts are printed below observed counts

            C1      C2     Total
    1        4      46       50
            9.33   40.67

    2       10      40       50
            9.33   40.67

    3       14      36       50
            9.33   40.67

Total       28     122      150

ChiSq =  3.048 +  0.699 +
         0.048 +  0.011 +
         2.333 +  0.536 = 6.674
df = 2

        TUMORS MALE
MTB > READ INTO C1 C2
DATA> 6 44
DATA> 4 46
DATA> 5 45
DATA> END

    3 ROWS READ
MTB > CHISQUARE C1 C2
```

(continued)

```
Expected counts are printed below observed counts

            C1       C2     Total
    1        6       44        50
          5.00    45.00

    2        4       46        50
          5.00    45.00

    3        5       45        50
          5.00    45.00

Total       15      135       150

ChiSq =  0.200 +  0.022 +
         0.200 +  0.022 +
         0.000 +  0.000 = 0.444
df = 2

MTB > STOP
```

16.28 The Congress was involved in a heated debate recently as to whether the federal government should continue to make funds available to those who cannot afford to pay for an abortion. The floor vote in the House of Representatives was as shown in the accompanying table. Use the chi-square test of independence to determine whether the variable "voting category" is independent of the variable "political party affiliation." If the two classifications are independent, then the Democrats and Republicans should have the same percentages of "yes" votes. Use $\alpha = .05$ for this test and draw a conclusion.

Party	Voting category Yes	No	Total
Democrat	181	137	318
Republican	101	111	212
Total	282	248	530

16.29 A novelty gift shop owner believes that 40% of the people who enter her store buy nothing, 35% buy one item, and 25% buy two or more items. Data for a one-week period were recorded and are displayed here:

Number of purchases	0	1	≥2
Number of customers	95	75	30

(a) Compute the percentage of customers who purchase 0, 1, or ≥ 2 items.

(b) Use these data to run a chi-square goodness-of-fit test. Draw a conclusion based on $\alpha = .05$.

16.30 In a survey of field agents for a major insurance company, the agents were asked to respond to questions concerning home office performance. Data are presented in the following table.

	Response				
Question	**Excellent**	**Good**	**Fair**	**Poor**	**Total**
How do you rate the speed of payment of agents' commissions?	16	48	14	12	90
Are home office personnel accessible?	18	36	16	20	90
Total	34	84	30	32	180

Is there sufficient evidence to conclude that the attitudes of the agents in the field toward the home office on commission payment and accessibility differ? Use $\alpha = .05$.

16.31 Chief marketing executives in major American corporations were asked to respond to a variety of questions in a survey conducted by a marketing professor. One of the questions in the survey was, "Do organizational conflicts and personnel problems prevent successful development and implementation of marketing plans?" The following table shows the responses of executives of manufacturing, retail trade, and financial institutions. Based on those data should one conclude that the responses differ according to industry class? Use $\alpha = .05$.

	Response class				
Industry	**Strongly agree**	**Mildly agree**	**Undecided**	**Mildly disagree**	**Strongly disagree**
Manufacturing	15	37	10	28	10
Retail trade	22	25	14	12	27
Finance	16	35	11	21	17

16.32 A survey of the customers of one of the very large petroleum retailers in the United States showed the following responses to the question, "How are the aesthetics of the service stations?"

Sex of respondent	Above average	Average	Below average	Total
Female	10	21	26	57
Male	8	24	11	43
Total	18	45	37	100

With $\alpha = .05$, test to see if the responses about the aesthetics of the firm's stations are independent of the sex of the respondent.

16.33 Student opinion on a resolution presented to the student council was surveyed to determine whether opinion was independent of fraternity and sorority affiliation. Two hundred students were interviewed, with the results as shown in the table. In a chi-square test of independence, how many degrees of freedom are there?

Student Opinion and Fraternity-Sorority Affiliation
for 200 Students Surveyed

	Status of affiliation		
Student opinion	**Fraternity**	**Sorority**	**Unaffiliated**
Favor	40	35	27
Opposed	18	25	55
Total	58	60	82

16.34 Use the data in Exercise 16.33 to determine whether or not there is sufficient evidence to indicate that student opinion on the resolution is independent of status (i.e., fraternity, sorority, or unaffiliated). Use $\alpha = .05$.

16.35 Refer to Exercise 16.33. Suppose we introduce a control variable, socioeconomic status of parents (families with incomes of $35,000 or more and with incomes below $35,000). The partial tables for the data are as given. Run separate chi-square tests of independence to determine whether the variable "student opinion" is independent of status (fraternity, sorority, unaffiliated) while controlling for socioeconomic status. Interpret your results. Use $\alpha = .05$.

	Status of affiliation					
Student opinion	**Income Below $35,000**			**Income $35,000 or More**		
	Fraternity	**Sorority**	**Unaffiliated**	**Fraternity**	**Sorority**	**Unaffiliated**
Favor	26	21	11	14	14	16
Opposed	8	15	39	10	10	16
Total	34	36	50	24	24	32

16.36 Use the results of Exercise 16.35 to conduct a pooled chi-square test of independence while controlling for the socioeconomic status of a student's family. What is the difference between the research hypothesis for this exercise and that for Exercise 16.34? Compare your conclusions to those of Exercise 16.34.

16.37 **Class Exercise** Sociologists and educators periodically point to the changing attitudes of college students, not only within an educational institution over a period of time but also between freshmen classes. For example, students during the 1950s were labeled complacent and likely to accept the pronouncements of those in authority without question; students of the 1960s were supposed to be the questioning, ever-seeking idealists, while the 1970s gave rise to the "me generation." And some of our colleagues profess now to see a distinct difference between the freshman and senior classes in their attitude toward study. All these pronouncements are subject to debate, and that leads us to a class problem associated with real data.

A current social question concerns the interpretation that should be given to equal rights and opportunity for minority groups. For example, should equal rights and opportunity imply that each human, regardless of race, religion, or sex, be accorded an equal opportunity for job employment and/or entrance to graduate or professional schools? Or does it mean that society should redress the wrongs of the past and accord members of minority groups a priority status in seeking jobs and/or educational opportunities? On this particular issue the attitudes of entering freshmen may very well differ from those of seniors, who will soon be entering the job market or competing for admission to professional schools.

Conduct a survey to determine whether there is a dependence between college class and the response to the question above. The twelve categories of the study appear as follows:

	College class			
Opinion	**Freshman**	**Sophomore**	**Junior**	**Senior**
Favor equal opportunity for each person				
Favor priority for minority groups				
No opinion, or neither of the above				

Use the student directory to randomly select 100 students from each college class. Contact each student by telephone to ascertain his or her response to the question. Analyze your data and determine whether the data provide sufficient evidence to indicate a dependence between college class and student attitude to the survey question.

Because seniors face the problems of admission to professional or graduate schools, competition for a limited number of good jobs, and imminent competition for advancement, their attitudes regarding the meaning of "equal opportunity" may differ from the others'. Collapse the 3×4 categorization of the data to a 3×2 categorization as follows:

Opinion	Seniors	Others
Favor equal opportunity for each person		
Favor priority for minority groups		
No opinion, or neither of the above		

Do the data provide sufficient evidence to indicate that the opinions of seniors differ from the opinions of others?

The calculations for this experiment can easily be accomplished on an electronic calculator, but you can also use packaged programs and a computer to perform the calculations. Useful packages include Minitab and SAS, mentioned in the "Using Computers" section, as well as the BMDP programs (see Dixon, 1983) and the SPSS^x software system (1984).

EXERCISES FROM THE DATABASE

16.38 Construct a 4×4 contingency table that categorizes patients from the clinical trials data base by *marital status* and by *treatment group*. Compute the percentages of males and females in each group. Do the groups appear to differ in their sex distributions? Test your intuition using a chi-square test of independence and give the p-value for your test.

16.39 Refer to the clinical trials data base and list other baseline comparisons that could be made among the treatment groups using a chi-square test of independence.

16.40 Refer to Exercise 16.39. Make the comparisons suggested in that exercise. Do the groups appear to be comparable?

16.41   Refer to the clinical trials data base.

(a) Construct a 4 × 5 contingency table to categorize patients by treatment group and therapeutic effect.

(b) Compute expected cell counts. Do the expected cell counts satisfy the criteria for running a chi-square test of independence?

(c) Run a chi-square test of independence on the table of (a), if appropriate, or on a "collapsed" 4 × 2 table.

Treatment group	Marked/moderate	Minimal/unchanged or worse
A		
B		
C		
D		

16.42   If the table is collapsed as suggested in Exercise 16.41, what information may be lost?

CHAPTER 17

NONPARAMETRIC STATISTICS

17.1 Introduction ▪ **17.2** A simple comparative test: The sign test ▪ **17.3** Wilcoxon's signed-rank test ▪ **17.4** Wilcoxon's rank sum test ▪ **17.5** Spearman's rank correlation coefficient ▪ **17.6** Using computers ▪ Summary ▪ Key terms ▪ Key formulas ▪ Supplementary exercises ▪ Exercise from the data base

---

17.1 INTRODUCTION

ordinal data

Some studies yield data identified by rank only **(ordinal data),** either because of the crudeness of the measuring scale employed or because of the inability of the investigator to further quantify the measurements. For example, in examining the effectiveness of an antihistamine in the treatment of ragweed allergy, we would be interested in measuring the change in symptomatology following treatment with the antihistamine. While it might be desirable to quantify the change in allergic symptoms, it is difficult to find objective measures of nose congestion, stuffiness, and so on. Thus we might be forced to measure improvement on an ordinal scale such as the following: worse, same, mild improvement, moderate improvement, marked improvement. Note that with this scale we cannot measure improvement in absolute terms, but rather we measure it in *relative* terms. For example, someone rated "mild" would have less improvement than someone rated "moderate," but there is no way to measure the difference in improvement using this scale.

Many experiments result in ordinal data. In the social sciences such variables as prestige, power, and alienation are measured by ordinal scales. In the behavioral sciences such variables as pain-threshold level, emotional stability, and drug reaction might be measured on an ordinal scale.

When the variable of interest is measured on an ordinal scale, the test procedures discussed thus far are inappropriate, and we must resort to **nonparametric statistical tests** to provide a means for analyzing these data. The word "nonparametric" evolves from the type of hypothesis usually tested when dealing with ordinal-level data. Most nonparametric tests

nonparametric statistical tests

do not involve inferences about parameters from the original distribution of measurements. For example, instead of hypothesizing that two populations have the same mean (as in Section 9.4), we could hypothesize that the two populations from which the samples were drawn are identical. Note that the practical implications of these two hypotheses are not the same. The first hypothesis is specific to a particular population parameter, whereas the second hypothesis addresses the question of equality (sameness) of the probability distributions. Two distributions of measurements could be different and still have the same mean.

In the next three sections we will discuss several nonparametric statistical tests for comparing two or more populations. Although they have been developed specifically for ordinal-level data, they are also appropriate for quantitative data when one or more of the assumptions underlying a particular parametric statistical test has been violated. For example, in conducting a t-test comparing two population means, we make the assumptions that the two populations are normal and have a common variance σ^2. If either of these assumptions is violated, the usual t-test for comparing independent samples is inappropriate.

This chapter provides useful alternatives to the parametric test procedures of previous chapters. The nonparametric tests presented here enable us to make statistical inferences even in experimental situations where the usual tests may be invalid.

17.2   A SIMPLE COMPARATIVE TEST: THE SIGN TEST

The sign test is a procedure for testing whether two populations have identical probability distributions. Since we will make no assumptions concerning the parameters of the original distributions of measurements, we refer to the sign test as a nonparametric statistical test.

Why use the sign test to make a comparison between two populations? First, some studies yield responses that are hard to quantify. For example, it would be hard to quantify a ranking of the moods of individuals who for the first time have been placed on welfare, or to evaluate the performance of the faculty within a large history department. The sign test works particularly well for these types of data because we do not need to know the exact value of each measurement, only whether one is larger or smaller than the other. Second, it's easy to perform a sign test. Third, we need make no assumptions about the form of the population probability distributions. For example, we need not assume that the populations are normal or mound-shaped.

The sign test is based on the differences in pairs of observations, one observation from each sample. We let x_1 denote an observation from sample 1 and x_2 denote an observation from sample 2, and we note the difference between observations x_1 and x_2 for each pair, showing the sign of the difference. If x_1 is greater than x_2, we show a plus sign; for x_1 less than x_2 we show a minus sign. We omit all pairs for which $x_1 = x_2$. Table 17.1 lists data and plus and minus signs for ten pairs of observations.

TABLE 17.1   Ten Pairs of Observations

Pair	Sample 1, x_1	Sample 2, x_2	Sign of $(x_1 - x_2)$
1	10.2	10.3	−
2	10.1	10.0	+
3	10.3	10.2	+
4	10.4	10.2	+
5	10.3	10.0	+
6	10.2	10.1	+
7	10.2	10.0	+
8	10.5	10.3	+
9	10.1	10.2	−
10	10.4	10.3	+

The sign test of the null hypothesis "there is no difference in the probability distribution for the two populations" utilizes the number of plus signs, x, for the pairs of observations from the two samples. If the null hypothesis is true, then for each pair the probability that x_1 is greater than x_2 is $\pi = .5$. That is, there is a 50:50 chance that x_1 will be greater than x_2, or vice versa. Testing the null hypothesis that the distributions are identical is equivalent to testing the hypothesis that a binomial parameter π is equal to .5. This test was discussed in detail in Section 10.3.

We illustrate the application of a binomial test of a proportion for the sign test in Example 17.1.

Example 17.1

Each of 20 young mothers was asked to compare two different approaches, 1 and 2, to socializing her young children. After two weeks of employing each approach, the mothers were asked to grade their satisfaction with each approach on a scale from 0 to 5. The results are given in Table 17.2. Determine if there is evidence to indicate that the two populations of ratings differ and hence whether approach 1 is preferred to approach 2 or vice versa. Use $\alpha = .05$.

TABLE 17.2   Measurements Made on a Random Sample of 20 Mothers

Mother	Approach 1, x_1	Approach 2, x_2	Sign of $(x_1 - x_2)$
1	3	2	+
2	4	2	+
3	3	5	−
4	5	4	+
5	4	3	+
6	3	2	+
7	3	4	−
8	4	3	+
9	3	2	+
10	4	2	+
11	5	4	+
12	3	4	−
13	2	1	+
14	3	2	+
15	5	3	+
16	5	4	+
17	5	3	+
18	2	3	−
19	4	2	+
20	4	3	+

Solution   Consider the pair of observations for each mother and let x be the number of times approach 1 has a higher score (satisfaction) than approach 2. The null hypothesis is that the two approaches are equally preferred (and hence the distributions of scores are identical), or, equivalently, the probability that x_1 exceeds x_2 for any pair is $\pi = .5$. The alternative hypothesis is that the two distributions are different; that is, π is greater than or less than .5. Thus we have

$$H_0: \pi = .5 \quad \text{and} \quad H_a: \pi \neq .5$$

The test statistic is

$$z = \frac{\hat{\pi} - \pi_0}{\sqrt{\dfrac{\pi_0(1 - \pi_0)}{n}}}$$

Substituting $\hat{\pi} = x/n$ and $\pi_0 = .5$, and rearranging terms, we can rewrite the formula for z as

$$z = \frac{x - .5n}{\sqrt{.25n}}$$

For $\alpha = .05$ we will reject the null hypothesis if z is greater than 1.96 or less than -1.96. Note that we are using a two-tailed test (as shown in Figure 17.1) because we wish to detect values of π that are either greater or smaller than $\pi = .5$.

FIGURE 17.1   Rejection Region for the Sign Test, Example 17.1

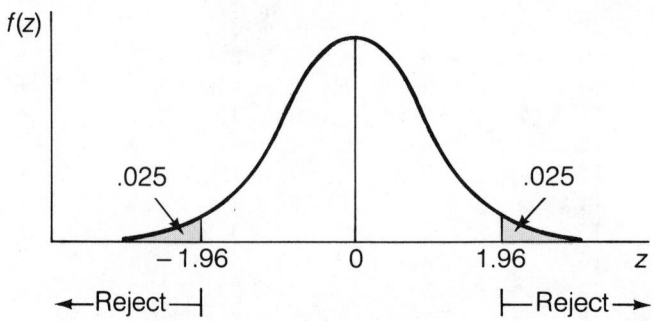

From Table 17.2 we see that the number of plus signs is $x = 16$, and n is 20. Thus the observed value of z is

$$z = \frac{x - .5n}{\sqrt{.25n}} = \frac{16 - 10}{2.24} = 2.7$$

Since this value is greater than 1.96, z falls in the rejection region. Hence we have evidence to indicate that the degree of satisfaction expressed by mothers differs for approaches 1 and 2. Further examination of the data suggests that mothers prefer approach 1.

One problem occasionally encountered when we use the sign test is that ties occur. That is, for one or more pairs $x_1 = x_2$. When this happens we omit the tied pair(s). For example, if you have 20 pairs and one pair results in a tie, you delete the tied pair and work with the remaining $n = 19$ pairs.

Example 17.2

Two psychiatrists were asked to rate (on a scale of 0–12) each of 12 prison inmates concerning their rehabilitative potential. Do the data shown in

Table 17.3 suggest a difference in the rating scales employed by the two psychiatrists? Use α = .05. (Note that a low reading on the scale indicates a low rehabilitative potential.)

TABLE 17.3 Data for Example 17.2

Inmate	Psychiatrist 1, x_1	Psychiatrist 2, x_2	Sign of $(x_1 - x_2)$
1	6	5	+
2	12	11	+
3	3	4	−
4	9	10	−
5	5	2	+
6	8	6	+
7	1	2	−
8	9	12	−
9	6	5	+
10	7	4	+
11	6	6	tie
12	9	8	+

Solution The four parts of our statistical test are as follows:

Null hypothesis: $\pi = .5$, where π is the probability that psychiatrist 1 will rate an inmate higher than psychiatrist 2.

Alt. hypothesis: $\pi \neq .5$

Test statistic: $z = \dfrac{x - .5n}{\sqrt{.25n}}$

Rejection region: For $\alpha = .05$ reject H_0 if $|z| > 1.96$

Substituting into the formula for z with $x = 7$ and $n = 11$ (note that there is one tie), we obtain

$$z = \frac{7 - 5.5}{\sqrt{.25(11)}} = .90$$

Since $z = .90$ does not fall in the rejection region, we have insufficient evidence to reject H_0: $\pi = .5$. That is, according to the sign test there is insufficient evidence to indicate that the population distributions of ratings are different for the two psychiatrists.

A summary of the elements of the sign test is presented next.

Sign Test for Comparing Two Populations

Null hypothesis: $\pi = .5$. (If π denotes the probability of observing a plus sign for a pair, then H_0: $\pi = .5$ implies that the two population distributions are identical.)

Alt. hypothesis: For a one-tailed test:
1. $\pi > .5$
2. $\pi < .5$

For a two-tailed test:
3. $\pi \neq .5$

Test statistic: $$z = \frac{x - .5n}{\sqrt{.25n}}$$

where n is the number of pairs, ignoring ties, and x is the number of pairs for which $x_1 > x_2$.

Rejection region: For $\alpha = .05$ (.01) and for a one-tailed test:
1. Reject H_0 if $z > 1.645$ (2.33)
2. Reject H_0 if $z < -1.645$ (-2.33)

For $\alpha = .05$ (.01) and for a two-tailed test:
3. Reject H_0 if $|z| > 1.96$ (2.58)

Note: This test is valid for $n \geq 10$.

To summarize, the advantages of the sign test are that it makes no assumptions concerning the nature of the original distributions of measurements, it can be applied to small samples ($n \geq 10$ pairs), and it is easy to apply.

Conversely, the disadvantage of the sign test is also clear. Since we use only the signs of the differences between x_1 and x_2 and not the actual values of the measurements, we sacrifice some of the information available in the experiment. Consequently, a sign test based on small samples may not detect a difference between two populations that other tests, which utilize information about the magnitudes of x_1 and x_2, may detect. Although this is a serious disadvantage it does not detract from the value of the sign test as a rapid method of detecting a difference between two populations.

The sign test is only one of a number of useful nonparametric statistical tests. Others are presented in Mendenhall (1987), Ott (1988), and Siegel (1956).

EXERCISES

$ 17.1 Fifty people were asked to rate each of two products, A and B. Use the sign test with $n = 50$ and $\alpha = .05$ to determine whether the distributions of ratings (for products A and B) differ if $x = 29$ people rated product A higher than product B.

$ 17.2 Refer to Exercise 17.1. State the level of significance for your test.

17.3 Two judges were asked to rate each of 22 male inmates on rehabilitative potential. The results are shown in the accompanying table. Use a sign test with $\alpha = .05$ to determine whether the distributions of scores are different for the two judges.

Inmate	Judge 1 rating	Judge 2 rating	Inmate	Judge 1 rating	Judge 2 rating
1	6	5	12	9	8
2	12	11	13	10	8
3	3	4	14	6	7
4	9	10	15	12	9
5	5	2	16	4	3
6	8	6	17	5	5
7	1	2	18	6	4
8	12	9	19	11	8
9	6	5	20	5	3
10	7	4	21	10	9
11	6	6	22	10	11

17.4 A test of manual dexterity was administered to 10th grade students. Prior to this, twelve pairs of students (one boy and one girl in each pair) were matched as nearly as possible according to manual dexterity skills. The students were then given the test. Use the results in the table shown here and the sign test to decide if the two groups have different distributions of scores. Use $\alpha = .05$.

Boys	78	57	59	63	65	79	58	75	63	81	72	65
Girls	85	62	58	76	82	72	61	74	69	70	80	73

17.5 Typical diets of twelve families subsisting on a poverty-level income were rated for their nutritional value. The ratings used were poor, fair, good, and excellent. The mothers of the families then participated in a series of classes on nutrition and preparation of balanced meals. A month later the diet of each family was rated again. The results are shown in the accompanying table.

Family	Before	After	Family	Before	After
1	Poor	Fair	7	Fair	Good
2	Poor	Fair	8	Good	Fair
3	Fair	Fair	9	Poor	Fair
4	Good	Excellent	10	Poor	Good
5	Fair	Good	11	Poor	Good
6	Poor	Good	12	Good	Excellent

Use the sign test with $\alpha = .05$ to decide whether the classes improved the quality of nutrition in the families' diets.

17.6 In a before-and-after study a social psychologist reported the accompanying data for ten small groups. Use the sign test to determine whether scores changed after the instruction. Set up all parts of the statistical test, using $\alpha = .05$.

Group	Before instruction	After instruction	Group	Before instruction	After instruction
1	8	9	6	3	5
2	10	13	7	5	8
3	7	9	8	10	9
4	4	7	9	12	13
5	6	8	10	11	15

17.7 The effect of Benzedrine on the heart rate of dogs (in beats per minute) was to be examined in an experiment. Fourteen dogs were chosen for the study, and each dog served as its own control. Half of the dogs were assigned to receive Benzedrine during the first study period while the other seven were given a placebo (saline solution). All dogs were examined to determine the heart rates after two hours on the medication. After two weeks in which no medication was given, the regimens for the dogs were switched for the second study period; that is, the dogs on Benzedrine were given the placebo while the others received Benzedrine. Again heart rates were measured two hours after administration of the medication. The accompanying sample data are not arranged in the order in which they were taken but are summarized by regimen. Use these data to test the alternative hypothesis H_a: $\pi > .5$, where π is the proportion of dogs for which the Benzedrine response was higher than the placebo response. Use a one-tailed test with $\alpha = .05$.

Dog	Placebo	Benzedrine	Dog	Placebo	Benzedrine
1	250	258	4	252	250
2	271	285	5	266	268
3	243	245	6	272	278

(continued)

Dog	Placebo	Benzedrine	Dog	Placebo	Benzedrine
7	293	280	11	310	320
8	296	305	12	286	293
9	301	319	13	306	305
10	298	308	14	309	313

17.3 WILCOXON'S SIGNED-RANK TEST

The Wilcoxon signed-rank test, which makes use of the sign and magnitude of the rank of the differences between pairs of measurements, provides a way to compare two populations when the variable of interest is measured on an ordinal scale. As with the sign test, Wilcoxon's signed-rank test also provides an alternative to the paired t-test. Utilizing the pairs of measurements with a nonzero difference, we rank the differences from lowest to highest, ignoring their signs. If two or more measurements have the same nonzero difference (ignoring sign), we assign each difference a rank equal to the average of the occupied ranks. The appropriate sign is then attached to the rank of each difference.

Before summarizing the Wilcoxon signed-rank test, we define the following notation:

n = number of pairs of observations with nonzero difference

T_+ = sum of the positive ranks; if there are no positive ranks $T_- = 0$

T_- = sum of the negative ranks; if there are no negative ranks $T_- = 0$

T = the smaller of T_+ and T_-, ignoring their signs

$$\mu_T = \frac{n(n + 1)}{4}$$

$$\sigma_T = \sqrt{\frac{n(n + 1)(2n + 1)}{24}}$$

The Wilcoxon signed-rank test is presented next. Note that we use a different test statistic and rejection region for large sample sizes ($n > 50$) than when $n \le 50$.

Wilcoxon Signed-Rank Test

Null hypothesis: The distribution of differences is symmetric about 0

(continued)

Wilcoxon Signed-Rank Test (continued)

Alt. hypothesis: For a one-tailed test:
1. The differences tend to be larger than 0
2. The differences tend to be smaller than 0

For a two-tailed test:
3. The differences tend to be shifted away from 0

Test statistic:
1. $T = |T_-|$
2. $T = T_+$
3. $T = $ smaller of $|T_-|$ and T_+

Rejection region: $(n \leq 50)$

For a specified value of α (one-tailed .05, .025, .01, or .005; two-tailed .10, .05, .02, or .01) and fixed number of nonzero differences n, reject H_0 if T is less than or equal to the appropriate entry in Table 10 of Appendix 3.

Compute the test statistic when $n > 50$:

$$z = \frac{T - \mu_T}{\sigma_T}$$

where $\mu_T = \dfrac{n(n + 1)}{4}$ and $\sigma_T = \sqrt{\dfrac{n(n + 1)(2n + 1)}{24}}$

for $\alpha = .05$ (.01)

1. Reject H_0 if $z > 1.645$ (2.33)
2. Reject H_0 if $z < -1.645$ (-2.33)
3. Reject H_0 if $|z| > 1.96$ (2.58)

Example 17.3

Consider the following experimental situation. Each of ten rats was weighed and then given a fixed diet supplemented with a dose of 10 mg/kg of a new drug product. After three months the rats were weighed individually again. Use the data in Table 17.4 and the Wilcoxon signed-rank test to decide whether there has been a change in weight for the three-month period. (Note: One reason for conducting such an experiment is to examine the drug product for potential toxicity. Toxicity of compounds is sometimes manifested through a weight loss over a three-month period.) Use $\alpha = .05$.

Solution This is clearly a two-tailed test. The parts of the test are shown here:

Null hypothesis:	The distribution of differences is symmetric about 0 (i.e., there is no change in weight over the three month period).
Alt. hypothesis:	The differences tend to be different from 0.
Test statistic:	$T = $ smaller of $|T_-|$ and T_+
Rejection region:	For $n \leq 50$, we refer to the appropriate two-tailed .05 value from Table 10 of Appendix 3.

TABLE 17.4 Data for Example 17.3

Rat	Predrug weight	Postdrug weight	Difference (postdrug − predrug)
1	66.3	122.0	55.7
2	75.9	128.4	52.5
3	84.8	142.7	57.9
4	79.8	137.8	58.0
5	81.6	136.8	55.2
6	87.2	143.0	55.8
7	66.3	110.4	44.1
8	80.2	125.9	45.7
9	79.6	133.6	54.0
10	78.2	74.6	− 3.6

To determine the test statistic T, we first must rank the ten (postdrug/predrug) differences from lowest to highest. These are shown in Table 17.5. Then the appropriate sign is attached to each rank. The sums for the positive and for the negative ranks are then

$$T_+ = -1$$
$$T_- = 2 + 3 + \cdots + 10 = 54$$

and T, the smaller of T_+ and T_-, ignoring the sign, is 1.

TABLE 17.5 Ranks for the Data of Table 17.4

Rat	Rank of difference	Rat	Rank of difference
1	7	6	8
2	4	7	2
3	9	8	3
4	10	9	5
5	6	10	− 1

The rejection region for a two-tailed test with $n = 10$ and $\alpha = .05$ is found by using Table 10 in Appendix 3: we will reject H_0 is T is less than or equal to 8. Since the observed value of T is 1, we reject H_0 and conclude that the distributions of predrug and postdrug weights are different. Practically, since all but one rat had a postdrug weight higher than the predrug weight, we can say that there appears to be an increase in weight for rats over the three months; there is no evidence of toxicity based on the weight data.

In Chapter 14 we discussed the paired t-test for comparing two population means based on dependent samples (paired data). There we made the assumption that the population of differences is roughly normal; if this is not true, then the Wilcoxon signed-rank test may be more appropriate. When in doubt about which test to use, do both of them. If the results do not agree seek out some help from a statistician to understand the peculiarities of the data that led to different results.

EXERCISES

17.8 Two different brands of fertilizer (A and B) were compared on each of ten different two-acre plots. Each plot was subdivided into one-acre subplots with brand A randomly assigned to one subplot and brand B to the other. Then fertilizers were applied to subplots at the rate of 60 lbs per acre. Barley yields (in bushels per acre) are listed by fertilizer and plot in the accompanying table. Use the Wilcoxon signed-rank test to test the null hypothesis that the distribution of differences is symmetrical about 0 against the alternative that the differences tend to be shifted away from 0. Use $\alpha = .05$.

Plot	Fertilizer A, x_1	Fertilizer B, x_2	Difference, $(x_1 - x_2)$
1	312	346	-34
2	333	372	-39
3	356	392	-36
4	316	351	-35
5	310	330	-20
6	352	364	-12
7	389	375	14
8	317	315	2
9	316	327	-11
10	346	378	-32

17.9 Refer to the data of Example 17.1. Use Wilcoxon's signed-rank test to compare the two different approaches. Use $\alpha = .05$. Are your conclusions similar to those obtained by using the sign test?

17.10 A single leaf was taken from each of eleven different tobacco plants. Each was divided in half: one-half was chosen at random and treated with preparation 1; the other half received preparation 2. The object of the experiment was to compare the effects of the two preparations for the control of mosaic virus as measured by the number of lesions on the half-leaves after a fixed period of time. The fewer the number of lesions, the better the preparation is. These data are recorded in the accompanying table. For $\alpha = .05$, use Wilcoxon's signed-rank test to examine the alternative hypothesis that the differences tend to be shifted away from 0.

Tobacco plant	Number of lesions on the half-leaf	
	Preparation 1	**Preparation 2**
1	18	14
2	20	15
3	9	6
4	14	12
5	38	32
6	26	30
7	15	9
8	10	2
9	25	18
10	7	3
11	13	6

17.11 Refer to Exercise 17.10. How does your test change if we want to detect whether preparation 2 is better than preparation 1?

17.12 Refer to Exercise 17.10. Conduct a sign test and compare your results to those of Exercise 17.10. Use $\alpha = .05$.

17.13 Refer to Exercise 17.7. Use Wilcoxon's signed-rank test to compare the distributions of heart rates for dogs on Benzedrine and on the placebo. Use a one-tailed test for the alternative hypothesis that the differences (placebo/Benzedrine) tend to be less than 0.

17.4 WILCOXON'S RANK SUM TEST

The Wilcoxon rank sum test (not to be confused with the Wilcoxon signed-rank test) provides a procedure for testing whether two populations are identical when *independent* random samples are selected from the two populations.

The *t*-test for comparing the means of two populations (Section 9.4) was based on two assumptions. We assumed that both populations were normal *and* that they had the same variance. the Wilcoxon rank sum test has weaker assumptions; the null hypothesis is that the two populations are identical (but not necessarily normal). Again for a given data set, if you have doubts as to which procedure to use (the *t*-test or Wilcoxon's rank sum test), do both tests. A statistician can help you interpret your data if the results of the two tests don't lead to the same conclusions.

There is another Wilcoxon nonparametric test that is appropriate for comparing two populations when an independent sample is obtained from each population. The test procedure is quite simple. If under the null hypothesis the two populations are assumed to be identical, then independent random samples from the respective populations should be similar. One way to measure the similarity between samples is to jointly rank (from lowest to highest) the measurements from the combined samples and examine the sum of the ranks for measurements in sample 1 (or, equivalently, sample 2). Let *T* denote the sum of the ranks for sample 1. Intuitively, we would have evidence to reject the null hypothesis that the two populations are identical if *T* is extremely small (or large).

The procedure is as follows. Independent random samples are obtained from the two populations. The combined sample data from the two populations are then jointly ranked. If there are ties among any measurements in the combined sample data, we assign each tied measurement a rank equal to the average of the occupied ranks. Then when the sample sizes n_1 and n_2 are both larger than 10, the sum of the ranks *T* for sample 1 will be approximately normally distributed, with mean and variance given by

$$\mu_T = \frac{n_1(n_1 + n_2 + 1)}{2}$$

$$\sigma_T^2 = \frac{n_1 n_2(n_1 + n_2 + 1)}{12}$$

The details of the test procedure are summarized next.

Wilcoxon's Rank Sum Test ($n_1 > 10$ and $n_2 > 10$)

Null hypothesis: The two populations are identical.

Alt. hypothesis: For a one-tailed test:

1. The distribution of measurements for population 1 is above (to the right of) the distribution of measurements for population 2.

(continued)

2. The distribution of measurements for population 1 is below (to the left of) the distribution of measurements for population 2.

For a two-tailed test:

3. The two populations are different.

Test statistic:

$$z = \frac{T - \mu_T}{\sigma_T}$$

where T is the sum of the ranks in sample 1,

$$\mu_T = n_1 (n_1 + n_2 + 1)/2$$

and

$$\sigma_T = \sqrt{\frac{n_1 n_2 (n_1 + n_2 + 1)}{12}}$$

Rejection region: For $\alpha = .05$ (.01) and for a one-tailed test:

1. Reject H_0 if $z > 1.645$ (2.33)
2. Reject H_0 if $z < -1.645$ (−2.33)

For $\alpha = .05$ (.01) and for a two-tailed test:

3. Reject H_0 if $|z| > 1.96$ (2.58)

Note: This is equivalent to the Mann-Whitney U-test (see Conover, 1980).

Example 17.4

Environmental engineers were interested in determining whether a cleanup project on a nearby lake was effective. Prior to initiation of the project, 12 samples of water had been obtained at random from the lake and analyzed for the amount of dissolved oxygen (in ppm). Due to diurnal fluctuations in the dissolved oxygen, all measurements were obtained at the 2 P.M. peak period. Six months after the cleanup project, 12 samples of measurements were obtained at the 2 P.M. peak period. The data are presented in Table 17.6; for convenience, the data are arranged in ascending order. Use $\alpha = .05$ to test the following hypothesis:

H_0: The distributions of measurements for before and six months after the cleanup project began are identical.

H_a: The distribution of dissolved oxygen measurements after the cleanup project is above (to the right of) the corresponding distribution of measurements before initiating the cleanup project. (Note that a cleanup project has been effective in one sense if the dissolved oxygen increases over a period of time.)

TABLE 17.6 Dissolved Oxygen Measurements (ppm)

Before cleanup		After cleanup	
10.2	10.8	11.0	11.6
10.3	10.8	11.2	11.7
10.4	10.9	11.2	11.8
10.6	11.1	11.2	11.9
10.6	11.1	11.4	11.9
10.7	11.3	11.5	12.1

Solution In order to compute the value of the test statistic z, first we must jointly rank the combined sample of 24 observations by assigning the rank of 1 to the smallest observation, the rank of 2 to the next smallest, and so on. When two or more measurements are the same, we assign all of them a rank equal to the average of the ranks they occupy. For example, there are two measurements equal to 10.6. Since these two measurements occupy the ranks 4 and 5, they are assigned a rank of 4.5, the average of the occupied ranks. The sample measurements and associated ranks (in parentheses) are listed in Table 17.7. Summing the ranks, we find that the "before cleanup" ranks have the smaller sum, which is $T = 84$.

TABLE 17.7 Dissolved Oxygen Measurements and Ranks

Before cleanup		After cleanup	
10.2 (1)	10.8 (7.5)	11.0 (10)	11.6 (19)
10.3 (2)	10.8 (7.5)	11.2 (14)	11.7 (20)
10.4 (3)	10.9 (9)	11.2 (14)	11.8 (21)
10.6 (4.5)	11.1 (11.5)	11.2 (14)	11.9 (22.5)
10.6 (4.5)	11.1 (11.5)	11.4 (17)	11.9 (22.5)
10.7 (6)	11.3 (16)	11.5 (18)	12.1 (24)

If we are trying to detect a shift to the right in the distribution for the measurements after the cleanup, we would expect the sum of the ranks for observations in sample 1 (before cleanup) to be small. Thus we will reject H_0 for $z < -1.645$. Substituting $n_1 = n_2 = 12$, we have

$$\mu_T = \frac{12(25)}{2} = 150$$

$$\sigma_T^2 = 12(25) = 300$$

$$\sigma_T = \sqrt{300} = 17.32$$

$$z = \frac{84 - 150}{17.32} = -3.81$$

Since the computed value of z is less than -1.645, the table value for $\alpha = .05$ and a one-tailed test, we reject H_0 and conclude that the population distribution of measurements after cleanup is shifted to the right of the corresponding population distribution of measurements before cleanup.

EXERCISES

17.14 Refer to Example 17.3. Ten rats were weighed and then placed on a fixed diet supplemented with 10 mg/kg of a new drug product for a period of three months. As part of the same experiment, an additional ten rats were placed in a control group. The same procedure was followed except that no drug product was mixed with the feed. The prestudy and poststudy weights for the ten drug-treated rats of Example 17.3 and the ten control rats are shown in the accompanying table. Use the Wilcoxon rank sum test to compare the weight gains for the two groups. Use $\alpha = .05$. Draw some conclusions.

Drug-treated rats			Control rats		
Prestudy weight	Poststudy weight	Difference (poststudy − prestudy)	Prestudy weight	Poststudy weight	Difference (poststudy − prestudy)
66.3	122.0	55.7	65.0	127.6	62.6
75.9	128.4	52.5	68.3	124.3	56.0
84.8	142.7	57.9	77.4	133.8	56.4
79.8	137.8	58.0	76.5	140.1	63.6
81.6	136.8	55.2	82.0	153.9	71.9
87.2	143.0	55.8	79.4	145.6	66.2
66.3	110.4	44.1	69.8	139.3	69.5
80.2	125.9	45.7	70.1	130.4	60.3
79.6	133.6	54.0	66.7	132.7	66.0
78.2	74.6	−3.6	71.3	144.2	72.9

17.15 Refer to the data of Exercise 17.14. Note the importance of using a control group in the experiment. Even though both groups of animals showed statistically significant weight gain (you can verify this with a sign test on the poststudy − prestudy differences in each group), the weight gain for the drug-treated rats was less than that for the corresponding control group. What conclusions could you draw if the control group was not included in the experiment? Can historical information on control animals substitute for the use of a concurrent control group?

17.16 Refer to Exercise 17.14. Are the two treatment groups comparable with regard to the pretreatment weights? Use Wilcoxon's rank sum test to compare the two groups. What problem might you encounter if the pretreatment weights were significantly different?

17.17 The accompanying data resulted from an experiment conducted to compare the heart rates (in beats per minute) for a group of twelve control animals and another group of twelve rats treated with an antihypertensive product. The response variable measured was the change in heart rate for each rat. Use the methods of Section 9.4 to conduct a two-sample t-test for comparing the mean change in heart rate for the two groups. Give the level of significance for your test.

Control		Treated	
9.0	− 17.0	59.0	51.0
12.0	36.0	44.0	− 9.0
36.0	42.0	63.0	75.5
77.5	65.0	87.5	28.0
− 7.5	30.5	30.5	82.0
32.5	45.5	57.5	65.0

17.18 Wilcoxon's rank sum test was performed for the data of Exercise 17.17. Use the computer output shown here to compare the results with those you obtained in Exercise 17.17. (Note: The quantity W for Minitab is the same as T, which we defined as the sum of the ranks in sample 1. Using $W = 118$, you will have to compute z and draw a conclusion.

```
MTB > READ INTO C1 C2
DATA>      9.0    59.0
DATA>     12.0    44.0
DATA>     36.0    63.0
DATA>     77.5    87.5
DATA>     -7.5    30.5
DATA>     32.5    57.5
DATA>    -17.0    51.0
DATA>     36.0    -9.0
DATA>     42.0    75.5
```

(continued)

```
DATA>     65.0   28.0
DATA>     30.5   82.0
DATA>     45.5   65.0
DATA> END
      12 ROWS READ

MTB > PRINT C1 C2

  ROW     C1     C2

    1     9.0    59.0
    2    12.0    44.0
    3    36.0    63.0
    4    77.5    87.5
    5    -7.5    30.5
    6    32.5    57.5
    7   -17.0    51.0
    8    36.0    -9.0
    9    42.0    75.5
   10    65.0    28.0
   11    30.5    82.0
   12    45.5    65.0

MTB > MANNWHITNEY C1 C2

Mann-Whitney Confidence Interval and Test

C1          N =  12    MEDIAN =      34.25
C2          N =  12    MEDIAN =      58.25
POINT ESTIMATE FOR ETA1-ETA2 IS     -23.00
95.4  PCT C.I. FOR ETA1-ETA2 IS (  -46.99,    1.49)
W =   118.0
TEST OF ETA1 = ETA2  VS.  ETA1 N.E. ETA2 IS SIGNIFICANT AT  0.0690

CANNOT REJECT AT ALPHA = 0.05

MTB > STOP
```

17.5 SPEARMAN'S RANK CORRELATION COEFFICIENT

The rank correlation coefficient provides a nonparametric procedure for measuring the strength of the relationship between two variables. It is used when we are working with the rankings of individual values for the two variables. In addition, it can be used to provide a general measure of the tendency for one variable to increase with another when the variables are not linearly related (and hence when the correlation coefficient of Section 12.4 is inappropriate). To see how it is used, we will look at an example.

Each of ten judges was asked to rate an experimental mattress according to both firmness and comfort, with scores to be assigned in the range from 0 to 7.0. Higher scores indicate greater firmness or comfort. The results of the study are presented in the second and fourth columns of Table 17.8.

TABLE 17.8 *Results of the Experimental Mattress Study*

Judge	Firmness	Rank, x	Comfort	Rank, y
1	2.5	2	5.0	7
2	3.0	4	4.7	6
3	5.0	8	3.0	1.5
4	4.0	7	4.2	4
5	3.5	6	4.5	5
6	2.0	1	3.0	1.5
7	3.3	5	5.9	9
8	5.2	9	5.5	8
9	2.8	3	3.2	3
10	5.8	10	6.1	10

The rank correlation coefficient is computed as follows. First, rank the measurements on each of the variables from smallest to largest. The score 2.0 is the smallest score for firmness; it receives rank 1. The measurement 2.5, which is the next smallest, receives rank 2, and so on. These ranks are denoted by x and are listed in the third column of Table 17.8. The smallest measurement for comfort is 3.0, but two judges gave this same rating. The procedure used for all tied scores is to give each of them a rank equal to the average of their occupied ranks. In this case the tied scores occupy ranks 1 and 2. Hence they both receive a rank of 1.5, the average of 1 and 2. Ranks assigned to comfort scores, denoted by y, are listed in the fifth column of Table 17.8.

Spearman rank correlation coefficient

The **Spearman rank correlation coefficient** $\hat{\rho}_s$ for the sample data can be calculated in the same way as the linear correlation coefficient of Section 12.4, with the exception that x and y now denote ranks rather than actual measurements. Thus

$$\hat{\rho}_s = \frac{S_{xy}}{\sqrt{S_{xx}S_{yy}}}$$

When no ties are present, the preceding complicated formula for $\hat{\rho}_s$ reduces to this simpler form:

$$\hat{\rho}_s = 1 - \frac{6(\Sigma\, d^2)}{n(n^2 - 1)}$$

where n is the number of pairs of observations and d represents the difference $(x - y)$ between a pair of ranks x and y. Even when ties occur, little error will result when using this formula if the number of ties is small in relation to the number of data points.

Example 17.5

Calculate the Spearman rank correlation coefficient $\hat{\rho}_s$ to measure the strength of the relationship between firmness and comfort for the sample data on experimental mattresses. (See Table 17.8.)

Solution   It is convenient to construct a table of ranks, squares, and cross-products to aid in our calculations; see Table 17.9.

TABLE 17.9   *Computations for Example 17.5*

Judge	Firmness	Rank, x	Comfort	Rank, y	x^2	y^2	xy
1	2.5	2	5.0	7	4	49	14
2	3.0	4	4.7	6	16	36	24
3	5.0	8	3.0	1.5	64	2.25	12
4	4.0	7	4.2	4	49	16	28
5	3.5	6	4.5	5	36	25	30
6	2.0	1	3.0	1.5	1	2.25	1.5
7	3.3	5	5.9	9	25	81	45
8	5.2	9	5.5	8	81	64	72
9	2.8	3	3.2	3	9	9	9
10	5.8	10	6.1	10	100	100	100
Total		55		55	385	384.5	335.5

Using the data in Table 17.9, we can determine the values of S_{xx}, S_{yy}, and S_{xy}, and hence we can obtain the value of $\hat{\rho}_s$ by using these values.

$$S_{xx} = \Sigma\, x^2 - \frac{(\Sigma\, x)^2}{n} = 385 - \frac{(55)^2}{10} = 82.5$$

$$S_{yy} = \Sigma\, y^2 - \frac{(\Sigma\, y)^2}{n} = 384.5 - \frac{(55)^2}{10} = 82.0$$

$$S_{xy} = \Sigma\, xy - \frac{(\Sigma\, x)(\Sigma\, y)}{n} = 335.5 - \frac{(55)(55)}{10} = 33.0$$

Thus for

$$\hat{\rho}_s = \frac{S_{xy}}{\sqrt{S_{xx}S_{yy}}}$$

we have

$$\hat{\rho}_s = \frac{33}{\sqrt{(82.5)\,(82)}} = .40$$

Other than observing the sign of $\hat{\rho}_s$ (which indicates a positive or a negative association) and the relative magnitude of $\hat{\rho}_s$, the only way for us to interpret $\hat{\rho}_s$ is to conduct a statistical test to see if the corresponding population rank correlation coefficient ρ_s is different from zero. But before introducing the statistical test, we will show the equivalence of the shortcut formula for $\hat{\rho}_s$ with an example.

Example 17.6

As indicated previously, the formula $\hat{\rho}_s = S_{xy}/\sqrt{S_{xx}S_{yy}}$ reduces to

$$\hat{\rho}_s = 1 - \frac{6\,\Sigma\,d^2}{n(n^2 - 1)}$$

when no ties occur. Even when a few ties occur, very little error will be introduced if we use the simpler formula. Compute $\hat{\rho}_s$ for the ranked data of Example 17.5 by using the simpler computational formula shown in Table 17.10.

TABLE 17.10 Computations for Calculating $\hat{\rho}_s$

x	y	$d = x - y$	d^2
2	7	−5	25
4	6	−2	4
8	1.5	6.5	42.25
7	4	3	9
6	5	1	1
1	1.5	−.5	.25
5	9	−4	16
9	8	1	1
3	3	0	0
10	10	0	0
Total		0	98.50

Solution   The data in Table 17.10 are useful for computing ρ_s. Substituting into the formula, with $n = 10$, we have

$$\hat{\rho}_s = 1 - \frac{6(98.50)}{10(100 - 1)} = 1 - .60 = .40$$

Note that this result is identical to that obtained in Example 17.5.

If the rank correlation coefficient ρ_s is computed from ranks of two variables, we can test whether or not there is a relationship (positive or negative) between the two variables by using the following procedure.

Large-Sample Rank Correlation Test for ρ_s

Null hypothesis: $\rho_s = 0$ (that is, no association between the ranks x and y).

Alt. hypothesis: For a one-tailed test:
1. $\rho_s > 0$
2. $\rho_s < 0$

For a two-tailed test:
3. $\rho_s \neq 0$

Test statistic: $z = \hat{\rho}_s \sqrt{n - 1}$

Rejection region: For $\alpha = .05$ (.01) and for a one-tailed test:
1. Reject H_0 if $z > 1.645$ (2.33)
2. Reject H_0 if $z < -1.645$ (-2.33)

For $\alpha = .05(.01)$ and for a two-tailed test:
3. Reject H_0 if $|z| > 1.96$ (2.58)

Note: The number n of pairs of ranks must be 10 or more.

Example 17.7

Test the hypothesis that there is a positive association between pairs of comfort and firmness scores for the experimental mattress data in Table 17.8. Use the rank order correlation coefficient. Use $\alpha = .05$.

Solution   The four parts of the statistical test are as follows:

Null hypothesis: $\rho_s = 0$
Alt. hypothesis: $\rho_s > 0$
Test statistic: $z = \hat{\rho}_s \sqrt{n - 1}$

Rejection region: For $\hat{\alpha} = .05$, reject H_0 if $z > 1.645$

Recall that the computed value of $\hat{\rho}_s$ for this problem was found to be .40. The test statistic z is then

$$z = (.40)\sqrt{10 - 1} = (.40)(3) = 1.20$$

Since $z = 1.20$, z does not fall in the rejection region, and we have insufficient evidence to conclude that there is a positive association between the ranks of firmness and comfort. Consequently, we have insufficient information to suggest that the actual firmness and comfort scores are related.

In this section we presented a nonparametric measure of the association between two variables, which is useful for data that can be ranked. No assumption is made concerning the distribution of the populations. This test is also useful when the two variables are related but not in a linear sense. The test statistic for a test of the hypothesis "no association" involves the rank correlation coefficient, which is merely the correlation coefficient for the ranked observations. A test, based on the Student's t-test, that is more exact for small samples is described in the text by Bradley (see the References at the end of the text).

17.6   USING COMPUTERS

The Wilcoxon rank sum test (also called the Mann-Whitney test) for comparing independent random samples from two populations can be done using Minitab and SAS. This has been illustrated in the output shown here for the data of Table 17.6. You should compare the output from the two systems to locate the quantities that we computed in Example 17.4. Additional Minitab output for a Wilcoxon rank sum test is shown in Exercise 17.18.

The Spearman rank correlation coefficient can also be computed using the SAS and Minitab software systems. In addition, using PROC CORR SPEARMAN in SAS, one gets the results of a t-test of H_0: $\rho_s = 0$. Note the similarities and differences in the rank correlation output from SAS and Minitab for the data of Table 17.8. Additional Minitab output is shown in Exercise 17.31.

You can refer to the programs presented here in order to analyze other data sets. The reference manuals for SAS and Minitab should be used to obtain more details about the various procedures and options for these two software systems.

Minitab Output

```
MTB > READ INTO C1 C2
DATA>  10.2    11.0
DATA>  10.3    11.2
DATA>  10.4    11.2
DATA>  10.6    11.2
DATA>  10.6    11.4
DATA>  10.7    11.5
DATA>  10.8    11.6
DATA>  10.8    11.7
DATA>  10.9    11.8
DATA>  11.1    11.9
DATA>  11.1    11.9
DATA>  11.3    12.1
DATA> END
     12 ROWS READ
MTB > PRINT C1 C2

  ROW    C1     C2

    1    10.2   11.0
    2    10.3   11.2
    3    10.4   11.2
    4    10.6   11.2
    5    10.6   11.4
    6    10.7   11.5
    7    10.8   11.6
    8    10.8   11.7
    9    10.9   11.8
   10    11.1   11.9
   11    11.1   11.9
   12    11.3   12.1


MTB > MANNWHITNEY C1 C2

Mann-Whitney Confidence Interval and Test

C1          N =  12     MEDIAN =       10.750
C2          N =  12     MEDIAN =       11.550
POINT ESTIMATE FOR ETA1-ETA2 IS       -0.800
95.4  PCT C.I. FOR ETA1-ETA2 IS (   -1.100,   -0.500)
W =      84.0
TEST OF ETA1 = ETA2  VS.  ETA1 N.E. ETA2 IS SIGNIFICANT AT  0.0002

MTB > STOP
```

SAS Output

```
OPTIONS LS=78 PS=60 NODATE NONUMBER;
DATA RAW;
  INPUT STATUS $ OXYGEN @@;
  LABEL OXYGEN = 'OXYGEN MEASUREMENT (PPM)';
```

(continued)

```
       CARDS;
       BEFORE 10.2    AFTER   11.0   BEFORE 10.3
       BEFORE 10.6    AFTER   11.2   BEFORE 10.6
       BEFORE 10.8    AFTER   11.6   BEFORE 10.8
       BEFORE 11.1    AFTER   11.9   BEFORE 11.1
       AFTER  11.2    BEFORE  10.4   AFTER  11.2
       AFTER  11.4    BEFORE  10.7   AFTER  11.5
       AFTER  11.7    BEFORE  10.9   AFTER  11.8
       AFTER  11.9    BEFORE  11.3   AFTER  12.1
    ;
    PROC PRINT N;
    TITLE1 'DISSOLVED OXYGEN MEASUREMENTS BEFORE AND AFTER CLEANUP';
    TITLE2 'LISTING OF THE DATA';

    PROC NPAR1WAY WILCOXON;
      VAR OXYGEN;
      CLASS STATUS;
    TITLE2 'EXAMPLE FOR PROC NPAR1WAY WITH THE WILCOXON OPTION';
    RUN;
```

```
             DISSOLVED OXYGEN MEASUREMENTS BEFORE AND AFTER CLEANUP
                           LISTING OF THE DATA


                     OBS     STATUS     OXYGEN

                       1     BEFORE      10.2
                       2     AFTER       11.0
                       3     BEFORE      10.3
                       4     BEFORE      10.6
                       5     AFTER       11.2
                       6     BEFORE      10.6
                       7     BEFORE      10.8
                       8     AFTER       11.6
                       9     BEFORE      10.8
                      10     BEFORE      11.1
                      11     AFTER       11.9
                      12     BEFORE      11.1
                      13     AFTER       11.2
                      14     BEFORE      10.4
                      15     AFTER       11.2
                      16     AFTER       11.4
                      17     BEFORE      10.7
                      18     AFTER       11.5
                      19     AFTER       11.7
                      20     BEFORE      10.9
                      21     AFTER       11.8
                      22     AFTER       11.9
                      23     BEFORE      11.3
                      24     AFTER       12.1

                           N = 24

             DISSOLVED OXYGEN MEASUREMENTS BEFORE AND AFTER CLEANUP
               EXAMPLE FOR PROC NPAR1WAY WITH THE WILCOXON OPTION

                  N P A R 1 W A Y   P R O C E D U R E
```

(continued)

```
        Wilcoxon Scores (Rank Sums) for Variable OXYGEN
                 Classified by Variable STATUS

                        Sum of      Expected       Std Dev       Mean
STATUS        N         Scores      Under H0       Under H0      Score

BEFORE       12          84.0        150.0       17.2903592      7.0
AFTER        12         216.0        150.0       17.2903592     18.0
                 Average Scores were used for Ties
        Wilcoxon 2-Sample Test (Normal Approximation)
        (with Continuity Correction of .5)

        S= 84.0000      Z= -3.78824      Prob > |Z| =    0.0002
        T-Test approx. Significance =       0.0010

        Kruskal-Wallis Test (Chi-Square Approximation)
        CHISQ= 14.571      DF= 1      Prob > CHISQ=      0.0001
```

Minitab Output

```
MTB > READ INTO C1 C3
DATA>   2.5      5.0
DATA>   3.0      4.7
DATA>   5.0      3.0
DATA>   4.0      4.2
DATA>   3.5      4.5
DATA>   2.0      3.0
DATA>   3.3      5.9
DATA>   5.2      5.5
DATA>   2.8      3.2
DATA>   5.8      6.1
DATA> END
     10 ROWS READ

MTB > LET C2 = RANKS (C1)
MTB > LET C4 = RANKS (C3)
MTB > PRINT C1-C4

ROW     C1      C2      C3      C4

 1      2.5      2      5.0     7.0
 2      3.0      4      4.7     6.0
 3      5.0      8      3.0     1.5
 4      4.0      7      4.2     4.0
 5      3.5      6      4.5     5.0
 6      2.0      1      3.0     1.5
 7      3.3      5      5.9     9.0
 8      5.2      9      5.5     8.0
 9      2.8      3      3.2     3.0
10      5.8     10      6.1    10.0

MTB > CORRELATION COEFFICIENT BETWEEN C2 AND C4

Correlation of C2 and C4 = 0.401

MTB > STOP
```

SAS Output

```
OPTIONS LS=78 PS=60 NODATE NONUMBER;
DATA RAW;
  INPUT JUDGE FIRMNESS COMFORT;
  CARDS;
  1    2.5    5.0
  2    3.0    4.7
  3    5.0    3.0
  4    4.0    4.2
  5    3.5    4.5
  6    2.0    3.0
  7    3.3    5.9
  8    5.2    5.5
  9    2.8    3.2
 10    5.8    6.1
;
PROC PRINT N;
TITLE1 'RESULTS OF THE EXPERIMENTAL MATTRESS STUDY';
TITLE2 'LISTING OF THE DATA';

PROC CORR SPEARMAN;
  VAR FIRMNESS COMFORT;
TITLE2 'EXAMPLE FOR PROC CORR WITH THE SPEARMAN OPTION';
RUN;
```

```
              RESULTS OF THE EXPERIMENTAL MATTRESS STUDY
                        LISTING OF THE DATA

              OBS     JUDGE     FIRMNESS     COMFORT

               1        1         2.5          5.0
               2        2         3.0          4.7
               3        3         5.0          3.0
               4        4         4.0          4.2
               5        5         3.5          4.5
               6        6         2.0          3.0
               7        7         3.3          5.9
               8        8         5.2          5.5
               9        9         2.8          3.2
              10       10         5.8          6.1

                        N = 10


              RESULTS OF THE EXPERIMENTAL MATTRESS STUDY
            EXAMPLE FOR PROC CORR WITH THE SPEARMAN OPTION

                        CORRELATION ANALYSIS

                2 'VAR' Variables:  FIRMNESS COMFORT
```

```
                    Simple Statistics

     Variable          N        Mean      Std Dev      Median

     FIRMNESS         10      3.71000     1.25914      3.40000
     COMFORT          10      4.51000     1.15897      4.60000

              Simple Statistics

     Variable       Minimum      Maximum

     FIRMNESS       2.00000      5.80000
     COMFORT        3.00000      6.10000


Spearman Correlation Coefficients / Prob > |R| under Ho: Rho=0 / N = 10

                           FIRMNESS          COMFORT

           FIRMNESS        1.00000          0.40122
                           0.0              0.2505

           COMFORT         0.40122          1.00000
                           0.2505           0.0
```

SUMMARY

In this chapter we have presented a few of the many nonparametric statistical procedures that are available to users of statistics. It is important to be aware of these techniques as potential alternatives to standard parametric procedures when one or more of the underlying assumptions of those procedures is violated.

The sign test, perhaps the simplest of all nonparametric statistical tests, can be used to compare two populations based on paired data. The null hypothesis tested is that the two populations have identical probability distributions. The test is based on the number of pairs for which the measurement in sample 1 is greater than the corresponding measurement in sample 2.

The Wilcoxon signed-rank test provides an alternative to the two-sample paired t-test of Section 14.3. In contrast, the Wilcoxon rank sum test is used as an alternative to the two-sample unpaired t-test of Section 9.4 when one or more of the underlying assumptions for the parametric test is violated. It, too, is easy to use and is based on the ranks of the measurements in the two samples.

Spearman's rank correlation coefficient $\hat{\rho}_s$ is an alternative to the correlation coefficient (of Section 12.4) and can be used to measure the strength of the relation between two variables based on the ranks of individual values. A test of the null hypothesis H_0: $\rho_s = 0$ can also be performed.

KEY TERMS

ordinal data
nonparametric statistical tests

Spearman rank correlation
  coefficient

KEY FORMULAS

1. Sign test

Test statistic: $z = \dfrac{x - .5n}{\sqrt{.25n}}$

2. Wilcoxon's signed-rank test

Test statistic: $T = |T_-|$ or T_+
(small sample)

Test statistic: $z = \dfrac{T - \mu_T}{\sigma_T}$
(large sample)

where $\mu_T = \dfrac{n(n + 1)}{4}$

and $\sigma_T = \sqrt{\dfrac{n(n + 1)(2n + 1)}{24}}$

3. Wilcoxon's rank sum test

Test statistic: $z = \dfrac{T - \mu_T}{\sigma_T}$

where $\mu_T = \dfrac{n_1(n_1 + n_2 + 1)}{2}$

and $\sigma_T = \sqrt{\dfrac{n_1 n_2 (n_1 + n_2 + 1)}{12}}$

4. Spearman's rank correlation coefficient

$$\hat{\rho}_s = \frac{S_{xy}}{\sqrt{S_{xx} S_{yy}}}, \text{ based on ranks}$$

$$\hat{\rho}_s = 1 - \frac{6\Sigma\, d^2}{n(n^2 - 1)}, \text{ when there are no ties}$$

Test for ρ_s

Test statistic: $z = \hat{\rho}_s \sqrt{n - 1}$

SUPPLEMENTARY EXERCISES

17.19 Ten sets of identical twins were administered drug products, and the increase in their pulse rates was observed. One twin of each set received drug A; the other received drug B. The purpose of the experiment was to determine if differences exist in the mean increases in pulse rate for the population of pulse rates for those administered drug A and the population corresponding to persons on drug B. The accompanying data were observed. Test the hypothesis "no difference in mean increase," using the sign test. (Note: We have no reason to believe one mean increase is greater than the other prior to conducting the test.) Use $\alpha = .05$.

Identical twin pair	Drug A	Drug B	Identical twin pair	Drug A	Drug B
1	12	19	6	12	15
2	14	13	7	13	10
3	8	6	8	15	18
4	11	24	9	18	21
5	14	12	10	17	22

17.20 Refer to Exercise 17.19. Conduct a Wilcoxon signed-rank test on these data and compare your results. What other test might be appropriate?

17.21 A panel of twelve people was asked to rate two new brands of coffee blends using a scale from 0 (terribly disagreeable) to 10 (superb).

Person	1	2	3	4	5	6	7	8	9	10	11	12
Blend 1	8	4	7	8	5	4	6	4	4	6	7	8
Blend 2	6	5	4	6	6	3	5	4	3	5	8	6

(a) Compute the differences (Blend 1/Blend 2).

(b) Use a sign test to compare the ratings for the two blends. Use $\alpha = .05$.

17.22 Refer to Exercise 17.21. Conduct a signed-rank test on the differences and compare your conclusions to those in Exercise 17.21.

§ 17.23   A panel of twelve taste testers was asked to compare a domestic and a foreign wine (denoted by A and B, respectively). Judges were asked to rate each wine on a seven-point scale (1, 2, . . . , 7), where 7 denotes the highest score. The results of the experiment are listed in the accompanying table. Use the sign test with $\alpha = .05$ to determine if the data indicate a difference in ratings for the domestic and the foreign wine.

Judge	Wine A	Wine B	Judge	Wine A	Wine B
1	7	4	7	4	4
2	5	4	8	6	4
3	4	3	9	5	3
4	7	6	10	4	5
5	6	7	11	7	3
6	5	2	12	6	5

17.24   Two judges rated each of twelve beauty pageant contestants on their poise, using a 10-point scale (1, 2, . . . , 10). The results of the judging are listed in the accompanying table.

	Contestant											
	1	2	3	4	5	6	7	8	9	10	11	12
Judge 1	5	4	3	10	3	9	10	1	8	6	3	4
Judge 2	7	8	4	6	5	8	10	3	7	5	8	4

(a) Calculate the rank correlation coefficient to measure the strength of the relationship between the scores given by the two judges.

(b) Conduct a test of the hypothesis that there is no relationship between the two judges' scores. Use $\alpha = .05$.

17.25   Ten pairs of identical twins participated in an experiment to investigate two methods of teaching children to read music. One child from each pair was taught to read music by method A; the other received instruction according to method B. The results of an examination given at the end of a six-week training period are presented in the accompanying table.

	Pair									
	1	2	3	4	5	6	7	8	9	10
Method A	75	80	67	73	93	88	70	95	84	92
Method B	73	76	65	70	95	82	65	85	83	95

(a) Replace the scores for the children taught by method A with ranks. Do the same for the scores from method B. Calculate the rank correlation coefficient to measure the strength of the relationship between methods A and B.

(b) Conduct a test of the alternative hypothesis H_a: $\rho_s \neq 0$. Use $\alpha = .05$.

17.26 Use a sign test on the original data of Exercise 17.25 to determine if there is a difference in the levels of achievement for the two methods of teaching children to read music ($\alpha = .05$).

17.27 Two deans rated each of fifteen fellowship applicants in terms of their expected academic success. The ratings, based on a scale of 1 to 5, are presented in the accompanying table. Do the data present sufficient evidence to indicate a difference in rating scales for the two deans? Base your conclusions on the sign test.

Applicant	Dean A	Dean B	Applicant	Dean A	Dean B
1	2	4	9	5	4
2	3	5	10	2	3
3	2	1	11	5	5
4	1	3	12	1	3
5	2	5	13	4	5
6	4	1	14	3	3
7	4	4	15	4	5
8	1	5			

17.28 Examine the accompanying computer output for the data of Example 17.4.

(a) Locate T in the output.

(b) Set up all parts of a test of the null hypothesis that the two populations are identical.

(c) Draw conclusions, using $\alpha = .05$.

```
WILCOXON RANK SUM TEST
                   X                Y
OBSERVATIONS    Before Cleanup    After Cleanup
                   10.20             11.00
                   10.30             11.20
                   11.40             11.20
                   10.60             11.20
                   10.60             11.40
                   10.70             11.50
                   10.80             11.60
```

(continued)

	10.80	11.70
	10.90	11.80
	11.10	11.90
	11.10	11.90
	11.30	12.10
NO. OBSERVATIONS	12.00	12.00
MEDIANS	10.75	11.50
SUM OF RANKS	84.00	216.00

17.29 **Class Exercise** Conduct a class experiment involving some aspect of marketing research or consumer preference sampling. For example, it might be of interest to conduct a consumer survey to determine whether there is a preference for one of two marketed products in a given category (brands of potato chips, beers, and so on). Alternatively, the class experiment could be involved with an election or local bond issue. For the problem the class should do the following:

(a) Develop a questionnaire (this could include a rating scale).

(b) Develop a procedure for identifying persons to be included in the survey.

(c) Collect the sample data.

(d) Analyze the data, using one or more of the nonparametric methods of this chapter.

(e) Draw conclusions and critique the procedure.

17.30 Fourteen persons were given tests on economics. Seven were randomly selected and given a test in an old room with obsolete dark furniture, a high ceiling, and poor lighting. After a short rest period, the group of seven was given a similar test in an attractive well-lit room with modern furnishings. For the other seven patrons, the procedure was reversed. The results of the tests are shown here.

Person	1	2	3	4	5	6	7	8	9	10	11	12	13	14
Old room	55	58	84	78	86	67	94	77	72	92	77	87	74	63
Modern room	63	50	83	92	89	78	93	78	74	93	77	93	78	60

Does the sample provide sufficient evidence to conclude that room environment had a positive effect on the exam scores? Use a sign test with $\alpha = .05$.

17.31 The data below show actual population changes for the years of 1970–1979 and predicted population changes for the years of 1980–1989 for a selected list of cities. The numerical values show percentages.

Area	Actual percentage change 1970–1979	Predicted percentage change 1980–1989
Las Vegas	69.1	31.3
Phoenix	55.6	23.7
Orlando	53.3	16.1
San Diego	37.0	22.6
Austin	47.8	28.4
Tampa	42.4	19.0
Sacramento	25.3	12.2
Seattle	12.4	20.1
Tulsa	23.6	21.8
Salt Lake City	32.6	21.5
Pittsburgh	−5.8	−6.2
Boston	−4.8	−1.4

Using Spearman's rank correlation and the Minitab output shown here, test the hypothesis that there is no correlation in the ranking of the cities for the actual percentage changes and the predicted percentage changes. Use $\alpha = .05$.

```
MTB > READ INTO C1 C3
DATA>   69.1    31.3
DATA>   55.6    23.7
DATA>   53.3    16.1
DATA>   37.0    22.6
DATA>   47.8    28.4
DATA>   42.4    19.0
DATA>   25.3    12.2
DATA>   12.4    20.1
DATA>   23.6    21.8
DATA>   32.6    21.5
DATA>   -5.8    -6.2
DATA>   -4.8    -1.4
     12 ROWS READ

MTB > LET C2 = RANKS (C1)
MTB > LET C4 = RANKS (C3)
MTB > PRINT C1-C4

 ROW    C1     C2     C3     C4

   1   69.1    12   31.3    12
   2   55.6    11   23.7    10
   3   53.3    10   16.1     4
   4   37.0     7   22.6     9
   5   47.8     9   28.4    11
   6   42.4     8   19.0     5
   7   25.3     5   12.2     3
```

(continued)

```
       8    12.4     3    20.1     6
       9    23.6     4    21.8     8
      10    32.6     6    21.5     7
      11    -5.8     1    -6.2     1
      12    -4.8     2    -1.4     2

MTB > CORRELATION COEFFICIENT BETWEEN C2 AND C4

Correlation of C2 and C4 = 0.706

MTB > STOP
```

17.32 A company is currently experimenting with two different procedures for adjusting folding machines used in greeting card processing. Observations were randomly selected for each procedure and the times of completion were recorded.

First technique	15	13	12	14	16	15	17	18	15	14	14	16	15
Second technique	13	13	14	14	15	12	14	16	16	13	14	13	12
	17	13											

Use the Wilcoxon rank-sum test to compare the two groups of adjustment times, with $\alpha = .05$. Interpret your conclusions.

17.33 Two students at a university were asked to rate rock artists on their levels of competencies, with 1 being assigned to the lowest-ranked artist and 10 being assigned to the highest-ranked artist. The ratings of the two students were as shown below.

Rock artist number	1	2	3	4	5	6	7	8	9	10
First student ranks	2	6	10	3	7	1	9	5	8	4
Second student ranks	5	4	1	9	3	2	8	6	10	7

Test the hypothesis that there is no association between the rankings of the two students. Use Spearman's rank correlation procedures when making your test. Set $\alpha = .05$.

EXERCISE FROM THE DATA BASE

17.34 Use the HAM-D sleep disturbance data and the HAM-D anxiety score data in the clinical data base (Appendix 1) to compute Spearman's rank correlation coefficient. Note that you will first have to rank the sleep disturbance and anxiety data separately. What can you conclude from the computed value of $\hat{\rho}_s$?

PART 6

STEP FOUR: REPORTING RESULTS

CHAPTER 18

COMMUNICATING THE RESULTS OF STATISTICAL ANALYSES

18.1 Introduction ▪ **18.2** Good communication is not easy ▪ **18.3** Communication hurdles: Graphical distortions ▪ **18.4** Communication hurdles: Biased samples ▪ **18.5** Communication hurdles: What's the sample size? ▪ **18.6** The statistical report ▪ Summary

18.1 INTRODUCTION

In Chapter 1 we introduced the subject of statistics as a study of making sense of data and identified the four major components of making sense of data, namely, data gathering, data summarization, data analysis, and communicating results. In this chapter, we will deal with methods or ways to communicating the results of statistical analyses. But rather than tell you what to write, which, of course, depends on the particular problem being discussed, the intended audience, and the form of the communication, we will consider some important elements of a statistical report. We will also discuss some of the potential pitfalls to effective communication.

18.2 GOOD COMMUNICATION IS NOT EASY

We've spent time throughout the text making sense of data; and the final step in this process is the communication of results. How might one communicate the results of a study or survey? The list of possibilities is almost endless, including all forms of verbal and written communication. There is quite a range of possibilities for verbal and written communication. For example, written communication within a company can vary from an informal short note or memo to a formal project report (Figure 18.1).

FIGURE 18.1 Forms of Written and Verbal Communication

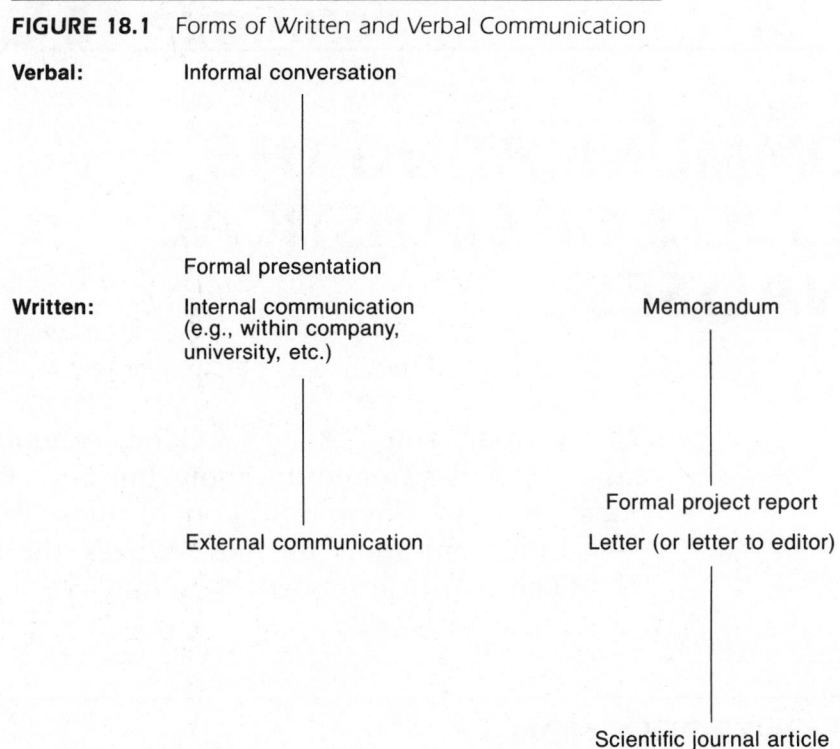

Verbal: Informal conversation

            Formal presentation

Written: Internal communication   Memorandum
            (e.g., within company,
            university, etc.)

                                                    Formal project report
            External communication   Letter (or letter to editor)

                                                    Scientific journal article

Communicating the results of a statistical analysis in concise, unambiguous terms is difficult. In fact, descriptions of most things are difficult. Try, for example, to describe the person sitting next to you so precisely that a stranger could select the individual from a group of others having similar physical characteristics. It is not an easy task. Fingerprints, voice-prints, and photographs—all pictorial descriptions—are the most precise methods of human identification. The description of a set of measurements is also a difficult task. But, like the description of a person, it can be accomplished more easily by using graphic or pictorial methods.

Cave drawings convey to us scattered bits of information about the life of prehistoric people. Similarly, vast quantities of knowledge about the ancient lives and cultures of the Babylonians, Egyptians, Greeks, and Romans are brought to life by means of drawings and sculpture. Art has been used to convey a picture of various life-styles, history, and culture in all ages. Not surprisingly, use of graphs and tables in conjunction with a written description can help to convey the meaning of a statistical analysis.

In reading the results of a statistical analysis and in communicating the results of our own analyses, we must be careful not to distort them because of the way the data and results are presented. You have all heard the expression, "It is easy to lie with statistics." The idea is *not* new. The famous British statesman Disraeli is quoted as saying, "There are three kinds of lies: lies, damned lies, and statistics." Where do things go wrong?

First of all, distortion of truth can occur only when we communicate. And since communication can be accomplished with graphs, pictures, sound, aroma, taste, words, numbers, or any other means devised to reach our senses, distortions can occur using any one or any combination of these methods of communication.

In this respect, statements that we make could be misleading to others because we might have omitted something in the explanation of the data-gathering stage or with the analyses done. For example, we might unintentionally fail to clearly explain the meaning of a numerical statement. Or, we might omit some background information that is necessary for a clear interpretation of the results. Even a correct statement may appear to be distorted if the reader lacks knowledge of elementary statistics. Thus a very clear expression of an inference using a 95% confidence interval is meaningless to a person who has not been exposed to the introductory concepts of statistics.

Now we will look at some potential hurdles to effective communication that must be carefully considered when we present the results of a statistical analysis—or when we try to interpret what someone else has presented.

18.3 COMMUNICATION HURDLES: GRAPHICAL DISTORTIONS

Pictures can easily distort the truth. We have seen clothing advertisements featuring beautiful people who are able to create an almost uncontrollable urge in us to dash out and buy whatever is modeled. Mail-order catalog sketches of products are frequently more attractive than the real thing, but we usually take this type of "distortion" for granted. Statistical pictures are the histograms, frequency polygons, pie charts, and bar graphs of Chapter 4. These drawings or displays of numerical results are difficult to combine with sketches of lovely women or handsome men and hence are secure from the most common form of graphic distortion. But other distortions are possible. One could inadvertently shrink or stretch the axes, thus distorting the actual results. The basic idea is that shallow and steep slopes are associated with small and large increases, respectively.

For example, suppose that the values of a leading consumer price index over the first six months of the year were 160, 165, 178, 189, 196, and 210. We might show the upward movement of this consumer price index by using the frequency polygon of Figure 18.2. In this graph the

---

FIGURE 18.2   Changes in a Consumer Price Index

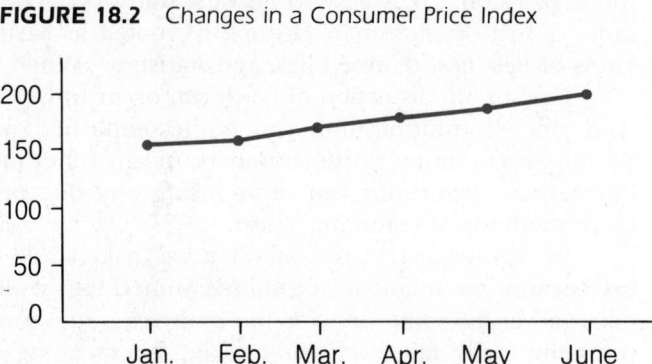

increase in the index is apparent, but it does not appear to be very great. On the other hand, we could present the sample data in a much different light as shown in Figure 18.3. For this graph the vertical axis is stretched and does not include zero. Note the impression of a substantial rise that is indicated by the steeper slope. Another way to achieve the same effect—to decrease or increase a slope—is to stretch or shrink the horizontal axis.

FIGURE 18.3   Changes in a Consumer Price Index

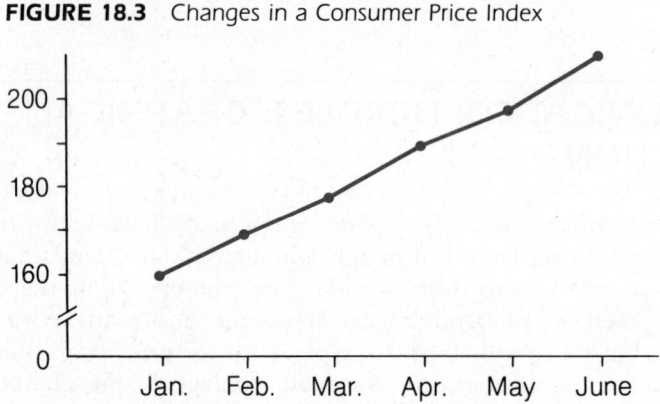

When we present data in the form of bar graphs, histograms, frequency polygons, or other figures we must be careful not to shrink or stretch axes because it will catch most readers off guard. Increases or decreases in responses should be judged large or small depending on the arbitrary importance to the observer of the change, not on the slopes shown in graphic representations. In reality, most people look only at the slopes in the "pictures."

18.4 COMMUNICATION HURDLES: BIASED SAMPLES

One of the most common statistical distortions occurs because the experimenter unwittingly or sometimes knowingly samples the wrong population. That is, he or she draws the sample from a set of measurements that is not the proper population of interest.

For example, suppose that we want to assess the reaction of taxpayers to a proposed park and recreation center for children. A random sample of households is selected, and interviewers are sent to those households in the sample. Unfortunately, no one is at home in 40% of the sample households, so we randomly select and substitute other households in the city to make up the deficit. The resulting sample is selected from the wrong population, and the sample is therefore said to be biased.

The specified population of interest in the household survey is the collection of opinions that would be obtained from the complete set of all households in the city. In contrast, the sample was drawn from a much smaller population or subset of this group—the set of opinions from householders who were at home when the sample was taken. It is possible that the fractions of householders favoring the park in these two populations are equal, and no damage was done by confining the sampling to those at home. But it is much more likely that those at home had small children and that this group would yield a higher fraction in favor of the park than would the city as a whole. Thus we have a biased sample because it is loaded in favor of families with small children. Perhaps a better way to see the difficulty is to note that we unwittingly selected the sample only from a special subset of the population of interest.

Biased samples frequently result from surveys that utilize mailed questionnaires. In a sense the investigator lets the selection and number of the sampling units depend on the interests, available time, and various other personal characteristics of the individuals who receive the questionnaires. Extremely busy and energetic people may drop the questionnaires in the nearest wastebasket; you rarely hear from those low-energy folk who are uninterested cr who are engrossed with other activities. Most often, the respondents are activists—those who are highly in favor, those who are opposed, or those who have something to gain from a certain outcome of the survey.

Although numerous newscasters and analysts utilize election results as an expression of public opinion on major issues, it is a well-known fact that voting results represent a biased sample of public opinion. Those who vote represent much less than half of the eligible voters; they are individuals who desire to exercise their rights and responsibilities as citizens or are individuals who have been specially motivated to participate. The resultant subset of voters is not representative of the interests and opinions of all eligible voters in the country.

Sampling the wrong population also occurs when people attempt to extrapolate experimental results from one population to another. Numerous experimental results have been published about the effect of various products (e.g., saccharin) in inducing cancer in moles, rats, the breasts of beagles, and so forth. These results are often used to imply that humans have a high risk of developing cancer after frequent or extended exposure to the product. These inferences are not always justified, because the experimental results were not obtained on humans. It is quite possible that humans are capable of resisting much higher doses than rats, or perhaps humans may be completely resistant for some reason. Drug induction of cancer in small mammals *does* indicate a need for concern and caution by humans, but it does not prove that the drug is definitely harmful to humans. Note that we are not criticizing experimentation in various species of animals, because it is frequently the only way we can obtain any information about potential toxicity in human beings. We simply point out that the experimenter is knowingly sampling a population that is only similar (and quite likely not too similar) to the one of interest.

Engineers also test "rats" instead of "humans." "Rats" in this context are miniature models or pilot plants of a new engineering system. Experiments on the models occasionally yield results that differ substantially from the results of the larger, real systems. So again we see a sampling from the wrong population, but it is the best the engineer can do because of the economics of the situation. Funds are not usually available to test a number of full-scale models prior to production.

Many other examples could be given of biased samples or of sampling from the wrong populations. The point is that when we communicate the results of a study or survey we should be clear about how the sample was drawn and whether it was *randomly* selected from the population of interest. If this information is not given in the published results of a survey or experiment, the reader should take the inferences with a grain of salt.

18.5 COMMUNICATION HURDLES: WHAT'S THE SAMPLE SIZE?

Distortions can occur when the sample size is not discussed. For example, suppose you read that a survey indicates that approximately 75% of a sample favor a new high-rise building complex. Further investigation might reveal that the investigator sampled only four people. When three out of the four favored his project, he decided to stop the survey. Of course, we exaggerate with this example; but we could also have revealed inconclusive results based on a sample of 25, even though many buyers would consider this sample size to be large enough. As you well know, very large samples are required to achieve adequate information in sampling binomial populations.

Fortunately, many publications are now providing more information about the sample size and how opinion surveys are conducted. Ten years ago, it was rare to find how many people were sampled, much less how they were sampled. Things are different now. In fact, sometimes the media have gone too far in an attempt to be completely open about how a survey was done. A case in point is the following article from the *Wall Street Journal* (Sept. 23, 1988). How many of us understand much more than the number of persons sampled and the approximate plus or minus (confidence interval)? It would take a person well trained in statistics and survey sampling to interpret what was done. Again, the moral of the story is, try to communicate in unambiguous terms.

How Poll Was Conducted

The Wall Street Journal/NBC News poll was based on nationwide telephone interviews conducted last Friday through Monday with 4159 adults age 18 or older. There were 2630 likely voters.

The sample was drawn from a complete list of telephone exchanges, chosen so that each region of the country was represented in proportion to its population. Households were selected by a method that gave all telephone numbers, listed and unlisted, a proportionate chance of being included. The results of the survey were weighted to adjust for variations in the sample relating to education, age, race, sex and region.

Chances are 19 of 20 that if all adults in the United States had been surveyed using the same questionnaire, the findings would differ from these poll results by no more than two percentage points in either direction. The margin of error for subgroups may be larger.

18.6 THE STATISTICAL REPORT

Now that you have seen the various ways to communicate the results of a statistical analysis and some of the more common hurdles to effectively communicating these results, let us address the content of a statistical report that would appear as part of an internal project report in a book or scientific journal article. Obviously companies and journals do not all abide by the same outline for a statistical report; however, based on what we present, you should have sufficient material to rearrange some sections and delete others to satisfy specific requirements.

A statistical report of the results of a study or experiment should clearly reflect all stages of making sense of data, much as we have emphasized these stages throughout the textbook. A general outline for a statistical report is shown in Table 18.1, along with a brief description of the content for each major section of the report.

TABLE 18.1 General Outline for a Statistical Report

Outline	Stage of making sense of data	Description
Summary		Sometimes this section comprises the abstract of a journal article or scientific paper. It is a short summary of the study results, preferably one page or less.
Introduction	Data-gathering stage	The introduction gives a brief summary of the background and rationale for the study (survey) run.
Study design and procedures	Data-gathering stage	The study design and procedures section gives the study (survey) objective, the study (survey) design, and a summary of procedures for conducting the study (survey), including details about the data-gathering stage.
Descriptive statistics	Data-summarization stage	This section includes the main descriptive techniques (such as histograms, scatter plots, means, standard errors, etc.) used to summarize the study (survey) data.
Statistical methodology	Data-analysis stage	This section gives a description of the methods used to analyze the study (survey) data. This would include a brief description of the statistical tests and estimation procedures used to address the objectives of the study (survey), including those on which the major conclusions are drawn. References for the tests and estimations procedures may

TABLE 18.1   (continued)

Outline	Stage of making sense of data	Description
		also be included, especially if the intended audience may not be aware of the techniques.
Results and conclusions	Data-analysis stage	In this section, the authors address the main results of the statistical analyses and the conclusions that can be drawn from these analyses in light of the study (survey) objectives.
Discussion		The discussion section of a statistical report provides an opportunity to interpret the results of the statistical analyses and to put the conclusions in the context of previous studies (surveys). Do the results confirm or contradict previous study (survey) results? If they contradict previous results, an attempt should be made to offer a viable explanation for the difference in results. The discussion section allows one to make recommendations for further studies or surveys, when appropriate.
Data listings (optional)		Sometimes it is appropriate to provide listings of the data upon which the summaries and analyses are based. This section could also include computer output for some (all) of the statistical analyses done in the data-analysis stage of the study.

SUMMARY

In this chapter we have discussed how to present the results of a statistical analysis of data and some of the troubles that we have in effectively communicating these results to the intended audience. The task is not easy.

Some of the obstacles or hurdles standing in the way of effectively communicating the results of a statistical analyses include graphical distortions, biased sampling, and omitting a discussion of the sampling technique and sample size. With some understanding of these obstacles, we can strive for more effective communication of the results of analyses so others don't misinterpret what we have found.

DATA BASE

DATA BASE: CLINICAL TRIAL DATA

The data presented here are from a clinical trial that was conducted to compare the safety and efficacy of three different compounds (A, B, and C) and a placebo (D) in the treatment of patients who exhibited characteristic signs and symptoms of depression. Certain predrug (baseline) determinations were made on each of the 100 patients to determine suitability for the study. Then each patient who qualified for entrance into the study was assigned at random to one of the four treatment groups and was dispensed medication for the duration of the study. Neither the investigator nor the patient knew which medication had been assigned.

At the end of the study, scores on numerous anxiety and depression scales were made. Data descriptions for the variables measured are shown next.

Variable descriptions

Below is a list of the variables and codes that are used in the accompanying output:

```
PATIENT  = patient number
AGE      = age (yrs)
MAR_STAT = marital status
           1) single
           2) married
           3) separated or divorced
           4) widowed
COFF_TEA = coffee/tea consumption (cups/day)
TOBACCO  = tobacco consumption
           0) none
           1) < 1 pack daily
           2) 1 pack daily
           3) > 1 pack daily
ALCOHOL  = alcohol consumption
           0) none
           1) social drinker (< 1 drink weekly)
           2) social drinker (1 to 2 drinks weekly)
           3) 1 to 2 drinks most days
           4) 3 or more drinks most days
```

TRT_EMOT = previous treatment for emotional problems
    1) psychiatrist
    2) non-psychiatrist physician
    3) both
    4) other

HOSPITAL = hospitalization for emotional problems
    0) no
    1) yes

PSY_DIAG = psychiatric diagnosis
    1) major depressive disorder, single episode
    2) major depressive disorder, recurrent episode
    3) bipolar affective disorder
    4) chronic depressive disorder
    5) atypical depressive disorder
    6) adjustment disorder with depressed mood

ANXIETY = HAM-D anxiety score
RETARDTN = HAM-D retardation score
SLEEP = HAM-D sleep disturbance score
TOTAL = HAM-D total score
OBRIST = HOPKINS OBRIST cluster total
APPETITE = appetite disturbance score

CHANGED = how much has the patient changed
    1) very much improved
    2) much improved
    3) minimally improved
    4) no change
    5) minimally worse
    6) much worse
    7) very much worse

THER_EFF = therapeutic effect
    1) marked
    2) moderate
    3) minimal
    4) unchanged
    5) worse

ADV_EFF = adverse effects
    1) none
    2) does not significantly interfere with patient's functioning
    3) significantly interferes with patient's functioning
    4) nullifies therapeutic effect

TREATMNT = drug treatment group
    A
    B
    C
    D-Placebo (control) group

Additional cross tabulations were generated for the clinical trial data using SAS. These tabulations should be helpful in solving some of the chapter exercises that refer to this data base.

EXAMPLE DATA SET

PATIENT	AGE	MAR_STAT	COFF_FTECA	TOBACCO	ALCOHOL	TRT_HOSP	HOSPITAL	PSY_DIAG	ANXIETY	RETARDTN	SLEEP	TOTAL	OBRIST	APPETITE	CHANGED	THER_EFF	ADV_EFF	TREATMNT
1	23	2	3	1	0	0	0	1	0.33	0.75	0.00	16	56	91.0	4	4	1	D
2	18	1	0	2	0	1	0	2	0.33	1.25	0.33	12	57	42.5	3	3	1	A
3	36	2	2	2	2	1	0	2	0.50	0.25	0.33	6	40	91.0	1	1	2	B
4	51	4	5	3	0	0	0	1	0.17	0.75	0.67	6	39	61.0	2	1	1	A
5	24	1	6	2	1	1	0	2	1.00	1.00	1.00	13	49	1.5	3	3	1	B
6	59	4	3	1	0	1	0	4	0.33	1.50	0.00	11	40	72.0	2	2	1	A
7	56	1	2	0	0	1	1	4	1.67	1.75	0.00	21	44	7.5	2	2	1	B
8	70	4	1	0	0	1	0	2	0.50	1.75	0.00	12	39	92.5	2	2	2	A
9	30	3	4	3	2	1	0	2	0.83	1.00	0.67	15	49	2.0	2	2	4	D
10	55	4	2	0	2	0	0	1	0.33	1.00	1.00	11	44	31.0	2	2	1	D
11	40	2	4	2	0	1	0	2	0.83	1.50	1.33	23	79	92.0	4	4	1	C
12	61	2	2	0	1	1	1	2	0.50	0.75	0.00	8	30	1.0	2	2	2	C
13	64	2	3	2	0	1	0	2	0.33	1.50	0.00	9	48	11.5	2	2	3	A
14	19	1	10	2	2	1	0	1	0.50	0.75	0.00	7	42	91.5	2	1	1	B
15	46	3	2	0	1	1	0	1	0.17	0.25	0.00	4	35	92.0	1	1	1	A
16	36	2	10	3	0	1	0	2	0.50	1.50	0.00	9	42	72.0	2	2	1	C
17	30	2	2	0	2	0	0	1	0.00	0.50	0.67	4	35	41.0	2	1	1	B
18	34	2	8	3	1	1	0	1	0.50	1.00	0.33	14	56	91.5	3	3	1	D
19	28	1	2	0	1	0	0	1	0.67	1.00	0.33	12	43	72.5	1	2	2	B
20	33	3	3	2	1	1	0	2	0.67	0.75	0.33	8	39	93.0	2	2	2	C
21	51	3	7	3	2	1	1	5	0.83	2.00	0.00	21	99	42.5	4	4	2	A
22	51	2	0	0	0	1	0	4	0.83	1.00	0.00	12	68	61.0	3	3	3	C
23	54	2	3	0	0	0	1	2	0.83	1.00	1.00	16	49	11.0	3	3	2	B
24	35	3	5	0	2	1	0	4	0.67	1.50	0.00	11	42	61.0	2	2	3	A
25	46	2	3	0	1	0	1	2	0.67	2.00	1.33	17	63	61.5	4	4	1	D
26	34	2	7	0	0	1	1	2	0.33	0.25	0.33	6	51	12.5	2	2	2	C
27	27	3	2	1	2	1	0	5	0.83	0.75	0.00	11	58	11.0	2	2	2	A
28	23	2	0	1	0	0	0	1	0.33	0.75	0.00	6	47	91.5	2	2	2	C
29	35	3	6	0	3	1	0	5	0.33	0.50	0.00	9	68	1.0	3	2	2	B
30	19	1	2	0	0	0	0	4	0.67	0.50	0.00	11	47	2.5	3	3	2	B
31	40	3	3	0	3	0	0	1	1.00	1.75	0.00	17	71	62.0	4	4	1	D
32	52	2	5	0	0	1	1	2	0.83	1.75	0.67	19	41	91.5	3	3	2	B
33	51	3	6	0	0	1	0	2	0.83	1.00	1.67	20	63	11.0	3	4	2	D
34	34	3	0	0	2	1	0	4	0.17	0.50	0.33	5	37	13.0	2	2	1	C
35	59	2	4	2	0	1	0	1	0.67	2.50	0.00	17	54	62.5	3	3	2	C
36	31	3	2	0	1	1	1	5	0.50	1.75	0.33	12	51	11.5	3	2	2	C
37	54	2	10	0	2	1	0	2	1.17	1.25	1.33	22	96	92.5	5	5	1	A
38	63	4	2	0	2	1	0	1	0.33	0.25	0.33	7	58	32.5	2	2	2	B
39	34	2	1	2	2	0	1	1	0.50	0.25	0.33	27	90	92.0	5	5	1	D
40	30	1	1	0	1	1	0	1	1.00	0.50	0.67	13	59	33.0	2	1	1	A
41	32	3	2	3	1	1	0	2	0.67	1.25	0.67	14	58	61.0	2	2	2	C
42	21	1	2	0	3	1	0	2	0.83	1.00	1.00	20	60	41.5	3	3	2	B
43	42	2	1	2	1	1	1	2	1.00	1.50	1.00	24	85	62.0	3	2	1	B
44	60	2	0	3	0	1	0	4	0.17	0.75	0.33	5	39	1.5	1	1	1	A
45	53	2	2	0	1	1	0	2	0.67	1.00	0.67	10	38	31.0	2	2	2	B
46	54	4	4	0	1	1	0	4	1.50	1.75	0.00	14	42	93.5	2	2	1	C
47	38	2	2	1	0	1	0	2	1.50	1.75	1.00	24	85	15.0	3	4	2	D
48	41	2	4	0	0	1	1	2	0.33	0.75	0.00	11	47	2.5	2	2	1	A
49	32	3	0	0	1	1	0	4	1.00	1.00	1.00	20	35	42.5	3	2	2	B
50	43	2	4	0	0	1	0	4	0.83	1.25	1.33	21	44	31.5	4	4	1	B

EXAMPLE DATA SET

PATIENT	AGE	MAR_STAT	COFF_TEA	TOBACCO	ALCOHOL	TRT_EMOT	HOSPITL	PSYDIAG	ANXIETY	RETARD	SLEEP	TOTAL	OBRITE	APPETITE	CHANGED	THER_EFF	ADV_EFF	TREATMNT
51	51	2	1	0	1	1	0	2	0.83	2.25	0.00	20	80	61.0	5	5	1	A
52	23	2	0	1	3	1	0	2	1.33	1.25	0.67	20	39	31.5	3	3	1	A
53	55	2	2	0	0	1	0	1	0.83	1.25	0.00	16	52	12.5	3	3	1	C
54	45	1	3	3	0	1	0	2	0.33	1.25	0.00	14	46	41.5	2	2	1	C
55	30	2	1	0	0	0	0	1	1.17	1.75	0.00	17	64	2.0	3	3	2	C
56	53	4	1	3	4	0	0	1	0.83	0.50	0.00	19	82	62.0	3	4	2	D
57	45	1	3	1	3	0	0	4	0.83	0.50	0.00	8	40	2.0	1	2	1	A
58	48	2	10	1	2	1	0	4	0.33	1.00	0.00	8	32	41.0	2	2	2	B
59	49	1	4	0	3	1	0	5	0.67	1.75	0.33	16	68	12.0	2	2	2	A
60	55	2	6	0	2	0	1	4	0.50	1.00	0.00	9	42	91.0	2	2	2	A
61	33	2	1	0	1	1	0	2	1.17	2.25	2.00	32	112	42.0	4	4	1	D
62	27	1	1	2	3	1	0	5	0.17	0.00	0.00	3	34	72.5	1	1	2	C
63	30	1	2	1	3	1	0	4	0.67	0.00	0.00	6	37	73.5	1	1	2	A
64	35	2	4	3	0	1	0	2	0.50	1.50	0.33	16	43	11.0	3	3	1	A
65	55	2	4	0	0	1	0	2	1.00	2.00	0.33	24	37	92.0	3	3	1	B
66	22	3	0	1	1	1	0	2	1.00	1.50	0.33	20	46	11.5	3	3	1	D
67	37	3	1	2	0	1	0	2	0.50	0.75	0.00	11	32	41.5	2	2	1	C
68	49	2	6	3	0	1	0	2	0.50	0.75	0.00	13	54	2.5	2	1	1	C
69	21	1	0	1	0	1	0	2	1.17	2.50	2.00	34	74	11.5	5	5	1	A
70	33	3	1	3	2	1	0	1	0.17	0.50	0.00	8	34	42.5	1	1	1	C
71	35	3	10	2	0	1	0	2	1.17	2.50	0.00	24	39	92.5	4	4	2	B
72	39	3	0	0	0	1	0	2	0.50	1.25	0.00	22	66	41.0	2	3	1	D
73	34	1	0	0	0	1	0	2	0.33	0.75	0.00	14	48	71.0	2	2	1	C
74	53	3	3	1	0	1	0	2	0.50	1.00	0.00	12	39	43.5	2	2	1	A
75	34	2	0	0	0	1	0	2	0.33	0.50	0.00	10	36	3.0	2	2	1	A
76	35	2	2	0	1	1	0	2	0.83	1.00	0.33	14	35	92.0	2	2	1	B
77	32	2	0	0	0	1	0	2	0.50	1.00	0.00	12	39	91.0	2	2	1	B
78	43	2	2	2	0	1	0	2	0.17	0.50	0.67	5	57	41.5	2	2	1	C
79	64	2	3	0	0	0	1	1	1.33	1.50	1.67	26	42	72.5	5	5	1	D
80	31	3	2	2	0	1	0	2	0.17	0.00	0.00	2	35	43.0	2	1	1	B
81	41	2	2	2	1	0	0	1	0.83	1.25	0.00	16	66	41.5	3	3	1	C
82	53	2	2	0	0	1	0	2	1.67	1.50	1.67	23	47	32.0	6	5	4	D
83	61	2	3	2	1	1	0	2	0.83	1.25	0.00	15	43	31.0	3	2	1	C
84	36	2	2	2	0	1	0	4	1.00	1.75	0.00	16	63	42.0	3	3	1	C
85	29	1	0	0	0	0	1	1	0.33	0.25	0.00	5	53	12.0	1	1	1	A
86	25	2	8	3	2	1	0	1	1.17	1.25	0.00	14	80	91.0	4	4	4	A
87	55	2	1	0	0	1	0	2	0.67	0.50	2.00	16	68	31.0	2	2	2	D
88	34	2	3	0	0	0	0	1	0.67	0.00	0.00	5	31	1.0	1	1	2	C
89	20	1	0	0	0	1	0	2	0.50	0.75	1.67	15	36	43.0	3	2	1	B
90	33	2	1	2	2	1	1	2	0.83	1.25	0.67	16	63	61.5	2	2	1	B
91	44	2	0	0	0	1	0	2	0.33	1.75	1.00	16	55	42.5	4	4	1	B
92	58	2	3	0	0	1	1	2	1.17	1.75	1.00	21	75	11.5	3	4	1	D
93	46	4	3	0	—2	1	0	1	0.83	1.25	1.00	16	83	13.5	3	3	2	D
94	31	2	2	0	1	1	0	4	1.17	1.50	0.00	21	65	61.5	4	4	2	D
95	29	3	2	2	2	0	0	1	0.50	1.00	0.67	20	92	32.5	4	4	2	D
96	50	3	3	0	2	1	0	1	0.83	0.25	0.67	18	50	91.0	2	1	2	D
97	27	2	0	0	0	1	0	2	0.67	1.00	2.00	17	64	62.5	3	3	2	D
98	31	3	5	0	1	1	0	2	0.50	0.75	0.67	19	74	61.5	3	3	1	D
99	69	4	4	1	2	1	0	1	1.67	2.25	1.33	26	87	63.5	5	5	3	D
100	41	2	0	0	0	1	0	2	0.00	0.50	0.00	3	37	92.5	2	2	1	D

EXAMPLE CLINICAL TRIAL
FREQUENCY COUNTS BY TREATMENT

TABLE OF TREATMNT BY MAR_STAT

TREATMNT   MAR_STAT   MARITAL STATUS

FREQUENCY	1	2	3	4	TOTAL
A	7	10	5	3	25
B	7	13	4	1	25
C	3	15	6	1	25
D	0	13	8	4	25
TOTAL	17	51	23	9	100

EXAMPLE CLINICAL TRIAL
FREQUENCY COUNTS BY TREATMENT

TABLE OF TREATMNT BY TOBACCO

TREATMNT   TOBACCO   TOBACCO DAILY CONSUMPTION

FREQUENCY	0	1	2	3	TOTAL
A	11	7	2	5	25
B	17	1	7	0	25
C	10	1	9	5	25
D	16	4	2	3	25
TOTAL	54	13	20	13	100

```
                    EXAMPLE CLINICAL TRIAL
                 FREQUENCY COUNTS BY TREATMENT

                 TABLE OF TREATMNT BY ALCOHOL

TREATMNT     ALCOHOL     HISTORY OF ALCOHOL USE


FREQUENCY |    0  |    1  |    2  |    3  |    4  | TOTAL
--------------- + ------- + ------- + ------- + ------- + ------- +
A         |   12  |    3  |    6  |    4  |    0  |   25
--------------- + ------- + ------- + ------- + ------- + ------- +
B         |   11  |    6  |    6  |    2  |    0  |   25
--------------- + ------- + ------- + ------- + ------- + ------- +
C         |   15  |    7  |    2  |    1  |    0  |   25
--------------- + ------- + ------- + ------- + ------- + ------- +
D         |   10  |    6  |    7  |    1  |    1  |   25
--------------- + ------- + ------- + ------- + ------- + ------- +
TOTAL         48      22      21       8       1     100
```

```
                    EXAMPLE CLINICAL TRIAL
                 FREQUENCY COUNTS BY TREATMENT

                 TABLE OF TREATMNT BY TRT_EMOT

TREATMNT     TRT_EMOT     PREVIOUS TRT FOR EMOTIONAL PROBLEMS

FREQUENCY |    0  |    1  | TOTAL
--------------- + ------- + ------- +
A         |    4  |   21  |   25
--------------- + ------- + ------- +
B         |    4  |   21  |   25
--------------- + ------- + ------- +
C         |    4  |   21  |   25
--------------- + ------- + ------- +
D         |    8  |   17  |   25
--------------- + ------- + ------- +
TOTAL         20      80     100
```

```
        EXAMPLE CLINICAL TRIAL
   FREQUENCY COUNTS BY TREATMENT

   TABLE OF TREATMNT BY HOSPITAL

TREATMNT      HOSPITAL    EVER HOSPITALIZED FOR EMOTIONAL PROBLEMS

FREQUENCY  |    0 |     1 | TOTAL
---------------  + ------- + ------- +
A          |   21 |    4 |    25
---------------  + ------- + ------- +
B          |   20 |    5 |    25
---------------  + ------- + ------- +
C          |   22 |    3 |    25
---------------  + ------- + ------- +
D          |   21 |    4 |    25
---------------  + ------- + ------- +
TOTAL           84       16      100
```

```
              EXAMPLE CLINICAL TRIAL
         FREQUENCY COUNTS BY TREATMENT

         TABLE OF TREATMNT BY PSY_DIAG

TREATMNT      PSY_DIAG    PSYCHIATRIC DIAGNOSIS

FREQUENCY  |    1 |    2 |    4 |    5 | TOTAL
---------------  + ------- + ------- + ------- + ------- +
A          |    5 |   11 |    6 |    3 |    25
---------------  + ------- + ------- + ------- + ------- +
B          |    4 |   15 |    5 |    1 |    25
---------------  + ------- + ------- + ------- + ------- +
C          |    7 |   12 |    4 |    2 |    25
---------------  + ------- + ------- + ------- + ------- +
D          |   11 |   13 |    1 |    0 |    25
---------------  + ------- + ------- + ------- + ------- +
TOTAL           27       51       16        6       100
```

EXAMPLE CLINICAL TRIAL
FREQUENCY COUNTS BY TREATMENT

TABLE OF TREATMNT BY CHANGED

TREATMENT   CHANGED   HOW MUCH HAS PATIENT CHANGED?

FREQUENCY	1	2	3	4	5	6	TOTAL
A	5	12	3	2	3	0	25
B	2	10	10	3	0	0	25
C	3	13	8	1	0	0	25
D	0	6	9	6	3	1	25
TOTAL	10	41	30	12	6	1	100

EXAMPLE CLINICAL TRIAL
FREQUENCY COUNTS BY TREATMENT

TABLE OF TREATMNT BY THER_EFF

TREATMNT   THER_EFF   THERAPEUTIC EFFECT

FREQUENCY	1	2	3	4	5	TOTAL
A	5	12	3	2	3	25
B	3	13	6	3	0	25
C	4	14	6	1	0	25
D	1	4	6	10	4	25
TOTAL	13	43	21	16	7	100

EXAMPLE CLINICAL TRIAL
FREQUENCY COUNTS BY TREATMENT

TABLE OF TREATMNT BY ADV_EFF

TREATMNT   ADV_EFF   SEVERITY OF ADVERSE EFFECTS

FREQUENCY	1	2	3	4	TOTAL
A	16	6	2	1	25
B	13	12	0	0	25
C	14	10	1	0	25
D	13	9	1	2	25
TOTAL	56	37	4	3	100

EXAMPLE CLINICAL TRIAL
MEANS BY TREATMENT

VARIABLE	LABEL	N	MEAN	STANDARD DEVIATION

--TREATMNT=A

VARIABLE	LABEL	N	MEAN	STANDARD DEVIATION
AGE		25	42.2400000	14.6380782
COFF_TEA	COFFEE OR TEA CUPS/DAY	25	2.9600000	2.7306898
ANXIETY	ANXIETY/SOMATIZATION FACTOR	25	0.6264000	0.3481245
RETARDTN	RETARDATION FACTOR	25	1.1300000	0.6419372
SLEEP	SLEEP DISTURBANCE FACTOR	25	0.2664000	0.4906193
TOTAL	TOTAL SCORE	25	12.7200000	6.7485801
OBRIST	OBRIST CLUSTER TOTAL	25	53.1600000	19.0999127
APPETITE	APPETITE DISTURBANCE	25	42.3400000	33.8334524
CHANGED	HOW MUCH HAS PATIENT CHANGED?	25	2.4400000	1.2609520

--TREATMNT=B

VARIABLE	LABEL	N	MEAN	STANDARD DEVIATION
AGE		25	37.6000000	12.8452326
COFF_TEA	COFFEE OR TEA CUPS/DAY	25	3.2000000	3.0550505
ANXIETY	ANXIETY/SOMATIZATION FACTOR	25	0.6928000	0.3595775
RETARDTN	RETARDATION FACTOR	25	1.0500000	0.5863020
SLEEP	SLEEP DISTURBANCE FACTOR	25	0.5332000	0.4911867
TOTAL	TOTAL SCORE	25	14.0400000	6.4838260
OBRIST	OBRIST CLUSTER TOTAL	25	45.9600000	12.6671491
APPETITE	APPETITE DISTURBANCE	25	50.0200000	32.0898738
CHANGED	HOW MUCH HAS PATIENT CHANGED?	25	2.5600000	0.8205689

--TREATMNT=C

VARIABLE	LABEL	N	MEAN	STANDARD DEVIATION
AGE		25	40.5200000	10.9549076
COFF_TEA	COFFEE OR TEA CUPS/DAY	25	2.6000000	2.3452079
ANXIETY	ANXIETY/SOMATIZATION FACTOR	25	0.6000000	0.3329289
RETARDTN	RETARDATION FACTOR	25	1.0400000	0.6110101
SLEEP	SLEEP DISTURBANCE FACTOR	25	0.1596000	0.3205137
TOTAL	TOTAL SCORE	25	11.4800000	4.9843087
OBRIST	OBRIST CLUSTER TOTAL	25	48.9000000	12.9131716
APPETITE	APPETITE DISTURBANCE	25	44.3000000	31.6497762
CHANGED	HOW MUCH HAS PATIENT CHANGED?	25	2.2800000	0.7371115

--TREATMNT=D

VARIABLE	LABEL	N	MEAN	STANDARD DEVIATION
AGE		25	42.0800000	12.9193653
COFF_TEA	COFFEE OR TEA CUPS/DAY	25	2.4800000	1.9390719
ANXIETY	ANXIETY/SOMATIZATION FACTOR	25	0.8400000	0.4243819
RETARDTN	RETARDATION FACTOR	25	1.1900000	0.5786118
SLEEP	SLEEP DISTURBANCE FACTOR	25	0.8536000	0.7015155
TOTAL	TOTAL SCORE	25	19.2000000	5.7445626
OBRIST	OBRIST CLUSTER TOTAL	25	66.6800000	18.6049277
APPETITE	APPETITE DISTURBANCE	25	49.5600000	29.5967059
CHANGED	HOW MUCH HAS PATIENT CHANGED?	25	3.3600000	1.1135529

MINIMUM VALUE	MAXIMUM VALUE	STD ERROR OF MEAN	SUM	VARIANCE
18.0000000	70.0000000	2.92761564	1056.00000	214.27333
0.0000000	10.0000000	0.54613796	74.00000	7.45667
0.1700000	1.3300000	0.06962490	15.66000	0.12119
0.0000000	2.5000000	0.12838743	28.25000	0.41208
0.0000000	2.0000000	0.09812387	6.66000	0.24071
4.0000000	34.0000000	1.34971602	318.00000	45.54333
35.0000000	99.0000000	3.81998255	1329.00000	364.80667
1.5000000	92.5000000	6.76669048	1058.50000	1144.70250
1.0000000	5.0000000	0.25219040	61.00000	1.59000
19.0000000	63.0000000	2.56904652	940.00000	165.00000
0.0000000	10.0000000	0.61101009	80.00000	9.33333
0.0000000	1.6700000	0.07191551	17.32000	0.12930
0.0000000	2.5000000	0.11726039	26.25000	0.34375
0.0000000	1.6700000	0.09823733	13.33000	0.24126
2.0000000	24.0000000	1.29676521	351.00000	42.04000
32.0000000	85.0000000	2.53342982	1149.00000	160.45667
1.0000000	92.5000000	6.41797476	1250.50000	1029.76000
1.0000000	4.0000000	0.16411378	64.00000	0.67333
23.0000000	61.0000000	2.19098152	1013.00000	120.01000
0.0000000	10.0000000	0.46904158	65.00000	5.50000
0.1700000	1.5000000	0.06658578	15.00000	0.11084
0.0000000	2.5000000	0.12220202	26.00000	0.37333
0.0000000	1.3300000	0.06410273	3.99000	0.10273
3.0000000	23.0000000	0.99686174	287.00000	24.84333
30.0000000	79.0000000	2.58263431	1222.50000	166.75000
1.0000000	93.5000000	6.32995524	1107.50000	1001.70833
1.0000000	4.0000000	0.14742230	57.00000	0.54333
22.0000000	69.000000	2.58387306	1052.00000	166.910000
0.0000000	8.000000	0.38781439	62.00000	3.760000
0.0000000	1.670000	0.08487638	21.00000	0.180100
0.2500000	2.250000	0.11572237	29.75000	0.334792
0.0000000	2.000000	0.14030310	21.34000	0.492124
3.0000000	32.000000	1.14891253	480.00000	33.000000
37.0000000	112.000000	3.72098553	1667.00000	346.143333
2.0000000	92.500000	5.91934118	1239.00000	875.965000
2.0000000	6.000000	0.22271057	84.00000	1.240000

USEFUL STATISTICAL TESTS AND CONFIDENCE INTERVALS

I. Inferences concerning the mean of a population.

    A. σ known

        1. Statistical test:

            Null hypothesis: $\mu = \mu_0$ (μ_0 is specified)

            Alt. hypothesis: For a one-tailed test:

                                    1. $\mu > \mu_0$
                                    2. $\mu < \mu_0$

                                  For a two-tailed test:

                                  3. $\mu \neq \mu_0$

            Test statistic: $z = \dfrac{\bar{x} - \mu_0}{\sigma_{\bar{x}}}$ where $\sigma_{\bar{x}} = \dfrac{\sigma}{\sqrt{n}}$

            Rejection region: For $\alpha = .05$ (or .01) and for a one-tailed test:

                                  1. Reject H_0 if $z > 1.645$ (or 2.33)
                                  2. Reject H_0 if $z < -1.645$ (or -2.33)

                                  For $\alpha = .05$ (or .01) and for a two-tailed test:

                                  3. Reject H_0 if $|z| > 1.96$ (or 2.58)

            Note: When $n \geq 30$ you may substitute s for σ in the formula for $\sigma_{\bar{x}}$.

        2. Confidence interval:

$$\bar{x} \pm z\sigma_{\bar{x}}$$

where $\sigma_{\bar{x}} = \sigma/\sqrt{n}$. Note: The values of z for a 90%, a 95%, or a 99% confidence interval for μ are 1.645, 1.96, or 2.58, respectively. When $n \geq 30$, you may substitute s for σ in the formula for $\sigma_{\bar{x}}$.

B. σ unknown, and the observations are nearly normally distributed.

   1. Statistical test:

Null hypothesis:	$\mu = \mu_0$ (μ_0 is specified)
Alter. hypothesis:	For a one-tailed test:

   1. $\mu > \mu_0$
   2. $\mu < \mu_0$

   For a two-tailed test:

   3. $\mu \neq \mu_0$

   Test statistic: $\quad t = \dfrac{\bar{x} - \mu_0}{s/\sqrt{n}}$

   Rejection region: For a specified value of α, df $= (n - 1)$, and for a one-tailed test:

   1. Reject H_0 if $t > t_\alpha$
   2. Reject H_0 if $t < -t_\alpha$

   For a specified value of α, df $= (n - 1)$, and for a two-tailed test:

   3. Reject H_0 if $|t| > t_{\alpha/2}$

   2. Confidence interval:

   $$\bar{x} \pm \frac{ts}{\sqrt{n}}$$

   The value of t corresponding to a 90%, a 95%, or a 99% confidence interval is found in Table 3 of Appendix 3 for df $=$ $(n - 1)$ and a $= .05, .025,$ or $.005$, respectively.

II. Inferences concerning the difference between the means of two populations.

   A. Assumptions:

   1. Population 1 is normally distributed with mean equal to μ_1 and variance equal to σ_1^2.

   2. Population 2 is normally distributed with mean equal to μ_2 and variance equal to σ_2^2.

   B. Some results:

   1. The sampling distribution of $(\bar{x}_1 - \bar{x}_2)$ is normal.

   2. The mean of the sampling distribution, $\mu_{\bar{x}_1 - \bar{x}_2}$, is equal to the difference between the populations means, $(\mu_1 - \mu_2)$.

3. The standard error of the sampling distribution is

$$\sigma_{\bar{x}_1 - \bar{x}_2} = \sqrt{\frac{\sigma_1^2}{n_1} + \frac{\sigma_2^2}{n_2}}$$

C. Statistical test:

Null hypothesis: $\mu_1 - \mu_2 = 0$

Alt. hypothesis: For a one-tailed test:

1. $\mu_1 - \mu_2 > 0$
2. $\mu_1 - \mu_2 < 0$

For a two-tailed test:

3. $\mu_1 - \mu_2 \neq 0$

Test statistic: $t = \dfrac{\bar{x}_1 - \bar{x}_2}{s_p \sqrt{\dfrac{1}{n_1} + \dfrac{1}{n_2}}}$

where

$$s_p = \sqrt{\frac{(n_1 - 1)s_1^2 + (n_2 - 1)s_2^2}{n_1 + n_2 - 2}}$$

Rejection region: For a specified value of α, for df $= (n_1 + n_2 - 2)$, and for a one-tailed test:

1. Reject H_0 if $t > t_\alpha$
2. Reject H_0 if $t < -t_\alpha$

For a specified value of α, for df $= (n_1 + n_2 - 2)$, and for a two-tailed test:

3. Reject H_0 if $|t| > t_{\alpha/2}$

D. Confidence interval:

$$\bar{x}_1 - \bar{x}_2 \pm t s_p \sqrt{\frac{1}{n_1} + \frac{1}{n_2}}$$

where

$$s_p = \sqrt{\frac{(n_1 - 1)s_1^2 + (n_2 - 1)s_2^2}{n_1 + n_2 - 2}}$$

The value of t corresponding to a 90%, a 95%, or a 99% confidence interval is found in Table 3 of Appendix 3 for df $= (n_1 + n_2 - 2)$ and a = .05, .025, or .005, respectively.

III. **Inferences about a population proportion π.**

A. Assumptions for a binomial experiment:

1. Experiment consists of n identical trials, each resulting in one of two outcomes, say, success and failure.

2. The probability of success is equal to π and remains the same from trial to trial.

3. The trials are independent of each other.

4. The variable measured is x, the number of successes observed during the n trials.

B. Results:

1. The estimator of π is $\hat{\pi} = x/n$

2. The mean of $\hat{\pi}$ is π

3. The variance of $\hat{\pi}$ is $\pi(1 - \pi)/n$

C. Statistical test:

Null hypothesis: $\quad \pi = \pi_0$ (π_0 is specified)

Alt. hypothesis: $\quad$ For a one-tailed test:

$\qquad$ 1. $\pi > \pi_0$
$\qquad$ 2. $\pi < \pi_0$

$\qquad$ For a two-tailed test:

$\qquad$ 3. $\pi \neq \pi_0$

Test statistic: $\quad z = \dfrac{\hat{\pi} - \pi_0}{\sigma_{\hat{\pi}}} \quad$ where $\quad \sigma_{\hat{\pi}} = \sqrt{\dfrac{\pi_0 (1 - \pi_0)}{n}}$

Rejection region: $\quad$ For $\alpha = .05$ (or .01) and for a one-tailed test:

$\qquad$ 1. Reject H_0 if $z > 1.645$ (or 2.33)
$\qquad$ 2. Reject H_0 if $z < -1.645$ (or -2.33)

$\qquad$ For $\alpha = .05$ (or .01) and for a two-tailed test:

$\qquad$ 3. Reject H_0 if $|z| > 1.96$ (or 2.58)

Note: This test is valid when $n\pi_0$ and $n(1 - \pi_0)$ are both 10 or more.

D. Confidence interval:

$$\hat{\pi} \pm z\sigma_{\hat{\pi}}$$

where

$$\sigma_{\hat{\pi}} = \sqrt{\frac{\pi(1 - \pi)}{n}}$$

Note: The z-values corresponding to a 90%, a 95%, or a 99% confidence interval are, respectively, 1.645, 1.96, or 2.58. If $n\hat{\pi}$ and $n(1 - \hat{\pi})$ are both 10 or greater, we may substitute $\hat{\pi}$ for π in the formula for $\sigma_{\hat{\pi}}$.

IV. Inferences comparing two population proportions π_1 and π_2.

A. Assumption: Independent random samples are drawn from each of two binomial populations.

	Population 1	Population 2
Probability of success:	π_1	π_2
Sample size:	n_1	n_2
Observed successes:	x_1	x_2

B. Results:

1. The estimated difference between π_1 and π_2 is

$$\hat{\pi}_1 - \hat{\pi}_2 = \frac{x_1}{n_1} - \frac{x_2}{n_2}$$

2. The mean of $(\hat{\pi}_1 - \hat{\pi}_2)$ is $(\pi_1 - \pi_2)$
3. The standard error of $(\hat{\pi}_1 - \hat{\pi}_2)$ is

$$\sqrt{\frac{\pi_1(1 - \pi_1)}{n_1} + \frac{\pi_2(1 - \pi_2)}{n_2}}$$

C. Statistical test:

Null hypothesis: $\pi_1 - \pi_2 = 0$

Alt. hypothesis: For a one-tailed test:

1. $\pi_1 - \pi_2 > 0$
2. $\pi_1 - \pi_2 < 0$

For a two-tailed test:

3. $\pi_1 - \pi_2 \neq 0$

Test statistic: $z = \dfrac{\hat{\pi}_1 - \hat{\pi}_2}{\sigma_{\hat{\pi}_1 - \hat{\pi}_2}}$

where

$$\sigma_{\hat{\pi}_1 - \hat{\pi}_2} = \sqrt{\pi(1 - \pi)\left(\frac{1}{n_1} + \frac{1}{n_2}\right)}$$

and π is approximated by

$$\hat{\pi} = \frac{x_1 + x_2}{n_1 + n_2}$$

Rejection region: For $\alpha = .05$ (or .01) and for a one-tailed test:

1. Reject H_0 if $z > 1.645$ (or 2.33)
2. Reject H_0 if $z < -1.645$ (or -2.33)

For $\alpha = .05$ (or .01) and for a two-tailed test:

3. Reject H_0 if $|z| > 1.96$ (or 2.58)

Note: $n\hat{\pi}$ and $n(1 - \hat{\pi})$ must be greater than or equal to 10 for both populations.

D. Confidence interval:

$$\hat{\pi}_1 - \hat{\pi}_2 \pm z\sigma_{\hat{\pi}_1 - \hat{\pi}_2}$$

where

$$\sigma_{\hat{\pi}_1 - \hat{\pi}_2} = \sqrt{\frac{\pi_1(1 - \pi_1)}{n_1} + \frac{\pi_2(1 - \pi_2)}{n_2}}$$

Substitute $z = 1.645$, 1.96, or 2.58 for a 90%, a 95%, or a 99% confidence interval, respectively. Also use $\hat{\pi}_1$ and $\hat{\pi}_2$ for the unknown parameters π_1 and π_2 in the formula for $\sigma_{\hat{\pi}_1 - \hat{\pi}_2}$. Very little error will result provided the sample sizes are large.

V. Inferences about σ^2 (or σ).

A. Assumption: The underlying population of measurements is normal.

B. Statistical test:

Null hypothesis: $\sigma^2 = \sigma_0^2$ (σ_0^2 is specified)

Alt. hypothesis: For a one-tailed test:

1. $\sigma^2 > \sigma_0^2$
2. $\sigma^2 < \sigma_0^2$

For a two-tailed test:

3. $\sigma^2 \neq \sigma_0^2$

Test statistic: $\chi^2 = \dfrac{(n - 1)s^2}{\sigma_0^2}$

Rejection region: For a specified value of α,

1. Reject H_0 if $\chi^2 > \chi_U^2$, the upper-tail value for $a = \alpha$ and df $= n - 1$
2. Reject H_0 if $\chi^2 < \chi_L^2$, the lower-tail value for $a = 1 - \alpha$ and df $= n - 1$
3. Reject H_0 if $\chi^2 > \chi_U^2$ based on $a = \alpha/2$ or $\chi^2 < \chi_L^2$ based on $a = 1 - \alpha/2$; df $= n - 1$

C. Confidence interval:

$$\frac{(n - 1)s^2}{\chi_U^2} < \sigma^2 < \frac{(n - 1)s^2}{\chi_L^2}$$

VI. Inferences about two population variances.

A. Assumptions:

1. Population 1 has a normal distribution, with mean μ_1 and variance σ_1^2.

2. Population 2 has a normal distribution, with mean μ_2 and variance σ_2^2.

3. Two independent random samples are drawn, n_1 measurements from population 1, n_2 from population 2.

B. Test for comparing two population variances:

Null hypothesis: $\sigma_1^2 = \sigma_2^2$

Alt. hypothesis: For a one-tailed test:

1. $\sigma_1^2 > \sigma_2^2$
2. $\sigma_1^2 < \sigma_2^2$

For a two-tailed test:

3. $\sigma_1^2 \neq \sigma_2^2$

Test statistic: $F = \dfrac{s_1^2}{s_2^2}$

Rejection region: For a given value of α, $\mathrm{df}_1 = (n_1 - 1)$, and $\mathrm{df}_2 = (n_2 - 1)$

1. Reject H_0 if $F > F_{\alpha,\,\mathrm{df}_1,\,\mathrm{df}_2}$
2. Reject H_0 if $F < 1/F_{\alpha,\,\mathrm{df}_2,\,\mathrm{df}_1}$
3. Reject H_0 if $F > F_{\alpha/2,\,\mathrm{df}_1,\,\mathrm{df}_2}$ or if $F < 1/F_{\alpha/2,\,\mathrm{df}_2,\,\mathrm{df}_1}$

C. Confidence interval:

$$\frac{s_1^2}{s_2^2} F_L < \frac{\sigma_1^2}{\sigma_2^2} < \frac{s_1^2}{s_2^2} F_U$$

where $F_L = 1/F_{\alpha/2,\,\mathrm{df}_1,\,\mathrm{df}_2}$ and $F_U = F_{\alpha/2,\,\mathrm{df}_2,\,\mathrm{df}_1}$

VII. Inference concerning β_1 in linear regression.

A. Assumptions:

1. The ϵs in $y = \beta_0 + \beta_1 x + \epsilon$ are normally distributed with mean 0 and variance σ_ϵ^2.

2. The ϵs are independent.

B. Statistical test:

Null hypothesis: $\beta_1 = 0$

Alt. hypothesis: For a one-tailed test:

1. $\beta_1 > 0$
2. $\beta_1 < 0$

For a two-tailed test:

3. $\beta_1 \neq 0$

Test statistic: $t = \dfrac{\hat{\beta}_1}{\sqrt{s_\epsilon^2/S_{xx}}}$

Rejection region: For a given value of α and df $= n - 2$:

1. Reject H_0 if $t > t_\alpha$
2. Reject H_0 if $t < -t_\alpha$
3. Reject H_0 if $|t| > t_{\alpha/2}$

C. Confidence interval:

$$\hat{\beta}_1 \pm t_{\alpha/2} \frac{s_\epsilon}{\sqrt{S_{xx}}}$$

APPENDIX 3

STATISTICAL TABLES

Table 1 Binomial probabilities, $P(x)$ for $n \le 20$ ■ Table 2 Normal curve areas ■ Table 3 Critical values of t ■ Table 4 Critical values of χ^2 ■ Table 5 Upper-tail values of F, a = .10 ■ Table 6 Upper-tail values of F, a = .05 ■ Table 7 Upper-tail values of F, a = .025 ■ Table 8 Upper-tail values of F, a = .01 ■ Table 9 Upper-tail values of F, a = .005 ■ Table 10 Wilcoxon signed-rank test ($n \le 50$) ■ Table 11 Random numbers ■ Table 12 $F_{max} = s^2_{max}/s^2_{min}$

TABLE 1 Binomial probabilities, $P(x)$ for $n \le 20$

$n = 2$ $x\downarrow$	0.05	0.10	0.15	0.20	π 0.25	0.30	0.35	0.40	0.45	0.50	
0	.9025	.8100	.7225	.6400	.5625	.4900	.4225	.3600	.3025	.2500	2
1	.0950	.1800	.2550	.3200	.3750	.4200	.4550	.4800	.4950	.5000	1
2	.0025	.0100	.0225	.0400	.0625	.0900	.1225	.1600	.2025	.2500	0
	0.95	0.90	0.85	0.80	0.75	0.70	0.65	0.60	0.55	0.50	$x\uparrow$

$n = 3$ $x\downarrow$	0.05	0.10	0.15	0.20	π 0.25	0.30	0.35	0.40	0.45	0.50	
0	.8574	.7290	.6141	.5120	.4219	.3430	.2746	.2160	.1664	.1250	3
1	.1354	.2430	.3251	.3840	.4219	.4410	.4436	.4320	.4084	.3750	2
2	.0071	.0270	.0574	.0960	.1406	.1890	.2389	.2880	.3341	.3750	1
3	.0001	.0010	.0034	.0080	.0156	.0270	.0429	.0640	.0911	.1250	0
	0.95	0.90	0.85	0.80	0.75	0.70	0.65	0.60	0.55	0.50	$x\uparrow$

(continued)

Table 1 ■ Binomial Probabilities 673

TABLE 1 continued

n = 4 *x* ↓	*0.05*	*0.10*	*0.15*	*0.20*	*π* *0.25*	*0.30*	*0.35*	*0.40*	*0.45*	*0.50*	
0	.8145	.6561	.5220	.4096	.3164	.2401	.1785	.1296	.0915	.0625	4
1	.1715	.2916	.3685	.4096	.4219	.4116	.3845	.3456	.2995	.2500	3
2	.0135	.0486	.0975	.1536	.2109	.2646	.3105	.3456	.3675	.3750	2
3	.0005	.0036	.0115	.0256	.0469	.0756	.1115	.1536	.2005	.2500	1
4	.0000	.0001	.0005	.0016	.0039	.0081	.0150	.0256	.0410	.0625	0
	0.95	0.90	0.85	0.80	0.75	0.70	0.65	0.60	0.55	0.50	*x* ↑

n = 5 *x* ↓	*0.05*	*0.10*	*0.15*	*0.20*	*π* *0.25*	*0.30*	*0.35*	*0.40*	*0.45*	*0.50*	
0	.7738	.5905	.4437	.3277	.2373	.1681	.1160	.0778	.0503	.0313	5
1	.2036	.3281	.3915	.4096	.3955	.3602	.3124	.2592	.2059	.1563	4
2	.0214	.0729	.1382	.2048	.2637	.3087	.3364	.3456	.3369	.3125	3
3	.0011	.0081	.0244	.0512	.0879	.1323	.1811	.2304	.2757	.3125	2
4	.0000	.0005	.0022	.0064	.0146	.0284	.0488	.0768	.1128	.1563	1
5	.0000	.0000	.0001	.0003	.0010	.0024	.0053	.0102	.0185	.0313	0
	0.95	0.90	0.85	0.80	0.75	0.70	0.65	0.60	0.55	0.50	*x* ↑

n = 6 *x* ↓	*0.05*	*0.10*	*0.15*	*0.20*	*π* *0.25*	*0.30*	*0.35*	*0.40*	*0.45*	*0.50*	
0	.7351	.5314	.3771	.2621	.1780	.1176	.0754	.0467	.0277	.0156	6
1	.2321	.3543	.3993	.3932	.3560	.3025	.2437	.1866	.1359	.0938	5
2	.0305	.0984	.1762	.2458	.2966	.3241	.3280	.3110	.2780	.2344	4
3	.0021	.0146	.0415	.0819	.1318	.1852	.2355	.2765	.3032	.3125	3
4	.0001	.0012	.0055	.0154	.0330	.0595	.0951	.1382	.1861	.2344	2
5	.0000	.0001	.0004	.0015	.0044	.0102	.0205	.0369	.0609	.0938	1
6	.0000	.0000	.0000	.0001	.0002	.0007	.0018	.0041	.0083	.0156	0
	0.95	0.90	0.85	0.80	0.75	0.70	0.65	0.60	0.55	0.50	*x* ↑

(continued)

TABLE 1   Binomial Probabilities, $P(x)$ for $n \leq 20$

$n = 7$					π						
$x\downarrow$	0.05	0.10	0.15	0.20	0.25	0.30	0.35	0.40	0.45	0.50	
0	.6983	.4783	.3206	.2097	.1335	.0824	.0490	.0280	.0152	.0078	7
1	.2573	.3720	.3960	.3670	.3115	.2471	.1848	.1306	.0872	.0547	6
2	.0406	.1240	.2097	.2753	.3115	.3177	.2985	.2613	.2140	.1641	5
3	.0036	.0230	.0617	.1147	.1730	.2269	.2679	.2903	.2918	.2734	4
4	.0002	.0026	.0109	.0287	.0577	.0972	.1442	.1935	.2388	.2734	3
5	.0000	.0002	.0012	.0043	.0115	.0250	.0466	.0774	.1172	.1641	2
6	.0000	.0000	.0001	.0004	.0013	.0036	.0084	.0172	.0320	.0547	1
7	.0000	.0000	.0000	.0000	.0001	.0002	.0006	.0016	.0037	.0078	0
	0.95	0.90	0.85	0.80	0.75	0.70	0.65	0.60	0.55	0.50	$x\uparrow$

$n = 8$					π						
$x\downarrow$	0.05	0.10	0.15	0.20	0.25	0.30	0.35	0.40	0.45	0.50	
0	.6634	.4305	.2725	.1678	.1001	.0576	.0319	.0168	.0084	.0039	8
1	.2793	.3826	.3847	.3355	.2670	.1977	.1373	.0896	.0548	.0313	7
2	.0515	.1488	.2376	.2936	.3115	.2965	.2587	.2090	.1569	.1094	6
3	.0054	.0331	.0839	.1468	.2076	.2541	.2786	.2787	.2568	.2188	5
4	.0004	.0046	.0185	.0459	.0865	.1361	.1875	.2322	.2627	.2734	4
5	.0000	.0004	.0026	.0092	.0231	.0467	.0808	.1239	.1719	.2188	3
6	.0000	.0000	.0002	.0011	.0038	.0100	.0217	.0413	.0703	.1094	2
7	.0000	.0000	.0000	.0001	.0004	.0012	.0033	.0079	.0164	.0313	1
8	.0000	.0000	.0000	.0000	.0000	.0001	.0002	.0007	.0017	.0039	0
	0.95	0.90	0.85	0.80	0.75	0.70	0.65	0.60	0.55	0.50	$x\uparrow$

$n = 9$					π						
$x\downarrow$	0.05	0.10	0.15	0.20	0.25	0.30	0.35	0.40	0.45	0.50	
0	.6302	.3874	.2316	.1342	.0751	.0404	.0207	.0101	.0046	.0020	9
1	.2985	.3874	.3679	.3020	.2253	.1556	.1004	.0605	.0339	.0176	8
2	.0629	.1722	.2597	.3020	.3003	.2668	.2162	.1612	.1110	.0703	7
3	.0077	.0446	.1069	.1762	.2336	.2668	.2716	.2508	.2119	.1641	6
4	.0006	.0074	.0283	.0661	.1168	.1715	.2194	.2508	.2600	.2461	5
5	.0000	.0008	.0050	.0165	.0389	.0735	.1181	.1672	.2128	.2461	4
6	.0000	.0001	.0006	.0028	.0087	.0210	.0424	.0743	.1160	.1641	3
7	.0000	.0000	.0000	.0003	.0012	.0039	.0098	.0212	.0407	.0703	2
8	.0000	.0000	.0000	.0000	.0001	.0004	.0013	.0035	.0083	.0176	1
9	.0000	.0000	.0000	.0000	.0000	.0000	.0001	.0003	.0008	.0020	0
	0.95	0.90	0.85	0.80	0.75	0.70	0.65	0.60	0.55	0.50	$x\uparrow$

(continued)

Table 1 ▪ Binomial Probabilities **675**

TABLE 1 continued

n = 10 x ↓	0.05	0.10	0.15	0.20	π 0.25	0.30	0.35	0.40	0.45	0.50	
0	.5987	.3487	.1969	.1074	.0563	.0282	.0135	.0060	.0025	.0010	10
1	.3151	.3874	.3474	.2684	.1877	.1211	.0725	.0403	.0207	.0098	9
2	.0746	.1937	.2759	.3020	.2816	.2335	.1757	.1209	.0763	.0439	8
3	.0105	.0574	.1298	.2013	.2503	.2668	.2522	.2150	.1665	.1172	7
4	.0010	.0112	.0401	.0881	.1460	.2001	.2377	.2508	.2384	.2051	6
5	.0001	.0015	.0085	.0264	.0584	.1029	.1536	.2007	.2340	.2461	5
6	.0000	.0001	.0012	.0055	.0162	.0368	.0689	.1115	.1596	.2051	4
7	.0000	.0000	.0001	.0008	.0031	.0090	.0212	.0425	.0746	.1172	3
8	.0000	.0000	.0000	.0001	.0004	.0014	.0043	.0106	.0229	.0439	2
9	.0000	.0000	.0000	.0000	.0000	.0001	.0005	.0016	.0042	.0098	1
10	.0000	.0000	.0000	.0000	.0000	.0000	.0000	.0001	.0003	.0010	0
	0.95	0.90	0.85	0.80	0.75	0.70	0.65	0.60	0.55	0.50	x ↑

n = 12 x ↓	0.05	0.10	0.15	0.20	π 0.25	0.30	0.35	0.40	0.45	0.50	
0	.5404	.2824	.1422	.0687	.0317	.0138	.0057	.0022	.0008	.0002	12
1	.3413	.3766	.3012	.2062	.1267	.0712	.0368	.0174	.0075	.0029	11
2	.0988	.2301	.2924	.2835	.2323	.1678	.1088	.0639	.0339	.0161	10
3	.0173	.0852	.1720	.2362	.2581	.2397	.1954	.1419	.0923	.0537	9
4	.0021	.0213	.0683	.1329	.1936	.2311	.2367	.2128	.1700	.1208	8
5	.0002	.0038	.0193	.0532	.1032	.1585	.2039	.2270	.2225	.1934	7
6	.0000	.0005	.0040	.0155	.0401	.0792	.1281	.1766	.2124	.2256	6
7	.0000	.0000	.0006	.0033	.0115	.0291	.0591	.1009	.1489	.1934	5
8	.0000	.0000	.0001	.0005	.0024	.0078	.0199	.0420	.0762	.1208	4
9	.0000	.0000	.0000	.0001	.0004	.0015	.0048	.0125	.0277	.0537	3
10	.0000	.0000	.0000	.0000	.0000	.0002	.0008	.0025	.0068	.0161	2
11	.0000	.0000	.0000	.0000	.0000	.0000	.0001	.0003	.0010	.0029	1
12	.0000	.0000	.0000	.0000	.0000	.0000	.0000	.0000	.0001	.0002	0
	0.95	0.90	0.85	0.80	0.75	0.70	0.65	0.60	0.55	0.50	x ↑

(continued)

TABLE 1 Binomial Probabilities, $P(x)$ for $n \leq 20$

n = 14 x ↓	0.05	0.10	0.15	0.20	π 0.25	0.30	0.35	0.40	0.45	0.50	
0	.4877	.2288	.1028	.0440	.0178	.0068	.0024	.0008	.0002	.0001	14
1	.3593	.3559	.2539	.1539	.0832	.0407	.0181	.0073	.0027	.0009	13
2	.1229	.2570	.2912	.2501	.1802	.1134	.0634	.0317	.0141	.0056	12
3	.0259	.1142	.2056	.2501	.2402	.1943	.1366	.0845	.0462	.0222	11
4	.0037	.0349	.0998	.1720	.2202	.2290	.2022	.1549	.1040	.0611	10
5	.0004	.0078	.0352	.0860	.1468	.1963	.2178	.2066	.1701	.1222	9
6	.0000	.0013	.0093	.0322	.0734	.1262	.1759	.2066	.2088	.1833	8
7	.0000	.0002	.0019	.0092	.0280	.0618	.1082	.1574	.1952	.2095	7
8	.0000	.0000	.0003	.0020	.0082	.0232	.0510	.0918	.1398	.1833	6
9	.0000	.0000	.0000	.0003	.0018	.0066	.0183	.0408	.0762	.1222	5
10	.0000	.0000	.0000	.0000	.0003	.0014	.0049	.0136	.0312	.0611	4
11	.0000	.0000	.0000	.0000	.0000	.0002	.0010	.0033	.0093	.0222	3
12	.0000	.0000	.0000	.0000	.0000	.0000	.0001	.0005	.0019	.0056	2
13	.0000	.0000	.0000	.0000	.0000	.0000	.0000	.0001	.0002	.0009	1
14	.0000	.0000	.0000	.0000	.0000	.0000	.0000	.0000	.0000	.0001	0
	0.95	0.90	0.85	0.80	0.75	0.70	0.65	0.60	0.55	0.50	x ↑

n = 16 x ↓	0.05	0.10	0.15	0.20	π 0.25	0.30	0.35	0.40	0.45	0.50	
0	.4401	.1853	.0743	.0281	.0100	.0033	.0010	.0003	.0001	.0000	16
1	.3706	.3294	.2097	.1126	.0535	.0228	.0087	.0030	.0009	.0002	15
2	.1463	.2745	.2775	.2111	.1336	.0732	.0353	.0150	.0056	.0018	14
3	.0359	.1423	.2285	.2463	.2079	.1465	.0888	.0468	.0215	.0085	13
4	.0061	.0514	.1311	.2001	.2252	.2040	.1553	.1014	.0572	.0278	12
5	.0008	.0137	.0555	.1201	.1802	.2099	.2008	.1623	.1123	.0667	11
6	.0001	.0028	.0180	.0550	.1101	.1649	.1982	.1983	.1684	.1222	10
7	.0000	.0004	.0045	.0197	.0524	.1010	.1524	.1889	.1969	.1746	9
8	.0000	.0001	.0009	.0055	.0197	.0487	.0923	.1417	.1812	.1964	8
9	.0000	.0000	.0001	.0012	.0058	.0185	.0442	.0840	.1318	.1746	7
10	.0000	.0000	.0000	.0002	.0014	.0056	.0167	.0392	.0755	.1222	6
11	.0000	.0000	.0000	.0000	.0002	.0013	.0049	.0142	.0337	.0667	5
12	.0000	.0000	.0000	.0000	.0000	.0002	.0011	.0040	.0115	.0278	4
13	.0000	.0000	.0000	.0000	.0000	.0000	.0002	.0008	.0029	.0085	3
14	.0000	.0000	.0000	.0000	.0000	.0000	.0000	.0001	.0005	.0018	2
15	.0000	.0000	.0000	.0000	.0000	.0000	.0000	.0000	.0001	.0002	1
	0.95	0.90	0.85	0.80	0.75	0.70	0.65	0.60	0.55	0.50	x ↑

(continued)

Table 1 ■ Binomial Probabilities 677

TABLE 1 continued

n = 18 x↓	0.05	0.10	0.15	0.20	π 0.25	0.30	0.35	0.40	0.45	0.50	
0	.3972	.1501	.0536	.0180	.0056	.0016	.0004	.0001	.0000	.0000	18
1	.3763	.3002	.1704	.0811	.0338	.0126	.0042	.0012	.0003	.0001	17
2	.1683	.2835	.2556	.1723	.0958	.0458	.0190	.0069	.0022	.0006	16
3	.0473	.1680	.2406	.2297	.1704	.1046	.0547	.0246	.0095	.0031	15
4	.0093	.0700	.1592	.2153	.2130	.1681	.1104	.0614	.0291	.0117	14
5	.0014	.0218	.0787	.1507	.1988	.2017	.1664	.1146	.0666	.0327	13
6	.0002	.0052	.0301	.0816	.1436	.1873	.1941	.1655	.1181	.0708	12
7	.0000	.0010	.0091	.0350	.0820	.1376	.1792	.1892	.1657	.1214	11
8	.0000	.0002	.0022	.0120	.0376	.0811	.1327	.1734	.1864	.1669	10
9	.0000	.0000	.0004	.0033	.0139	.0386	.0794	.1284	.1694	.1855	9
10	.0000	.0000	.0001	.0008	.0042	.0149	.0385	.0771	.1248	.1669	8
11	.0000	.0000	.0000	.0001	.0010	.0046	.0151	.0374	.0742	.1214	7
12	.0000	.0000	.0000	.0000	.0002	.0012	.0047	.0145	.0354	.0708	6
13	.0000	.0000	.0000	.0000	.0000	.0002	.0012	.0045	.0134	.0327	5
14	.0000	.0000	.0000	.0000	.0000	.0000	.0002	.0011	.0039	.0117	4
15	.0000	.0000	.0000	.0000	.0000	.0000	.0000	.0002	.0009	.0031	3
16	.0000	.0000	.0000	.0000	.0000	.0000	.0000	.0000	.0001	.0006	2
17	.0000	.0000	.0000	.0000	.0000	.0000	.0000	.0000	.0000	.0001	1
	0.95	0.90	0.85	0.80	0.75	0.70	0.65	0.60	0.55	0.50	x↑

n = 20 x↓	0.05	0.10	0.15	0.20	π 0.25	0.30	0.35	0.40	0.45	0.50	
0	.3585	.1216	.0388	.0115	.0032	.0008	.0002	.0000	.0000	.0000	20
1	.3774	.2702	.1368	.0576	.0211	.0068	.0020	.0005	.0001	.0000	19
2	.1887	.2852	.2293	.1369	.0669	.0278	.0100	.0031	.0008	.0002	18
3	.0596	.1901	.2428	.2054	.1339	.0716	.0323	.0123	.0040	.0011	17
4	.0133	.0898	.1821	.2182	.1897	.1304	.0738	.0350	.0139	.0046	16
5	.0022	.0319	.1028	.1746	.2023	.1789	.1272	.0746	.0365	.0148	15
6	.0003	.0089	.0454	.1091	.1686	.1916	.1712	.1244	.0746	.0370	14
7	.0000	.0020	.0160	.0545	.1124	.1643	.1844	.1659	.1221	.0739	13
8	.0000	.0004	.0046	.0222	.0609	.1144	.1614	.1797	.1623	.1201	12
9	.0000	.0001	.0011	.0074	.0271	.0654	.1158	.1597	.1771	.1602	11
10	.0000	.0000	.0002	.0020	.0099	.0308	.0686	.1171	.1593	.1762	10
11	.0000	.0000	.0000	.0005	.0030	.0120	.0336	.0710	.1185	.1602	9
12	.0000	.0000	.0000	.0001	.0008	.0039	.0136	.0355	.0727	.1201	8
13	.0000	.0000	.0000	.0000	.0002	.0010	.0045	.0146	.0366	.0739	7
14	.0000	.0000	.0000	.0000	.0000	.0002	.0012	.0049	.0150	.0370	6
15	.0000	.0000	.0000	.0000	.0000	.0000	.0003	.0013	.0049	.0148	5
16	.0000	.0000	.0000	.0000	.0000	.0000	.0000	.0003	.0013	.0046	4
17	.0000	.0000	.0000	.0000	.0000	.0000	.0000	.0000	.0002	.0011	3
18	.0000	.0000	.0000	.0000	.0000	.0000	.0000	.0000	.0000	.0002	2
	0.95	0.90	0.85	0.80	0.75	0.70	0.65	0.60	0.55	0.50	x↑

TABLE 2 Normal curve areas

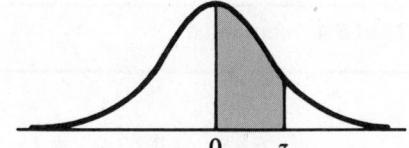

z	.00	.01	.02	.03	.04	.05	.06	.07	.08	.09
0.0	.0000	.0040	.0080	.0120	.0160	.0199	.0239	.0279	.0319	.0359
0.1	.0398	.0438	.0478	.0517	.0557	.0596	.0636	.0675	.0714	.0753
0.2	.0793	.0832	.0871	.0910	.0948	.0987	.1026	.1064	.1103	.1141
0.3	.1179	.1217	.1255	.1293	.1331	.1368	.1406	.1443	.1480	.1517
0.4	.1554	.1591	.1628	.1664	.1700	.1736	.1772	.1808	.1844	.1879
0.5	.1915	.1950	.1985	.2019	.2054	.2088	.2123	.2157	.2190	.2224
0.6	.2257	.2291	.2324	.2357	.2389	.2422	.2454	.2486	.2517	.2549
0.7	.2580	.2611	.2642	.2673	.2704	.2734	.2764	.2794	.2823	.2852
0.8	.2881	.2910	.2939	.2967	.2995	.3023	.3051	.3078	.3106	.3133
0.9	.3159	.3186	.3212	.3238	.3264	.3289	.3315	.3340	.3365	.3389
1.0	.3413	.3438	.3461	.3485	.3508	.3531	.3554	.3577	.3599	.3621
1.1	.3643	.3665	.3686	.3708	.3729	.3749	.3770	.3790	.3810	.3830
1.2	.3849	.3869	.3888	.3907	.3925	.3944	.3962	.3980	.3997	.4015
1.3	.4032	.4049	.4066	.4082	.4099	.4115	.4131	.4147	.4162	.4177
1.4	.4192	.4207	.4222	.4236	.4251	.4265	.4279	.4292	.4306	.4319
1.5	.4332	.4345	.4357	.4370	.4382	.4394	.4406	.4418	.4429	.4441
1.6	.4452	.4463	.4474	.4484	.4495	.4505	.4515	.4525	.4535	.4545
1.7	.4554	.4564	.4573	.4582	.4591	.4599	.4608	.4616	.4625	.4633
1.8	.4641	.4649	.4656	.4664	.4671	.4678	.4686	.4693	.4699	.4706
1.9	.4713	.4719	.4726	.4732	.4738	.4744	.4750	.4756	.4761	.4767
2.0	.4772	.4778	.4783	.4788	.4793	.4798	.4803	.4808	.4812	.4817
2.1	.4821	.4826	.4830	.4834	.4838	.4842	.4846	.4850	.4854	.4857
2.2	.4861	.4864	.4868	.4871	.4875	.4878	.4881	.4884	.4887	.4890
2.3	.4893	.4896	.4898	.4901	.4904	.4906	.4909	.4911	.4913	.4916
2.4	.4918	.4920	.4922	.4925	.4927	.4929	.4931	.4932	.4934	.4936
2.5	.4938	.4940	.4941	.4943	.4945	.4946	.4948	.4949	.4951	.4952
2.6	.4953	.4955	.4956	.4957	.4959	.4960	.4961	.4962	.4963	.4964
2.7	.4965	.4966	.4967	.4968	.4969	.4970	.4971	.4972	.4973	.4974
2.8	.4974	.4975	.4976	.4977	.4977	.4978	.4979	.4979	.4980	.4981
2.9	.4981	.4982	.4982	.4983	.4984	.4984	.4985	.4985	.4986	.4986
3.0	.4987	.4987	.4987	.4988	.4988	.4989	.4989	.4989	.4990	.4990

Source: This table is abridged from Table 1 of *Statistical Tables and Formulas,* by A. Hald (New York: John Wiley & Sons, 1952). Reprinted by permission of A. Hald and the publishers, John Wiley & Sons.

Table 3 ▪ Critical Values of *t*   **679**

TABLE 3   Critical values of *t*

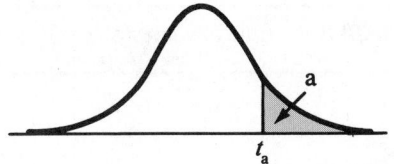

df	a = .10	a = .05	a = .025	a = .010	a = .005
1	3.078	6.314	12.706	31.821	63.657
2	1.886	2.920	4.303	6.965	9.925
3	1.638	2.353	3.182	4.541	5.841
4	1.333	2.132	2.776	3.747	4.604
5	1.476	2.015	2.571	3.365	4.032
6	1.440	1.943	2.447	3.143	3.707
7	1.415	1.895	2.365	2.998	3.499
8	1.397	1.860	2.306	2.896	3.355
9	1.383	1.833	2.262	2.821	3.250
10	1.372	1.812	2.228	2.764	3.169
11	1.363	1.796	2.201	2.718	3.106
12	1.356	1.782	2.179	2.681	3.055
13	1.350	1.771	2.160	2.650	3.012
14	1.345	1.761	2.145	2.624	2.977
15	1.341	1.753	2.131	2.602	2.947
16	1.337	1.746	2.120	2.583	2.921
17	1.333	1.740	2.110	2.567	2.898
18	1.330	1.734	2.101	2.552	2.878
19	1.328	1.729	2.093	2.539	2.861
20	1.325	1.725	2.086	2.528	2.845
21	1.323	1.721	2.080	2.518	2.831
22	1.321	1.717	2.074	2.508	2.819
23	1.319	1.714	2.069	2.500	2.807
24	1.318	1.711	2.064	2.492	2.797
25	1.316	1.708	2.060	2.485	2.787
26	1.315	1.706	2.056	2.479	2.779
27	1.314	1.703	2.052	2.473	2.771
28	1.313	1.701	2.048	2.467	2.763
29	1.311	1.699	2.045	2.462	2.756
30	1.310	1.697	2.042	2.457	2.750
40	1.303	1.684	2.021	2.423	2.704
60	1.296	1.671	2.000	2.390	2.660
120	1.289	1.658	1.980	2.358	2.617
inf.	1.282	1.645	1.960	2.326	2.576

Source: From "Table of Percentage Points of the *t*-Distribution." Computed by Maxine Merrington, *Biometrika*, Vol. 32 (1941), p. 300. Reproduced by permission of the *Biometrika* Trustees.

TABLE 4 Critical values of χ^2

df	a = .995	a = .990	a = .975	a = .950	a = .900
1	0.0000393	0.0001571	0.0009821	0.0039321	0.0157908
2	0.0100251	0.0201007	0.0506356	0.102587	0.210720
3	0.0717212	0.114832	0.215795	0.351846	0.584375
4	0.206990	0.297110	0.484419	0.710721	1.063625
5	0.411740	0.554300	0.831211	1.145476	1.61031
6	0.675727	0.872085	1.237347	1.63539	2.20413
7	0.989265	1.239043	1.68987	2.16735	2.83311
8	1.344419	1.646482	2.17973	2.73264	3.48954
9	1.734926	2.087912	2.70039	3.32511	4.16816
10	2.15585	2.55821	3.24697	3.94030	4.86518
11	2.60321	3.05347	3.81575	4.57481	5.57779
12	3.07382	3.57056	4.40379	5.22603	6.30380
13	3.56503	4.10691	5.00874	5.89186	7.04150
14	4.07468	4.66043	5.62872	6.57063	7.78953
15	4.60094	5.22935	6.26214	7.26094	8.54675
16	5.14224	5.81221	6.90766	7.96164	9.31223
17	5.69724	6.40776	7.56418	8.67176	10.0852
18	6.26481	7.01491	8.23075	9.39046	10.8649
19	6.84398	7.63273	8.90655	10.1170	11.6509
20	7.43386	8.26040	9.59083	10.8508	12.4426
21	8.03366	8.89720	10.28293	11.5913	13.2396
22	8.64272	9.54249	10.9823	12.3380	14.0415
23	9.26042	10.19567	11.6885	13.0905	14.8479
24	9.88623	10.8564	12.4011	13.8484	15.6587
25	10.5197	11.5240	13.1197	14.6114	16.4734
26	11.1603	12.1981	13.8439	15.3791	17.2919
27	11.8076	12.8786	14.5733	16.1513	18.1138
28	12.4613	13.5648	15.3079	16.9279	18.9392
29	13.1211	14.2565	16.0471	17.7083	19.7677
30	13.7867	14.9535	16.7908	18.4926	20.5992
40	20.7065	22.1643	24.4331	26.5093	29.0505
50	27.9907	29.7067	32.3574	34.7642	37.6886
60	35.5346	37.4848	40.4817	43.1879	46.4589
70	43.2752	45.4418	48.7576	51.7393	55.3290
80	51.1720	53.5400	57.1532	60.3915	64.2778
90	59.1963	61.7541	65.6466	69.1260	73.2912
100	67.3276	70.0648	74.2219	77.9295	82.3581

Table 4 ▪ Critical Values of χ^2 681

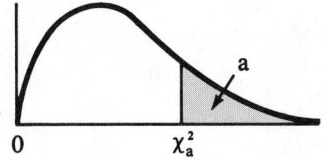

a = .10	a = .05	a = .025	a = .010	a = .005	df
2.70554	3.84146	5.02389	6.63490	7.87944	1
4.60517	5.99147	7.37776	9.21034	10.5966	2
6.25139	7.81473	9.34840	11.3449	12.8381	3
7.77944	9.48773	11.1433	13.2767	14.8602	4
9.23635	11.0705	12.8325	15.0863	16.7496	5
10.6446	12.5916	14.4494	16.8119	18.5476	6
12.0170	14.0671	16.0128	18.4753	20.2777	7
13.3616	15.5073	17.5346	20.0902	21.9550	8
14.6837	16.9190	19.0228	21.6660	23.5893	9
15.9871	18.3070	20.4831	23.2093	25.1882	10
17.2750	19.6751	21.9200	24.7250	26.7569	11
18.5494	21.0261	23.3367	26.2170	28.2995	12
19.8119	22.3621	24.7356	27.6883	29.8194	13
21.0642	23.6848	26.1190	29.1413	31.3193	14
22.3072	24.9958	27.4884	30.5779	32.8013	15
23.5418	26.2962	28.8454	31.9999	34.2672	16
24.7690	27.5871	30.1910	33.4087	35.7185	17
25.9894	28.8693	31.5264	34.8053	37.1564	18
27.2036	30.1435	32.8523	36.1908	38.5822	19
28.4120	31.4104	34.1696	37.5662	39.9968	20
29.6151	32.6705	35.4789	38.9321	41.4010	21
30.8133	33.9244	36.7807	40.2894	42.7956	22
32.0069	35.1725	38.0757	41.6384	44.1813	23
33.1963	36.4151	39.3641	42.9798	45.5585	24
34.3816	37.6525	40.6465	40.3141	46.9278	25
35.5631	38.8852	41.9232	45.6417	48.2899	26
36.7412	40.1133	43.1944	46.9630	49.6449	27
37.9159	41.3372	44.4607	48.2782	50.9933	28
39.0875	42.5569	45.7222	49.5879	52.3356	29
40.2560	43.7729	46.9792	50.8922	53.6720	30
51.8050	55.7585	59.3417	63.6907	66.7659	40
63.1671	67.5048	71.4202	76.1539	79.4900	50
74.3970	79.0819	83.2976	88.3794	91.9517	60
85.5271	90.5312	95.0231	100.425	104.215	70
96.5782	101.879	106.629	112.329	116.321	80
107.565	113.145	118.136	124.116	128.299	90
118.498	124.342	129.561	135.807	140.169	100

Source: From "Tables of the Percentage Points of the χ^2-Distribution," *Biometrika*, Vol. 32 (1941), pp. 188–189, by Catherine M. Thompson. Reproduced by permission of the *Biometrika* Trustees.

TABLE 5   Upper-tail values of *F*,   a = .10

	df₁								
df₂	*1*	*2*	*3*	*4*	*5*	*6*	*7*	*8*	*9*
1	39.86	49.50	53.59	55.83	57.24	58.20	58.91	59.44	59.86
2	8.53	9.00	9.16	9.24	9.29	9.33	9.35	9.37	9.38
3	5.54	5.46	5.39	5.34	5.31	5.28	5.27	5.25	5.24
4	4.54	4.32	4.19	4.11	4.05	4.01	3.98	3.95	3.94
5	4.06	3.78	3.62	3.52	3.45	3.40	3.37	3.34	3.32
6	3.78	3.46	3.29	3.18	3.11	3.05	3.01	2.98	2.96
7	3.59	3.26	3.07	2.96	2.88	2.83	2.78	2.75	2.72
8	3.46	3.11	2.92	2.81	2.73	2.67	2.62	2.59	2.56
9	3.36	3.01	2.81	2.69	2.61	2.55	2.51	2.47	2.44
10	3.29	2.92	2.73	2.61	2.52	2.46	2.41	2.38	2.35
11	3.23	2.86	2.66	2.54	2.45	2.39	2.34	2.30	2.27
12	3.18	2.81	2.61	2.48	2.39	2.33	2.28	2.24	2.21
13	3.14	2.76	2.56	2.43	2.35	2.28	2.23	2.20	2.16
14	3.10	2.73	2.52	2.39	2.31	2.24	2.19	2.15	2.12
15	3.07	2.70	2.49	2.36	2.27	2.21	2.16	2.12	2.09
16	3.05	2.67	2.46	2.33	2.24	2.18	2.13	2.09	2.06
17	3.03	2.64	2.44	2.31	2.22	2.15	2.10	2.06	2.03
18	3.01	2.62	2.42	2.29	2.20	2.13	2.08	2.04	2.00
19	2.99	2.61	2.40	2.27	2.18	2.11	2.06	2.02	1.98
20	2.97	2.59	2.38	2.25	2.16	2.09	2.04	2.00	1.96
21	2.96	2.57	2.36	2.23	2.14	2.08	2.02	1.98	1.95
22	2.95	2.56	2.35	2.22	2.13	2.06	2.01	1.97	1.93
23	2.94	2.55	2.34	2.21	2.11	2.05	1.99	1.95	1.92
24	2.93	2.54	2.33	2.19	2.10	2.04	1.98	1.94	1.91
25	2.92	2.53	2.32	2.18	2.09	2.02	1.97	1.93	1.89
26	2.91	2.52	2.31	2.17	2.08	2.01	1.96	1.92	1.88
27	2.90	2.51	2.30	2.17	2.07	2.00	1.95	1.91	1.87
28	2.89	2.50	2.29	2.16	2.06	2.00	1.94	1.90	1.87
29	2.89	2.50	2.28	2.15	2.06	1.99	1.93	1.89	1.86
30	2.88	2.49	2.28	2.14	2.05	1.98	1.93	1.88	1.85
40	2.84	2.44	2.23	2.09	2.00	1.93	1.87	1.83	1.79
60	2.79	2.39	2.18	2.04	1.95	1.87	1.82	1.77	1.74
120	2.75	2.35	2.13	1.99	1.90	1.82	1.77	1.72	1.68
∞	2.71	2.30	2.08	1.94	1.85	1.77	1.72	1.67	1.63

Table 5 ▪ Upper-Tail Values of *F*, a = .10   **683**

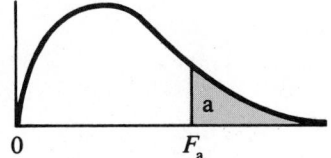

					df₁						
10	**12**	**15**	**20**	**24**	**30**	**40**	**60**	**120**	**∞**	**df₂**
60.19	60.71	61.22	61.74	62.00	62.26	62.53	62.79	63.06	63.33	1
9.39	9.41	9.42	9.44	9.45	9.46	9.47	9.47	9.48	9.49	2
5.23	5.22	5.20	5.18	5.18	5.17	5.16	5.15	5.14	5.13	3
3.92	3.90	3.87	3.84	3.83	3.82	3.80	3.79	3.78	3.76	4
3.30	3.27	3.24	3.21	3.19	3.17	3.16	3.14	3.12	3.10	5
2.94	2.90	2.87	2.84	2.82	2.80	2.78	2.76	2.74	2.72	6
2.70	2.67	2.63	2.59	2.58	2.56	2.54	2.51	2.49	2.47	7
2.54	2.50	2.46	2.42	2.40	2.38	2.36	2.34	2.32	2.29	8
2.42	2.38	2.34	2.30	2.28	2.25	2.23	2.21	2.18	2.16	9
2.32	2.28	2.24	2.20	2.18	2.16	2.13	2.11	2.08	2.06	10
2.25	2.21	2.17	2.12	2.10	2.08	2.05	2.03	2.00	1.97	11
2.19	2.15	2.10	2.06	2.04	2.01	1.99	1.96	1.93	1.90	12
2.14	2.10	2.05	2.01	1.98	1.96	1.93	1.90	1.88	1.85	13
2.10	2.05	2.01	1.96	1.94	1.91	1.89	1.86	1.83	1.80	14
2.06	2.02	1.97	1.92	1.90	1.87	1.85	1.82	1.79	1.76	15
2.03	1.99	1.94	1.89	1.87	1.84	1.81	1.78	1.75	1.72	16
2.00	1.96	1.91	1.86	1.84	1.81	1.78	1.75	1.72	1.69	17
1.98	1.93	1.89	1.84	1.81	1.78	1.75	1.72	1.69	1.66	18
1.96	1.91	1.86	1.81	1.79	1.76	1.73	1.70	1.67	1.63	19
1.94	1.89	1.84	1.79	1.77	1.74	1.71	1.68	1.64	1.61	20
1.92	1.87	1.83	1.78	1.75	1.72	1.69	1.66	1.62	1.59	21
1.90	1.86	1.81	1.76	1.73	1.70	1.67	1.64	1.60	1.57	22
1.89	1.84	1.80	1.74	1.72	1.69	1.66	1.62	1.59	1.55	23
1.88	1.83	1.78	1.73	1.70	1.67	1.64	1.61	1.57	1.53	24
1.87	1.82	1.77	1.72	1.69	1.66	1.63	1.59	1.56	1.52	25
1.86	1.81	1.76	1.71	1.68	1.65	1.61	1.58	1.54	1.50	26
1.85	1.80	1.75	1.70	1.67	1.64	1.60	1.57	1.53	1.49	27
1.84	1.79	1.74	1.69	1.66	1.63	1.59	1.56	1.52	1.48	28
1.83	1.78	1.73	1.68	1.65	1.62	1.58	1.55	1.51	1.47	29
1.82	1.77	1.72	1.67	1.64	1.61	1.57	1.54	1.50	1.46	30
1.76	1.71	1.66	1.61	1.57	1.54	1.51	1.47	1.42	1.38	40
1.71	1.66	1.60	1.54	1.51	1.48	1.44	1.40	1.35	1.29	60
1.65	1.60	1.55	1.48	1.45	1.41	1.37	1.32	1.26	1.19	120
1.60	1.55	1.49	1.42	1.38	1.34	1.30	1.24	1.17	1.00	∞

Source: From "Tables of Percentage Points of the Inverted Beta (*F*)-Distribution," *Biometrika*, Vol. 33 (1943), pp. 73–88, by Maxine Merrington and Catherine M. Thompson. Reproduced by permission of the *Biometrika* Trustees.

Table 6 Upper-tail values of F, $a = .05$

					df_1				
df_2	1	2	3	4	5	6	7	8	9
1	161.4	199.5	215.7	224.6	230.2	234.0	236.8	238.9	240.5
2	18.51	19.00	19.16	19.25	19.30	19.33	19.35	19.37	19.38
3	10.13	9.55	9.28	9.12	9.01	8.94	8.89	8.85	8.81
4	7.71	6.94	6.59	6.39	6.26	6.16	6.09	6.04	6.00
5	6.61	5.79	5.41	5.19	5.05	4.95	4.88	4.82	4.77
6	5.99	5.14	4.76	4.53	4.39	4.28	4.21	4.15	4.10
7	5.59	4.74	4.35	4.12	3.97	3.87	3.79	3.73	3.68
8	5.32	4.46	4.07	3.84	3.69	3.58	3.50	3.44	3.39
9	5.12	4.26	3.86	3.63	3.48	3.37	3.29	3.23	3.18
10	4.96	4.10	3.71	3.48	3.33	3.22	3.14	3.07	3.02
11	4.84	3.98	3.59	3.36	3.20	3.09	3.01	2.95	2.90
12	4.75	3.89	3.49	3.26	3.11	3.00	2.91	2.85	2.80
13	4.67	3.81	3.41	3.18	3.03	2.92	2.83	2.77	2.71
14	4.60	3.74	3.34	3.11	2.96	2.85	2.76	2.70	2.65
15	4.54	3.68	3.29	3.06	2.90	2.79	2.71	2.64	2.59
16	4.49	3.63	3.24	3.01	2.85	2.74	2.66	2.59	2.54
17	4.45	3.59	3.20	2.96	2.81	2.70	2.61	2.55	2.49
18	4.41	3.55	3.16	2.93	2.77	2.66	2.58	2.51	2.46
19	4.38	3.52	3.13	2.90	2.74	2.63	2.54	2.48	2.42
20	4.35	3.49	3.10	2.87	2.71	2.60	2.51	2.45	2.39
21	4.32	3.47	3.07	2.84	2.68	2.57	2.49	2.42	2.37
22	4.30	3.44	3.05	2.82	2.66	2.55	2.46	2.40	2.34
23	4.28	3.42	3.03	2.80	2.64	2.53	2.44	2.37	2.32
24	4.26	3.40	3.01	2.78	2.62	2.51	2.42	2.36	2.30
25	4.24	3.39	2.99	2.76	2.60	2.49	2.40	2.34	2.28
26	4.23	3.37	2.98	2.74	2.59	2.47	2.39	2.32	2.27
27	4.21	3.35	2.96	2.73	2.57	2.46	2.37	2.31	2.25
28	4.20	3.34	2.95	2.71	2.56	2.45	2.36	2.29	2.24
29	4.18	3.33	2.93	2.70	2.55	2.43	2.35	2.28	2.22
30	4.17	3.32	2.92	2.69	2.53	2.42	2.33	2.27	2.21
40	4.08	3.23	2.84	2.61	2.45	2.34	2.25	2.18	2.12
60	4.00	3.15	2.76	2.53	2.37	2.25	2.17	2.10	2.04
120	3.92	3.07	2.68	2.45	2.29	2.17	2.09	2.02	1.96
∞	3.84	3.00	2.60	2.37	2.21	2.10	2.01	1.94	1.88

Table 6 ▪ Upper-Tail Values of F, a = .05   **685**

					df$_1$						
10	*12*	*15*	*20*	*24*	*30*	*40*	*60*	*120*	∞	*df$_2$*
241.9	243.9	245.9	248.0	249.1	250.1	251.1	252.2	253.3	254.3	1
19.40	19.41	19.43	19.45	19.45	19.46	19.47	19.48	19.49	19.50	2
8.79	8.74	8.70	8.66	8.64	8.62	8.59	8.57	8.55	8.53	3
5.96	5.91	5.86	5.80	5.77	5.75	5.72	5.69	5.66	5.63	4
4.74	4.68	4.62	4.56	4.53	4.50	4.46	4.43	4.40	4.36	5
4.06	4.00	3.94	3.87	3.84	3.81	3.77	3.74	3.70	3.67	6
3.64	3.57	3.51	3.44	3.41	3.38	3.34	3.30	3.27	3.23	7
3.35	3.28	3.22	3.15	3.12	3.08	3.04	3.01	2.97	2.93	8
3.14	3.07	3.01	2.94	2.90	2.86	2.83	2.79	2.75	2.71	9
2.98	2.91	2.85	2.77	2.74	2.70	2.66	2.62	2.58	2.54	10
2.85	2.79	2.72	2.65	2.61	2.57	2.53	2.49	2.45	2.40	11
2.75	2.69	2.62	2.54	2.51	2.47	2.43	2.38	2.34	2.30	12
2.67	2.60	2.53	2.46	2.42	2.38	2.34	2.30	2.25	2.21	13
2.60	2.53	2.46	2.39	2.35	2.31	2.27	2.22	2.18	2.13	14
2.54	2.48	2.40	2.33	2.29	2.25	2.20	2.16	2.11	2.07	15
2.49	2.42	2.35	2.28	2.24	2.19	2.15	2.11	2.06	2.01	16
2.45	2.38	2.31	2.23	2.19	2.15	2.10	2.06	2.01	1.96	17
2.41	2.34	2.27	2.19	2.15	2.11	2.06	2.02	1.97	1.92	18
2.38	2.31	2.23	2.16	2.11	2.07	2.03	1.98	1.93	1.88	19
2.35	2.28	2.20	2.12	2.08	2.04	1.99	1.95	1.90	1.84	20
2.32	2.25	2.18	2.10	2.05	2.01	1.96	1.92	1.87	1.81	21
2.30	2.23	2.15	2.07	2.03	1.98	1.94	1.89	1.84	1.78	22
2.27	2.20	2.13	2.05	2.01	1.96	1.91	1.86	1.81	1.76	23
2.25	2.18	2.11	2.03	1.98	1.94	1.89	1.84	1.79	1.73	24
2.24	2.16	2.09	2.01	1.96	1.92	1.87	1.82	1.77	1.71	25
2.22	2.15	2.07	1.99	1.95	1.90	1.85	1.80	1.75	1.69	26
2.20	2.13	2.06	1.97	1.93	1.88	1.84	1.79	1.73	1.67	27
2.19	2.12	2.04	1.96	1.91	1.87	1.82	1.77	1.71	1.65	28
2.18	2.10	2.03	1.94	1.90	1.85	1.81	1.75	1.70	1.64	29
2.16	2.09	2.01	1.93	1.89	1.84	1.79	1.74	1.68	1.62	30
2.08	2.00	1.92	1.84	1.79	1.74	1.69	1.64	1.58	1.51	40
1.99	1.92	1.84	1.75	1.70	1.65	1.59	1.53	1.47	1.39	60
1.91	1.83	1.75	1.66	1.61	1.55	1.50	1.43	1.35	1.25	120
1.83	1.75	1.67	1.57	1.52	1.46	1.39	1.32	1.22	1.00	∞

Source: From "Tables of Percentage Points of the Inverted Beta (*F*)-Distribution," *Biometrika*, Vol. 33 (1943), pp. 73–88, by Maxine Merrington and Catherine M. Thompson. Reproduced by permission of the *Biometrika* Trustees.

TABLE 7   Upper-tail values of F, $a = .025$

					df_1				
df_2	*1*	*2*	*3*	*4*	*5*	*6*	*7*	*8*	*9*
1	647.8	799.5	864.2	899.6	921.8	937.1	948.2	956.7	963.3
2	38.51	39.00	39.17	39.25	39.30	39.33	39.36	39.37	39.39
3	17.44	16.04	15.44	15.10	14.88	14.73	14.62	14.54	14.47
4	12.22	10.65	9.98	9.60	9.36	9.20	9.07	8.98	8.90
5	10.01	8.43	7.76	7.39	7.15	6.98	6.85	6.76	6.68
6	8.81	7.26	6.60	6.23	5.99	5.82	5.70	5.60	5.52
7	8.07	6.54	5.89	5.52	5.29	5.12	4.99	4.90	4.82
8	7.57	6.06	5.42	5.05	4.82	4.65	4.53	4.43	4.36
9	7.21	5.71	5.08	4.72	4.48	4.32	4.20	4.10	4.03
10	6.94	5.46	4.83	4.47	4.24	4.07	3.95	3.85	3.78
11	6.72	5.26	4.63	4.28	4.04	3.88	3.76	3.66	3.59
12	6.55	5.10	4.47	4.12	3.89	3.73	3.61	3.51	3.44
13	6.41	4.97	4.35	4.00	3.77	3.60	3.48	3.39	3.31
14	6.30	4.86	4.24	3.89	3.66	3.50	3.38	3.29	3.21
15	6.20	4.77	4.15	3.80	3.58	3.41	3.29	3.20	3.12
16	6.12	4.69	4.08	3.73	3.50	3.34	3.22	3.12	3.05
17	6.04	4.62	4.01	3.66	3.44	3.28	3.16	3.06	2.98
18	5.98	4.56	3.95	3.61	3.38	3.22	3.10	3.01	2.93
19	5.92	4.51	3.90	3.56	3.33	3.17	3.05	2.96	2.88
20	5.87	4.46	3.86	3.51	3.29	3.13	3.01	2.91	2.84
21	5.83	4.42	3.82	3.48	3.25	3.09	2.97	2.87	2.80
22	5.79	4.38	3.78	3.44	3.22	3.05	2.93	2.84	2.76
23	5.75	4.35	3.75	3.41	3.18	3.02	2.90	2.81	2.73
24	5.72	4.32	3.72	3.38	3.15	2.99	2.87	2.78	2.70
25	5.69	4.29	3.69	3.35	3.13	2.97	2.85	2.75	2.68
26	5.66	4.27	3.67	3.33	3.10	2.94	2.82	2.73	2.65
27	5.63	4.24	3.65	3.31	3.08	2.92	2.80	2.71	2.63
28	5.61	4.22	3.63	3.29	3.06	2.90	2.78	2.69	2.61
29	5.59	4.20	3.61	3.27	3.04	2.88	2.76	2.67	2.59
30	5.57	4.18	3.59	3.25	3.03	2.87	2.75	2.65	2.57
40	5.42	4.05	3.46	3.13	2.90	2.74	2.62	2.53	2.45
60	5.29	3.93	3.34	3.01	2.79	2.63	2.51	2.41	2.33
120	5.15	3.80	3.23	2.89	2.67	2.52	2.39	2.30	2.22
∞	5.02	3.69	3.12	2.79	2.57	2.41	2.29	2.19	2.11

Table 7 ▪ Upper-Tail Values of *F*, a = .025   **687**

					df$_1$						
10	*12*	*15*	*20*	*24*	*30*	*40*	*60*	*120*	∞	*df$_2$*
968.6	976.7	984.9	993.1	997.2	1001	1006	1010	1014	1018	1
39.40	39.41	39.43	39.45	39.46	39.46	39.47	39.48	39.49	39.50	2
14.42	14.34	14.25	14.17	14.12	14.08	14.04	13.99	13.95	13.90	3
8.84	8.75	8.66	8.56	8.51	8.46	8.41	8.36	8.31	8.26	4
6.62	6.52	6.43	6.33	6.28	6.23	6.18	6.12	6.07	6.02	5
5.46	5.37	5.27	5.17	5.12	5.07	5.01	4.96	4.90	4.85	6
4.76	4.67	4.57	4.47	4.42	4.36	4.31	4.25	4.20	4.14	7
4.30	4.20	4.10	4.00	3.95	3.89	3.84	3.78	3.73	3.67	8
3.96	3.87	3.77	3.67	3.61	3.56	3.51	3.45	3.39	3.33	9
3.72	3.62	3.52	3.42	3.37	3.31	3.26	3.20	3.14	3.08	10
3.53	3.43	3.33	3.23	3.17	3.12	3.06	3.00	2.94	2.88	11
3.37	3.28	3.18	3.07	3.02	2.96	2.91	2.85	2.79	2.72	12
3.25	3.15	3.05	2.95	2.89	2.84	2.78	2.72	2.66	2.60	13
3.15	3.05	2.95	2.84	2.79	2.73	2.67	2.61	2.55	2.49	14
3.06	2.96	2.86	2.76	2.70	2.64	2.59	2.52	2.46	2.40	15
2.99	2.89	2.79	2.68	2.63	2.57	2.51	2.45	2.38	2.32	16
2.92	2.82	2.72	2.62	2.56	2.50	2.44	2.38	2.32	2.25	17
2.87	2.77	2.67	2.56	2.50	2.44	2.38	2.32	2.26	2.19	18
2.82	2.72	2.62	2.51	2.45	2.39	2.33	2.27	2.20	2.13	19
2.77	2.68	2.57	2.46	2.41	2.35	2.29	2.22	2.16	2.09	20
2.73	2.64	2.53	2.42	2.37	2.31	2.25	2.18	2.11	2.04	21
2.70	2.60	2.50	2.39	2.33	2.27	2.21	2.14	2.08	2.00	22
2.67	2.57	2.47	2.36	2.30	2.24	2.18	2.11	2.04	1.97	23
2.64	2.54	2.44	2.33	2.27	2.21	2.15	2.08	2.01	1.94	24
2.61	2.51	2.41	2.30	2.24	2.18	2.12	2.05	1.98	1.91	25
2.59	2.49	2.39	2.28	2.22	2.16	2.09	2.03	1.95	1.88	26
2.57	2.47	2.36	2.25	2.19	2.13	2.07	2.00	1.93	1.85	27
2.55	2.45	2.34	2.23	2.17	2.11	2.05	1.98	1.91	1.83	28
2.53	2.43	2.32	2.21	2.15	2.09	2.03	1.96	1.89	1.81	29
2.51	2.41	2.31	2.20	2.14	2.07	2.01	1.94	1.87	1.79	30
2.39	2.29	2.18	2.07	2.01	1.94	1.88	1.80	1.72	1.64	40
2.27	2.17	2.06	1.94	1.88	1.82	1.74	1.67	1.58	1.48	60
2.16	2.05	1.94	1.82	1.76	1.69	1.61	1.53	1.43	1.31	120
2.05	1.94	1.83	1.71	1.64	1.57	1.48	1.39	1.27	1.00	∞

Source: From "Tables of Percentage Points of the Inverted Beta (*F*)-Distribution," *Biometrika,* Vol. 33 (1943), pp. 73–88, by Maxine Merrington and Catherine M. Thompson. Reproduced by permission of the *Biometrika* Trustees.

TABLE 8   Upper-tail values of F, $a = .01$

					df$_1$				
df$_2$	1	2	3	4	5	6	7	8	9
1	4052	4999.5	5403	5625	5764	5859	5928	5982	6022
2	98.50	99.00	99.17	99.25	99.30	99.33	99.36	99.37	99.39
3	34.12	30.82	29.46	28.71	28.24	27.91	27.67	27.49	27.35
4	21.20	18.00	16.69	15.98	15.52	15.21	14.98	14.80	14.66
5	16.26	13.27	12.06	11.39	10.97	10.67	10.46	10.29	10.16
6	13.75	10.92	9.78	9.15	8.75	8.47	8.26	8.10	7.98
7	12.25	9.55	8.45	7.85	7.46	7.19	6.99	6.84	6.72
8	11.26	8.65	7.59	7.01	6.63	6.37	6.18	6.03	5.91
9	10.56	8.02	6.99	6.42	6.06	5.80	5.61	5.47	5.35
10	10.04	7.56	6.55	5.99	5.64	5.39	5.20	5.06	4.94
11	9.65	7.21	6.22	5.67	5.32	5.07	4.89	4.74	4.63
12	9.33	6.93	5.95	5.41	5.06	4.82	4.64	4.50	4.39
13	9.07	6.70	5.74	5.21	4.86	4.62	4.44	4.30	4.19
14	8.86	6.51	5.56	5.04	4.69	4.46	4.28	4.14	4.03
15	8.68	6.36	5.42	4.89	4.56	4.32	4.14	4.00	3.89
16	8.53	6.23	5.29	4.77	4.44	4.20	4.03	3.89	3.78
17	8.40	6.11	5.18	4.67	4.34	4.10	3.93	3.79	3.68
18	8.29	6.01	5.09	4.58	4.25	4.01	3.84	3.71	3.60
19	8.18	5.93	5.01	4.50	4.17	3.94	3.77	3.63	3.52
20	8.10	5.85	4.94	4.43	4.10	3.87	3.70	3.56	3.46
21	8.02	5.78	4.87	4.37	4.04	3.81	3.64	3.51	3.40
22	7.95	5.72	4.82	4.31	3.99	3.76	3.59	3.45	3.35
23	7.88	5.66	4.76	4.26	3.94	3.71	3.54	3.41	3.30
24	7.82	5.61	4.72	4.22	3.90	3.67	3.50	3.36	3.26
25	7.77	5.57	4.68	4.18	3.85	3.63	3.46	3.32	3.22
26	7.72	5.53	4.64	4.14	3.82	3.59	3.42	3.29	3.18
27	7.68	5.49	4.60	4.11	3.78	3.56	3.39	3.26	3.15
28	7.64	5.45	4.57	4.07	3.75	3.53	3.36	3.23	3.12
29	7.60	5.42	4.54	4.04	3.73	3.50	3.33	3.20	3.09
30	7.56	5.39	4.51	4.02	3.70	3.47	3.30	3.17	3.07
40	7.31	5.18	4.31	3.83	3.51	3.29	3.12	2.99	2.89
60	7.08	4.98	4.13	3.65	3.34	3.12	2.95	2.82	2.72
120	6.85	4.79	3.95	3.48	3.17	2.96	2.79	2.66	2.56
∞	6.63	4.61	3.78	3.32	3.02	2.80	2.64	2.51	2.41

Table 8 ▪ Upper-Tail Values of *F*,   a = .01   **689**

					df$_1$						
10	**12**	**15**	**20**	**24**	**30**	**40**	**60**	**120**	**∞**	**df$_2$**
6056	6106	6157	6209	6235	6261	6287	6313	6339	6366	1
99.40	99.42	99.43	99.45	99.46	99.47	99.47	99.48	99.49	99.50	2
27.23	27.05	26.87	26.69	26.60	26.50	26.41	26.32	26.22	26.13	3
14.55	14.37	14.20	14.02	13.93	13.84	13.75	13.65	13.56	13.46	4
10.05	9.89	9.72	9.55	9.47	9.38	9.29	9.20	9.11	9.02	5
7.87	7.72	7.56	7.40	7.31	7.23	7.14	7.06	6.97	6.88	6
6.62	6.47	6.31	6.16	6.07	5.99	5.91	5.82	5.74	5.65	7
5.81	5.67	5.52	5.36	5.28	5.20	5.12	5.03	4.95	4.86	8
5.26	5.11	4.96	4.81	4.73	4.65	4.57	4.48	4.40	4.31	9
4.85	4.71	4.56	4.41	4.33	4.25	4.17	4.08	4.00	3.91	10
4.54	4.40	4.25	4.10	4.02	3.94	3.86	3.78	3.69	3.60	11
4.30	4.16	4.01	3.86	3.78	3.70	3.62	3.54	3.45	3.36	12
4.10	3.96	3.82	3.66	3.59	3.51	3.43	3.34	3.25	3.17	13
3.94	3.80	3.66	3.51	3.43	3.35	3.27	3.18	3.09	3.00	14
3.80	3.67	3.52	3.37	3.29	3.21	3.13	3.05	2.96	2.87	15
3.69	3.55	3.41	3.26	3.18	3.10	3.02	2.93	2.84	2.75	16
3.59	3.46	3.31	3.16	3.08	3.00	2.92	2.83	2.75	2.65	17
3.51	3.37	3.23	3.08	3.00	2.92	2.84	2.75	2.66	2.57	18
3.43	3.30	3.15	3.00	2.92	2.84	2.76	2.67	2.58	2.49	19
3.37	3.23	3.09	2.94	2.86	2.78	2.69	2.61	2.52	2.42	20
3.31	3.17	3.03	2.88	2.80	2.72	2.64	2.55	2.46	2.36	21
3.26	3.12	2.98	2.83	2.75	2.67	2.58	2.50	2.40	2.31	22
3.21	3.07	2.93	2.78	2.70	2.62	2.54	2.45	2.35	2.26	23
3.17	3.03	2.89	2.74	2.66	2.58	2.49	2.40	2.31	2.21	24
3.13	2.99	2.85	2.70	2.62	2.54	2.45	2.36	2.27	2.17	25
3.09	2.96	2.81	2.66	2.58	2.50	2.42	2.33	2.23	2.13	26
3.06	2.93	2.78	2.63	2.55	2.47	2.38	2.29	2.20	2.10	27
3.03	2.90	2.75	2.60	2.52	2.44	2.35	2.26	2.17	2.06	28
3.00	2.87	2.73	2.57	2.49	2.41	2.33	2.23	2.14	2.03	29
2.98	2.84	2.70	2.55	2.47	2.39	2.30	2.21	2.11	2.01	30
2.80	2.66	2.52	2.37	2.29	2.20	2.11	2.02	1.92	1.80	40
2.63	2.50	2.35	2.20	2.12	2.03	1.94	1.84	1.73	1.60	60
2.47	2.34	2.19	2.03	1.95	1.86	1.76	1.66	1.53	1.38	120
2.32	2.18	2.04	1.88	1.79	1.70	1.59	1.47	1.32	1.00	∞

Source: From "Tables of Percentage Points of the Inverted Beta (*F*)-Distribution," *Biometrika*, Vol. 33 (1943), pp. 73–88, by Maxine Merrington and Catherine M. Thompson. Reproduced by permission of the *Biometrika* Trustees.

TABLE 9   Upper-tail values of F, $a = .005$

df_2	1	2	3	4	5	6	7	8	9
1	16211	20000	21615	22500	23056	23437	23715	23925	24091
2	198.5	199.0	199.2	199.2	199.3	199.3	199.4	199.4	199.4
3	55.55	49.80	47.47	46.19	45.39	44.84	44.43	44.13	43.88
4	31.33	26.28	24.26	23.15	22.46	21.97	21.62	21.35	21.14
5	22.78	18.31	16.53	15.56	14.94	14.51	14.20	13.96	13.77
6	18.63	14.54	12.92	12.03	11.46	11.07	10.79	10.57	10.39
7	16.24	12.40	10.88	10.05	9.52	9.16	8.89	8.68	8.51
8	14.69	11.04	9.60	8.81	8.30	7.95	7.69	7.50	7.34
9	13.61	10.11	8.72	7.96	7.47	7.13	6.88	6.69	6.54
10	12.83	9.43	8.08	7.34	6.87	6.54	6.30	6.12	5.97
11	12.23	8.91	7.60	6.88	6.42	6.10	5.86	5.68	5.54
12	11.75	8.51	7.23	6.52	6.07	5.76	5.52	5.35	5.20
13	11.37	8.19	6.93	6.23	5.79	5.48	5.25	5.08	4.94
14	11.06	7.92	6.68	6.00	5.56	5.26	5.03	4.86	4.72
15	10.80	7.70	6.48	5.80	5.37	5.07	4.85	4.67	4.54
16	10.58	7.51	6.30	5.64	5.21	4.91	4.69	4.52	4.38
17	10.38	7.35	6.16	5.50	5.07	4.78	4.56	4.39	4.25
18	10.22	7.21	6.03	5.37	4.96	4.66	4.44	4.28	4.14
19	10.07	7.09	5.92	5.27	4.85	4.56	4.34	4.18	4.04
20	9.94	6.99	5.82	5.17	4.76	4.47	4.26	4.09	3.96
21	9.83	6.89	5.73	5.09	4.68	4.39	4.18	4.01	3.88
22	9.73	6.81	5.65	5.02	4.61	4.32	4.11	3.94	3.81
23	9.63	6.73	5.58	4.95	4.54	4.26	4.05	3.88	3.75
24	9.55	6.66	5.52	4.89	4.49	4.20	3.99	3.83	3.69
25	9.48	6.60	5.46	4.84	4.43	4.15	3.94	3.78	3.64
26	9.41	6.54	5.41	4.79	4.38	4.10	3.89	3.73	3.60
27	9.34	6.49	5.36	4.74	4.34	4.06	3.85	3.69	3.56
28	9.28	6.44	5.32	4.70	4.30	4.02	3.81	3.65	3.52
29	9.23	6.40	5.28	4.66	4.26	3.98	3.77	3.61	3.48
30	9.18	6.35	5.24	4.62	4.23	3.95	3.74	3.58	3.45
40	8.83	6.07	4.98	4.37	3.99	3.71	3.51	3.35	3.22
60	8.49	5.79	4.73	4.14	3.76	3.49	3.29	3.13	3.01
120	8.18	5.54	4.50	3.92	3.55	3.28	3.09	2.93	2.81
∞	7.88	5.30	4.28	3.72	3.35	3.09	2.90	2.74	2.62

Table 9 ▪ Upper-tail values of F, a = .005   691

	df_1										
10	*12*	*15*	*20*	*24*	*30*	*40*	*60*	*120*	*∞*	*df_2*
24224	24426	24630	24836	24940	25044	25148	25253	25359	25465	1
199.4	199.4	199.4	199.4	199.5	199.5	199.5	199.5	199.5	199.5	2
43.69	43.39	43.08	42.78	42.62	42.47	42.31	42.15	41.99	41.83	3
20.97	20.70	20.44	20.17	20.03	19.89	19.75	19.61	19.47	19.32	4
13.62	13.38	13.15	12.90	12.78	12.66	12.53	12.40	12.27	12.14	5
10.25	10.03	9.81	9.59	9.47	9.36	9.24	9.12	9.00	8.88	6
8.38	8.18	7.97	7.75	7.65	7.53	7.42	7.31	7.19	7.08	7
7.21	7.01	6.81	6.61	6.50	6.40	6.29	6.18	6.06	5.95	8
6.42	6.23	6.03	5.83	5.73	5.62	5.52	5.41	5.30	5.19	9
5.85	5.66	5.47	5.27	5.17	5.07	4.97	4.86	4.75	4.64	10
5.42	5.24	5.05	4.86	4.76	4.65	4.55	4.44	4.34	4.23	11
5.09	4.91	4.72	4.53	4.43	4.33	4.23	4.12	4.01	3.90	12
4.82	4.64	4.46	4.27	4.17	4.07	3.97	3.87	3.76	3.65	13
4.60	4.43	4.25	4.06	3.96	3.86	3.76	3.66	3.55	3.44	14
4.42	4.25	4.07	3.88	3.79	3.69	3.58	3.48	3.37	3.26	15
4.27	4.10	3.92	3.73	3.64	3.54	3.44	3.33	3.22	3.11	16
4.14	3.97	3.79	3.61	3.51	3.41	3.31	3.21	3.10	2.98	17
4.03	3.86	3.68	3.50	3.40	3.30	3.20	3.10	2.99	2.87	18
3.93	3.76	3.59	3.40	3.31	3.21	3.11	3.00	2.89	2.78	19
3.85	3.68	3.50	3.32	3.22	3.12	3.02	2.92	2.81	2.69	20
3.77	3.60	3.43	3.24	3.15	3.05	2.95	2.84	2.73	2.61	21
3.70	3.54	3.36	3.18	3.08	2.98	2.88	2.77	2.66	2.55	22
3.64	3.47	3.30	3.12	3.02	2.92	2.82	2.71	2.60	2.48	23
3.59	3.42	3.25	3.06	2.97	2.87	2.77	2.66	2.55	2.43	24
3.54	3.37	3.20	3.01	2.92	2.82	2.72	2.61	2.50	2.38	25
3.49	3.33	3.15	2.97	2.87	2.77	2.67	2.56	2.45	2.33	26
3.45	3.28	3.11	2.93	2.83	2.73	2.63	2.52	2.41	2.29	27
3.41	3.25	3.07	2.89	2.79	2.69	2.59	2.48	2.37	2.25	28
3.38	3.21	3.04	2.86	2.76	2.66	2.56	2.45	2.33	2.21	29
3.34	3.18	3.01	2.82	2.73	2.63	2.52	2.42	2.30	2.18	30
3.12	2.95	2.78	2.60	2.50	2.40	2.30	2.18	2.06	1.93	40
2.90	2.74	2.57	2.39	2.29	2.19	2.08	1.96	1.83	1.69	60
2.71	2.54	2.37	2.19	2.09	1.98	1.87	1.75	1.61	1.43	120
2.52	2.36	2.19	2.00	1.90	1.79	1.67	1.53	1.36	1.00	∞

Source: From "Tables of Percentage Points of the Inverted Beta (*F*)-Distribution," *Biometrika*, Vol. 33 (1943), pp. 73–88, by Maxine Merrington and Catherine M. Thompson. Reproduced by permission of the *Biometrika* Trustees.

Table 10 Critical values for the Wilcoxon signed-rank test
$(n = 5(1)50)$

One-sided	Two-sided	$n = 5$	$n = 6$	$n = 7$	$n = 8$	$n = 9$	$n = 10$	$n = 11$	$n = 12$
$\alpha = .05$	$\alpha = .10$	1	2	4	6	8	11	14	17
$\alpha = .025$	$\alpha = .05$		1	2	4	6	8	11	14
$\alpha = .01$	$\alpha = .02$			0	2	3	5	7	10
$\alpha = .005$	$\alpha = .01$				0	2	3	5	7

One-sided	Two-sided	$n = 13$	$n = 14$	$n = 15$	$n = 16$	$n = 17$	$n = 18$	$n = 19$	$n = 20$
$\alpha = .05$	$\alpha = .10$	21	26	30	36	41	47	54	60
$\alpha = .025$	$\alpha = .05$	17	21	25	30	35	40	46	52
$\alpha = .01$	$\alpha = .02$	13	16	20	24	28	33	38	43
$\alpha = .005$	$\alpha = .01$	10	13	16	19	23	28	32	37

One-sided	Two-sided	$n = 21$	$n = 22$	$n = 23$	$n = 24$	$n = 25$	$n = 26$	$n = 27$	$n = 28$
$\alpha = .05$	$\alpha = .10$	68	75	83	92	101	110	120	130
$\alpha = .025$	$\alpha = .05$	59	66	73	81	90	98	107	117
$\alpha = .01$	$\alpha = .02$	49	56	62	69	77	85	93	102
$\alpha = .005$	$\alpha = .01$	43	49	55	61	68	76	84	92

One-sided	Two-sided	$n = 29$	$n = 30$	$n = 31$	$n = 32$	$n = 33$	$n = 34$	$n = 35$	$n = 36$
$\alpha = .05$	$\alpha = .10$	141	152	163	175	188	201	214	228
$\alpha = .025$	$\alpha = .05$	127	137	148	159	171	183	195	208
$\alpha = .01$	$\alpha = .02$	111	120	130	141	151	162	174	186
$\alpha = .005$	$\alpha = .01$	100	109	118	128	138	149	160	171

One-sided	Two-sided	$n = 37$	$n = 38$	$n = 39$	$n = 40$	$n = 41$	$n = 42$	$n = 43$	$n = 44$
$\alpha = .05$	$\alpha = .10$	242	256	271	287	303	319	336	353
$\alpha = .025$	$\alpha = .05$	222	235	250	264	279	295	311	327
$\alpha = .01$	$\alpha = .02$	198	211	224	238	252	267	281	297
$\alpha = .005$	$\alpha = .01$	183	195	208	221	234	248	262	277

One-sided	Two-sided	$n = 45$	$n = 46$	$n = 47$	$n = 48$	$n = 49$	$n = 50$		
$\alpha = .05$	$\alpha = .10$	371	389	408	427	446	466		
$\alpha = .025$	$\alpha = .05$	344	361	379	397	415	434		
$\alpha = .01$	$\alpha = .02$	313	329	345	362	380	398		
$\alpha = .005$	$\alpha = .01$	292	307	323	339	356	373		

Source: From *Some Rapid Approximate Statistical Procedures* (Revised) by Frank Wilcoxon and Roberta A. Wilcox (Pearl River, NY: Lederle Laboratories, 1964), Table 2. Reproduced by permission of Lederle Laboratories, a division of American Cyanamid Company.

Table 11 ■ Random Numbers   **693**

TABLE 11   Random numbers

Line	Column									
	1	2	3	4	5	6	7	8	9	10
1	75029	50152	25648	02523	84300	83093	39852	91276	88988	12439
2	73741	30492	19280	41255	74008	72750	70420	67769	72837	27098
3	07049	98408	27011	76385	15212	03806	85928	81312	14514	55277
4	01033	08705	42934	79257	89138	21506	26797	67223	62165	67981
5	48399	78564	35787	07647	23794	73938	29477	11420	03228	16586
6	70459	73480	06740	79124	14078	72352	07410	93292	93057	18715
7	74770	80185	08181	27417	90866	98444	72870	51219	51481	47916
8	24167	13753	65011	66288	12633	79199	61497	56186	83643	96184
9	24316	80240	62592	53393	57028	61626	56508	84407	97873	27571
10	84565	59254	94435	33322	50014	00180	50954	04099	66005	59141
11	60794	32497	47830	94509	36576	68874	84062	84503	50454	42199
12	99104	14833	97062	48867	19645	78069	91602	46991	57523	22219
13	15604	93654	21487	86036	22827	62637	70378	58539	17827	80108
14	20204	00253	19678	15789	17628	63667	23348	67083	92361	50413
15	71233	73676	00958	42662	47344	00104	74530	46238	06655	23791
16	82846	82954	52107	66054	27358	69664	71760	03577	75622	21536
17	48613	97858	49627	17036	55574	80116	80533	62146	48083	29177
18	42313	91287	66900	79817	76803	42462	63542	99089	22655	44130
19	60879	68102	60700	51281	61386	06782	88214	68246	15552	79093
20	34593	95713	62942	16236	30933	39470	58423	95304	46017	18364
21	96033	10917	01205	08978	43021	77321	76736	64527	96534	98457
22	21932	45476	75464	43497	81807	99369	59945	65349	52588	27386
23	91019	99635	78638	75114	42943	81629	03283	85036	80666	18675
24	86053	48238	14952	55565	98821	92843	67663	70387	13356	46650
25	59700	38346	92770	11506	34101	01051	99390	86884	26788	78768

TABLE 12   Percentage Points of $F_{max} = s^2_{max}/s^2_{min}$

Upper 5% points

df_2	t 2	3	4	5	6	7	8	9	10	11	12
2	39.0	87.5	142	202	266	333	403	475	550	626	704
3	15.4	27.8	39.2	50.7	62.0	72.9	83.5	93.9	104	114	124
4	9.60	15.5	20.6	25.2	29.5	33.6	37.5	41.1	44.6	48.0	51.4
5	7.15	10.8	13.7	16.3	18.7	20.8	22.9	24.7	26.5	28.2	29.9
6	5.82	8.38	10.4	12.1	13.7	15.0	16.3	17.5	18.6	19.7	20.7
7	4.99	6.94	8.44	9.70	10.8	11.8	12.7	13.5	14.3	15.1	15.8
8	4.43	6.00	7.18	8.12	9.03	9.78	10.5	11.1	11.7	12.2	12.7
9	4.03	5.34	6.31	7.11	7.80	8.41	8.95	9.45	9.91	10.3	10.7
10	3.72	4.85	5.67	6.34	6.92	7.42	7.87	8.28	8.66	9.01	9.34
12	3.28	4.16	4.79	5.30	5.72	6.09	6.42	6.72	7.00	7.25	7.48
15	2.86	3.54	4.01	4.37	4.68	4.95	5.19	5.40	5.59	5.77	5.93
20	2.46	2.95	3.29	3.54	3.76	3.94	4.10	4.24	4.37	4.49	4.59
30	2.07	4.20	2.61	2.78	2.91	3.02	3.12	3.21	3.29	3.36	3.39
60	1.67	1.85	1.96	2.04	2.11	2.17	2.22	2.26	2.30	2.33	2.36
∞	1.00	1.00	1.00	1.00	1.00	1.00	1.00	1.00	1.00	1.00	1.00

Upper 1% points

df_2	t 2	3	4	5	6	7	8	9	10	11	12
2	199	448	729	1036	1362	1705	2063	2432	2813	3204	3605
3	47.5	85	120	151	184	21(6)	24(9)	28(1)	31(0)	33(7)	36(1)
4	23.2	37	49	59	69	79	89	97	106	113	120
5	14.9	22	28	33	38	42	46	50	54	57	60
6	11.1	15.5	19.1	22	25	27	30	32	34	36	37
7	8.89	12.1	14.5	16.5	18.4	20	22	23	24	26	27
8	7.50	9.9	11.7	13.2	14.5	15.8	16.6	17.9	18.9	19.8	21
9	6.54	8.5	9.9	11.1	12.1	13.1	13.9	14.7	15.3	16.0	16.6
10	5.85	7.4	8.6	9.6	10.4	11.1	11.8	12.4	12.9	13.4	13.9
12	4.91	6.1	6.9	7.6	8.2	8.7	9.1	9.5	9.9	10.2	10.6
15	4.07	4.9	5.5	6.0	6.4	6.7	7.1	7.3	7.5	7.8	8.0
20	3.32	3.8	4.3	4.6	4.9	5.1	5.3	5.5	5.6	5.8	5.9
30	2.63	3.0	3.3	3.4	3.6	3.7	3.8	3.9	4.0	4.1	4.2
60	1.96	2.2	2.3	2.4	2.4	2.5	2.5	2.6	2.6	2.7	2.7
∞	1.00	1.0	1.0	1.0	1.0	1.0	1.0	1.0	1.0	1.0	1.0

s^2_{max} is the largest and s^2_{min} the smallest in a set of t independent mean squares, each based on $df_2 = n - 1$ df. Values in the column $t = 2$ and in the rows $df_2 = 2$ and ∞ are exact. Elsewhere the third digit may be in error by a few units for the 5% points and several units for the 1% points. The third-digit figures in parentheses for $df_2 = 3$ are the most uncertain.

From *Biometrika Tables for Statisticians*, 3rd ed., Vol. 1, edited by E. S. Pearson and H. O. Hartley (New York: Cambridge University Press, 1966), Table, p. 202. Reproduced by permission of the *Biometrika Trustees*.

GLOSSARY OF COMMON STATISTICAL TERMS

acceptance region   Set of values of a test statistic that imply acceptance of the null hypothesis.

alternative hypothesis   Hypothesis to be accepted if null hypothesis is rejected.

analysis of variance   A procedure for comparing more than two population means. There are many applications of analysis of variance beyond those discussed in the text.

arithmetic mean   Average.

bar chart   A graphical method for showing how data fall into a group of categories.

binomial experiment   An experiment involving n identical independent trials. (See Chapter 10 for an exact description of a binomial experiment.)

binomial random variable   Discrete random variable representing the number of successes x in n identical independent trials. (For an exact definition of a binomial experiment, see Chapter 10.)

box-and-whiskers plot   *See* box plot.

box plot   A graphical method for describing data concerned with the symmetry and central tendency of a set of measurements.

Central Limit Theorem   Theorem stating that the sampling distribution of the sample mean (or sum) will be approximately normal when certain conditions are satisfied. (See Chapter 7.)

chi-square   Test statistic used to test the null hypothesis of independence for the two classifications of a contingency table. Also has many other statistical applications not discussed in this text.

chi-square distribution   Distribution of a chi-square statistic.

class boundary   The dividing point between two cells in a frequency histogram.

classes   Cells of a frequency histogram.

class frequency   The number of observations falling in a class (referring to a frequency histogram).

classical interpretation of probability   If there are N possible outcomes in an experiment and N_E result in event E, then

$$P(E) = \frac{N_E}{N}$$

class interval   Each sub-interval of a frequency table (or histogram) is called a class interval.

class interval width   The width of each class interval in a frequency table (or histogram).

cluster sampling   Sampling so that a simple random sample of groups (clusters) is selected; all items within a selected cluster are then sampled.

coefficient of determination   The square of the sample correlation coefficient, $\hat{\rho}$.

complement   The complement of an event A is the event that A does not occur.

completely randomized design   An experimental design where treatments are assigned to experimental units—"at random."

conditional probability (A given B):   The probability that A occurs given that B has occurred.

confidence coefficient   The probability that an interval estimate (a confidence interval) will enclose the parameter of interest.

confidence interval   Two numbers, computed from sample data, that form an interval estimate for some parameter.

contingency table   A two-way table constructed for classifying count data. The entries in the table show the number of observations falling in the cells. The objective of an analysis is to determine whether the two directions of classification are dependent (contingent) upon one another.

continuous random variable   A quantitative random variable that can assume any one of a countless number of values on a line interval.

correlation coefficient   A measure of linear dependence between two random variables.

degrees of freedom   A parameter of Student's t-, the F-, and the chi-square probability distributions. Degrees of freedom measure the quantity of information available in normally distributed data for estimating the population variance σ^2.

deviation   The difference between a measurement and the mean of the set of measurements from which the measurement was drawn.

deviation from the mean   Distance between a sample observation and the sample mean $\bar{x}$.

direction observation   A method of data collection whereby the surveyor observes the event(s) of interest.

discrete random variable   A random variable that can assume only a countable number of values.

dot diagram   A way to illustrate the variability in a small set of measurements.

Empirical Rule   A rule that describes the variability of data that possess a mound-shaped frequency distribution. (See Chapter 5.)

estimate   A number computed from sample data, used to approximate a population parameter.

estimation   One of two approaches to making inferences about parameters; included in this approach are point estimates and interval estimates (confidence intervals).

event   A collection of outcomes.

expected value of y   For the model $y = \beta_0 + \beta_1 x + \epsilon$, the expected value of y, $E(y)$, is $E(y) = \beta_0 + \beta_1 x$

expected values of $\hat{\beta}_0$ and $\hat{\beta}_1$ The expected values of $\hat{\beta}_0$ and $\hat{\beta}_1$ for the linear regression model $y = \beta_0 + \beta_1 x + \epsilon$ are, respectively, β_0 and β_1.

experiment   The process by which an observation is obtained.

exploratory data analysis   A relatively new area of applied statistics dealing with data description.

factorial experiment   An experiment where the response is observed at each factor-level combination of the independent variables.

F-distribution   Distribution of an F-statistic.

freehand regression line   A trend line "fit" by eye.

frequency   Number of observations falling in some cell or in some classification category.

frequency polygon   One of several methods for graphing frequency data.

frequency table   A table used to summarize how many measurements in a set fall into each of the sub-intervals (or classes).

F-statistic   Test statistic used to compare variances from two normal populations. Used in the analysis of variance.

hinges   Quantities similar to the upper and lower quartiles of a data set.

histogram   A graphical method for describing a set of data. (See Chapter 4.)

hypothesis testing   *See* statistical test.

independent events   Event A is independent of event B if $P(A|B) = P(A)$ (or if $P(B|A) = P(B)$).

interaction   The failure of a variable to exert the same effect on a response at different levels of one or more other independent variables.

interval estimate   Two numbers computed from the sample data. The interval formed by the numbers should enclose some parameter of interest. An interval estimate is usually called a confidence interval.

latin square design   An experimental design that employs blocking in two dimensions.

leading digit   The first digit of measurements in a stem-and-leaf plot.

least squares   A method of curve fitting that selects as the best-fitting curve the one that minimizes the sum of the squares of deviations of the data points from the fitted curve. (See Chapter 12.)

level of significance   Refers to the outcome of a specific statistical test of a hypothesis. The level of significance of the test is the probability of drawing a value of the test statistic that is as contradictory, or more contradictory, to the null hypothesis than the value observed, assuming that the null hypothesis is true.

linear correlation   Dependence between two random variables.

linear equation   An equation of the form $y = \beta_0 + \beta_1 x$.

lower boundary   The lower limit of a class interval.

lower confidence limit   The smaller of the two numbers that form a confidence interval.

lower quartile   The 25th percentile of a set of measurements.

mean   The average of a set of measurements. The symbols $\bar{x}$ and μ denote the means of a sample and a population, respectively.

measures of central tendency   Numerical descriptive measures that locate the center or central value in a set.

median   The middle measurement when a set is ordered according to numerical value. (See Chapter 5.)

method of least squares   A procedure for finding the estimates $\hat{\beta}_0$ and $\hat{\beta}_1$ for the model $y = \beta_0 + \beta_1 x + \epsilon$.

mode   The measurement in a set that occurs with greatest frequency.

mound-shaped frequency distribution   A symmetrical, single-peaked frequency distribution.

mutually exclusive events A and B are mutually exclusive events if the occurrence of one precludes the occurrence of the other.

nonparametric methods   Usually refer to statistical tests of hypotheses about population probability distributions but not about specific parameters of the distributions. (See Chapter 17.)

normal curve   Smooth, bell-shaped curve known as the normal distribution.

normal distribution   A bell-shaped probability distribution. The curve possesses a specific mathematical formula.

null hypothesis   The hypothesis under test in a statistical test of a hypothesis.

numerical descriptive measures   Quantities used to describe a set of measurements.

observational study   A study where the conditions under which observation are obtained are not fixed (controlled).

one-at-a-time approach   A popular, but inefficient, way to evaluate the effect of each variable in a problem involving more than one independent variable.

outcome   The result of an experiment.

paired-difference test   A statistical test for the comparison of two population means. The test is based on paired observations, one from each of the two populations.

parameter of a population   A numerical descriptive measure of a population.

parameters   Numerical descriptive measures for a population.

parametric methods   Statistical methods for estimating parameters or testing hypotheses about population parameters.

percentiles   See Chapter 5 for definition.

personnel interviews   Interviews conducted in person.

pie chart   A graphical method for describing data. (See Chapter 4.)

point estimate   *See* estimate.

population   The set of measurements, existing or conceptual, that is of interest to the experimenter. Samples are selected from the population.

predicted value of y   A value of y computed from the prediction equation $\hat{y} = \hat{\beta}_0 + \hat{\beta}_1 x$.

prediction equation   An equation relating a dependent variable y to one or more independent variables.

probability   As a practical matter, we think of the probability of an event as a measure of one's belief that the event will occur when the experiment is conducted once. The exact definition, giving a quantitative measure of this belief, is subject to debate. The relative frequency concept is most widely accepted.

probability distribution   A formula, table, or graph used to display the probabilities for a discrete random variable.

probability distribution, continuous   A smooth curve that gives the theoretical frequency distribution for the continuous random variable. An area under the curve over an interval is proportional to the probability that the random variable will fall in the interval.

probability distribution, discrete   A listing, a mathematical formula, or a histogram that gives the probability associated with each value of the random variable.

p-value   Level of significance of a statistical test.

qualitative random variable   A random variable with qualitative outcomes.

qualitative variable   A variable that has qualitative observations.

quantitative random variable   A random variable with quantitative outcomes.

quantitative variable   A variable that has quantitative observations.

randomized block design   An experimental design whereby treatments are assigned to experimental units "at random" within each block.

random sample   A sample of n measurements selected in such a way that every different sample of n elements in the population has an equal probability of being selected.

random variable   A random variable is associated with an experiment. Its values are numerical events that cannot be predicted with certainty.

range of a set of measurements   The difference between the largest and smallest members of the set.

rank correlation coefficient   A rank correlation coefficient is a coefficient of linear correlation between two random variables that is based on the ranks of the measurements, not on their actual values. *See also* correlation coefficient.

ratio estimation   A way in survey sampling to use information on an auxiliary variable in the estimation of a population parameter.

regression line   Line fit to data points, using the method of least squares.

rejection region   Set of values of a test statistic that indicates rejection of the null hypothesis.

relative frequency   Class frequency divided by the total number of measurements.

relative frequency concept of probability   If an experiment is repeated n times and n_E result in event E, then

$$P(E) \approx \frac{n_E}{n}$$

relative frequency histogram   A histogram displaying the fraction of measurements falling in each class.

residual The difference between a value of y and its predicted value $\hat{y}$: residual $= y - \hat{y}$.

sample A subset of measurements selected from a population.

sample correlation coefficient The correlation coefficient computed from a sample of pairs (x,y).

sample mean Sample average

sampling distribution The probability distribution for a sample statistic.

scatterplot A graphical method for displaying bivariate data.

self-administered questionnaire A method of collecting data in a sample survey whereby those surveyed complete a questionnaire.

significance level *See* level of significance.

sign test A nonparametric statistical test used to compare two populations. (See Chapter 17.)

simple random sampling Sampling so that each sample of size n has an equal chance of being selected.

skewed distribution An asymmetric distribution of measurements trailing off to the right or left.

slope The term β_1 in the linear equation $y = \beta_0 + \beta_1 x$.

software Computer programs.

Spearman's rank correlation coefficient One of several correlation coefficients based on the ranks of the two random variables.

standard deviation A measure of data variation. (See Chapter 5.)

standardized normal distribution A normal distribution with mean and standard deviation equal to 0 and 1, respectively. The standardized normal variable is denoted by the symbol z.

standard normal random variable A random variable having a normal distribution with $\mu = 0$, $\sigma = 1$.

statistical software systems Computer programs to summarize and statistically analyze data that are integrated into a single system of programs.

statistical test A procedure for making an inference about one or more population parameters by using information from sample data. The procedure is based on the concept of proof by contradiction.

statistics Numerical, descriptive measures for a sample.

stem-and-leaf plot A graphical method for displaying data from exploratory data analysis.

stratified random sampling Sampling so that a simple random sample is selected within each of the strata.

Student's *t*-test A test statistic used for small-sample tests of means.

Student's *t*-distribution A particular symmetric mound-shaped distribution that possesses more spread than the standard normal probability distribution.

subjective probability One's personal estimate of the probability of an event.

sum of squares about the mean The sum of the squared differences about the mean, $\Sigma(y - \bar{y})^2$.

sum of squares due to regression The sum of the squared deviations of the predicted values from the mean, $\Sigma(\hat{y} - \bar{y})^2$.

sum of squares for error   The sum of the squared residuals, $\text{SSE} = \Sigma(y - \hat{y})^2$

systematic sample   An economical way to sample a population when the elements of the population are arranged in a list.

test statistic   A function of the sample measurements, used as a decision maker in a test of a hypothesis.

trailing digits   The remaining digits after the leading digits of measurements in a stem-and-leaf plot.

two-tailed test   A statistical test with a two-sided alternative hypothesis.

type I error   Rejecting the null hypothesis when it is true.

type II error   Accepting the null hypothesis when it is false and the alternative hypothesis is true.

unconditional probability   The probability of an event.

upper confidence limit   The larger of the two numbers that form a confidence interval.

upper quartile   The 75th percentile of a set of measurements.

variable   A phenomenon where observations vary from trial to trial.

variance   A measure of data variation. (See Chapter 5.)

y-intercept   The term β_0 in the linear equation $y = \beta_0 + \beta_1 x$.

z-score   A standardized score formed by subtracting the mean and dividing by the standard deviation.

z-statistic   A standardized normal random variable that is frequently used as a test statistic.

REFERENCES

Bureau of Labor Statistics Handbook of Methods (1982). Washington, D.C.: U.S. Government Printing Office.

Bradley, J. V. (1968). *Distribution-Free Statistical Tests.* Englewood Cliffs, N.J.: Prentice-Hall.

Conover, W. J. (1980). *Practical Nonparametric Statistics,* 2nd ed. New York: John Wiley.

Dixon, W. J. (1981). *BMDP Statistical Software.* Berkeley, CA.: University of California Press.

Dixon, W. J. and M. B. Brown, eds. (1978). *Biomedical Computer Programs,* rev. ed. Los Angeles: University of California Press.

Hald, A. (1952). *Statistical Tables and Formulas.* New York: John Wiley.

Handbook of Tables for Probability and Statistics, 2nd ed. (1968). Cleveland, Ohio: The Chemical Rubber Co.

Helwig, J. T. (1977). *SAS Supplementary Library User's Guide.* Raleigh, N.C.: SAS Institute.

Hicks, C. R. (1973). *Fundamental Concepts in the Design of Experiments,* 2nd ed. New York: Holt, Rinehart and Winston.

Hildebrand, D. and L. Ott (1987). *Statistical Thinking for Managers,* 2nd ed. Boston: Duxbury Press.

Hollander, M. and D. A. Wolfe (1973). *Nonparametric Statistical Methods.* New York: John Wiley.

Huntsberger, D. V. and P. Billingsley (1977). *Elements of Statistical Inference,* 4th ed. Boston: Allyn and Bacon.

Lefkowitz, J. M. (1984). *Introduction to Statistical Software Packages.* Boston: Duxbury Press.

Marshall, K. P. (1974). "A Study of Job Satisfaction Among 219 Nurses in the Southeast." Unpublished paper.

Mendenhall, W. (1968). *An Introduction to Linear Models and the Design and Analysis of Experiments.* Belmont, CA: Wadsworth.

Mendenhall, W. (1987). *Introduction to Probability and Statistics,* 7th ed. Boston: Duxbury Press.

Mendenhall, W., R. Scheaffer, and D. Wackerly (1981). *Mathematical Statistics with Applications,* 2nd ed. Boston: Duxbury Press.

Merrington, M. (1941). "Table of percentage points of the *t*-distribution." *Biometrika* 32:300.

Merrington, M. and C. M. Thompson (1943). "Points of the inverted beta (*F*)-distribution." *Biometrika* 33:73–88.

Minitab Reference Manual. Release 6.1 (1988). State College, PA: Minitab, Inc.

National Bureau of Standards (1949). *Tables of the Binomial Probability Distribution.* Washington, D.C.: U.S. Government Printing Office.

Newman, R. W. and R. M. White (1951). *Reference Anthropometry of Army Men.* Report No. 180, Lawrence, MA.: Environmental Climatic Research Laboratory.

Nie, N., C. H. Hull, J. G. Jenkins, K. Steinbrenner, and D. H. Bent (1975). *Statistical Package for the Social Sciences,* 2nd ed. New York: McGraw-Hill.

Omstead, P. S. and J. W. Tukey (1947). "A corner test for association." *Annals of Mathematical Statistics* 18:495–513.

Ott, L. (1988). *An Introduction to Statistical Methods and Data Analysis,* 3rd ed. Boston: Duxbury Press.

Ott, L., R. Larson, and W. Mendenhall (1987). *Statistics: A Tool for the Social Sciences,* 4th ed. Boston: Duxbury Press.

Ryan, A. A., ed. (1985). *SAS User's Guide: Statistics.* Cary, N.C.: SAS Institute, Inc.

Ryan, T. A., B. L. Joiner, and B. F. Ryan (1985). *Minitab Student Handbook,* 2nd ed. Boston: Duxbury Press.

SAS Institute (1988). *SAS Introductory Guide,* Release 6.03 Edition. Cary, N.C.: SAS Institute.

Scheaffer, R. L., W. Mendenhall, and L. Ott (1986). *Elementary Survey Sampling,* 3rd ed. Boston: Duxbury Press.

Siegel, S. (1956). *Nonparametric Statistics for the Behavioral Sciences.* New York: McGraw-Hill.

SPSS, Inc. Staff (1984). *SPSS-X User's Guide,* 2nd ed. New York: McGraw-Hill.

Thompson, C. M. (1941). "Tables of the percentage points of the χ^2-distribution." *Biometrika* 32:188–189.

Tukey, J. W. (1977). *Exploratory Data Analysis.* Reading, MA: Addison-Wesley.

U.S. Bureau of the Census (1980). *Statistical Abstract of the United States.* Washington, D.C.: U.S. Government Printing Office.

U.S. Department of Justice (1980). *Uniform Crime Reports for the United States.* Washington, D.C.: U.S. Government Printing Office, pp. 60–86.

SELECTED ANSWERS

Chapter 1

1.1 (a) The population is the set of weights of all shrimp on the diet.
(b) The sample is the set of weights of the 100 shrimp selected from the pond.
(c) A single weight that typifies the collection of weights contained in the population, for example, the "average" or mean weight.
(d) A measure of reliability is needed so that you will know how much faith you can place in your inference.

1.3 (a) The population is the set of numbers of children in all households that receive welfare support in the city.
(b) The sample is the set of numbers of children corresponding to the 400 households selected from the welfare roles.
(c) The characteristic of interest would be a number that typifies the number of children per welfare household—for example, the "average" or mean number per household.
(d) Same as the answer to 1.1(d).

Chapter 4

4.1   **4.3**

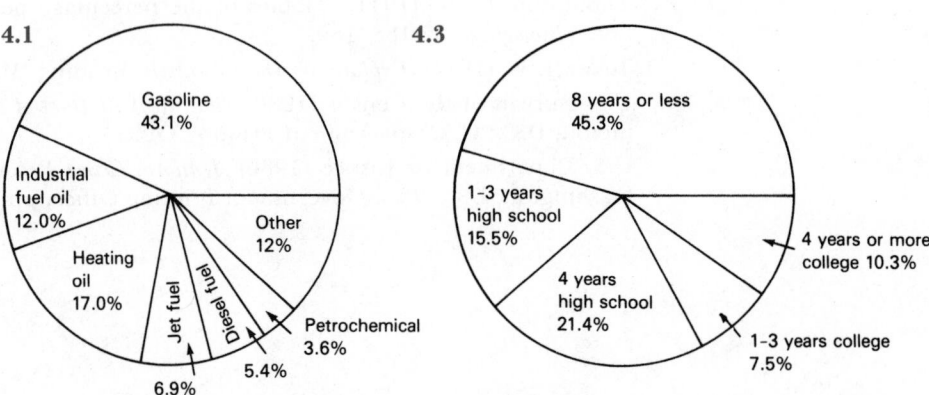

4.5 No, we can't add the percentages of males and females for each category.

4.7

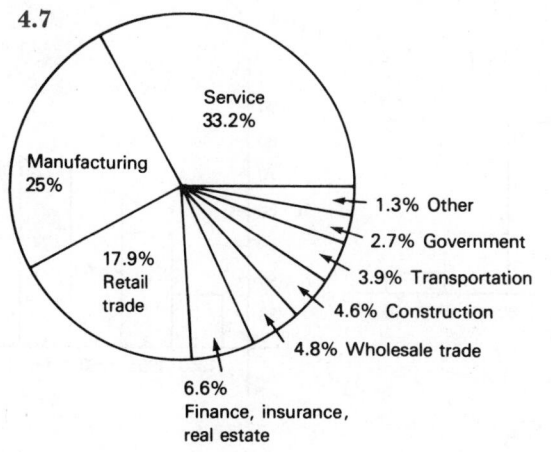

4.9

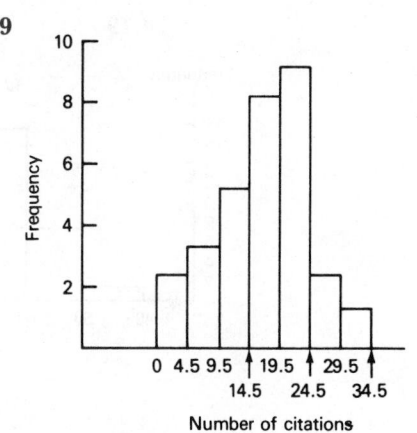

4.11

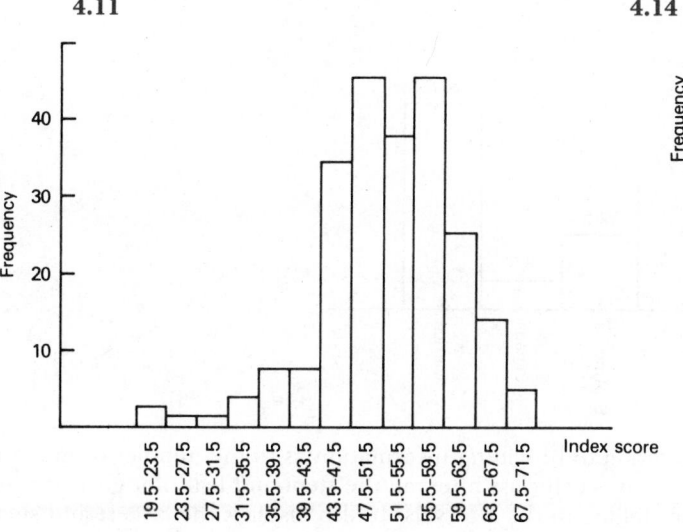

4.14

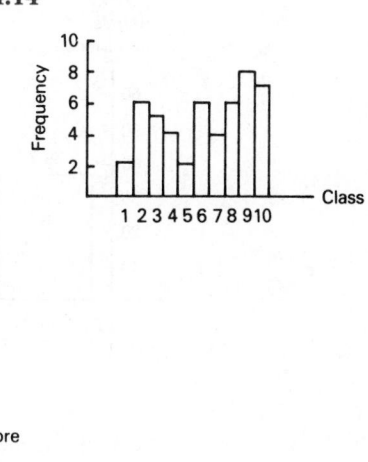

4.15

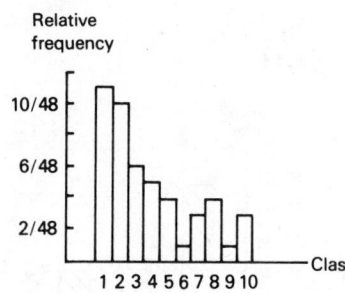

4.17

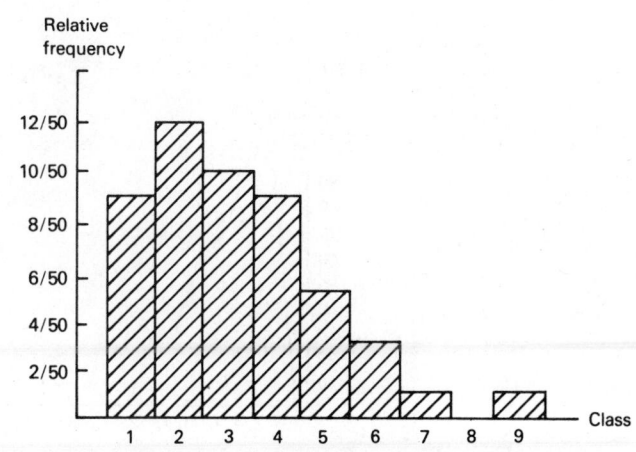

4.19

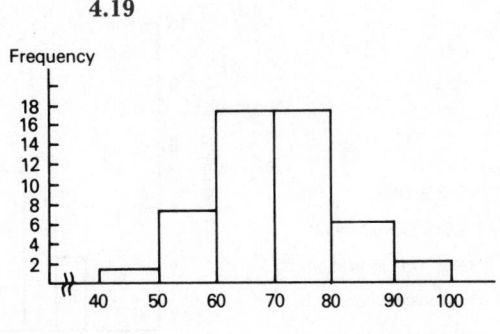

4.34

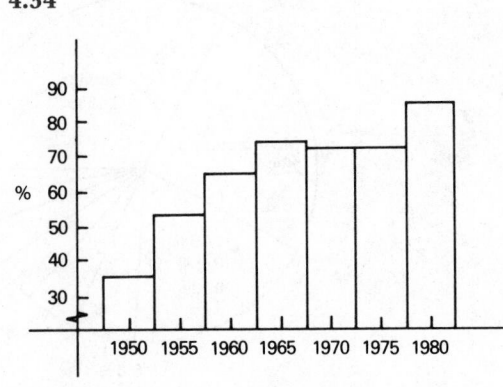

4.36

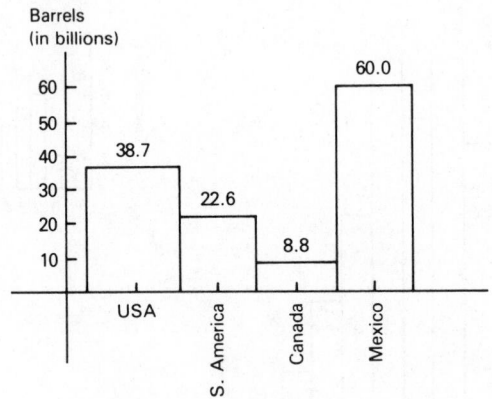

4.38 In addition to providing information as to the number of measurements falling into each class interval, the stem-and-leaf plot gives the actual values falling in the intervals. It also displays the largest and smallest measurements in the set.

4.40

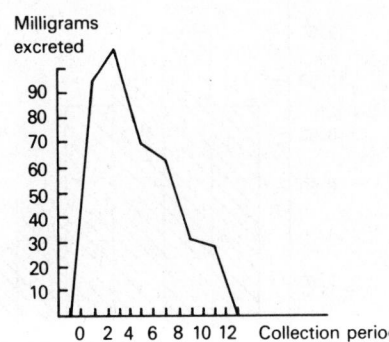

4.44

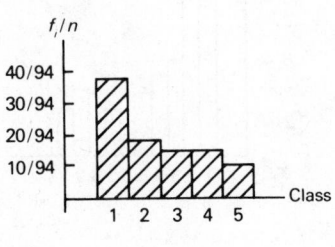

4.46

Age group	Frequency	Fraction of total	Sector angle
20–29	32	.0175	6.300
30–39	175	.0958	34.488
40–49	438	.2397	86.292
50 and over	1182	.6470	232.920
Total	1827	1.0000	360.000

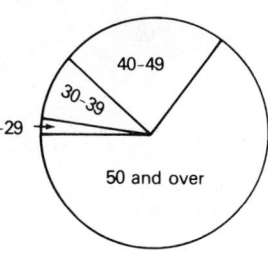

4.48

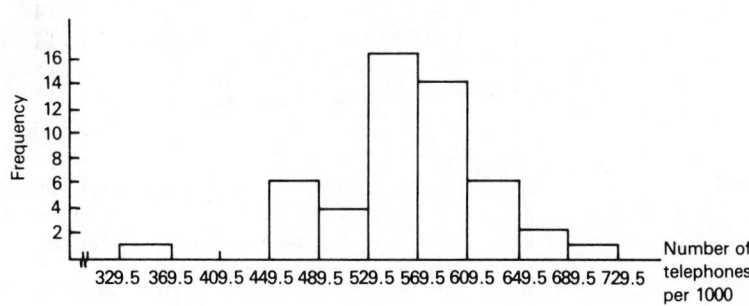

Chapter 5

5.1   Mean: 9   Median: 9   Mode: 9

5.3   Mean: 109   Median: 103   Mode would remain the same.

5.5   Mean: 64.2   Median: 64   Mode: there is none since each measurement appears once.

5.7   Mean: 14.51   Median: 15   Mode: 16

5.13 50% of all workers earn more than $8.48 per hour; 50% earn less than $8.48.

5.15 Range: 14   Mean: 8   Variance: $s^2 = 23$
Standard deviation: $s = 4.80$

5.19 Approximately 68% will lie between .7 and 1.9. Approximately 95% will lie between .1 and 2.5. All or nearly all will lie between $-.5$ and 3.1.

5.21 (b) $\Sigma x = 15, \Sigma x^2 = 55$
(c) $s^2 = 2.50, s = 1.58$

5.23 (b) $\Sigma x = 25, \Sigma x^2 = 85, \Sigma (x - \bar{x})^2 = 22.5$
(c) $s^2 = 2.5, s = 1.58$

5.25

k	$\bar{x} - ks$ to $\bar{x} + ks$	Percentage in interval
1	11.95 to 28.69	70%
2	3.58 to 37.06	96%
3	-4.79 to 45.43	100%

5.27 (a)

Class	Class boundaries	f_i	f_i/n
1	1.5–2.5	1	.0143
2	2.5–3.5	1	.0143
3	3.5–4.5	3	.0429
4	4.5–5.5	5	.0714
5	5.5–6.5	5	.0714
6	6.5–7.5	12	.1714
7	7.5–8.5	18	.2571
8	8.5–9.5	15	.2143
9	9.5–10.5	6	.0857
10	10.5–11.5	3	.0429
11	11.5–12.5	0	.0000
12	12.5–13.5	1	.0143
Total		70	1.0000

(b) $\bar{x} = 7.73$

(c) $\Sigma (x - \bar{x})^2 = 271.8429$; $s^2 = 3.9398$; $s = 1.98$

Interval	Percentage	Empirical Rule
$\bar{x} \pm s$: 5.75–9.71	71	68%
$\bar{x} \pm 2s$: 3.77–11.69	96	95%
$\bar{x} \pm 3s$: 1.79–13.67	100	almost all

5.31 (a) $s \approx$ range/4 $= 2.0$
(b) $s = 2.02$

5.33 (a) $s \approx 9.50$
(b) $s = 8.37$

5.35 (a) .348   (b) $s \approx .07$ (c) $s = .095$

5.43 Mean: 111.56   Median: 112.67   Mode: 117

5.45 (a) False   (b) False   (c) True   (d) False
(e) True   (f) True

5.53 The value 17 is *the median*.

5.55 According to the Empirical Rule, 68%.

5.57 A measurement of 20% lies exactly one standard deviation ($s = 6\%$) away from the mean. $\bar{x} = 14\%$. The proportion of cities giving percentages of unlisted numbers exceeding 20% would probably be in the neighborhood of 16%.

5.59 $\bar{x} = 2.1$; $s^2 = 2.6211$; $s = 1.62$; all the measurements fall within the interval $\bar{x} \pm 2s$, or -1.14 to 5.34.

5.61

Class (× 1000)	Frequency	Midpoint x_i	$f_i \times x_i$	f_i/n	cum f_i	cum f_i/n
0–399.5	26	199.5	5187	.52	26	.52
399.5–799.5	14	599.5	8393	.28	40	.80
799.5–1199.5	4	999.5	3998	.08	44	.88
1199.5–1599.5	3	1399.5	4198.5	.06	47	.94
1599.5–1999.5	0	1799.5	0	0	47	.94
1999.5–2399.5	2	2199.5	4399	.04	49	.98
2399.5–2799.5	1	2599.5	2599.5	.02	50	1.00
	50		28775			

5.63 The data are skewed to the right, with a few extreme values far to the right.

5.65 $\bar{x} = 102.2$, $s^2 = 141.7$, $s = 11.90$
Yes; a value of 140 is more than 3 standard deviations away from the sample mean, $\bar{x} = 102.2$

5.67

Class	Class boundaries	f_i	f_i/n
1	81.9–91.1	7	.2187
2	91.1–100.3	11	.3437
3	100.3–109.5	5	.1562
4	109.5–118.7	6	.1875
5	118.7–127.9	1	.0312
6	127.9–137.1	2	.0625
Total		32	1.0000

5.78 (a) 20.32   (b) 19.5

(c)

Class boundaries	f_i	f_i/n
5.5–10.5	4	.08
10.5–15.5	12	.24
15.5–20.5	11	.22
20.5–25.5	11	.22
25.5–30.5	6	.12
30.5–35.5	4	.08
35.5–40.5	0	.00
40.5–45.5	2	.04

(d) no

5.80 $\bar{x} = 4214.29$. This distribution is mound-shaped, so the Empirical Rule would not be appropriate. The median is more representative. Third quartile: $3000

5.82 $\bar{x} = 6847.14$, $s = 6232.68$
The range approximation for s is $(21,010 - 379)/4 = 5158$; this is reasonably close for an approximation.

5.84

DESCRIPTIVE STATISTICS BY TREATMENT GROUP FOR THE FOLLOWING VARIABLES
ANXIETY, RETARDATION, SLEEP, AND TOTAL HAMILTON SCORE

VARIABLE	LABEL	N	MEAN	MINIMUM VALUE	MAXIMUM VALUE	STANDARD DEVIATION
--TREATMNT = A --						
ANXIETY	ANXIETY/SOMATIZATION FACTOR	25	0.62640000	0.17000000	1.33000000	0.34812450
RETARDTN	RETARDATION FACTOR	25	1.13000000	0.00000000	2.50000000	0.64193717
SLEEP	SLEEP DISTURBANCE FACTOR	25	0.26640000	0.00000000	2.00000000	0.49061934
TOTAL	TOTAL SCORE	25	12.72000000	4.00000000	34.00000000	6.74858010
--TREATMNT = B --						
ANXIETY	ANXIETY/SOMATIZATION FACTOR	25	0.69280000	0.00000000	1.67000000	0.35957753
RETARDTN	RETARDATION FACTOR	25	1.05000000	0.00000000	2.50000000	0.58630197
SLEEP	SLEEP DISTURBANCE FACTOR	25	0.53320000	0.00000000	1.67000000	0.49118666
TOTAL	TOTAL SCORE	25	14.04000000	2.00000000	24.00000000	6.48382603
--TREATMNT = C --						
ANXIETY	ANXIETY/SOMATIZATION FACTOR	25	0.60000000	0.17000000	1.50000000	0.33292892
RETARDTN	RETARDATION FACTOR	25	1.04000000	0.00000000	2.50000000	0.61101009
SLEEP	SLEEP DISTURBANCE FACTOR	25	0.15960000	0.00000000	1.33000000	0.32051365
TOTAL	TOTAL SCORE	25	11.48000000	3.00000000	23.00000000	4.98430871
--TREATMNT = D --						
ANXIETY	ANXIETY/SOMATIZATION FACTOR	25	0.84000000	0.00000000	1.67000000	0.42438190
RETARDTN	RETARDATION FACTOR	25	1.19000000	0.25000000	2.25000000	0.57861184
SLEEP	SLEEP DISTURBANCE FACTOR	25	0.85360000	0.00000000	2.00000000	0.70151550
TOTAL	TOTAL SCORE	25	19.20000000	3.00000000	32.00000000	5.74456265

Chapter 6

6.1 (a) Subjective probability   (b) Relative frequency concept
(c) Classical interpretation   (d) Relative frequency concept
(e) Relative frequency concept   (f) Classical interpretation

6.3

	Result on	
Toss 1	Toss 2	Toss 3
H	H	H
H	H	T
H	T	H
T	H	H
H	T	T
T	H	T
T	T	H
T	T	T

6.5 (a) $\overline{A}$ = {HHH, HHT, HTH, THH, TTT}
$\overline{B}$ = {TTT}
$\overline{C}$ = {HHH, HHT, HTH, HTT, THH, THT, TTH}
(b) A and B are not mutually exclusive.

6.7 A and B are not independent.
A and C are not independent.
B and C are not independent.

6.9 (a) $P(A) = \frac{1}{6}$

(b) $P(A) = \frac{1}{2}$

(c) $P(A) = \frac{2}{3}$

(d) $P(A) = \frac{1}{3}$

6.14   (a) $P(\text{accept}) = .7435$, $P(\text{reject}) = .2565$
(b) $P(\text{two-career marriage}) = .46$, $P(\text{one-career marriage}) = .37$

6.20   Yes, assuming only two possible outcomes.

6.22   Yes. The estimate will tend to be biased.

6.24   (a) $P(x = 10) = .0000$, $P(x = 6) = .0368$, $P(x \geq 6) = .0473$,
$P(x = 0) = .0282$
(b) The formula is

$$P(x \leq 100) = \sum_{k=0}^{100} \frac{1000!}{k!\,(1000 - k)} (.3)^k\, (.7)^{1000-k}$$

6.26   .729   .972

6.28   With $n = 3$ and $\pi = \frac{1}{2}$, we have $P(0) = \frac{1}{8}$, $P(1) = \frac{3}{8}$, $P(2) = \frac{3}{8}$, $P(3) = \frac{1}{8}$.

6.30   (a) .4032   (b) .4713

6.32   (a) .4015   (b) .2794

6.34   (a) .0542   (b) .1857   **6.36**   .8729

6.38 $z_0 = 1.96$ **6.40** $z_0 = 1.645$

6.42   (a) .5   (b) .1056   (c) .9699   (d) .8664   (e) .3413

6.44   (a) .1056   (b) .1056   (c) .0062   (d) .7888

6.46   (a) $x = 628$ (b) 475. We call this the 40th percentile.

6.48   (d) All of the above are true.   **6.62**   (a) .3125   (b) .5000

6.64   (a) $\sigma \approx 55$ (b) The probability is approximately .0344.

6.66   (a) $P = .1736$ (b) $P = .3783$ (c) $P = .4481$

6.68   Not a binomial experiment.

6.70   No, $x = 1373$ is more than 15 standard deviations above the mean.

6.72   (a) The 5 characteristics of a binomial experiment have been met.
(b) Inferences can only be made about soldiers stationed in the U.S. since only U.S. bases were sampled.

6.74   (a) .9938   (b) 1490.7   (c) μ is less than 40.

6.76   (a) .4332   (b) .4641   **6.78**   (a) .2881   (b) .5762

6.80   (a) .5119   (b) .0267   **6.82** $z = 1.645$

6.84 $z = -.50$ **6.86**   .4649   **6.93**   Approximately 70 minutes

6.95 $\mu = 6$; $\sigma = 2.19$ **6.97**   (a) $\sigma = 30$ (b) $P = .50$ (c) 6.68%

6.99 $\mu = 6400$; $\sigma = 35.78$. Yes, $x = 6200$ is more than 5 standard deviations away from the mean.

6.101   (a) .3069   (b) .4074   (c) .4126   (d) They are not independent.

6.103   (a) $16 for the new book   (b) 72.89%

6.107   (a) .3783

Chapter 7

7.1   A random sample of n measurements from a population is one in which every different subset of size n from the population has an equal probability of being selected. It is probably not possible to draw a truly random sample.

7.3   The process is not likely to approximate random sampling. For example, if the sampling is conducted late in the day, a disproportionately large number of discarded homeowners might be those without children and those who might have negative views on expenditures for public schools.

7.7 (a) 10 samples   (b) $\mu_{\bar{x}} = 3.4$ (c) $\sigma_{\bar{x}} = 1.98$

7.9 According to the Central Limit Theorem, the sampling distribution of $\bar{x}$ will be approximately normal, with $\mu_{\bar{x}} = \mu = 55$ and $\sigma_{\bar{x}} = \sigma/\sqrt{n} = 2.0$.

7.11 The distribution for Σx will be approximately normal, with mean $60(192) = 11,520$ and standard error $43\sqrt{60} = 333.08$.

7.19 (a) .0548

(b) If the mean oxygen content really were 6.0 ppm, the probability that a sample mean would exceed 6.5 ppm is very small (.0548). This suggests that the mean oxygen content is not 6.0 ppm. The data suggest that it is larger than 6.0 ppm.

7.21 About 1.00   **7.23** (a) .2033   (b) .0668   **7.25** .2076

7.27 (a) The sampling distribution will be approximately normal with mean .02 and standard deviation .000081.

(b) $P(x \geq 3\%) = $ approximately 0

  $P(x \geq 4\%) = $ approximately 0

7.29 (a) 0   (b) 0   **7.31** .3821

7.42 If the store were open 6 days per week, the mean of the four-week (24 days) shrinkage would be $24(\$320) = \$7680$. The standard deviation would be $\sigma\sqrt{n} = \$80\sqrt{24} = \$391.92$. The sampling distribution of the total shrinkage would be approximately normal.

7.44 .0793

7.46 No; $x = 2600$ lies 27.8 standard deviations below $\mu = 3500$. This is so improbable it suggests that μ is less than 3500 and hence that the proportion of incorrect returns this year is less than 70%.

7.53 A highly improbable result has occurred, assuming $\pi = .05$. This might lead us to believe that the fraction defective exceeds .05.

7.55 .0125   **7.57** .3844   **7.61** $P(\hat{\pi} \leq .70) = 0$

7.63 $P(z \geq 3.43) = 0$, so this is not a likely outcome.

Chapter 8

8.1 (a) The set of all registered voters in the state

(b) Divide the state into a series of districts. A random sample of districts could be selected; and within the selected districts, a sample of voting precincts could be selected. The voting records within the sampled precincts could be used to obtain information about the percentage (or proportion) of registered voters who have voted at least once over the past two years.

8.3 (a) The perceptual set of lifetimes for all fuses of that particular type manufactured by the company

(b) Testing a hypothesis about a population mean

8.5 $35 \pm 1.645 (7.5/\sqrt{100}$ or 33.77 to 36.23

8.7 $6.5 \pm 1.96(2.6/\sqrt{42})$, or 5.71 to 7.29

8.9 $3300 \pm 1.96(500/\sqrt{30})$, or 3121.08 to 3478.92

8.11 $3.27 \pm 1.96(.23/\sqrt{100})$, or 3.22 to 3.32

8.18 $z = \dfrac{63.7 - 68}{14.2/\sqrt{50}} = -2.14$. Since $-2.14 < -1.645$, reject H_0.

8.20 $H_0: \mu = 525$
      $H_a: \mu > 525$
      T.S.: $z = \dfrac{\bar{x} - \mu_0}{\sigma/\sqrt{n}}$
      R.R.: For $\alpha = .05$, reject H_0 when $z > 1.645$.

8.22 $H_0: \mu = 0$
      $H_a: \mu > 0$
      T.S.: $z = \dfrac{\bar{x} - \mu_0}{\sigma/\sqrt{n}}$
      R.R.: For $\alpha = .05$, reject H_0 if $z > 1.645$.

8.24 $H_0: \mu = 8.2$
      $H_a: \mu < 8.2$

8.26 $H_0: \mu = .3$
      $H_a: \mu > .3$
      T.S.: $z = \dfrac{.7 - .3}{.4/\sqrt{60}} = 7.75$
      R.R.: For $\alpha = .05$, reject H_0 if $z > 1.645$. Reject H_0.

8.28 $z = 1.74$; level of significance: .0409

8.30 The distribution of the test statistic $\dfrac{\bar{x} - \mu_0}{s/\sqrt{n}}$ for $n < 30$ is no longer normal but possesses a t-distribution.

8.32 (a) Reject H_0 if $t < -2.624$.   (b) Reject H_0 if $t > 2.819$.
      (c) Reject H_0 if $t > 3.365$.

8.34 $H_0: \mu = 650$
      $H_a: \mu < 650$
      T.S.: $t = \dfrac{638.6 - 650}{12.74/\sqrt{5}} = -2.00$
      R.R.: Reject H_0 if $t < -2.132$. Insufficient evidence to reject H_0.

8.36 $3.78 \pm 2.262 \,(.181/\sqrt{10})$ or 3.65 to 3.91

8.43 Estimates are unnecessary if the sample constitutes the population, since the estimates (statistics) become parameters and there is no element of uncertainty in the sampling process.

8.53 $9.2 \pm 1.645 \,(12.4/\sqrt{100})$ or 7.16 to 11.24

8.55 $3.2 \pm 1.645 \,(.3/\sqrt{50})$ or 3.13 to 3.27

8.57 $p < .005$ (based on t-values)
      $p < .001$ (based on z-values)

8.59 $t = 2.95$; reject H_0

8.61 $t = -3.21$; reject H_0

8.63 $t = -2.71$; reject H_0

8.65 $t = -2.74$; reject H_0

8.67 $p < .01$

8.69 $t = 1.08$; insufficient evidence to reject H_0

8.71 $t = 0.58$; insufficient evidence to reject H_0

8.73 $2.28 \pm 1.895 \, (.25/\sqrt{8})$ or 2.11 to 2.45

8.75 $8.898 < \mu < 9.035$; we are 95% confident that the mean potency for the batch lies in the interval 8.898 to 9.035

8.77 Statements (a), (b), and (c) are true

8.79 $t = -1.25$; insufficient evidence to reject H_0

8.81 $t = 5.99$; reject H_0

8.83 $2567 \pm 1.658 \, (4230/\sqrt{45})$ or 1521.51 to 3612.49

8.85 $t = -5.92$; reject H_0

8.87 $5.07 \pm 2.131 \, (2.55/\sqrt{16})$ or 3.71 to 6.43

8.104 $53.16 \pm 1.711 \, (19.10/\sqrt{25})$ or 46.62 to 59.70

Chapter 9

9.1 -1.22 to 11.22   9.3 .14 to 3.86   9.5 -3.66 to 7.66

9.7 T.S.: $t = -2.67$ R.R.: Reject H_0 if $t < -1.703$

9.9 $t = -18.51$; reject H_0

9.11 T.S.: $t = -4.54$ R.R.: Reject H_0 if $|t| > 1.96$; reject H_0

9.19 2.68 to 9.72

9.21 $-212.67 \pm 1.812 \, (19.90) \sqrt{\dfrac{1}{6} + \dfrac{1}{6}}$ or -233.49 to -191.85

9.23 (a) $t = -2.16$; $p = .056$ (b) $t = 1.689$; $p = .122$

9.26 $t = 6.90$; reject H_0

9.28 $t = 2.46$; insufficient evidence to reject H_0

9.30 $3.83 \pm 2.132 \, (0.68) \sqrt{\dfrac{1}{3} + \dfrac{1}{3}}$ or 2.65 to 5.01

9.32 $t = -3.33$; reject H_0

9.37 (a) H_0: $\mu_2 - \mu_1 = 0$ (b) H_a: $\mu_2 - \mu_1 > 0$
(c) Reject H_0 if $t > 1.860$ (d) $t = 2.00$ (e) Reject H_0

9.39 $t = 2.00$; $0.025 < p < .05$

9.41 H_0: $\mu_1 - \mu_2 = 0$ H_a: $\mu_1 - \mu_2 > 0$ T.S.: $t = .90$
R.R.: Reject H_0 if $t > 1.746$; insufficient evidence to conclude the ghetto prices are higher than the suburban prices.
A Type II error might have been made.

9.43 (a) $t = 4.92$; $p < .001$
(b) (i) Reject H_0 and conclude $\mu_E - \mu_C > 0$
(ii) There is strong evidence that the mean IQ gain for students taught by teachers who expect such a gain is higher than for students taught by teachers who have no such expectations.

9.45 We must assume normality of the two populations and equality of the two population variances. Equality of the population variances may be a problem.

9.47 Yes, no, no, yes   9.49 -0.78 to 3.18

9.51 3252.59 to 2146.59

9.53 $t = -1.11$; insufficient evidence to indicate a difference in teaching effectiveness

9.57 For H_0: $\mu_D - \mu_B = 0$, $t = 2.98$; based on a one-tailed test with df = 24, $p < .005$

For H_a: $\mu_D - \mu_C = 0$, $t = 5.08$; based on a one-tailed test with df = 24, $p < .005$

Treatment A appears to have the lowest mean HAM-D total score.

9.59 H_0: $\mu_B - \mu_D = 0$ H_a: $\mu_B - \mu_D \neq 0$

T.S.: $t = \dfrac{\bar{x}_B - \bar{x}_D}{s_p \sqrt{\dfrac{1}{25} + \dfrac{1}{25}}} = -1.23$

R.R.: For a two-tailed test with $\alpha = .05$, df = 48, we will reject H_0 if $|t| > t_{.025}$ based on df = 48. Since Table 3 of Appendix 3 does not list t-values for df = 48, use the .025 value based on the nearest df (df = 40). Hence we will reject H_0 if $|t| > 2.021$.

Conclusion: Since $|-1.23|$ is not greater than 2.021, we have insufficient evidence to reject H_0.

Discussion: It is important that the treatment groups have similar ages, so that differences between the treatment groups that are based on the HAM-D total score (or some other measure of depression) are due to differences in treatment effects of the prescribed medication.

Chapter 10

10.1 $.68 \pm 1.645 \dfrac{}{\sqrt{.68(.32)/1000}}$ or .66 to .70   **10.3** .10 to .42

10.5 .056 to .084   **10.7** .64 to .74

10.9 (b) No, sample size unknown

10.15 $\hat{\pi} = .55$, $z = 1.41$; insufficient evidence to reject H_0

10.17 $\hat{\pi} = .56$, $z = 3.79$; reject H_0

10.19 H_0: $\pi = .05$ H_a: $\pi > .05$ T.S.: $z = 3.08$
R.R.: For $\alpha = .05$, reject H_0 if $z > 1.645$; reject H_0

10.21 $x = 18$, $z = 2.67$; reject H_0

10.24 $\hat{\pi}_1 = .36$, $\hat{\pi}_2 = .25$, $-.14$ to .36

10.26 $\hat{\pi}_1 = .58$, $\hat{\pi}_2 = .46$, $z = 1.70$; $p = .0446$ **10.28** $-.01$ to .21

10.30 .15 to .43

10.35 .455 to .511; we are 95% confident that the proportion of church-goers (π) lies between .455 and .511.

10.37 $z = -1.15$; $p = .1251$; insufficient evidence to reject H_0: $\pi = .75$

10.39 $z = -3.67$; reject H_0: $\pi = .90$

10.43 (a) .370 to .430   (b) No

10.45 (a) Agree   (b) -1.645 (c) Agree

10.47 $z = 3.33$; reject H_0 **10.49** $z = -2.81$; $p = .005$

10.51 $z = 1.36$; insufficient evidence to reject H_0

10.53 $z = -10.69$; reject H_0

10.55 (1) The 55 human volunteers may or may not constitute a random sample, although all appeared to be clear of infection.

(2) Exact sample sizes should be given since 55 is not evenly divisible by four. Then we could test

$$H_0: \pi_1 - \pi_2$$

for the two groups (no drops and no juice vs nose drops and juice).

(3) For the groups receiving the virus infection, we should know the percent of volunteers having respiratory symptoms so a test as shown in (z) could be performed.

10.57 .036 to .058   **10.59** Class exercise   **10.61** .078 to .122

10.63 .48 to .62   **10.65** .400 to .560   **10.67** .06 to .14

10.69 Class exercise   **10.71** $H_0: \pi = .5$ $H_a: \pi \neq .5$

T.S.: $z = \dfrac{\hat{\pi} - .5}{\sigma_{\hat{\pi}}}$ where $\sigma_{\hat{\pi}} = \sqrt{.5(.5)/100} = .05$

R.R.: For $\alpha = .05$, reject H_0 if $|z| > 1.96$ **10.73** .0278

10.75 $H_0: \pi = .5$ $H_a: \pi > .5$ $z = 4.67; p < .001$

10.77 $\hat{\pi} = .07, 2 = 1.30$; insufficient evidence to reject H_0 **10.79** 0.17

10.81 $\hat{\pi} = .16$, $z = -4.09$;   reject H_0 **10.83** .247 to .353

10.85 $z = 3.81$; reject H_0 **10.87** $z = 4.81$; reject $H_0: \pi = .50$

10.94 $\pi_A - \pi_D$: .12 ± .27 $\pi_B - \pi_D$: 0 ± .28
$\pi_C - \pi_D$: .04 ± .28

Chapter 11

11.1 .0017 to .0038   **11.3** 2.043 to 15.702

11.5 (a) 2.215   (b) 2.48   (c) 1.96   (d) 2.70   (e) 2.79

11.7 No; insufficient evidence to reject H_0; $F = 3.15$ does not exceed $F_{.05, 7, 7} = 3.79$

11.9 Insufficient evidence to reject H_0; $F = 2.73$ does not exceed 2.98

11.11 $F = 3.33$; reject H_0

11.13 (a) 1.212 to 111.62   (b) 1.10 to 10.57

11.15 $\chi^2 = 8.510; p > .10$ **11.17** (c) $F = 3.95$; reject $H_0: \sigma_1^2 = \sigma_2^2$

11.19 (a) $F = 2.27$; reject H_0
(b) $F = 2.27$; insufficient evidence to reject H_0

11.21 $F = 3.57$; reject H_0 **11.23** $F = 3.13$; reject H_0

11.25 $F = 3.62$; reject H_0 **11.27** $F = 2.35; p > .10$

11.29 $\bar{d} = -.33, s_{\bar{d}} = 3.35$

11.31 (a) Equality of the population variances   (b) No; $F = 5.90$

11.33

Treatment	Mean	Std. dev.	n
A	0.626	0.348	25
B	0.693	0.360	25
C	0.600	0.333	25
D	0.840	0.424	25

(a) $H_0: \sigma_A^2 = \sigma_D^2$
    $F = .67$; insufficient evidence to reject H_0
(b) $H_0: \sigma_B^2 = \sigma_D^2$
    $F = .72$; insufficient evidence to reject H_0

(c) H_0: $\sigma_C^2 = \sigma_D^2$
$F = .62$; insufficient evidence to reject H_0

11.35 (a) $.89 < \sigma_B^2/\sigma_C^2 < 6.22$ (b) $.38 < \sigma_B^2/\sigma_A^2 < 2.66$

Chapter 12

12.3 $y = 7.8$ **12.7** $\hat{y} = .50 + 1.70x$ **12.9** $\hat{y} = 2.47 + 1.63x$

12.11 $\hat{y} = -1089.67 + 0.34x$; $\hat{y} = 1486.33$ **12.18** $\hat{\rho} = .993$

12.20 $\hat{\rho} = .809$; $\hat{\rho}^2 = .655$; that is, 65.5% of the variability in the y-values is accounted for by variability in the x-values

12.24 $\hat{\rho}^2$ is R-SQUARE $= 0.9438$
$\hat{\rho}$ is $\sqrt{.9438} = -.972$ (the negative sign must be used since $\hat{\beta}_1 < 0$)
This would indicate a strong negative relationship with the variability in x-values accounting for 94.38% of the variability in the y-values.

12.26 $\hat{\rho} = .907$ **12.28** $\hat{\rho} = .99$; yes **12.30** $\hat{\rho} = .63$

12.32 (a) $\hat{y} = 81.9 + 1.26x_1 - 0.425x_2 - 1.19x_3 + 0.0018x_4$
(b) $\hat{y} = 37.955$
(c) Since the only large t-value is for β_1, this regression equation may not fit the data well.

12.34 (b) $\hat{\rho} = .988$
(c) There is a strong positive correlation between scores. Since $\hat{\rho}^2 = .976$, the perceived self-image score accounts for 97.6% of the variability in the social adjustment scores.

12.40 (a) Strong positive relationship (b) Little or no relationship
(c) Perfect negative relationship

12.42 (a) $y = 2 + x$ (b) $y = 2 - x$ (c) $y = 1 + x$
(d) $y = -2 + x$

12.44 ENGLACT VS MATHACT, $\hat{\rho} = .595$
ENGLACT VS SOCSACT, $\hat{\rho} = .668$
MATHACT VS SOCSACT, $\hat{\rho} = .567$

12.46 $\hat{\rho} = .935$ **12.48** (b) $\hat{\rho} = .854$ **12.55** $\hat{\rho} = .720$ **12.57** $\hat{\rho} = .130$

Chapter 13

13.1 (a) $\hat{y} = 9.275 - .505x$ (b) $t = 7.83$; reject H_0

13.6 (b) $\hat{y} = 62.81 + 4.39x$

13.8 The 95% confidence interval for β_1 is 2.226 to 6.553. Since this interval does not include 0 as a possible value for β_1, we arrive at the same conclusion as found in the two-tailed test of Exercise 13.7.

13.14 $t = -13.08$; reject H_0 **13.16** $t = 16.92$; reject H_0

13.18 0.742 to 2.030 **13.20** 41.930 to 69.218

13.22 $t = -7.86$; reject H_0

13.24 $t = 8.73$; reject H_0

13.26 $4.9995 \pm .0640$; a 90% confidence interval would not be as wide; a 95% confidence interval for $E(y)$ when $x = 9$ would be wider since $x = 9$ is further away from $\bar{x} = 5.5$ than $x = 5$

13.28 9.39368 ± 1.45122 **13.30** $t = 5.35$; reject H_0

13.32 (b) $\hat{y} = 8.412234 + 5.026596x$
(c) $x = 4$ is observation 5; the predicted y value is 28.519

13.34 $\hat{y} = 36.10 + 0.384x$; since $t = 1.13$ for $H_0: \beta_1 = 0$ there is insufficient evidence to reject H_0, hence it does not seem to make sense to predict y based on x

13.36 $\hat{y} = 738.07 + 4.70x$. $t = 12.31$; reject H_0

13.42 $\hat{\rho} = .615, t = 7.72$; reject H_0

Chapter 14

14.1 312   **14.3** 35   **14.5** 385   **14.14** $t = 5.45$; reject H_0

14.16 Will likely lose information   **14.18** $-.14$ to 2.90

14.20 (1) The variation in the population (noise)
(2) The size of the sample (volume)

14.22 Paired-difference experiment

14.24 $t = 1.75$; insufficient evidence to reject H_0

14.26 (a) No, because the two observations in each pair are not independent.
(b) If no pairing occurred in the design, then a paired-difference analysis will result in a loss of information (degrees of freedom for the t-test).

14.28 $t = -1.45$; $p = 0.099$

14.30 (a) Probably experiment B, because men and women were matched (paired) according to IQ.
(b) A had 48 degrees of freedom; B had 24 degrees of freedom.

14.32 78   **14.34** 55

14.36 (b) $t = 2.39$; insufficient evidence to reject H_0

14.38 $t = 2.12$; $p = 0.067$. We don't have strong evidence to reject H_0

14.40 1849

14.42 (a) $H_0: \mu_d = 0$ ($\mu_d = \mu_{\text{Location 1}} - \mu_{\text{Location 2}}$), $H_a: \mu_d \neq 0$ (b) t (c) 9
(d) $\mu_d > 0$; i.e., location 1 gets more traffic on the average than location 2

14.44 865   **14.46** 24   **14.48** 62   **14.50** $t = 3.89$; reject H_0 **14.52** 45

Chapter 15

15.3 .86 to 2.88   **15.5** $F = 2.51$; insufficient evidence to reject H_0

15.7 $t = -2.424$; reject H_0 **15.9** -4.03 to 22.03

15.11 (b) 1.732   (c) $F = 6.055$; $p = .022$

15.13 (a) $H_0: \mu_1 = \mu_2 = \mu_3 = \mu_4$, i.e., the means of the four different preconditioning groups are the same.
(b) Reject H_0; at least one of the means differs from the rest. There is no F-value listed in Appendix 3 Tables 5–9 for $df_2 = 84$. Taking the next-lower tabled value for df_2, namely $df_2 = 60$, we have $F_{.01, 3, 60} = 4.13$.
(c) Hypothetical populations of all possible students who could have been tested under these four preconditioning states. If we had been working with non-random samples, the differences we detected using the analysis of variance could not be attributed to differences among the preconditioning states.

15.15 $F = 17.69$; reject H_0

15.17

Source	SS	df	MS	F
Between	8.7704	2	4.3852	14.730
Within	2.679	9	0.2977	
Total	11.4494	11		

15.21   (a) 6.63 to 13.81   (b) -1.41 to 5.77

15.23 $F = 64.50$; reject H_0

15.25   (b), (c) $F = 2.64$; insufficient evidence to reject H_0
(d) Locations 1 and 4, but this difference was not detected in the ANOVA

15.29

Source	SS	df	MS	F
Between	1485.7468	2	742.8734	19.49
Within	571.66	15	38.11	
Total	2057.4068	17		

Reject H_0

15.31

Source	SS	df	MS	F
Between	.02973	3	.00991	10.30
Within	.03463	36	.000962	
Total	.06436	39		

Reject H_0

15.38

Source	SS	df	MS	F
Between	6297.71	3	2099.24	8.09
Within	24912.40	96	259.50	
Total	31210.11	99		

Reject H_0

15.40

Source	SS	df	MS	F
Between	347.15	3	115.72	0.69
Within	15988.64	96	166.55	
Total	16335.79	99		

Insufficient evidence to reject H_0

Chapter 16

16.1 $\chi^2 = 10$; reject H_0 **16.3** $\chi^2 = 64.64$; reject H_0: all $\pi_i = .20$

16.5   (a) 2   (b) 4   **16.7** $\chi^2 = 6.46$; reject H_0

16.9   (a) $\chi^2 = 6.091$ (b) $.025 < p < .05$

16.11 $\chi^2 = 4.00$; insufficient evidence to reject H_0

16.13 $\chi^2 = 33.62$; reject H_0 **16.15** $\chi^2 = 17.43$; reject H_0

16.17 $\chi^2 = .773$; insufficient evidence to reject H_0

16.19 $\chi^2 = 1.43$; insufficient evidence to reject H_0

16.21 $\chi^2 = 4.36$; insufficient evidence to reject H_0

16.23   (a) .323, .248, .167, .167   (b) $\chi^2 = 10.71$; reject H_0
(c) The percentage of accidents with fatalities differs depending on the probable cause of the accident

16.25 (a) .10, .14, .19   (b) $\chi^2 = 3.31$; insufficient evidence to reject H_0

16.29 $\chi^2 = 5.585$; insufficient evidence to reject H_0

16.31 $\chi^2 = 35.246$; reject H_0 **16.37** Class exercise

16.39 We could make any number of comparisons; more important would be comparisons of the treatment groups for other variables (e.g., age, tobacco, alcohol, psychiatric diagnosis) which would help establish the comparability of the treatment groups with respect to these other variables prior to treatment.

16.41 $\chi^2 = 17.86$; reject H_0. Groups A–C have a much higher percentage of patients showing a marked or moderate therapeutic effect.

Chapter 17

17.1 $z = 1.13$; insufficient evidence to reject H_0

17.3 $z = 2.236$; reject H_0 **17.5** $z = -2.71$; reject H_0

17.7 $z = 2.14$; reject H_0 **17.9** $T = 40$; reject H_0

17.11 $T = 4$; reject H_0

17.13 $T = 16$; reject H_0 for $\alpha = .05$

17.17 $t = 2.05$; $.05 < p < .10$

17.19 $z = -.63$; insufficient evidence to reject H_0

17.21 $z = 1.51$; insufficient evidence to reject H_0

17.23 $z = 2.11$; reject H_0

17.25 (a) $\hat{\rho}_s = .945$ (b) $z = 2.84$; reject H_0

17.27 $z = -1.73$; insufficient evidence to reject H_0

17.29 Class exercise   **17.31** $z = 2.34$; reject H_0

17.33 $z = .054$; insufficient evidence to reject H_0

INDEX

TABLE 2 Normal curve areas

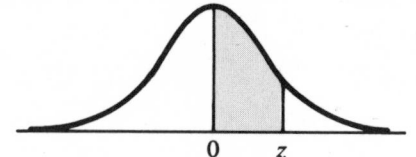

z	.00	.01	.02	.03	.04	.05	.06	.07	.08	.09
0.0	.0000	.0040	.0080	.0120	.0160	.0199	.0239	.0279	.0319	.0359
0.1	.0398	.0438	.0478	.0517	.0557	.0596	.0636	.0675	.0714	.0753
0.2	.0793	.0832	.0871	.0910	.0948	.0987	.1026	.1064	.1103	.1141
0.3	.1179	.1217	.1255	.1293	.1331	.1368	.1406	.1443	.1480	.1517
0.4	.1554	.1591	.1628	.1664	.1700	.1736	.1772	.1808	.1844	.1879
0.5	.1915	.1950	.1985	.2019	.2054	.2088	.2123	.2157	.2190	.2224
0.6	.2257	.2291	.2324	.2357	.2389	.2422	.2454	.2486	.2517	.2549
0.7	.2580	.2611	.2642	.2673	.2704	.2734	.2764	.2794	.2823	.2852
0.8	.2881	.2910	.2939	.2967	.2995	.3023	.3051	.3078	.3106	.3133
0.9	.3159	.3186	.3212	.3238	.3264	.3289	.3315	.3340	.3365	.3389
1.0	.3413	.3438	.3461	.3485	.3508	.3531	.3554	.3577	.3599	.3621
1.1	.3643	.3665	.3686	.3708	.3729	.3749	.3770	.3790	.3810	.3830
1.2	.3849	.3869	.3888	.3907	.3925	.3944	.3962	.3980	.3997	.4015
1.3	.4032	.4049	.4066	.4082	.4099	.4115	.4131	.4147	.4162	.4177
1.4	.4192	.4207	.4222	.4236	.4251	.4265	.4279	.4292	.4306	.4319
1.5	.4332	.4345	.4357	.4370	.4382	.4394	.4406	.4418	.4429	.4441
1.6	.4452	.4463	.4474	.4484	.4495	.4505	.4515	.4525	.4535	.4545
1.7	.4554	.4564	.4573	.4582	.4591	.4599	.4608	.4616	.4625	.4633
1.8	.4641	.4649	.4656	.4664	.4671	.4678	.4686	.4693	.4699	.4706
1.9	.4713	.4719	.4726	.4732	.4738	.4744	.4750	.4756	.4761	.4767
2.0	.4772	.4778	.4783	.4788	.4793	.4798	.4803	.4808	.4812	.4817
2.1	.4821	.4826	.4830	.4834	.4838	.4842	.4846	.4850	.4854	.4857
2.2	.4861	.4864	.4868	.4871	.4875	.4878	.4881	.4884	.4887	.4890
2.3	.4893	.4896	.4898	.4901	.4904	.4906	.4909	.4911	.4913	.4916
2.4	.4918	.4920	.4922	.4925	.4927	.4929	.4931	.4932	.4934	.4936
2.5	.4938	.4940	.4941	.4943	.4945	.4946	.4948	.4949	.4951	.4952
2.6	.4953	.4955	.4956	.4957	.4959	.4960	.4961	.4962	.4963	.4964
2.7	.4965	.4966	.4967	.4968	.4969	.4970	.4971	.4972	.4973	.4974
2.8	.4974	.4975	.4976	.4977	.4977	.4978	.4979	.4979	.4980	.4981
2.9	.4981	.4982	.4982	.4983	.4984	.4984	.4985	.4985	.4986	.4986
3.0	.4987	.4987	.4987	.4988	.4988	.4989	.4989	.4989	.4990	.4990

Source: This table is abridged from Table 1 of *Statistical Tables and Formulas*, by A. Hald (New York: John Wiley & Sons, 1952). Reprinted by permission of A. Hald and the publishers, John Wiley & Sons.